AF443111

HIGHER-DIMENSIONAL GEOMETRY
OVER FINITE FIELDS

NATO Science for Peace and Security Series

This Series presents the results of scientific meetings supported under the NATO Programme:
Science for Peace and Security (SPS).

The NATO SPS Programme supports meetings in the following Key Priority areas: (1) Defence
Against Terrorism; (2) Countering other Threats to Security and (3) NATO, Partner and
Mediterranean Dialogue Country Priorities. The types of meeting supported are generally
"Advanced Study Institutes" and "Advanced Research Workshops". The NATO SPS Series
collects together the results of these meetings. The meetings are co-organized by scientists from
NATO countries and scientists from NATO's "Partner" or "Mediterranean Dialogue" countries.
The observations and recommendations made at the meetings, as well as the contents of the
volumes in the Series, reflect those of participants and contributors only; they should not
necessarily be regarded as reflecting NATO views or policy.

Advanced Study Institutes (ASI) are high-level tutorial courses to convey the latest
developments in a subject to an advanced-level audience.

Advanced Research Workshops (ARW) are expert meetings where an intense but informal
exchange of views at the frontiers of a subject aims at identifying directions for future action.

Following a transformation of the programme in 2006 the Series has been re-named and re-
organised. Recent volumes on topics not related to security, which result from meetings
supported under the programme earlier, may be found in the NATO Science Series.

The Series is published by IOS Press, Amsterdam, and Springer Science and Business Media,
Dordrecht, in conjunction with the NATO Public Diplomacy Division.

Sub-Series

A.	Chemistry and Biology	Springer Science and Business Media
B.	Physics and Biophysics	Springer Science and Business Media
C.	Environmental Security	Springer Science and Business Media
D.	Information and Communication Security	IOS Press
E.	Human and Societal Dynamics	IOS Press

http://www.nato.int/science
http://www.springer.com
http://www.iospress.nl

Sub-Series D: Information and Communication Security – Vol. 16   ISSN 1874-6268

Higher-Dimensional Geometry over Finite Fields

Edited by

Dmitry Kaledin
Steklov Institute, Moscow, Russia

and

Yuri Tschinkel
Courant Institute, NYU and Mathematisches Institut, Göttingen, Germany

IOS Press

Amsterdam • Berlin • Oxford • Tokyo • Washington, DC

Published in cooperation with NATO Public Diplomacy Division

Proceedings of the NATO Advanced Study Institute on Higher-Dimensionals Geometry over
Finite Fields
Göttingen, Germany
25 June – 6 July 2007

ISBN 978-1-58603-855-7
Library of Congress Control Number: 2008924572

Publisher
IOS Press
Nieuwe Hemweg 6B
1013 BG Amsterdam
Netherlands
fax: +31 20 687 0019
e-mail: order@iospress.nl

Distributor in the UK and Ireland
Gazelle Books Services Ltd.
White Cross Mills
Hightown
Lancaster LA1 4XS
United Kingdom
fax: +44 1524 63232
e-mail: sales@gazellebooks.co.uk

Distributor in the USA and Canada
IOS Press, Inc.
4502 Rachael Manor Drive
Fairfax, VA 22032
USA
fax: +1 703 323 3668
e-mail: iosbooks@iospress.com

Higher-Dimensional Geometry over Finite Fields
D. Kaledin and Y. Tschinkel (Eds.)
IOS Press, 2008

v

Preface

Number systems based on a finite collection of symbols, such as the 0s and 1s of computer circuitry, are ubiquitous in the modern age. Finite fields are the most important such number systems, playing a vital role in military and civilian communications through coding theory and cryptography. These disciplines have evolved over recent decades, and where once the focus was on algebraic curves over finite fields, recent developments have revealed the increasing importance of higher-dimensional algebraic varieties over finite fields.

These are the proceedings of the NATO Advanced Study Institute "Higher-dimensional geometry over finite fields" held at the University of Göttingen in June–July 2007. They introduce the reader to recent developments in algebraic geometry over finite fields with particular attention to applications of geometric techniques to the study of rational points on varieties over finite fields of dimension at least 2.

Dmitry Kaledin
Yuri Tschinkel

Contents

Finite Field Experiments

Institut für Algebraische Geometrie
Leibniz Universität Hannover
Welfengarten 1
D-30167 Hannnover

e-mail: bothmer@math.uni-hannover.de

Hans-Christian GRAF V. BOTHMER

Abstract. We show how to use experiments over finite fields to gain information about the solution set of polynomial equations in characteristic zero.

Introduction

Let X be a variety defined over $\mathbb{Z}$. According to Grothendieck we can picture X as a family of varieties X_p over $\operatorname{Spec} \mathbb{Z}$ with fibers over closed points of $\operatorname{Spec} \mathbb{Z}$ corresponding to reductions modulo p and the generic fiber over (0) corresponding to the variety $X_{\mathbb{Q}}$ defined by the equations of X over $\mathbb{Q}$.

The generic fiber is related to the special fibers by semicontinuity theorems. For example, the dimension of X_p is upper semicontinuous with

$$\dim X_{\mathbb{Q}} = \min_{p>0} \dim X_p.$$

This allows us to gain information about $X_{\mathbb{Q}}$ by investigating X_p which is often computationally much simpler.

Even more surprising is the relation between the geometry of X_p and the number of $\mathbb{F}_p$ rational points of X_p discovered by Weil:

Theorem 0.1. *Let* $X_p \subset \mathbb{P}^n_{\mathbb{F}_p}$ *be a smooth curve of genus* g, *and* N *be the number of* $\mathbb{F}_p$-*rational points of* X_p. *Then*

$$|1 - N + p| \leq 2g\sqrt{p}.$$

He conjectured even more precise relations for varieties of arbitrary dimension which were proved by Deligne using l-adic cohomology.

In this tutorial we will use methods which are inspired by Weil's ideas, but are not nearly as deep. Rather we will rely on some basic probabilistic estimates which

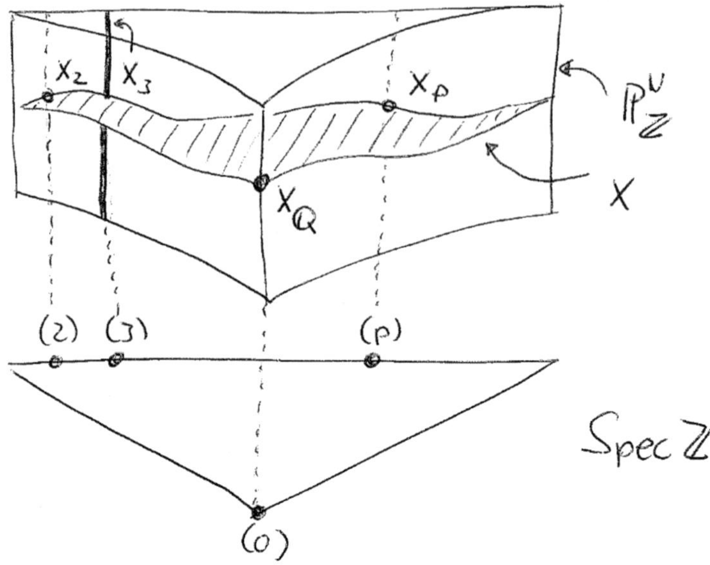

Figure 1. A variety over $\operatorname{Spec}\mathbb{Z}$

are nevertheless quite useful. I have learned these ideas from my advisor Frank Schreyer, but similar methods have been used independently by other people, for example Joachim von zur Gathen and Igor Shparlinski [1], Oliver Labs [2] and Noam Elkies [3].

The structure of these notes is as follows: We start in Section 1 by evaluating the polynomials defining a variety X at random points. This can give some heuristic information about the codimension c of X and about the number d of codimension-c components of X.

In Section 2 we refine this method by looking at the tangent spaces of X in random points. This gives a way to also estimate the number of components in every codimension. As an application we show how this can be applied to gain new information about the Poincaré center problem.

In Section 3 we explain how it is often possible to prove that a solution found over $\mathbb{F}_p$ actually lifts to $\overline{\mathbb{Q}}$. This is applied to the construction of new surfaces in $\mathbb{P}^4$.

Often one would like not only to prove the existence of a lift, but explicitly find one. It is explained in Section 4 how this can be done if the solution set is zero dimensional.

We close in Section 5 with a beautiful application of these lifting techniques found by Oliver Labs, showing how he constructed a new septic with 99 real nodes in $\mathbb{P}^3_{\mathbb{R}}$.

For all experiments in this tutorial we have used the computer algebra system Macaulay 2 [4]. The most important Macaulay 2 commands used are explained in Appendix A, for more detailed information we refer to the online help of Macaulay

2 [4]. In Appendix B Stefan Wiedmann provides a MAGMA translation of the Macaulay 2 scripts in this tutorial. All scripts are available online at [5]. We would like to include translations to other computer algebra packages, so if you are for example a **Singular**-expert, please contact us.

Finally I would like to thank the referee for many valuable suggestions.

1. Guessing

We start by considering the most simple case, namely that of a hypersurface $X \subset \mathbb{A}^n$ defined by a single polynomial $f \in \mathbb{F}_p[x_1, \ldots, x_n]$. If $a \in \mathbb{A}^n$ is a point we have

$$f(a) = \begin{cases} 0 & \text{one possibility} \\ \neq 0 & (p-1) \text{ possibilities} \end{cases}$$

Naively we would therefore expect that we obtain zero for about $\frac{1}{p}$ of the points.

Experiment 1.1. We evaluate a given polynomial in 700 random points, using Macaulay 2:

```
R = ZZ[x,y,z,w]                    -- work in AA^4
F = x^23+1240*y*z+w+129269698      -- a Polynomial
K = ZZ/7                           -- work over F_7
L = apply(700,                     -- substitute 700
      i->sub(F,random(K^1,K^4)))   -- random points
tally L                            -- count the results
```

obtaining:

```
o5 = Tally{-1 => 100}
           -2 => 108
           -3 => 91
           0 => 98
           1 => 102
           2 => 101
           3 => 100
```

Indeed, all elements of $\mathbb{F}_7$ occur about $700/7 = 100$ times as one would expect naively.

If $f = g \cdot h \in \mathbb{F}_p[x_1, \ldots, x_n]$ is a reducible polynomial we have

$$f(a) = g(a)h(a) = \begin{cases} 0 \cdot 0 & 1 \text{ possibility} \\ * \cdot 0 & (p-1) \text{ possibilities} \\ 0 \cdot * & (p-1) \text{ possibilities} \\ * \cdot * & (p-1)^2 \text{ possibilities} \end{cases}$$

so one might expect a zero for about $\frac{2p-1}{p^2} \approx \frac{2}{p}$ of the points.

Experiment 1.2. We continue Experiment 1.1 and evaluate a product of two polynomials in 700 random points:

```
G = x*y*z*w+z^25-938493+x-z*w      -- a second polynomial
tally apply(700,                   -- substitute 700
    i->sub(F*G,random(K^1,K^4)))   -- random points & count
```

This gives:

```
o8 = Tally{-1 => 86}
            -2 => 87
            -3 => 77
            0 => 198
            1 => 69
            2 => 84
            3 => 99
```

Indeed, the value 0 now occurs about twice as often, i.e. $198 \approx \frac{2}{7} \cdot 700$.

Repeating Experiments 1.1 and 1.2 for 100 random polynomials and 100 random products we obtain Figure 2.

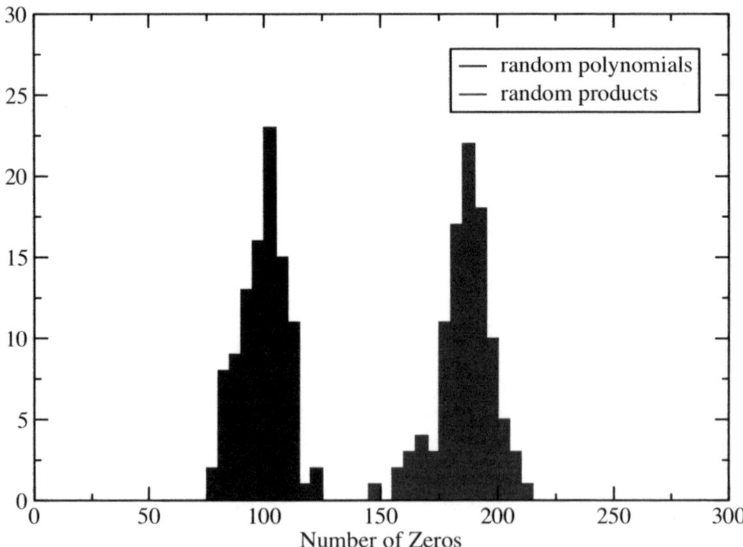

Figure 2. Evaluating 100 random polynomials and 100 random products at 700 points each.

Observe that the results for irreducible and reducible polynomials do not overlap. Evaluating a polynomial at random points might therefore give some indication on the number of its irreducible factors. For this we will make the above naive observations more precise.

Definition 1.3. If $f \in \mathbb{F}_p[x_1, \ldots, x_n]$ is a polynomial, we call the map

$$f|_{\mathbb{F}_p^n} : \mathbb{F}_p^n \to \mathbb{F}_p$$
$$a \mapsto f(a)$$

the corresponding *polynomial function*. We denote by

$$V_p := \{f \colon \mathbb{F}_p^n \to \mathbb{F}_p\}$$

the vector space of all polynomial functions on $\mathbb{F}_p^n$.

Being a polynomial function is nothing special:

Lemma 1.4 (Interpolation). *Let* $\phi \colon \mathbb{F}_p^n \to \mathbb{F}_p$ *be any function. Then there exists a polynomial* $f \in \mathbb{F}_p[x_1, \ldots, x_n]$ *such that* $\phi = f|_{\mathbb{F}_p^n}$.

Proof. Notice that $(1 - x^{p-1}) = 0 \iff x \neq 0$. For every $a \in \mathbb{F}_p^n$ we define

$$f_a(x) := \prod_{i=1}^{n}(1 - (x_i - a_i)^{p-1})$$

and obtain

$$f_a(x) = \begin{cases} 1 \text{ if } x = a \\ 0 \text{ if } x \neq a. \end{cases}$$

Since $\mathbb{F}_p^n$ is finite we can consider $f := \sum_{a \in \mathbb{F}_p^n} \phi(a) f_a$ and obtain $f(x) = \phi(x)$ for all $x \in \mathbb{F}_p^n$. $\square$

Remark 1.5. From Lemma 1.4 it follows that

   (i) V_p is a vector space of dimension p^n.
   (ii) V_p is a finite set with p^{p^n} elements.
   (iii) Two distinct polynomials can define the same polynomial function, for example x^p and x. More generally if $F \colon \mathbb{F}_p \to \mathbb{F}_p$ is the *Frobenius endomorphism* then $f(a) = f(F(a))$ for all polynomials f and all $a \in \mathbb{F}_p^n$.

This makes it easy to count polynomial functions:

Proposition 1.6. *The number of polynomial functions* $f \in V_p$ *with k zeros is*

$$\binom{p^n}{k} \cdot 1^k \cdot (p-1)^{p^n - k}.$$

Proof. Since V_p is simply the set of all functions $f \colon \mathbb{F}_p^n \to \mathbb{F}_p$, we can enumerate the ones with k zeros as follows: First choose k points and assign the value 0 and then chose any of the other $(p-1)$ values for the remaining $p^n - k$ points. $\square$

Corollary 1.7. *The average number of zeros for polynomial functions $f \in V_p$ is*

$$\mu = p^{n-1}$$

and the standard deviation of the number of zeros in this set is

$$\sigma = \sqrt{p^n \left(\frac{1}{p}\right)\left(\frac{p-1}{p}\right)} < \sqrt{\mu}.$$

Proof. Standard facts about binomial distributions. $\qquad\qquad\square$

Remark 1.8. Using the normal approximation of the binomial distribution, we can estimate that more than 99% of all $f \in V_p$ satisfy

$$|\#V(f) - \mu| \le 2.58\sqrt{\mu}$$

For products of polynomials we have

Proposition 1.9. *The number of pairs $(f, g) \in V_p \times V_p$ whose product has k zeros is*

$$\#\{(f, g)\,|\,\#V(f \cdot g) = k\} = \binom{p^n}{k} \cdot (2p-1)^k \cdot ((p-1)^2)^{p^n - k}.$$

In particular, the average number of zeros in this set is

$$\mu' = p^n \left(\frac{2p-1}{p^2}\right) \approx 2\mu$$

and the standard deviation is

$$\sigma' = \sqrt{p^n \left(\frac{2p-1}{p^2}\right)\left(\frac{(p-1)^2}{p^2}\right)} < \sqrt{\mu'}$$

Proof. As in the proof of Proposition 1.6 we first choose k points. For each of these points x we choose either the value of $f(x) = 0$ and $g(x) \ne 0$ or $f(x) \ne 0$ and $g(x) = 0$ or $f(x) = g(x) = 0$. This gives $2p-1$ possibilities. For the remaining $p^n - k$ we choose f and q nonzero. For this we have $(p-1)^2$ possibilities. The formulas then follow again from standard facts about binomial distributions. $\quad\square$

Remark 1.10. It follows that more than 99% of pairs $(f, g) \in V_p \times V_p$ satisfy

$$|\#V(f \cdot g) - \mu'| \le 2.58\sqrt{\mu'}.$$

In particular, if a polynomial f has a number of zeros that lies outside of this range one can reject the hypothesis that f is a product of two irreducible with 99% confidence.

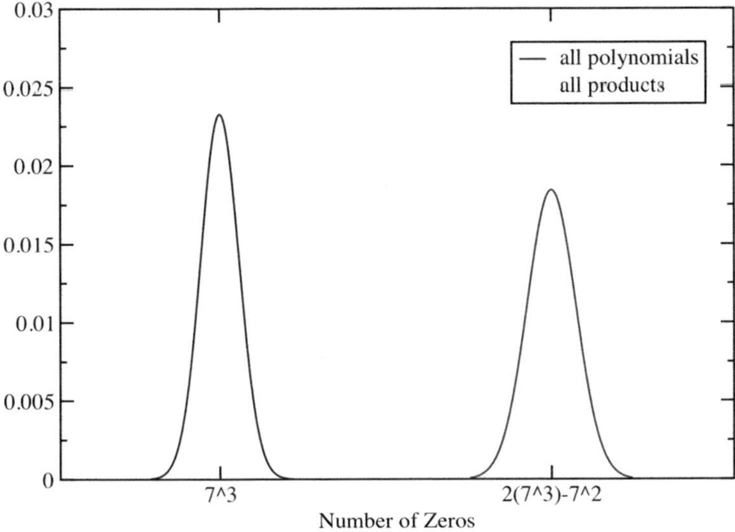

Figure 3. Distribution of the number of zeros on hypersurfaces in $\mathbb{A}^4$ in characteristic 7.

Even for small p the distributions of Proposition 1.6 and Proposition 1.9 differ substantially (see Figure 3).

Remark 1.11. For plane curves we can compare our result to the Weil conjectures. Weil shows that 100% of smooth pane curves of genus g in $\mathbb{P}^2_{\mathbb{F}_p}$ satisfy

$$|N - (p+1)| \le 2g\sqrt{p}$$

while we proved that 99% of the polynomial functions on $\mathbb{A}^2$ satisfy

$$|N - p| \le 2.58\sqrt{p}.$$

Of course Weil's theorem is much stronger. If $p > 4g^2$ Weil's theorem implies for example that every smooth curve of genus g over $\mathbb{F}_p$ has a rational point, while no such statement can be derived from our results. If on the other hand one is satisfied with approximate results, our estimates have the advantage that they are independent of the genus g. In Figure 4 we compare the two results with an experiment in the case of plane quartics. (Notice that smooth plane quartics have genus 3.)

For big n it is very time consuming to count all $\mathbb{F}_p$-rational points on $V(f) \subset \mathbb{A}^n$. We can avoid this problem by using a statistical approach once again.

Definition 1.12. Let $X \subset \mathbb{A}^n$ be a variety over $\mathbb{F}_p$. Then

$$\gamma_p(X) := \frac{\#X(\mathbb{F}_p)}{\#\mathbb{A}^n(\mathbb{F}_p)}$$

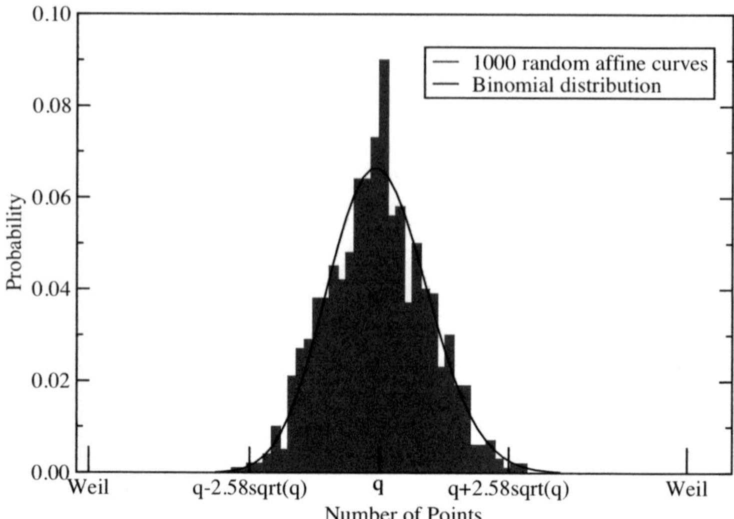

Figure 4. Number of points on 1000 affine quartics in characteristic $q = 37$ compared to the corresponding binomial distribution and Weil's bound.

is called the *fraction of zeros* of X. If furthermore $x_1, \ldots, x_m \in \mathbb{A}^n(\mathbb{F}_p)$ are points then

$$\hat{\gamma}_p(X) := \frac{\#\{i \mid x_i \in X\}}{m}$$

is called an *empirical fraction of zeros* of X.

Remark 1.13. If we choose the points x_i randomly and independently, the probability that $x_i \in X$ is $\gamma_p(X)$. Therefore we have the following:

(i) $\mu(\hat{\gamma}_p) = \gamma_p$, i.e. for large m we expect $\hat{\gamma}_p(X) \approx \gamma_p(X)$

(ii) $\sigma(\hat{\gamma}_p) \approx \sqrt{\frac{\gamma_p}{m}}$, i.e. the quadratic mean of the error $|\gamma_p - \hat{\gamma}_p|$ decreases with $\sqrt{m}$.

(iii) Since for hypersurfaces $\gamma_p \approx \frac{1}{p}$ the average error depends neither on the number of variables n nor on the degree of X.

(iv) Using the normal approximation again one can show that it is usually enough to test about $100 \cdot p$ points to distinguish between reducible and irreducible polynomials (for more precise estimates see [6]).

Experiment 1.14. Consider quadrics in $\mathbb{P}^3$ and let

$$\Delta := \{\text{singular quadric}\} \subset \{\text{all quadrics}\} \cong \mathbb{P}^9$$

be the subvariety of singular quadrics in the space of all quadrics. Since having a singularity is a codimension 1 condition for surfaces in $\mathbb{P}^3$ we expect Δ to be a hypersurface. Is Δ irreducible? Using our methods we obtain a heuristic answer using Macaulay 2:

```
-- work in characteristic 7
K = ZZ/7
-- the coordinate Ring of IP^3
R = K[x,y,z,w]
-- look at 700 quadrics
tally apply(700, i->codim singularLocus(ideal random(2,R)))
```

giving

```
o12 = Tally{2 => 5  }
            3 => 89
            4 => 606.
```

We see $95 = 89 + 5$ of our 700 quadrics were singular, i.e. $\hat{\gamma}(\Delta) = \frac{95}{700}$. Since this is much closer to $\frac{1}{7}$, then it is to $\frac{2}{7}$ we guess that Δ is irreducible. Notice that we have not even used the equation of Δ to obtain this estimate.

Let's now consider an irreducible variety $X \subset \mathbb{A}^n$ of codimension $c > 1$. Projecting $\mathbb{A}^n$ to a subspace $\mathbb{A}^{n-c+1}$ we obtain a projection $X' \subset \mathbb{A}^{n-c+1}$ of X (see Figure 5). Generically X' is a hypersurface, so by our arguments above X' has approximately p^{n-c} points. Generically most points of X' have only one preimage in X so we obtain the following very rough heuristic:

Heuristic 1.15. Let $X \subset \mathbb{A}^n_{\mathbb{F}_p}$ be a variety of codimension c and d the number of components of codimension c, then

$$\hat{\gamma}_p(X) \approx \frac{d}{p^c}$$

Remark 1.16. A more precise argument for this heuristic comes from the Weil Conjectures. Indeed, the number of $\mathbb{F}_p$-rational points on an absolutely irreducible projective variety X is

$$p^{\dim X} + \text{lower order terms,}$$

so $\gamma_p \approx \frac{1}{p^{\mathrm{codim}\, X}}$. Our elementary arguments still work in the case of complete intersections and determinantal varieties [6].

Remark 1.17. Notice that Heuristic 1.15 involves two unknowns: c and d. To determine these one has to measure over several primes of good reduction.

Experiment 1.18. As in Experiment 1.14 we look at quadrics in $\mathbb{P}^3$. These are given by their 10 coefficients and form a $\mathbb{P}^9$. This time we are interested in the variety $X \subset \mathbb{P}^9$ of quadrics whose singular locus is at least one dimensional. For this we first define a function that looks at random quadrics over $\mathbb{F}_p$ until it has found at least k examples whose singular locus has codimension at most c. It then returns the number of trials needed to do this.

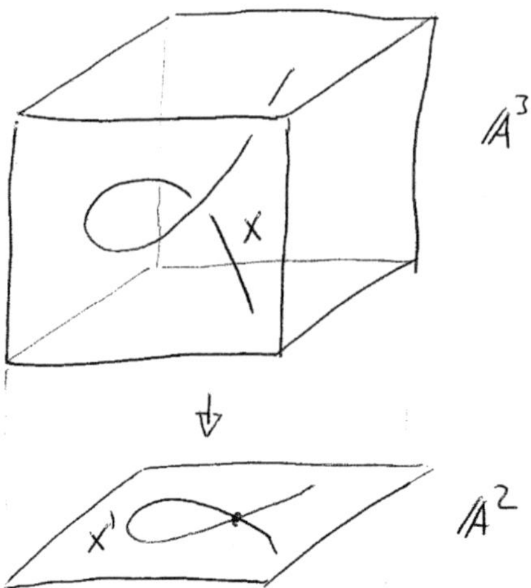

Figure 5. The projection of a curve in $\mathbb{A}^3$ is a hypersurface in $\mathbb{A}^2$ and most points have only one preimage.

```
findk = (p,k,c) -> (
    K := ZZ/p;
    R := K[x,y,z,w];
    trials := 0;
    found := 0;
    while found < k do (
      Q := ideal random(2,R);
      if c>=codim (Q+ideal jacobian Q) then (
            found = found + 1;
            print found;
            );
      trials = trials + 1;
      );
    trials
    )
```

Here we use `(Q+ideal jacobian Q)` instead of `singularLocus(Q)`, since the second option quickly produces a memory overflow.

The function `findk` is useful since the error in estimating γ from $\hat{\gamma}$ depends on the number of singular quadrics found. By searching until a given number of singular quadrics is found make sure that the error estimates will be small enough.

We now look for quadrics that have singularities of dimension at least one

```
k=50; time L1 = apply({5,7,11},q->(q,time findk(q,k,2)))
```

obtaining

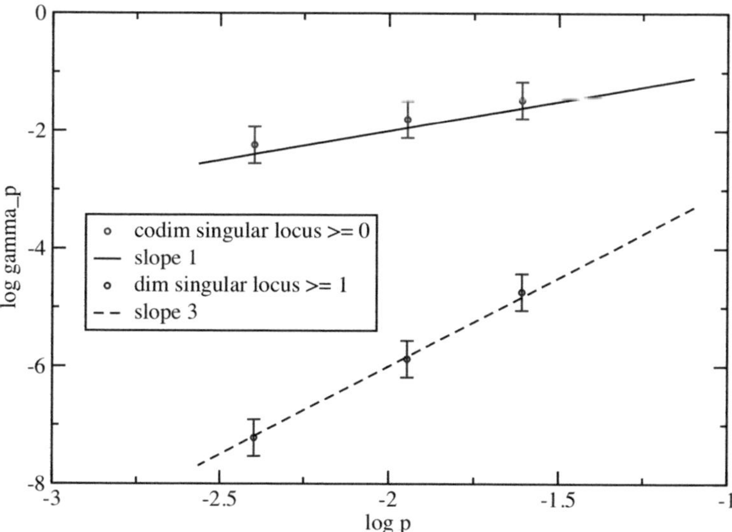

Figure 6. Measuring the codimension of quadrics with zero and one dimensional singular loci. Here we compare the measurements with lines of the correct slope 1 and 3.

```
{(5, 5724), (7, 17825), (11, 68349)}
```

i.e. $\gamma_5 \approx \frac{50}{5724}$, $\gamma_7 \approx \frac{50}{17825}$ and $\gamma_{11} \approx \frac{50}{68349}$. The codimension c of X can be interpreted as the negative slope in a log-log plot of $\gamma_p(X)$ since Heuristic 1.15 gives

$$\hat{\gamma}_p(X) \approx \frac{d}{p^c} \iff \log(\hat{\gamma}_p(X)) \approx \log(d) - c\log(p).$$

This is illustrated in Figure 6.

By using `findk` with $k = 50$ the errors of all our measurements are of the same magnitude. We can therefore use regression to calculate the slope of a line fitting these measurements:

```
-- calculate slope of regression line by
-- formula from [2] p. 800
slope = (L) -> (
     xbar := sum(apply(L,1->1#0))/#L;
     ybar := sum(apply(L,1->1#1))/#L;
     sum(apply(L,1->(1#0-xbar)*(1#1-ybar))))/
       sum(apply(L,1->(1#0-xbar)^2))
     )

-- slope for dim 1 singularities
slope(apply(L1,1->(log(1/1#0),log(k/1#1))))

o5 = 3.13578
```

The codimension of X is indeed 3 as can be seen by the following geometric argument: Each quadric with a singular locus of dimension 1 is a union of two hyperplanes. Since the family $\hat{\mathbb{P}}^3$ of all hyperplanes in $\mathbb{P}^3$ is 3-dimensional, we obtain $\dim X = 6$ which has codimension 3 in the $\mathbb{P}^9$ of all quadrics.

The approach presented in this section measures the number of components of minimal codimension quite well. At the same time it is very difficult to see components of larger codimension. One reason is that the rough approximations that we have made introduce errors in the order of $\frac{1}{p^{c+1}}$.

We will see in the next section how one can circumvent these problems.

2. Using Tangent Spaces

If $X \subset \mathbb{A}^n$ has components of different dimensions, the guessing method of Section 1 does not detect the smaller components.

If for example X is the union of a curve and a surface in $\mathbb{A}^3$, we expect the surface to have about p^2 points while the curve will have about p points (see Figure 7).

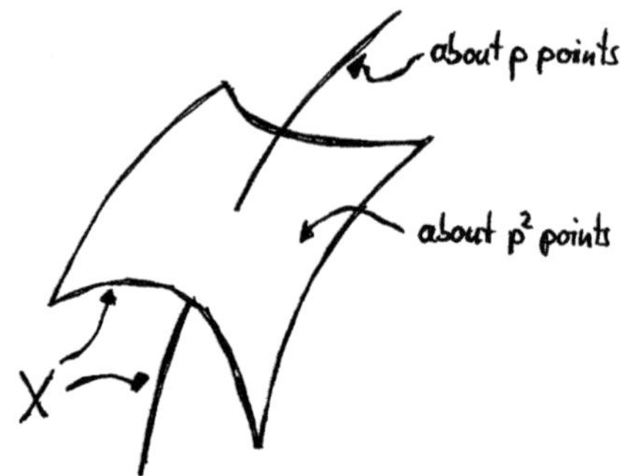

Figure 7. Expected number of $\mathbb{F}_p$ rational points on a union of a curve and a surface.

Using Heuristic 1.15 we obtain

$$\gamma_p = \frac{p^2 + p}{p^3} \approx \frac{1}{p}$$

indicating that X has 1 component of codimension 1. The codimension 2 component remains invisible.

Experiment 2.1. Let's check the above reasoning in an experiment. First define a function that produces a random inhomogeneous polynomial of given degree:

```
randomAffine = (d,R) -> sum apply(d+1,i->random(i,R))
```

with this we choose random polynomials F, G and H in 6 variables

```
n=6
R=ZZ[x_1..x_n];
F = randomAffine(2,R)
G = randomAffine(6,R);
H = randomAffine(7,R);
```

and consider the ideal $I = (FG, FH)$

```
I = ideal(F*G,F*H);
```

Finally, we evaluate the polynomials of I in 700 points of characteristic 7 and count how many of them lie in $X = V(I)$:

```
K = ZZ/7
t = tally apply(700,i->(
    0 == sub(I,random(K^1,K^n))
    ))
```

This yields

```
o9 = Tally{false => 598}
           true => 102
```

i.e. $\hat{\gamma}_7(X) = \frac{102}{700}$ which is very close to $\frac{1}{7}$. Consequently we would conclude that X has one component of codimension 1. The codimension 2 component given by $G = H = 0$ remains invisible.

To improve this situation we will look at tangent spaces. Let $a \in X \subset \mathbb{A}^n$ be a point and $T_{X,a}$ the tangent space of X in a. If $I_X = (f_1, \ldots, f_m)$, let

$$J_X = \begin{pmatrix} \frac{df_1}{dx_1} & \cdots & \frac{df_1}{dx_n} \\ \vdots & & \vdots \\ \frac{df_m}{dx_1} & \cdots & \frac{df_m}{dx_n} \end{pmatrix}$$

be the Jacobian matrix. We know from differential geometry that

$$T_{X,a} = \ker J_X(a) = \{v \in K^n \mid J_X(a)v = 0\}.$$

We can use tangent spaces to estimate the dimension of components of X:

Proposition 2.2. *Let $a \in X \subset \mathbb{A}^n$ be a point and $X' \subset X$ a component containing a. Then $\dim X' \leq \dim T_{X,a}$ with equality holding in smooth points of X.*

Proof. [8, II.1.4. Theorem 3]　　　　　　　　　　　　　　　　　□

In particular, we can use the dimension of the tangent space in a point $a \in X$ to separate points that lie on different dimensional components, at least if these components are non reduced (see Figure 8). For each of these sets we use Heuristic 1.15 to obtain

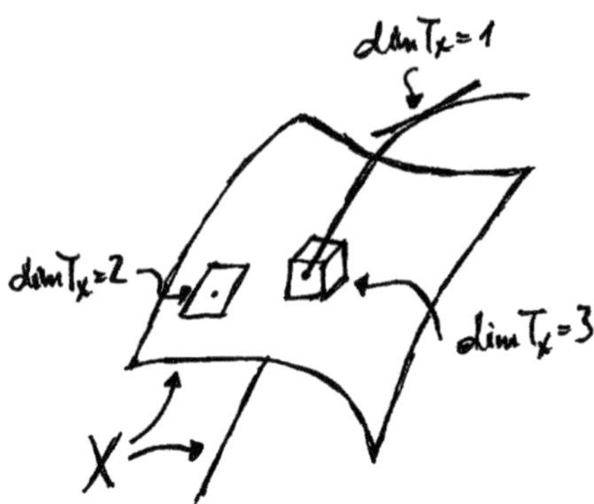

Figure 8. Dimension of tangent spaces in $\mathbb{F}_p$ rational points on a union of a curve and a surface.

Heuristic 2.3. Let $X \subset \mathbb{A}^n$ be a variety . If J_X is the Jacobian matrix of X and $a_1, \ldots, a_m \in \mathbb{A}^n$ are points, then the number of codimension c components of X is approximately

$$\frac{\#\{i \mid a_i \in X \text{ and } \operatorname{rank} J_X(a_i) = c\} \cdot p^c}{m}$$

Experiment 2.4. Let's test this heuristic by continuing Experiment 2.1. For this we first calculate the Jacobian matrix of the ideal I

```
J = jacobian I;
```

Now we check again 700 random points, but when we find a point on $X = V(I)$ we also calculate the rank of the Jacobian matrix in this point:

```
K=ZZ/7
time t = tally apply(700,i->(
    point := random(K^1,K^n);
    if sub(I,point) == 0 then
        rank sub(J,point)
))
```

The result is

```
o12 = Tally{0    => 2      }
            1    => 106
            2    => 14
            null => 578
```

Indeed, we find that there are about $\frac{106 \cdot 7^1}{700} = 1.06$ components of dimension 1 and about $\frac{14 \cdot 7^2}{700} = 0.98$ components of codimension 2. For codimension 0 the result is $\frac{2 \cdot 7^0}{700} \approx 0.003$ consistent with the fact that there are no components of codimension 0.

Remark 2.5. It is a little dangerous to give the measurements as in Experiment 2.4 without error bounds. Using the Poisson approximation of binomial distributions with small success probability we obtain

$$\sigma(\text{number of points found}) \approx \sqrt{\text{number of points found}}.$$

In the above experiment this gives

$$\# \text{ codim 1 components} = \frac{(106 \pm 2.58\sqrt{106}) \cdot 7^1}{700} = 1.06 \pm 0.27$$

and

$$\# \text{ codim 2 components} = \frac{(14 \pm 2.58\sqrt{14}) \cdot 7^2}{700} = 0.98 \pm 0.68.$$

where the error terms denote the 99% confidence interval. Notice that the measurement of the codimension 2 components is less precise. As a rule of thumb good error bounds are obtained if one searches until about 50 to 100 points of interest are found.

Remark 2.6. This heuristic assumes that the components do not intersect. If components do have high dimensional intersections, the heuristic might give too few components, since intersection points are singular and have lower codimensional tangent spaces.

In more involved examples calculating and storing the Jacobian matrix J_X can use a lot of time and space. Fortunately one can calculate $J_X(a)$ directly without calculating J_X first:

Proposition 2.7. *Let* $f \in \mathbb{F}_p[x_1, \ldots, x_n]$ *be a polynomial,* $a \in \mathbb{F}_p^n$ *a point and* $b \in \mathbb{F}_p^n$ *a vector. Then*

$$f(a + b\varepsilon) = f(a) + d_b f(a)\varepsilon \in \mathbb{F}_p^n[\varepsilon]/(\varepsilon^2).$$

with $d_b f$ *denoting the derivative of* f *in direction of* b. *In particular, if* $e_i \in \mathbb{F}_p^n$ *is the* i-*th unit vector, we have*

$$f(a + e_i\varepsilon) = f(a) + \frac{df}{dx_i}(a)\varepsilon.$$

Proof. Use the Taylor expansion. $\square$

Example 2.8. $f(x) = x^2 \implies f(1 + \varepsilon) = (1 + \varepsilon)^2 = 1 + 2\varepsilon = f(1) + \varepsilon f'(1)$

Experiment 2.9. To compare the two methods of calculating derivatives, we consider the determinant of a random matrix with polynomial entries. First we create a random matrix

```
K = ZZ/7                          -- characteristic 7
R = K[x_1..x_6]                   -- 6 variables
M = random(R^{5:0},R^{5:-2})      -- a random 5x5 matrix with
                                  -- quadratic entries
```

calculate the determinant

```
time F = det M;
-- used 13.3 seconds
```

and its derivative with respect to x_1.

```
time F1 = diff(x_1,F);
-- used 0.01 seconds
```

Now we substitute a random point:

```
point = random(K^1,K^6)
time sub(F1,point)
-- used 0. seconds
```

```
o7 = 2
```

By far the most time is used to calculate the determinant. With the ε-method this can be avoided. We start by creating a vector in the direction of x_1:

```
T = K[e]/(e^2)                    -- a ring with e^2=0
e1 = matrix{{1,0,0,0,0,0}}        -- the first unit vector
point1 = sub(point,T) + e*sub(e1,T)  -- point with direction
```

Now we first evaluate the matrix M in this vector

```
time M1 = sub(M,point1)
-- used 0. seconds
```

and only then take the determinant

```
time det sub(M,point1)
-- used 0. seconds
```

```
o12 = 2e + 1
```

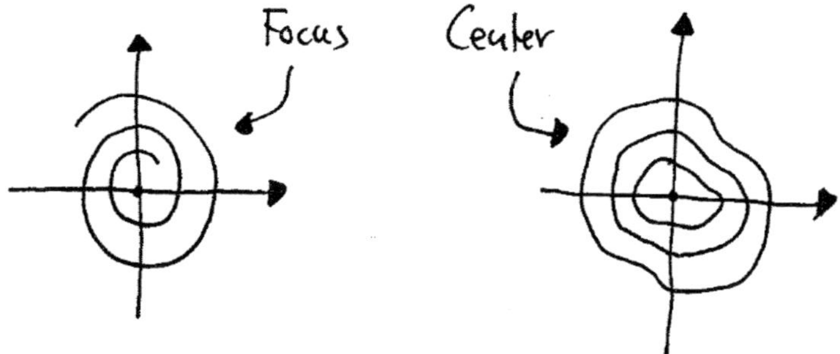

Figure 9. A focus and a center.

Indeed, the coefficient of e is the derivative of the determinant in this point. This method is too fast to measure by the **time** command of Macaulay 2. To get a better time estimate, we calculate the derivative of the determinant at 5000 random points:

```
time apply(5000,i->(
     point := random(K^1,K^6);      -- random point
        point1 := sub(point,T)+e*sub(e1,T); -- tangent direction
    det sub(M,point1);              -- calculate derivative
     ));
-- used 12.76 seconds
```

Notice that this is still faster than calculating the complete determinant once.

Remark 2.10. The ε-method is most useful if there exists a fast algorithm for evaluating the polynomials of interest. The determinant of an $n \times n$ matrix for example has $n!$ terms, so the time to evaluate it directly is proportional to $n!$. If we use Gauss elimination on the matrix first, the time needed drops to n^3.

For the remainder of this section we will look at an application of these methods to the Poincaré center problem. We start by considering the well known system of differential equations

$$\dot{x} = -y$$

$$\dot{y} = x$$

whose integral curves are circles around the origin. Let's now disturb these equations with polynomials P and Q whose terms have degree at least 2:

$$\dot{x} = -y + P$$

$$\dot{y} = x + Q.$$

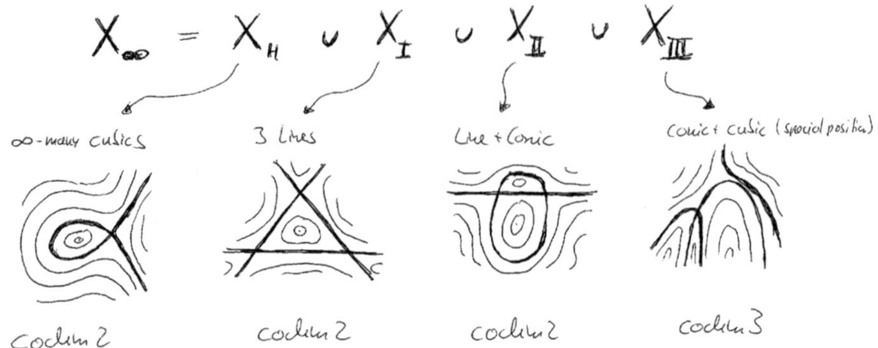

Figure 10. Geometric interpretation of the components of the center variety in the case $d = 2$.

Near zero the integral curves of the disturbed system are either closed or not. In the second case one says that the equations have a *focus* in $(0,0)$ while in the first case they have a *center* (see Figure 9).

The condition of having a center is closed in the space of all (P, Q):

Theorem 2.11 (Poincaré). *There exists an infinite series of polynomials f_i in the coefficients of P and Q such that*

$$\begin{matrix} \dot{x} = -y + P \\ \dot{y} = x + Q \end{matrix} \quad \text{has a center} \iff f_i(P, Q) = 0 \text{ for all } i.$$

We call $f_i(P, Q)$ the i-th focal value of (P, Q).

If the terms of P and Q have degree at most d then the f_i describe an algebraic variety X_∞ in the finite-dimensional space of pairs (P, Q). This variety is called the center variety.

Remark 2.12. By Hilbert's Basis Theorem $I_\infty := (f_0, f_1, \dots)$ is finitely generated. Unfortunately, Hilbert's Basis Theorem is not constructive, so it is a priory unknown how many generators I_∞ has. It is therefore useful to consider the i-th partial center varieties $X_i = V(f_0, \dots, f_i)$.

The following is known:

Theorem 2.13. *If $d = 2$ then the center variety has four components*

$$X_\infty = X_H \cup X_{III} \cup X_{II} \cup X_I \subset \mathbb{A}^6,$$

three of codimension 2 and one of codimension 3. Moreover $X_\infty = X_3$.

Proof. Decompose $I_3 = (f_1, f_2, f_3)$ with a computer algebra system and show that all solutions do have a center [9], [10]. □

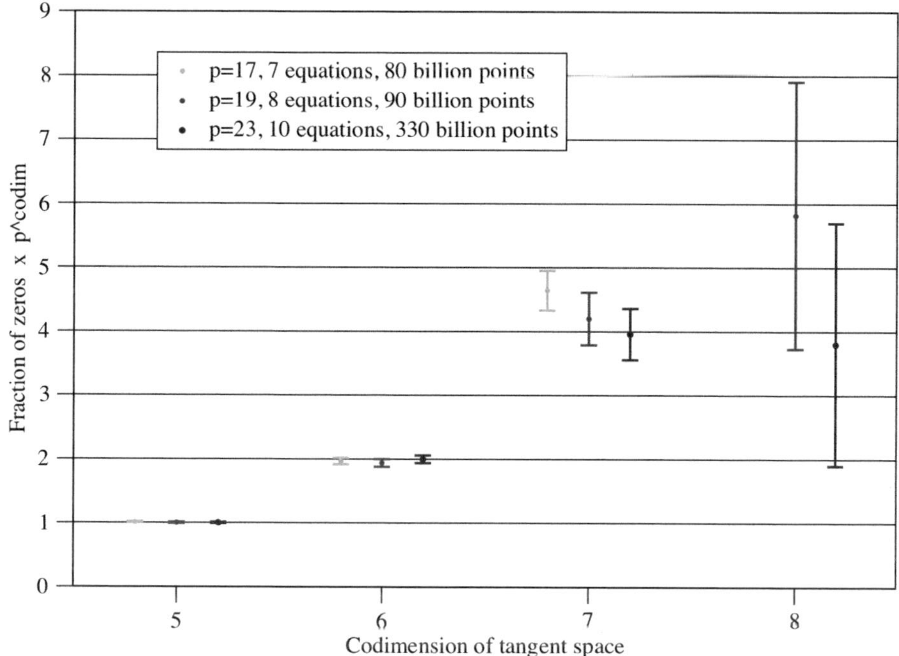

Figure 11. Measurements for the Poincaré center problem with $d = 3$

Looking at algebraic integral curves one even obtains a geometric interpretation of the components in this case (see Figure 10).

For $d = 3$ almost nothing is known. The best results so far are lists of centers given by Zoladec[11], [12]. The problem from a computer algebra perspective is that the f_i are too large to be handled, already f_5 has 5348 terms and it is known that $X_\infty \neq X_{10}$.

Experiment 2.14. Fortunately for our method, Frommer [9] has devised an algorithm to calculate $f_i(P, Q)$ for given (P, Q). A closer inspection shows that Frommer's Algorithm works over finite fields and will also calculate $f_i(P+\varepsilon P', Q+\varepsilon Q')$. So we have all ingredients to use Heuristic 2.3. Using a fast C++ implementation of Frommer's Algorithm by Martin Cremer and Jacob Kröker [13] we first check our method on the known degree 2 case. For this we evaluate $f_1, \ldots, f_{10}$ for $d = 2$ at 1.000.000 random points in characteristic 23. This gives

```
codim tangent space = 0: 5
codim tangent space = 1: 162
codim tangent space = 2: 5438
codim tangent space = 3: 88
```

Heuristic 2.3 translates this into

```
codim 0 components: 0.00 +/- 0.00
```

Table 1. Known families of cubic centers that could have codimension below 8 in $\mathbb{A}^{14}$ [12].

Type	Name	Codimension
Darboux	CD_1	5
Darboux	CD_2	6
Darboux	CD_3	7
Darboux	CD_4	7
Darboux	CD_5	7
Reversible	CR_1	≥ 6
Reversible	CR_5	≥ 7
Reversible	CR_7	≥ 7
Reversible	CR_{11}	≥ 7
Reversible	CR_{12}	≥ 7
Reversible	CR_{16}	≥ 7

```
codim 1 components: 0.00 +/- 0.00
codim 2 components: 2.87 +/- 0.10
codim 3 components: 1.07 +/- 0.29
```

This agrees well with Theorem 2.11.

For $d = 3$ we obtain the measurements in Figure 11. One can check these results against Zoladec's lists as depicted in Table 1. Here the measurements agree in codimension 5 and 6. In codimension 7 there seem to be 8 known families while we only measure 4. Closer inspection of the known families reveals that CR_5 and CR_7 are contained in CD_4 and that CR_{12} and CR_{16} are contained in CD_2 [14]. After accounting for this our measurement agrees with Zoladec's results and we conjecture that Zoladec's lists are complete up to codimension 7.

3. Existence of a Lift to Characteristic Zero

Often one is not interested in characteristic p solutions, but in solutions over $\mathbb{C}$. Unfortunately, not all solutions over $\mathbb{F}_p$ lift to characteristic 0.

Example 3.1. Consider the variety $X = V(3x) \subset \mathbb{P}^1_{\mathbb{Z}}$ over $\operatorname{Spec}\mathbb{Z}$. As depicted in Figure 12, X decomposes into two components: $V(3) = \mathbb{P}^1_{\mathbb{F}_3}$ which lives only over $\mathbb{F}_3$ and $V(x) = \{(0 : 1)\}$ which has fibers over all of $\operatorname{Spec}\mathbb{Z}$. In particular, the point $(1 : 0) \in \mathbb{P}^1_{\mathbb{F}_3} \subset X$ does not lift to characteristic 0.

To prove that a given solution point over $\mathbb{F}_p$ does lift to characteristic zero the following tool is very helpful:

Proposition 3.2 (Existence of a Lifting). *Let* $X \subset Y \subset \mathbb{A}^n_{\mathbb{Z}}$ *be varieties with* $\dim Y_{\mathbb{F}_p} = \dim Y_{\mathbb{Z}} - 1$ *for all p and $X \subset Y$ determinantal, i.e. there exists a vector bundle morphism*

$$\phi \colon E \to F$$

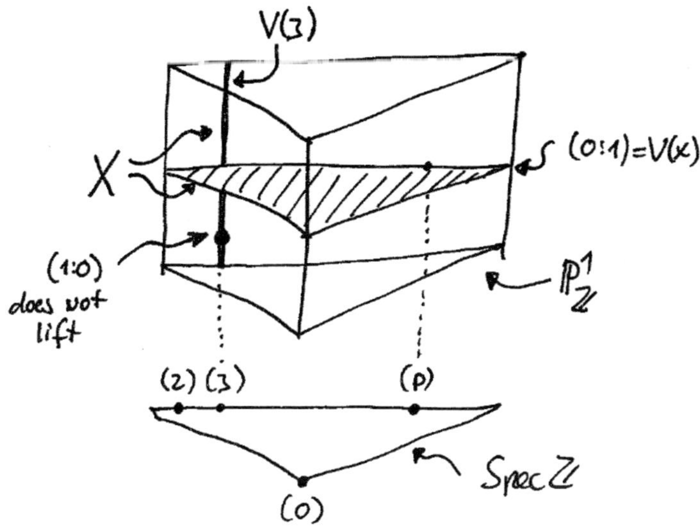

Figure 12. The vanishing set of $3x$ in $\mathbb{P}^1_\mathbb{Z}$ over $\operatorname{Spec}\mathbb{Z}$

on Y and a number $r \leq \min(\operatorname{rank} E, \operatorname{rank} F)$ such that $X = X_r(\phi)$ is the locus where ϕ has rank at most r. If $x \in X_{\mathbb{F}_n}$ is a point with

$$\dim T_{X_{\mathbb{F}_p},x} = \dim Y_{\mathbb{F}_p} - (\operatorname{rank} E - r)(\operatorname{rank} F - r)$$

then X is smooth in x and there exists a component Z of $X_\mathbb{Z}$ containing x and having a nonzero fiber over (0).

Proof. Set $d = \dim Y_{\mathbb{F}_p} - (\operatorname{rank} E - r)(\operatorname{rank} F - r)$. Since $X_{\mathbb{F}_p}$ is determinantal, we have

$$\dim Z_{\mathbb{F}_p} \geq d$$

for every irreducible component $Z_{\mathbb{F}_p}$ of $X_{\mathbb{F}_p}$ and d is the expected dimension of $Z_{\mathbb{F}_p}$ [15, Ex. 10.9, p. 245]. If $Z_{\mathbb{F}_p}$ contains the point x we obtain

$$d \leq \dim Z_{\mathbb{F}_p} \leq \dim T_{Z_{\mathbb{F}_p},x} \leq \dim T_{X_{\mathbb{F}_p},x} = d$$

by our assumptions. So $Z_{\mathbb{F}_p}$ is of dimension d and smooth in x. Let now $Z_\mathbb{Z}$ be a component of $X_\mathbb{Z}$ that contains $Z_{\mathbb{F}_p}$ and x. Since $X_\mathbb{Z}$ is determinantal in $Y_\mathbb{Z}$ and $\dim Y_\mathbb{Z} = \dim Y_{\mathbb{F}_p} + 1$ we have

$$\dim Z_\mathbb{Z} \geq d + 1.$$

Since $\dim Z_{\mathbb{F}_p} = d$ the fiber of $Z_\mathbb{Z}$ over p cannot contain all of $Z_\mathbb{Z}$. Indeed, in this case we would have $Z_{\mathbb{F}_p} = Z_\mathbb{Z}$ since both are irreducible, but $\dim Z_{\mathbb{F}_p} \neq \dim Z_\mathbb{Z}$. It follows that $Z_\mathbb{Z}$ has nonempty fibers over an open subset of $\operatorname{Spec} Z_\mathbb{Z}$ and therefore also over (0) [16], [17]. $\qquad\square$

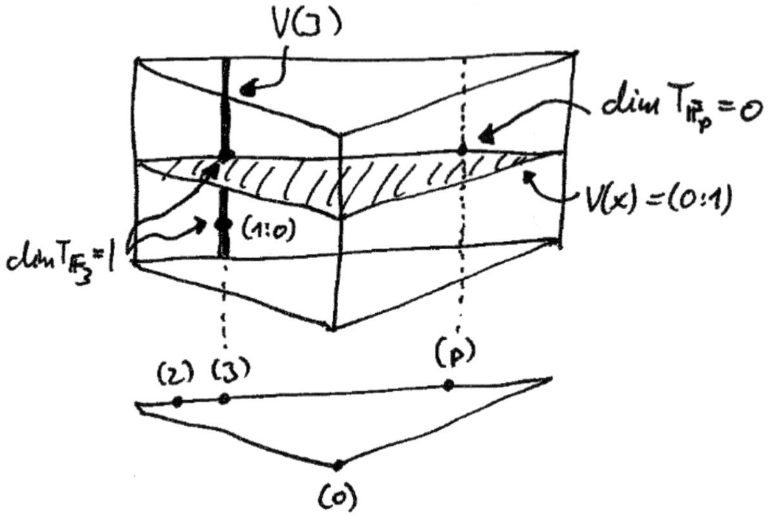

Figure 13. Tangent spaces in several points of $X = V(3x) \subset \mathbb{P}^1_\mathbb{Z}$ over $\mathrm{Spec}\,\mathbb{Z}$

Example 3.3. The variety $X = V(3x)$ is determinantal on $Y = \mathbb{P}^1_\mathbb{Z}$ since it is the rank 0 locus of the vector bundle morphism

$$\phi \colon \mathcal{O}_{\mathbb{P}^1_\mathbb{Z}} \xrightarrow{\;3x\;} \mathcal{O}_{\mathbb{P}^1_\mathbb{Z}}(1).$$

Furthermore $\dim \mathbb{P}^1_{\mathbb{F}_p} = 1 = \dim \mathbb{P}^1_\mathbb{Z} - 1$ for all p. The expected dimension of $X_{\mathbb{F}_p}$ is therefore $1 - (1-0) \cdot (1-0) = 0$. As depicted in Figure 13 we have three typical examples:

   (i) $x = (0 : 1)$ over $\mathbb{F}_p$ with $p \neq 3$. Here the tangent space over $\mathbb{F}_p$ is zero dimensional and the point lifts according to Proposition 3.2.
  (ii) $x = (0 : 1)$ over $\mathbb{F}_3$. Here the tangent space is 1-dimensional and Proposition 3.2 does not apply. Even though the point does lift.
 (iii) $x = (1 : 0)$ over $\mathbb{F}_3$. Here the tangent space is also 1-dimensional and Proposition 3.2 does not apply. In this case the point does not lift.

This method has been used first by Frank Schreyer [16] to construct new surfaces in $\mathbb{P}^4$ which are not of general type. The study of such surfaces started in 1989 when Ellingsrud and Peskine showed that their degree is bounded [18] and therefore only finitely many families exist. Since then the degree bound has been sharpened by various authors, most recently by [19] to 52. On the other hand a classification is only known up to degree 10 and examples are known up to degree 15 (see [19] for an overview and references).

Here I will explain how Cord Erdenberger, Katharina Ludwig and I found a new family of rational surfaces S of degree 11 and sectional genus 11 in $\mathbb{P}^4$ with finite field experiments.

Our plan is to realize S as a blowup of $\mathbb{P}^2$. First we consider some restrictions on the linear system that embeds S into $\mathbb{P}^4$:

Proposition 3.4. *Let* $S = \mathbb{P}^2_{\mathbb{C}}(p_1, \ldots, p_l)$ *be the blowup of* $\mathbb{P}^2_{\mathbb{C}}$ *in* l *distinct points. We denote by* $E_1, \ldots, E_l$ *the corresponding exceptional divisors and by* L *the pullback of a general line in* $\mathbb{P}^2_{\mathbb{C}}$ *to* S. *Let* $|aL - \sum_{i=1}^{l} b_i E_i|$ *be a very ample linear system of dimension four and set* $\beta_j = \#\{i \,|\, b_i = j\}$. *Then*

$$d = a^2 - \sum_j \beta_j j^2$$

$$\pi = \binom{a-1}{2} - \sum_j \beta_j \binom{j}{2}$$

$$K^2 = 9 - \sum_j \beta_j.$$

where d *is the degree,* π *the sectional genus and* K *the canonical divisor of* S.

Proof. Intersection theory on S [17, Corollary 4.1]. $\qquad\qquad\square$

By the double point formula for surfaces in $\mathbb{P}^4$ [20, Appendix A, Example 4.1.3] a rational surface of degree 11 and sectional genus 11 must satisfy $K^2 = -11$. For fixed a the equations above can be solved by integer programming, using for example the algorithm described in Chapter 8 of [21].

In the case $a < 9$ we find that there are no solutions. For $a = 9$ the only solution is $\beta_3 = 1$, $\beta_2 = 14$ and $\beta_1 = 5$. Our first goal is therefore to find 5 simple points, 14 double points and one triple point in $\mathbb{P}^2$ such that the ideal of the union of these points contains 5 polynomials of degree 9.

To make the search fast, we would like to use characteristic 2. The difficulty here is that $\mathbb{P}^2$ contains only 7 rational points, while we need 20. Our solution to this problem was to choose

$$P \in \mathbb{P}^2(\mathbb{F}_2) \qquad\qquad Q \in \mathbb{P}^2(\mathbb{F}_{2^{14}}) \qquad\qquad R \in \mathbb{P}^2(\mathbb{F}_{2^5})$$

such that the Frobenius orbit of Q and R are of length 14 and 5 respectively. The ideals of the orbits are then defined over $\mathbb{F}_2$.

```
-- define coordinate ring of P^2 over F_2
F2 = GF(2)
S2  = F2[x,y,z]

-- define coordinate ring of P^2 over F_2^14 and F_2^5
St  = F2[x,y,z,t]
use St; I14 = ideal(t^14+t^13+t^11+t^10+t^8+t^6+t^4+t+1);
        S14 = St/I14
use St; I5 = ideal(t^5+t^3+t^2+t+1); S5 = St/I5
```

```
-- the random points
use S2; P = matrix{{0_S2, 0_S2, 1_S2}}
use S14;Q = matrix{{t^(random(2^14-1)), t^(random(2^14-1)), 1_S14}}
use S5; R = matrix{{t^(random 31), t^(random 31), 1_S5}}

-- their ideals
IP = ideal ((vars S2)*syz P)
IQ = ideal ((vars S14)_{0..2}*syz Q)
IR = ideal ((vars S5)_{0..2}*syz R)

-- their orbits
f14 = map(S14/IQ,S2); Qorbit = ker f14
degree Qorbit    -- hopefully degree = 14

f5 = map(S5/IR,S2); Rorbit = ker f5
degree Rorbit    -- hopefully degree = 5
```

If Q and R have the correct orbit length we calculate $|9H - 3P - 2Q - R|$

```
-- ideal of 3P
P3 = IP^3;

-- orbit of 2Q
f14square = map(S14/IQ^2,S2); Q2orbit = ker f14square;

-- ideal of 3P + 2Qorbit + 1Rorbit
I = intersect(P3,Q2orbit,Rorbit);

-- extract 9-tics
H = super basis(9,I)
rank source H    -- hopefully affine dimension = 5
```

If at this point we find 5 sections, we check that there are no unassigned base points

```
-- count basepoints (with multiplicities)
degree ideal H    -- hopefully degree = 1x6+14x3+1x5 = 53
```

If this is the case, the next difficulty is to check if the corresponding linear system is very ample. On the one hand this is an open condition, so it should be satisfied by most examples, on the other hand we are in characteristic 2, so exceptional loci can have very many points. An irreducible divisor for example already contains approximately half of the rational points.

```
-- construct map to P^4
T = F2[x0,x1,x2,x3,x4]
```

```
fH = map(S2,T,H);

-- calculate the ideal of the image
Isurface = ker fH;

-- check invariants
betti res coker gens Isurface
codim Isurface      -- codim = 2
degree Isurface    -- degree = 11
genera Isurface    -- genera = {0,11,10}

-- check smoothness
J = jacobian Isurface;
mJ = minors(2,J) + Isurface;
codim mJ   -- hopefully codim = 5
```

Indeed, after about 100.000 trials one comes up with the points

```
use S14;Q = matrix{{t^11898, t^137, 1_S14}}
use S5; R = matrix{{t^6, t^15, 1_S5}}
```

These satisfy all of the above conditions and prove that rational surfaces of degree 11 and sectional genus 11 in $\mathbb{P}^4$ exist over $\mathbb{F}_2$.

As a last step we have to show that this example lifts to char 0. For this we consider the morphism

$$\tau_k : H^0(\mathcal{O}_{\mathbb{P}^2_{\mathbb{Z}}}(a)) \to \mathcal{O}_{\mathbb{P}^2_{\mathbb{Z}}}(a) \oplus 3\mathcal{O}_{\mathbb{P}^2_{\mathbb{Z}}}(a-1) \oplus \cdots \oplus \binom{k+2}{2} \mathcal{O}_{\mathbb{P}^2_{\mathbb{Z}}}(a-k)$$

on $\mathbb{P}^2_{\mathbb{Z}}$ that associates to each polynomial of degree a the coefficients of its Taylor expansion up to degree k in a given point P.

Lemma 3.5. *If $a > k$ then the image of τ_k is a vector bundle $\mathcal{F}_k$ of rank $\binom{k+2}{2}$ over* $\operatorname{Spec} \mathbb{Z}$.

Proof. In each point we consider an affine 2-dimensional neighborhood where we can choose the $\binom{k+2}{2}$ coefficients of the affine Taylor expansion independently. This shows that the image has at least this rank everywhere. If follows from the Euler relation for homogeneous polynomials

$$x\frac{df}{dx} + y\frac{df}{dy} + z\frac{df}{dz} = (\deg f) \cdot f$$

that this is also the maximal rank. $\qquad\square$

Now set $Y_{\mathbb{Z}} = \text{Hilb}_{1,\mathbb{Z}} \times \text{Hilb}_{14,\mathbb{Z}} \times \text{Hilb}_{5,\mathbb{Z}}$ where $\text{Hilb}_{k,\mathbb{Z}}$ denotes the Hilbert scheme of k points in $\mathbb{P}^2_{\mathbb{Z}}$ over $\text{Spec}\,\mathbb{Z}$, and let

$$X_{\mathbb{Z}} = \{(p, q, r) \mid h^0(9L - 3p - 2q - 1r) \geq 5\} \subset Y_{\mathbb{Z}}$$

be the subset where the linear system of nine-tics with a triple point in p, double points in q and single base points in r is at least of projective dimension 4.

Proposition 3.6. *There exist vector bundles E and F of ranks 55 and 53 respectively on $Y_{\mathbb{Z}}$ and a morphism*

$$\phi\colon E \to F$$

such that $X_{50}(\phi) = X_{\mathbb{Z}}$.

Proof. On the Cartesian product

$$\begin{array}{ccc}
\text{Hilb}_{d,\mathbb{Z}} \times \mathbb{P}^2_{\mathbb{Z}} & \xrightarrow{\;\pi_2\;} & \mathbb{P}^2_{\mathbb{Z}} \\
\Big\downarrow{\scriptstyle \pi_1} & & \\
\text{Hilb}_{d,\mathbb{Z}} & &
\end{array}$$

we have the morphisms

$$\pi_2^* \tau_k\colon H^0(\mathcal{O}_{\mathbb{P}^2_{\mathbb{Z}}}(9)) \otimes \mathcal{O}_{\text{Hilb}_{d,\mathbb{Z}} \times \mathbb{P}^2_{\mathbb{Z}}} \to \pi_2^* \mathcal{F}_k.$$

Let now $P_d \subset \text{Hilb}_{d,\mathbb{Z}} \times \mathbb{P}^2_{\mathbb{Z}}$ be the universal set of points. Then P_d is a flat family of degree d over $\text{Hilb}_{d,\mathbb{Z}}$ and

$$\mathcal{G}_k := (\pi_1)_*((\pi_2^* \mathcal{F}_k)|_{P_d})$$

is a vector bundle of rank $d\binom{k+2}{2}$ over $\text{Hilb}_{d,\mathbb{Z}}$. On

$$Y_{\mathbb{Z}} = \text{Hilb}_{1,\mathbb{Z}} \times \text{Hilb}_{14,\mathbb{Z}} \times \text{Hilb}_{5,\mathbb{Z}}$$

the induced map

$$\phi\colon H^0(\mathcal{O}_{\mathbb{P}^2_{\mathbb{Z}}}(9)) \otimes \mathcal{O}_{X_{\mathbb{Z}}} \xrightarrow{\;\tau_2 \oplus \tau_1 \oplus \tau_0\;} \sigma_1^* \mathcal{G}_2 \oplus \sigma_{14}^* \mathcal{G}_1 \oplus \sigma_5^* \mathcal{G}_0$$

has the desired properties, where σ_d denotes the projection to $\text{Hilb}_{d,\mathbb{Z}}$. $\qquad\square$

So we have to show that the tangent space of $X_{\mathbb{F}_2}$ in our base locus has codimension $(55 - 50)(53 - 50) = 15$. This can be done by explicitly calculating the differential of ϕ in our given base scheme using the ε-method. The script is too long for this paper, but can be downloaded at [22]. Indeed, we find that the codimension of the tangent space is 15, so this shows that our example lies on an irreducible component that is defined over an open subset of $\text{Spec}\,\mathbb{Z}$.

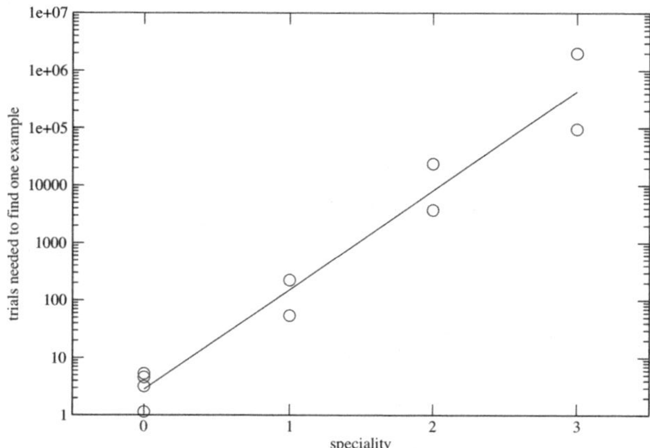

Figure 14. The difficulty of finding a surface grows exponentially with the speciality

Remark 3.7. The overall time to find smooth surfaces that lift to characteristic zero can be substantially reduced if one calculates the tangent space of a given point (P, Q, R) in the Hilbert scheme $X_{\mathbb{Z}}$ directly after establishing $|9H - 3P - 2Q - R| = \mathbb{P}^4$. One then needs to check very ampleness only for smooth points of $X_{\mathbb{Z}}$. This is useful since the tangent space calculation is just a linear question, while the check for very ampleness requires Gröbner bases. We use a very fast C-implementation by Jakob Kröker to do the whole search algorithm up to checking smoothness. Only the (very few) remaining examples are then checked for very ampleness using Macaulay 2.

Experiment 3.8. We also tried to reconstruct the other known rational surfaces in $\mathbb{P}^4$ with our program. The number of trials needed is depicted in Figure 14. The expected codimension of X in the corresponding Hilbert scheme turns out to be 5 times the speciality $h^1(\mathcal{O}_X(1))$ of the surface. As expected, the logarithm of the number of trials needed to find a surface is proportional to the codimension of X.

Remark 3.9. We could not reconstruct all known families. The reason for this is that we only look at examples where the base points of a given multiplicity form an irreducible Frobenius orbit. In some cases such examples do not exist for geometric reasons.

Experiment 3.10. Looking at the linear system $|14H - 4P - 3Q - R|$ with $\deg P = 8$, $\deg Q = 6$ and $\deg R = 2$, we find rational surfaces of degree 12 and sectional genus 12 in $\mathbb{P}^4$ with this method (not published) .

4. Finding a Lift

In some good cases characteristic p methods even allow one to find a solution over $\mathbb{Q}$ quickly. Basically this happens when the solution set is zero dimensional with two different flavors.

The first good situation, depicted in Figure 15, arises when $X = V(f_1, \ldots, f_m) \subset \mathbb{A}^n_{\mathbb{Z}}$ has a unique solution over $\bar{\mathbb{Q}}$, maybe with high multiplicity. In this case it follows that the solution is defined over $\mathbb{Q}$.

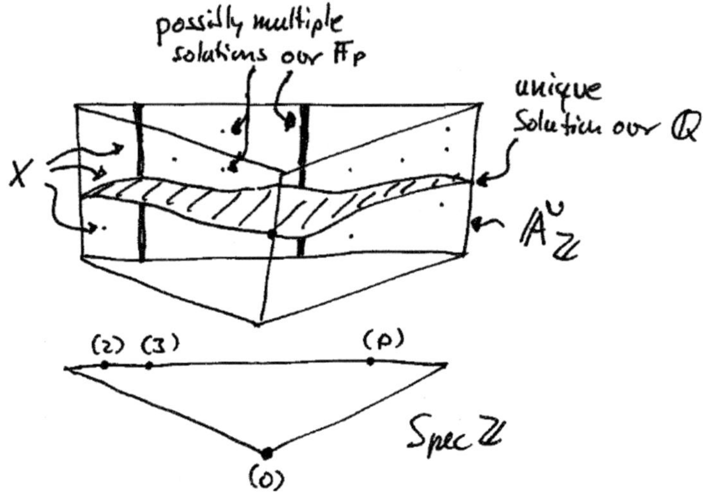

Figure 15. A scheme over $\operatorname{Spec}\mathbb{Z}$ with a *unique* solution over $\bar{\mathbb{Q}}$, possibly with high multiplicity

Algorithm 4.1. If the coordinates of the unique solution over $\mathbb{Q}$ are even in $\mathbb{Z}$ one can find this solution as follows:

- (i) Reduce mod p_i and test *all* points in $\mathbb{F}_{p_i}^n$
- (ii) Find many primes p_i with a unique solution in $\mathbb{F}_{p_i}^n$
- (iii) Use Chinese remaindering to find a solution mod $\prod_i p_i \gg 0$.
- (iv) Test if this is a solution over $\mathbb{Z}$. If not, find more primes p_i with unique solutions over $\mathbb{F}_{p_i}$.

Remark 4.2. Even if the solution y over $\mathbb{Q}$ is unique, there can be several solutions over $\mathbb{F}_p$. Since the codimension of points in $\mathbb{A}^n$ is n we expect that the probability of a random point $x \in \mathbb{A}^n$ to satisfy $x \in X$ is $\frac{1}{p^n}$ by Heuristic 1.15. We therefore expect that the probability of $x \notin X$ for all points $x \neq y$ is

$$\left(1 - \frac{1}{p^n}\right)^{p^n - 1} \approx \frac{1}{e}.$$

Experiment 4.3. Let's use this Algorithm 4.1 to solve

$$-8x^2 - xy - 7y^2 + 5238x - 11582y - 7696 = 0$$

$$4xy - 10y^2 - 2313x - 16372y - 6462 = 0$$

For this we need a function that looks at all points over a given prime:

```
allPoints = (I,p) -> (
   K = ZZ/p;
   flatten apply(p,i->
       flatten apply(p,j->
           if (0==codim sub(I,matrix{{i*1_K,j*1_K}}))
           then {(i,j)}
           else {}
       ))
   )
```

With this we look for solutions of our equations over the first nine primes.

```
R = ZZ[x,y]
-- the equations
I = ideal (-8*x^2-x*y-7*y^2+5238*x-11582*y-7696,
    4*x*y-10*y^2-2313*x-16372*y-6462)

-- look for solutions
tally apply({2,3,5,7,11,13,17,19,23},p->(p,time allPoints(I,p)))
```

We obtain:

```
o8 = Tally{(2, {(0, 0)}) => 1
          (3, {(0, 2), (1, 0), (2, 0)}) => 1
          (5, {(4, 1)}) => 1
          (7, {(2, 3), (5, 5)}) => 1
          (11, {(2, 7), (8, 1)}) => 1
          (13, {(3, 4), (12, 6)}) => 1
          (17, {(10, 8)}) => 1
          (19, {(1, 3), (1, 17), (18, 5), (18, 18)}) => 1
          (23, {(15, 8)}) => 1
```

As expected for the intersection of two quadrics we find at most 4 solutions. Over four primes we find unique solutions, which is reasonably close to the expected number $9/e \approx 3.31$. We now combine the information over these four primes using the Chinese remainder Theorem.

```
-- Chinese remaindering
-- given solutions mod m and n find
-- a solution mod m*n
-- sol1 = (n,solution)
-- sol2 = (m,solution)
chinesePair = (sol1, sol2) -> (
     n = sol1#0;an = sol1#1;
     m = sol2#0;am = sol2#1;
     drs = gcdCoefficients(n,m);
```

```
      -- returns {d,r,s} so that a*r + b*s is the
      -- greatest common divisor d of a and b.
      r = drs#1;
      s = drs#2;
      amn = s*m*an+r*n*am;
      amn = amn - (round(amn/(m*n)))*(m*n);
      if (drs#0) == 1 then (m*n,amn)
                          else print "m and n not coprime"
      )
```

```
-- take a list {(n_1,s_1),...,(n_k,s_k)}
-- and return (n,a) such that
-- n = n_1 * ... * n_k    and
-- s_i = a mod n_i
chineseList = (L) -> (fold(L,chinesePair))
```

```
-- x coordinate
chineseList({(2,0),(5,4),(17,10),(23,15)})
-- y coordinate
chineseList({(2,0),(5,1),(17,8),(23,8)})
```

This gives

```
o11 = (3910, 1234)
o12 = (3910, -774)
```

i.e. $(1234, -774)$ is the unique solution mod $3910 = 2 \cdot 5 \cdot 17 \cdot 23$. Substituting this into the original equations over $\mathbb{Z}$ shows that this is indeed a solution over $\mathbb{Z}$.

```
sub(I,matrix{{1234,-774}})
```

```
o13 = ideal (0, 0)
```

If the unique solution does not have $\mathbb{Z}$ but $\mathbb{Q}$ coordinates then one can find the solution using the extended Euclidean Algorithm [23, Section 5.10].

Example 4.4. Let's try to find a small solution to the equation

$$\frac{r}{s} \equiv 7 \mod 37.$$

Each solution satisfies

$$r = 7s + 37t$$

with s and t in $\mathbb{Z}$. Using the extended Euclidean Algorithm

	r	s	t	r/s
	37	0	1	
-5	7	1	0	$7/1$
-3	2	-5	1	$-2/5$
	1	16	-3	$1/16$

we find the solution

$$1 = \gcd(7, 37) = 7 \cdot 16 + 37 \cdot (-3)$$

to our linear equation. Observe, however, that the intermediate step in the Euclidean Algorithm also gives solutions, most of them with small coefficients. Indeed, $r/s = -2/5$ is a solution with $r, s \leq \sqrt{37}$ which is the best that we can expect.

If we find a small solution by this method, we even can be sure that it is the only one satisfying the congruence:

Proposition 4.5. *There exist at most two solutions (r, s) of*

$$r \equiv as + bt \mod m$$

that satisfy $r, s \leq \sqrt{m}$. If a solution satisfies $r, s \leq \frac{1}{2}\sqrt{m}$, then this solution is unique.

Proof. [23, Section 5.10] $\qquad\qquad\qquad\qquad\qquad\qquad\qquad\qquad\qquad\qquad\square$

Experiment 4.6. Let's find a solution to

$$176x^2 + 148xy + 301y^2 - 742x + 896y + 768 = 0$$

$$-25xy + 430y^2 + 33x + 1373y + 645 = 0$$

As in Experiment 4.3 we search for primes with unique solutions

```
I = ideal (176*x^2+148*x*y+301*y^2-742*x+896*y+768,
      -25*x*y+430*y^2+33*x+1373*y+645)
tally apply({2,3,5,7,11,13,17,19,23,29,31,37,41},
      p->(p,time allPoints(I,p)))
```

and obtain

```
o10 = Tally{(2, {(1, 0)}) => 1
        (3, {(0, 0), (0, 1), (2, 0)}) => 1
        (5, {(3, 2), (4, 1)}) => 1
        (7, {(2, 6), (4, 0)}) => 1
        (11, {}) => 1
        (13, {(5, 10)}) => 1
```

```
(17, {(5, 4), (9, 13), (11, 16), (12, 12)}) => 1
(19, {(3, 15), (8, 6), (13, 15), (17, 1)}) => 1
(23, {(15, 18), (19, 12)}) => 1
(29, {(26, 15), (28, 9)}) => 1
(31, {(7, 22)}) => 1
(37, {(14, 18)}) => 1
(41, {(0, 23)}) => 1
```

Notice that there is no solution mod 11. If there is a solution over $\mathbb{Q}$ this means that 11 has to divide at least one of the denominators. Chinese remaindering gives a solution mod $2 \cdot 13 \cdot 31 \cdot 37 \cdot 41 = 1222702$:

```
-- x coordinate
chineseList({(2,1),(13,5),(31,7),(37,14),(41,0)})
o11 = (1222702, 138949)
-- y coordinate
chineseList({(2,0),(13,10),(31,22),(37,18),(41,23)})
o12 = (1222702, -526048)
```

Substituting this into the original equations gives

```
sub(I,matrix{{138949,-526048}})
o13 = ideal (75874213835186, 120819022681578)
```

so this is not a solution over $\mathbb{Z}$. To find a small possible solution over $\mathbb{Q}$ we use an implementation of the extended Euclidean Algorithm from [23, Section 5.10].

```
-- take (a,n) and calculate a solution to
--    r =  as mod n
-- such that r,s < sqrt(n).
-- return (r/s)
recoverQQ = (a,n) -> (
   r0:=a;s0:=1;t0:=0;
   r1:=n;s1:=0;t1:=1;
   r2:=0;s2:=0;t2:=0;
   k := round sqrt(r1*1.0);
   while k <= r1 do (
       q = r0//r1;
       r2 = r0-q*r1;
       s2 = s0-q*s1;
       t2 = t0-q*t1;
       --print(q,r2,s2,t2);
       r0=r1;s0=s1;t0=t1;
       r1=r2;s1=s2;t1=t2;
       );
   (r2/s2)
   )
```

This yields

```
-- x coordinate
recoverQQ(138949,2*13*31*37*41)

        123
o21 = ---
         22
```

Notice that Macaulay reduced 246/44 to 123/22 in this case. Therefore this is not a solution mod 2. Indeed, no solution mod 2 exists, since the denominator of the x coordinate is divisible by 2. For the y coordinate we obtain

```
-- y coordinate
recoverQQ(-526048,2*13*31*37*41)

         77
o22 = - --
         43
```

As a last step we substitute this $\mathbb{Q}$-point into the original equations.

```
sub(I, matrix{{123/22,-77/43}})
o24 = ideal (0, 0)
```

This shows that we have indeed found a solution over $\mathbb{Q}$. Notice also that as argued above one of the denominators is divisible by 2 and the other by 11.

Remark 4.7.

(i) The assumption that we have a unique solution over $\mathbb{Q}$ is not as restrictive as it might seem. If we have for example 2 solutions, then at least the line through them is unique. More generally, if the solution set over $\mathbb{Q}$ lies on k polynomials of degree d then the corresponding point in the Grassmannian $\mathbb{G}(k, \binom{d+n}{n})$ is unique.

(ii) Even if we do not have isolated solutions, we can use this method to find the polynomials of $\mathrm{rad}(I(X))$, at least if the polynomials are of small degree.

(iii) For this method we do not need explicit equations, rather an algorithm that decides whether a point lies on X is enough. This is indeed an important distinction. It is for example easy to check whether a given hypersurface is singular, but very difficult to give an explicit discriminant polynomial in the coefficients of the hypersurface that vanishes if and only if it is singular.

Before we finish this tutorial by looking at a very nice application of this method by Oliver Labs, we will look briefly at a second situation in which we can find explicit solutions over $\mathbb{Q}$. I learned this method from Noam Elkies in his talk at the Clay Mathematics Institute Summer School "Arithmetic geometry" 2006.

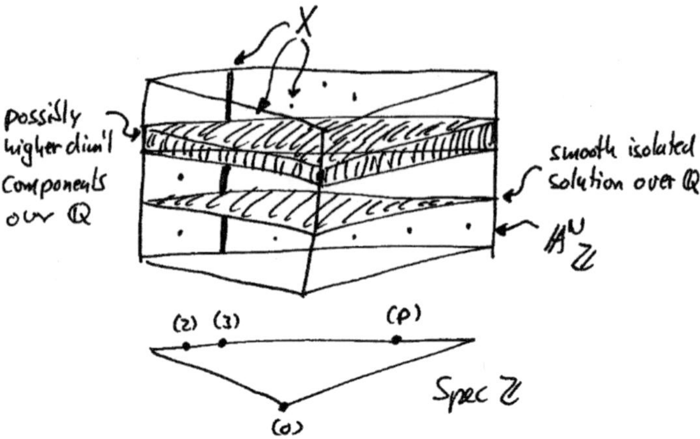

Figure 16. A scheme X over $\operatorname{Spec}\mathbb{Z}$ with a smooth isolated solution over $\mathbb{Q}$.

Algorithm 4.8. Assume that X has a *smooth* point x over $\mathbb{Q}$ that is isolated over $\bar{\mathbb{Q}}$ as depicted in Figure 16, and that p is a prime that does not divide the denominators of the coordinates of x. Then we can find this point as follows:

 (i) Reduce mod p and test all points.
 (ii) Calculate the tangent spaces at the found points. If the dimension of such a tangent space is 0 then the corresponding point is smooth and isolated.
 (iii) Lift the point mod p^k with k large using p-adic Newton iteration, as explained in Prop 4.9.

Proposition 4.9. *Let $a \in \mathbb{A}^n_{\mathbb{Z}}$ be a solution of*

$$f_1(a) = \cdots = f_n(a) = 0 \quad \mod p^k$$

and assume that the Jacobian matrix $J = \left(\frac{df_i}{dx_j}\right)$ is invertible at $a \mod p$. Then

$$a' = a - (f_1(a), \dots, f_n(a))J(a)^{-1}$$

is a solution mod p^{2k}.

Proof. Use the Taylor expansion as in the proof of Newton iteration. $\qquad\square$

Experiment 4.10. Let's solve the equations of Experiment 4.3 using p-adic Newton iteration. For this we need some functions for modular calculations:

```
-- calculate reduction of a matrix M mod n
modn = (M,n) -> (
    matrix apply(rank target M, i->
        apply(rank source M,j-> M_j_i-round(M_j_i/n)*n)))

-- divide a matrix of integers by an integer
```

```
-- (in our application this division will not have a remainder)
divn = (M,n) -> (
    matrix apply(rank target M, i->
        apply(rank source M,j-> M_j_i//n)))


-- invert number mod n
invn = (i,n) -> (
    c := gcdCoefficients(i,n);
    if c#0 == 1 then c#1 else "error"
    )


-- invert a matrix mod n
-- M a square matrix over ZZ
-- (if M is not invertible mod n, then 0 is returned)
invMatn = (M,n) -> (
    Mn  := modn(M,n);
    MQQ := sub(Mn,QQ);
    detM = sub(det Mn,QQ);
    modn(invn(sub(detM,ZZ),n)*sub(detM*MQQ^-1,ZZ),n)
    )
```

With this we can implement Newton iteration. We will represent a point by a pair (P, eps) with P a matrix of integers that is a solution modulo eps.

```
-- (P,eps)  an approximation mod eps (contains integers)
-- M         affine polynomials (over ZZ)
-- J         Jacobian matrix (over ZZ)
-- returns an approximation (P,eps^2)
newtonStep = (Peps,M,J) -> (
    P  := Peps#0;
    eps := Peps#1;
    JPinv := invMatn(sub(J,P),eps);
    correction := eps*modn(divn(sub(M,P)*JPinv,eps),eps);
    {modn(P-correction,eps^2),eps^2}
    )


-- returns an approximation mod Peps^(2^num)
newton = (Peps,M,J,num) -> (
  i := 0;
  localPeps := Peps;
  while i < num do (
    localPeps = newtonStep(localPeps,M,J);
    print(localPeps);
    i = i+1;
  );
  localPeps
)
```

We now consider equations of Example 4.3

```
I = ideal (-8*x^2-x*y-7*y^2+5238*x-11582*y-7696,
    4*x*y-10*y^2-2313*x-16372*y-6462)
```

their Jacobian matrix

```
J = jacobian(I)
```

and their solutions over $\mathbb{F}_7$:

```
apply(allPoints(I,7),Pseq -> (
    P := matrix {toList Pseq};
    (P,0!=det modn(sub(J,P),7))
  ))

o25 = {(| 2 3 |, true), (| 5 5 |, true)}
```

Both points are isolated and smooth over $\mathbb{F}_7$ so we can apply p-adic Newton iteration to them. The first one lifts to the solution found in Experiment 4.3:

```
newton((matrix{{2,3}},7),gens I, J,4)

{| 9 10 |, 49}
{| -1167 -774 |, 2401}
{| 1234 -774 |, 5764801}
{| 1234 -774 |, 33232930569601}
```

while the second point probably does not lift to $\mathbb{Z}$:

```
newton((matrix{{5,5}},7),gens I, J,4)

{| 5 -9 |, 49}
{| -926 334 |, 2401}
{| 359224 -66894 |, 5764801}
{| 11082657337694 -9795607574104 |, 33232930569601}
```

Remark 4.11. Noam Elkies has used this method to find interesting elliptic fibrations over $\mathbb{Q}$. See for example [3, Section III, p. 11].

Remark 4.12. The Newton method is much faster than lifting by Chinese remaindering, since we only need to find one smooth point in one characteristic. Unfortunately, it does not work if we cannot calculate tangent spaces. An application where this happens is discussed in the next section.

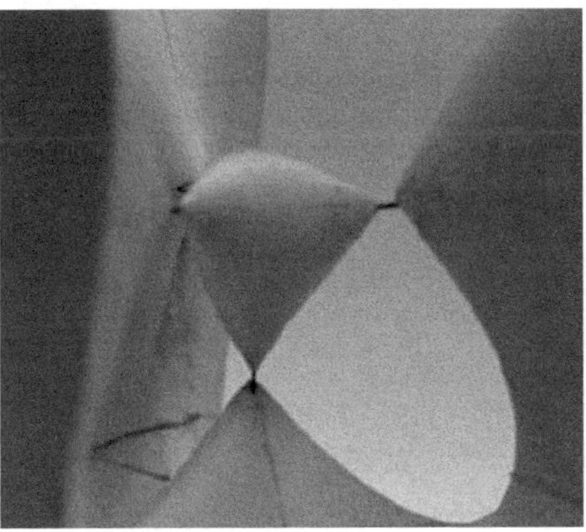

Figure 17. Historic plaster model of the Cayley Cubic as displayed in the mathematical instute of the university of Göttingen

5. Surfaces with Many Real Nodes

A very nice application of finite field experiments with beautiful characteristic zero results was done by Oliver Labs in his thesis [2]. We look at his ideas and results in this section.

Consider an algebraic surface $X \subset \mathbb{P}^3_{\mathbb{R}}$ of degree d and denote by $N(X)$ the number of real nodes of X. A classical question of real algebraic geometry is to determine the maximal number of nodes a surface of degree d can have. We denote this number by

$$\mu(d) := \max\{N(X) \mid X \subset \mathbb{P}^3_{\mathbb{R}} \wedge \deg X = d\}.$$

Moreover one would like to find explicit equations for surfaces X that do have $\mu(d)$ real nodes. The cases $\mu(1) = 0$ and $\mu(2) = 1$, i.e the plane and the quadric cone, have been known since antiquity.

Cayley [24] and Schäfli [25] solved $\mu(3) = 4$, while Kummer proved $\mu(4) = 16$ in [26]. Plaster models of a Cayley-Cubic and a Kummer-Quartic are on display in the Göttingen Mathematical Institute as numbers 124 and 136, see Figure 17 and 18. These and many other pictures are available at

http://www.uni-math.gwdg.de/modellsammlung.

For the case $d = 5$, Togliatti proved in [27] that quintic surfaces with 31 nodes exist. One such surface is depicted in Figure 20. It took 40 years before Beauville [28] finally proved that 31 is indeed the maximal possible number.

In 1994 Barth [29] found the beautiful sextic with the icosahedral symmetry and 65 nodes shown in Figure 21. Jaffe and Rubermann proved in [30] that no sextics with 66 or more nodes exist.

For $d = 7$ the problem is still open. By works of Chmutov [31], Breske/Labs/van Straten [32] and Varchenko [33] we only know $93 \leq \mu(7) \leq 104$. For large d Chmutov and Breske/Labs/van Straten show

$$\mu(d) \geq \frac{5}{12}d^3 + \text{lower order terms,}$$

while Miyaoka [34] proves

$$\mu(d) \leq \frac{4}{9}d^3 + \text{lower order terms.}$$

Here we explain how Oliver Labs found a new septic with many nodes, using finite field experiments [35].

Experiment 5.1. The most naive approach to find septics with many nodes is to look at random surfaces of degree 7 in some small characteristic:

```
-- Calculate milnor number for hypersurfaces in IP^3
-- (for nonisolated singularities and smooth surfaces 0
-- is returned)
mu = (f) -> (
      J := (ideal jacobian ideal f)+ideal f;
      if 3==codim J then degree J else 0
      )

K = ZZ/5           -- work in char 5
R = K[x,y,z,w]  -- coordinate ring of IP^3

-- look at 100 random surfaces
time tally apply(100, i-> mu(random(7,R)))
```

After about 18 seconds we find

```
o4 = Tally{0 => 69}
           1 => 24
           2 => 5
           3 => 1
           4 => 1
```

which is still far from 93 nodes. Since having an extra node is a codimension-one condition, a rough estimation gives that we would have to search $5^{89} \approx 1.6 \times 10^{62}$ times longer to find 89 more nodes in characteristic 5.

One classical idea to find surfaces with many nodes, is to use symmetry. If for example we only look at mirror symmetric surfaces, we obtain singularities in pairs, as depicted in Figure 19.

Figure 18. A Kummer surface with 16 nodes.

Experiment 5.2. We look at 100 random surfaces that are symmetric with respect to the $x = 0$ plane

```
-- make a random f mirror symmetric
sym = (f) -> f+sub(f,{x=>-x})

time tally apply(100, i-> mu(sym(random(7,R))))

o6 = Tally{0 => 57}
           1 => 10
           2 => 11
           3 => 9
           4 => 4
           5 => 3
           6 => 3
           7 => 1
           9 => 1
          13 => 1
```

Indeed, we obtain more singularities, but not nearly enough.

The symmetry approach works best if we have a large symmetry group. In the $d = 7$ case Oliver Labs used the D_7 symmetry of the 7-gon. If D_7 acts on $\mathbb{P}^3$ with symmetry axis $x = y = 0$ one can use representation theory to find a 7-dimensional family of D_7-invariant 7-tic we use in the next experiment.

Experiment 5.3. Start by considering the cone over a 7-gon given by

$$P = 2^6 \prod_{j=0}^{6} \left(\cos\left(\frac{2\pi j}{7} \right) x + \sin\left(\frac{2\pi j}{7} \right) y - z \right),$$

which can be expanded to

Figure 19. A mirror symmetric cubic

```
P = x*(x^6-3*7*x^4*y^2+5*7*x^2*y^4-7*y^6)+
    7*z*((x^2+y^2)^3-2^3*z^2*(x^2+y^2)^2+2^4*z^4*(x^2+y^2))-
    2^6*z^7
```

Now parameterize D_7 invariant septics U that contain a double cubic.

```
S = K[a1,a2,a3,a4,a5,a6,a7]
RS = R**S  -- tensor product of rings
U = (z+a5*w)*
    (a1*z^3+a2*z^2*w+a3*z*w^2+a4*w^3+(a6*z+a7*w)*(x^2+y^2))^2
```

We will look at random sums of the form $P + U$ using

```
randomInv = () -> (
    P-sub(U,vars R|random(R^{0},R^{7:0}))
    )
```

Let's try 100 of these

```
time tally apply(100, i-> mu(randomInv()))
```

```
o9 = Tally{63 => 48}
          64 => 6
          65 => 4
             . . .
          136 => 1
          140 => 1
```

Unfortunately, this looks better than it is, since many of the surfaces with high Milnor numbers have singularities that are not ordinary nodes. We can detect this by looking at the Hessian matrix which has rank ≥ 3 only at smooth points and

Figure 20. A Togliatti quintic

ordinary nodes. The following function returns the number of nodes of $X = V(f)$ if all nodes are ordinary and 0 otherwise.

```
numA1 = (f) -> (
   -- singularities of f
   singf := (ideal jacobian ideal f)+ideal f;
   if 3==codim singf then (
           -- calculate Hessian
       Hess := diff(transpose vars R,diff(vars R,f));
         ssf := singf + minors(3,Hess);
         if 4==codim ssf then degree singf else 0
         )
       else 0
   )
```

With this we test another 100 examples:

```
time tally apply(100, i-> numA1(randomInv()))

o12 = Tally{0 => 28 }
            63 => 51
            64 => 13
            65 => 1
            70 => 6
            72 => 1
```

which takes about 30 seconds. Notice that most surfaces have $N(X)$ a multiple of 7 as expected from the symmetry.

To speed up these calculations Oliver Labs intersects the surfaces $X = V(P + U)$ with the hyperplane $y = 0$ see Figure 22. Since the operation of D_7 moves this hyperplane to 7 different positions, every singularity of the intersection curve

Figure 21. The Barth sextic

C that does not lie on the symmetry axis corresponds to 7 singularities of X. Singular points on C that do lie on the symmetry axis contribute only one node to the singularities of X. Using the symmetry of the construction one can show that for surfaces X with only ordinary double points all singularities are obtained this way [36, p. 18, Cor. 2.3.10], [35, Lemma 1].

Experiment 5.4. We now look at 10000 random D_7-invariant surfaces and their intersection curves with $y = 0$. We estimate the number of nodes on X from the number of nodes on C and return the point in the parameter space of U if this number is large enough.

```
use R
time tally apply(10000,i-> (
     r := random(R^{0},R^{7:0});
     f := sub(P-sub(U,vars R|r),y_R=>0);
     singf := ideal f + ideal jacobian ideal f;
     if 2 == codim singf then (
          -- calculate Hessian
        Hess := diff(transpose vars R,diff(vars R,f));
         ssf := singf + minors(2,Hess);
         if 3==codim ssf then (
               d := degree singf;
               -- points on the line x=0
            singfx := singf+ideal(x);
             dx := degree singfx;
             if 2!=codim singfx then dx=0;
              d3 = (d-dx)*7+dx;
             (d,d-dx,dx,d3,if d3>=93 then r)
         )
         else -1
     )
  ))
```

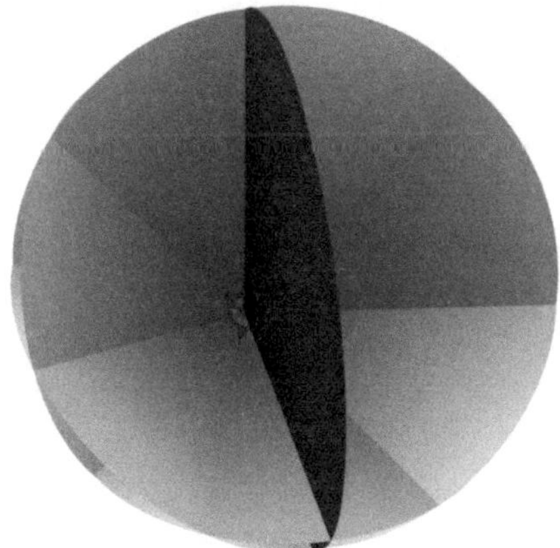

Figure 22. Intersection of the 7-gon with a perpendicular hypersurface

In this way we find

```
o16 = Tally{(9, 9, 0, 63, ) => 5228
            (10, 9, 1, 64, ) => 731
              . . . . .
            (16, 14, 2, 100, | 1 2 2 1 1 0 1 |) => 1
            -1 => 3071
            null => 8
```

It remains to check whether the found U really gives rise to surfaces with 100 nodes

```
f = P-sub(U,vars R|sub(matrix{{1,2,2,1,1,0,1}},R))
numA1(f)

o18 = 100
```

This proves that there exists a surface with 100 nodes over $\mathbb{F}_5$.

Looking at other fields one finds that $\mathbb{F}_5$ is a special case. In general one only finds surfaces with 99 nodes. To lift these examples to characteristic zero, Oliver Labs analyzed the geometry of the intersection curves of the 99-nodal examples and found that

(i) All such intersection curves decompose into a line and a 6-tic.
(ii) The singularities of the intersection curves are in a special position that can be explicitly described (see [35] for details)

These geometric properties imply (after some elimination) that there exists an α such that

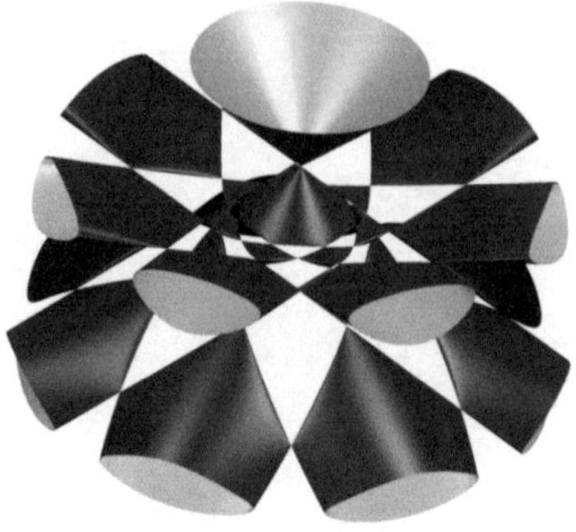

Figure 23. The Labs septic

$$\alpha_1 = \alpha^7 + 7\alpha^5 - \alpha^4 + 7\alpha^3 - 2\alpha^2 - 7\alpha - 1$$

$$\alpha_2 = (\alpha^2 + 1)(3\alpha^5 + 14\alpha^4 - 3\alpha^2 + 7\alpha - 2)$$

$$\alpha_3 = (\alpha^1 + 1)^2(3\alpha^3 + 7\alpha - 3)$$

$$\alpha_4 = (\alpha(1 + \alpha^2) - 1)(1 + \alpha^2)^2$$

$$\alpha_5 = -\frac{\alpha^2}{1 + \alpha^2}$$

$$\alpha_6 = \alpha_7 = 1$$

It remained to determine which α lead to 99-nodal septics. Experiments over many primes show that there are at most 3 such α. Over primes with exactly 3 solutions, Oliver Labs represented them as zeros of a degree 3 polynomial. By using the Chinese remaindering method, he lifted the coefficients of this polynomial to characteristic 0 and obtained

$$7\alpha^3 + 7\alpha + 1 = 0.$$

This polynomial has exactly one real solution, and with this α one can calculate this time over $\mathbb{Q}(\alpha)$ that the resulting septic has indeed 99 real nodes. Figure 23 shows the inner part of this surface.

A movie of this and many other surfaces in this section can by found on my home page

$$\text{www.iag.uni-hannover.de/~bothmer/goettingen.php,}$$

on the home page of Oliver Labs

$$\text{http://www.algebraicsurface.net/,}$$

or on youTube.com

http://www.youtube.com/profile?user=bothmer.

The movies and the surfaces in this article were produced using the public domain programs `surf` by Stefan Endraß [37] and `surfex` by Oliver Labs [38].

A. Selected Macaulay Commands

Here we review some `Macaulay 2` commands used in this tutorial. Lines starting with "i" are input lines, while lines starting with "o" are output lines. For more detailed explanations we refer to the online help of `Macaulay2` [4].

A.1. apply

This command applies a function to a list. In Macaulay 2 this is often used to generate loops.

```
i1 : apply({1,2,3,4},i->i^2)
o1 = {1, 4, 9, 16}
o1 : List
```

The list $\{0, 1, \ldots, n-1\}$ can be abbreviated by n:

```
i2 : apply(4,i->i^2)
o2 = {0, 1, 4, 9}
o2 : List
```

A.2. map

With `map(R,S,m)` a map from S to R is produced. The matrix m over S contains the images of the variables of R:

```
i1 : f = map(ZZ,ZZ[x,y],matrix{{2,3}})
o1 = map(ZZ,ZZ[x,y],{2, 3})
o1 : RingMap ZZ <--- ZZ[x,y]

i2 : f(x+y)
o2 = 5
```

If no matrix is given, all variables to variables of the same name or to zero.

```
i3 : g = map(ZZ[x],ZZ[x,y])
o3 = map(ZZ[x],ZZ[x,y],{x, 0})
o3 : RingMap ZZ[x] <--- ZZ[x,y]

i4 : g(x+y+1)
o4 = x + 1
o4 : ZZ[x]
```

A.3. random

This command can be used either to construct random matrices

```
i1 : K = ZZ/3

o1 = K
o1 : QuotientRing

i2 : random(K^2,K^3)

o2 = | 1 0  -1 |
     | 1 -1 1  |
              2        3
o2 : Matrix K  <--- K
```

or to construct random homogeneous polynomials of given degree

```
i3 : R = K[x,y]

o3 = R
o3 : PolynomialRing

i4 : random(2,R)

          2          2
o4 = x  + x*y - y
o4 : R
```

A.4. sub

This command is used to substitute values for the variables of a ring:

```
i1 : K = ZZ/3

o1 = K
o1 : QuotientRing
```

```
i2 : R = K[x,y]

o2 = R
o2 : PolynomialRing

i3 : f = x*y

o3 = x*y
o3 : R

i4 : sub(f,matrix{{2,3}})

o4 = 6
```

Another application is the transfer a polynomial, ideal or matrix from one ring R to another ring S that has some variables in common with R

```
i5 : S = K[x,y,z]

o5 = S
o5 : PolynomialRing

i6 : sub(f,S)

o6 = x*y
o6 : S
```

A.5. syz

The command is used here to calculate a presentation for the kernel of a matrix:

```
i1 : M = matrix{{1,2,3},{4,5,6}}

o1 = | 1 2 3 |
     | 4 5 6 |
                  2           3
o1 : Matrix ZZ   <--- ZZ

i2 : syz M

o2 = | -1 |
     | 2  |
     | -1 |
                  3           1
o2 : Matrix ZZ   <--- ZZ
```

A.6. tally

With `tally` one can count how often an element appears in a list:

```
i1 : tally{1,2,1,3,2,2,17}

o1 = Tally{1 => 2 }
            2 => 3
            3 => 1
            17 => 1
o1 : Tally
```

B. Magma Scripts (by Stefan Wiedmann)

Stefan Wiedmann [5] has translated the Macaulay 2 scripts of this article to Magma. Here they are:

Experiment B.1.1. Evaluate a given polynomial in 700 random points.

```
K := FiniteField(7);                //work over F_7
R<x,y,z,w> := PolynomialRing(K,4);  //Polynomialring in 4 variables
                                    //over F_7
K4:=CartesianPower(K,4);            //K^4
F := x^23+1248*y*z*w+129269698;     //a polynomial

M := [Random(K4): i in [1..700]];   //random points
T := {*Evaluate(F,s): s in M*};

Multiplicity(T,0);                  //Results with muliplicity
```

Experiment B.1.2. Evaluate a product of two polynomials in 700 random points

```
K := FiniteField(7);                //work over F_7
R<x,y,z,w> := PolynomialRing(K,4);  //AA^4 over F_7
K4:=CartesianPower(K,4);            //K^4

F := x^23+1248*y*z*w+129269698;     //a polynomial
G := x*y*z*w+z^25-938493+x-z*w;     //a second polynomial
H := F*G;

M := [Random(K4): i in [1..700]];   //random points
T := {*Evaluate(H,s): s in M*};
T;
Multiplicity(T,0);
```

Experiment B.1.14. Count singular quadrics.

```
K := FiniteField(7);
R<X,Y,Z,W> := PolynomialRing(K,4);
{* Dimension(JacobianIdeal(Random(2,R,0))) : i in [1..700]*};
```

Experiment B.1.18. Count quadrics with dim > 0 singular locus

```
function findk(n,p,k,c)
    //Search until k singular examples of codim at most c are found,
    //p prime number, n dimension
    K :=  FiniteField(p);
    R := PolynomialRing(K,n);
    trials := 0;
    found := 0;
    while found lt k do
      Q := Ideal([Random(2,R,0)]);
      if c ge n - Dimension(Q+JacobianIdeal(Basis(Q))) then
        found := found + 1;
      else
        trials := trials + 1;
      end if;
    end while;
    print "Trails:",trials;
    return trials;
end function;

k := 50;

time L1 := [[p,findk(4,p,k,2)] : p in [5,7,11]];
L1;

time findk(4,5,50,2);
time findk(4,7,50,2);
time findk(4,11,50,2);

function slope(L)
    //calculate slope of regression line by
    //formula form [2] p. 800
    xbar := &+[L[i][1] : i in [1..#L]]/#L;
    ybar := &+[L[i][2] : i in [1..#L]]/#L;
    return &+[(L[i][1]-xbar)*(L[i][2]-ybar): i in [1..3]]/
      &+[(L[i][1]-xbar)^2 : i in [1..3]];
end function;

//slope for dim 1 singularities
slope([[Log(1/x[1]), Log(k/x[2])] : x in L1]);
```

Experiment B.2.1. Count points on a reducible variety.

```
K := FiniteField(7);
V := CartesianPower(K,6);
R<x1,x2,x3,x4,x5,x6> := PolynomialRing(K,6);

//random affine polynomial of degree d
randomAffine := func< d | &+[ Random(i,R,7) : i in [0..d]]>;

//some polynomials
F := randomAffine(2);
G := randomAffine(6);
H := randomAffine(7);

//generators of I(V(F) \cup V(H,G))
I := Ideal([F*G,F*H]);

//experiment
null := [0 : i in [1..#Basis(I)]];

t := {**};

for j in [1..700] do
    point := Random(V);
    Include(~t, null eq [Evaluate(Basis(I)[i],point) :
            i in [1..#Basis(I)]]);
end for;

//result
t;
```

Experiment B.2.4. Count points and tangent spaces on a reducible variety.

```
K := FiniteField(7); //charakteristik 7
R<x1,x2,x3,x4,x5,x6> := PolynomialRing(K,6); //6 variables
V := CartesianPower(K,6);

//random affine polynomial of degree d
randomAffine := func< d | &+[ Random(i,R,7) : i in [0..d]]>;

//some polynomials
F := randomAffine(2);
G := randomAffine(6);
H := randomAffine(7);

//generators of I(V(F) \cup V(H,G))
I := Ideal([F*G,F*H]);

null := [0 : i in [1..#Basis(I)]];
```

```
//the Jacobi-Matrix
J := JacobianMatrix(Basis(I));
size := [NumberOfRows(J),NumberOfColumns(J)];


A := RMatrixSpace(R,size[1],size[2]);
B := KMatrixSpace(K,size[1],size[2]);


t := {**};


time
for j in [1..700] do
  point := Random(V);
  substitude := map< A -> B | x :->
                    [Evaluate(t,point): t in ElementToSequence(x)]>;
  if null eq [Evaluate(Basis(I)[i], point) : i in [1..#Basis(I)]]
    then Include(~t, Rank(substitude(J)));
  else
    Include(~t,-1);
  end if;
end for;


//result
t;
```

Experiment B.2.9.

```
K := FiniteField(7);                          //charakteristik 7
R<x1,x2,x3,x4,x5,x6> := PolynomialRing(K,6);  //6 variables
V := CartesianPower(K,6);

//consider an 5 x 5 matrix with degree 2 entries
r := 5;
d := 2;
Mat := MatrixAlgebra(R,r);


//random matrix
M := Mat![Random(d,R,7) : i in [1..r^2]];

//calculate determinant and derivative w.r.t x1
time F := Determinant(M);
time F1 := Derivative(F,1);

//substitute a random point
point := Random(V);
time Evaluate(F1,point);
```

```
//calculate derivative with epsilon

Ke<e> := AffineAlgebra<K,e|e^2>; //a ring with e^2 = 0

Mate := MatrixAlgebra(Ke,r);

//the first unit vector
e1 := <>;
for i in [1..6] do
    if i eq 1 then
      Append(~e1,e);
    else
      Append(~e1,0);
    end if;
end for;

//point with direction
point1 := < point[i]+e1[i] : i in [1..6]>;

time Mate![Evaluate(x,point1): x in ElementToSequence(M)];
time Determinant(Mate![Evaluate(x,point1):
                 x in ElementToSequence(M)]);

//determinant at 5000 random points

time
for i in [1..5000] do
    point := Random(V);                      //random point
    point1 := <point[i]+e1[i] : i in [1..6]>;  //tangent direction
    //calculate derivative
    _:=Determinant(Mate![Evaluate(x,point1):
                 x in ElementToSequence(M)]);
end for;
```

Experiment B.4.3.

```
R<x,y> := PolynomialRing(IntegerRing(),2); //two variables

//the equations
F := -8*x^2-x*y-7*y^2+5238*x-11582*y-7696;
G := 4*x*y-10*y^2-2313*x-16372*y-6462;
I := Ideal([F,G]);

//now lets find the points over F_p
function allPoints(I,p)
    M := [];
    K := FiniteField(p);
```

```
    A := AffineSpace(K,2);
    R := CoordinateRing(A);
    for pt in CartesianPower(K,2) do
        Ipt := Ideal([R|Evaluate(Basis(I)[k],pt) :
                    k in [1..#Basis(I)]]);
        SIpt := Scheme(A,Ipt);
        if Codimension(SIpt) eq 0 then;
            Append(~M,pt);
        end if;
    end for;
    return M;
end function;

for p in PrimesUpTo(23) do
    print p, allPoints(I,p);
end for;

/*Chinese remaindering
given solutions mod m and n find
a solution mod m*n
sol1 = [n,solution]
sol2 = [m,solution]*/
function chinesePair(sol1,sol2)
    n := sol1[1];
    an := sol1[2];
    m := sol2[1];
    am := sol2[2];
    d,r,s :=  Xgcd(n,m);
    //returns d,r,s so that a*r + b*s is
    //the greatest common divisor d of a and b.
    amn := s*m*an+r*n*am;
    amn := amn - (Round(amn/(m*n)))*(m*n);
    if d eq 1 then
        return [m*n,amn];
    else
        print "m and n not coprime";
        return false;
    end if;
end function;

/*take a list {(n_1,s_1),...,(n_k,s_k)}
and return (n,a) such that
n=n_1* ... * n_k and
s_i = a mod n_i*/
function chineseList(L)
    //#L >= 2
    erg := L[1];
```

```
    for i in [2..#L] do
        erg := chinesePair(L[i],erg);
    end for;
    return erg;
end function;

//x coordinate
chineseList([[2,0],[5,4],[17,10],[23,15]]);
//y coordinate
chineseList([[2,0],[5,1],[17,8],[23,8]]);

//test the solution
Evaluate(F,[1234,-774]);
Evaluate(G,[1234,-774]);
```

Experiment B.4.6. Rational recovery, as suggested in von zur Gathen in [23, Section 5.10]. Uses the functions **allPoints** and **chineseList** from Experiment B.4.3.

```
R<x,y> := PolynomialRing(IntegerRing(),2); //two variables

//equations
F := 176*x^2+148*x*y+301*y^2-742*x+896*y+768;
G := -25*x*y+430*y^2+33*x+1373*y+645;
I := Ideal([F,G]);

for p in PrimesUpTo(41) do
    print p, allPoints(I,p);
end for;

// x coordinate
chineseList([[2,1],[13,5],[31,7],[37,14],[41,0]]);
// y coordinate
chineseList([[2,0],[13,10],[31,22],[37,18],[41,23]]);

//test the solution
Evaluate(F,[138949,-526048]);
Evaluate(G,[138949,-526048]);

/*take (a,n) and calculate a solution to
r = as mod n
such that r,s < sqrt(n).
return (r/s)*/
function recoverQQ(a,n)
    r0:=a;
    s0:=1;
    t0:=0;
```

```
    r1:=n;
    s1:=0;
    t1:=1;
    r2:=0;
    s2:=0;
    t2:=0;
    k  := Round(Sqrt(r1*1.0));
    while k le r1 do
        q  := r0 div r1;
        r2 := r0-q*r1;
        s2 := s0-q*s1;
        t2 := t0-q*t1;
        r0:=r1;
        s0:=s1;
        t0:=t1;
        r1:=r2;
        s1:=s2;
        t1:=t2;
    end while;
    return (r2/s2);
end function;

//x coordinate
recoverQQ(138949,2*13*31*37*41);
//y coordinate
recoverQQ(-526048,2*13*31*37*41);

//test the solution
Evaluate(F,[123/22,-77/43]);
Evaluate(G,[123/22,-77/43]);
```

Experiment B.4.10. Lifting solutions using p-adic Newtoniteration (as suggested
by N.Elkies). Uses the function `allPoints` from Example B.4.3.

```
//calculate reduction of a matrix M mod n
function modn(M,n)
    return Matrix(Nrows(M),Ncols(M),[x - Round(x/n)*n :
                x in Eltseq(M)]);
end function;

//divide a matrix of integer by an integer
//(in our application this division will not have a remainder)
function divn(M,n)
    return Matrix(Nrows(M),Ncols(M),[x div n : x in Eltseq(M)]);
end function;

// invert number mod n
```

```
function invn(i,n)
    a,b := Xgcd(i,n);
    if a eq 1 then
        return b;
    else return false;
    end if;
end function;


//invert a matrix mod n
//M a square matrix over ZZ
//(if M is not invertible mod n, then 0 is returned)
function invMatn(M,n)
    Mn := modn(M,n);
    MQQ := MatrixAlgebra(RationalField(),Nrows(M))!Mn;
    detM := Determinant(Mn);
    if Type(invn(detM,n)) eq BoolElt then
        return 0;
    else
        return
            (MatrixAlgebra(IntegerRing(),Nrows(M))!
            (modn(invn(detM,n)*detM*MQQ^(-1),n)));
    end if;
end function;


//(P,eps) an approximation mod eps (contains integers)
//M        affine polynomials (over ZZ)
//J        Jacobian matrix (over ZZ)
//returns an approximation (P,eps^2)
function newtonStep(Peps,M,J)
    P := Peps[1];
    eps := Peps[2];
    JatP:=Matrix(Ncols(J),Nrows(J),[Evaluate(x,Eltseq(P)) :
                                x in Eltseq(J)]);
    JPinv := invMatn(JatP,eps);
    MatP:= Matrix(1,#M,[Evaluate(x,Eltseq(P)) : x in Eltseq(M)]);
    correction := eps*modn(divn(MatP*Transpose(JPinv),eps),eps);
    return <modn(P-correction,eps^2),eps^2>;
end function;


//returns an approximation mod Peps^(2^num)
function newton(Peps,M,J,num)
    localPeps := Peps;
    for i in [1..num] do
    localPeps := newtonStep(localPeps,M,J);
    print localPeps;
    end for;
    return localPeps;
```

```
end function;

//c.f. example 4.3
R<x,y> := PolynomialRing(IntegerRing(),2); //two variables

//the equations
F := -8*x^2-x*y-7*y^2+5238*x-11582*y-7696;
G := 4*x*y-10*y^2-2313*x-16372*y-6462;
I := Ideal([F,G]);
J := JacobianMatrix(Basis(I));

Ap := allPoints(I,7);
for x in Ap do
    MatP := Matrix(Nrows(J),Ncols(J),[Evaluate(j,x):
                                      j in Eltseq(J)]);
    print x, (0 ne Determinant(MatP));
end for;

Peps := <Matrix(1,2,[2,3]),7>;
newton(Peps,Basis(I),J,4);

Peps := <Matrix(1,2,[5,5]),7>;
newton(Peps,Basis(I),J,4);
```

Experiment B.5.1. Count singularities of random surfaces over $\mathbb{F}_5$.

```
K := FiniteField(5);                     //work in char 5
A := AffineSpace(K,4);
R<x,y,z,w> :=CoordinateRing(A);    //coordinate ring of IP^3

//Calculate milnor number
//(For nonisolated singularities and smooth surfaces 0 is returned)
function mu(f)
    SJ := Scheme(A,(Ideal([f])+JacobianIdeal(f)));
    if 3 eq Codimension(SJ) then
        return Degree(ProjectiveClosure(SJ));
    else
        return 0;
    end if;
end function;

//look at 100 random surfaces
M := {**};
time
for i in [1..100] do
    f := Random(7,CoordinateRing(A),5);
    Include(~M,mu(f));
```

```
end for;

print "M:", M;
```

Experiment B.5.2. Count singularities of mirror symmetic random surfaces. Uses the function mu from Example B.5.1.

```
K := FiniteField(5);                    //work in char 5
A := AffineSpace(K,4);
R<x,y,z,w> :=CoordinateRing(A);    //coordinate ring of IP^3

//make a random f mirror symmetric
function mysym(f)
    return (f + Evaluate(f,x,-x));
end function;

//look at 100 random surfaces
M := {**};
time
for i in [1..100] do
f:=R!mysym(Random(7,CoordinateRing(A),5));
Include(~M,mu(f));
end for;

print "M:", M;
```

Experiment B.5.3. Count $A1$-singularities of D_7 invariant surfaces. Uses the function mu from Experiment B.5.1.

```
K := FiniteField(5);                    //work in char 5
A := AffineSpace(K,4);
R<X,Y,Z,W> :=CoordinateRing(A);    //coordinate ring of IP^3
RS<x,y,z,w,a1,a2,a3,a4,a5,a6,a7> := PolynomialRing(K,11);

//the 7-gon
P := X*(X^6-3*7*X^4*Y^2+5*7*X^2*Y^4-7*Y^6)
    +7*Z*((X^2+Y^2)^3-2^3*Z^2*(X^2+Y^2)^2+2^4*Z^4*(X^2+Y^2))
    -2^6*Z^7;

//parametrising invariant 7 tics with a double cubic
U := (z+a5*w)*(a1*z^3+a2*z^2*w+a3*z*w^2+a4*w^3
    +(a6*z+a7*w)*(x^2+y^2))^2;

//random invariant 7-tic
function randomInv()
    return (P - Evaluate(U,[X,Y,Z,W] cat [Random(K):i in [1..7]]));
end function;
```

```
//test with 100 examples
M1 := {**};
time
for i in [1..100] do
Include(~M1,mu(randomInv()));
end for;

print "M1:", M1;

//singularities of f
function numA1(f)
    singf := Ideal([f])+JacobianIdeal(f);
    Ssingf := Scheme(A,singf);
    if 3 eq Codimension(Ssingf) then
        T := Scheme(A,Ideal([f]));
        //calculate Hessian
        Hess := HessianMatrix(T);
        ssf := singf + Ideal(Minors(Hess,3));
        if 4 eq Codimension(Scheme(A,ssf)) then
            return Degree(ProjectiveClosure(Ssingf));
        else
            return 0;
        end if;
    else
        return 0;
    end if;
end function;

//test with 100 examples
M2 := {**};
time
for i in [1..100] do
Include(~M2,numA1(randomInv()));
end for;

print "M2:", M2;
```

Experiment B.5.4. Estimate number of $A1$-singularities by looking at $y = 0$. Uses the function **numA1** from Experiment B.5.3.

```
K := FiniteField(5); //work in char 5
A := AffineSpace(K,4);
R<X,Y,Z,W> := CoordinateRing(A); //coordinate ring of IP^3
RS<x,y,z,w,a1,a2,a3,a4,a5,a6,a7> := PolynomialRing(K,11);

//the 7-gon
```

```
P  := X*(X^6-3*7*X^4*Y^2+5*7*X^2*Y^4-7*Y^6)
    +7*Z*((X^2+Y^2)^3-2^3*Z^2*(X^2+Y^2)^2+2^4*Z^4*(X^2+Y^2))
    -2^6*Z^7;

//parametrising invariant 7 tics with a double cubic
U := (z+a5*w)*(a1*z^3+a2*z^2*w+a3*z*w^2+a4*w^3
    +(a6*z+a7*w)*(x^2+y^2))^2;

//estimate number of nodes
function numberofsing()
    r := [Random(K): i in [1..7]];
    f := Evaluate(P - Evaluate(U,[X,Y,Z,W] cat r),Y,0);
    singf := Ideal([f])+ JacobianIdeal(f);
    Ssingf := Scheme(A,singf);
    if 2 eq Codimension(Ssingf) then
        //calculate Hessian
        S := Scheme(A,Ideal([f]));
        Hess := HessianMatrix(S);
        ssf := singf + Ideal(Minors(Hess,2));
        Sssf := Scheme(A,ssf);
        if 3 eq Codimension(Sssf) then
            d := Degree(ProjectiveClosure(Ssingf));
            //points on the line x=0
            singfx := singf + Ideal([X]);
            Ssingfx :=  Scheme(A,singfx);
            dx := Degree(ProjectiveClosure(Ssingfx));
            if 2 ne Codimension(Ssingfx) then
                dx := 0;
            end if;
            d3 := (d-dx)*7+dx;
            if d3 ge 93 then
                return <d, d-dx, dx, d3>, r;
            else
                return <d, d-dx, dx, d3>, _;
            end if;
        else
            return <-1,0,0,0>,_;
        end if;
    else return <0,0,0,0>,_;
    end if;
end function;

M1 := {**};
M1hit := {**};
time
for i in [1..10000] do
    a,b := numberofsing();
```

```
   if assigned(b) then
        Include(~M1hit,<a,b>);
   else
        Include(~M1,a);
   end if;
end for;

print "M1:", M1;
print "M1hit:", M1hit;

//test
f  :=  P-Evaluate(U,[X,Y,Z,W,1,2,2,1,1,0,1]);
numA1(f);
```

References

[1] J. von zur Gathen and I. Shparlinski. Computing components and projections of curves over finite fields. *SIAM J. Comput.*, 28(3):822–840 (electronic), 1999.

[2] O. Labs. *Hypersurfaces with Many Singularities*. PhD thesis, Johannes Gutenberg Universität Mainz, 2005. available from www.OliverLabs.net.

[3] N. D. Elkies. Three lectures on elliptic surfaces and curves of high rank 2007, arXiv:0709.2908v1 [math.NT].

[4] D. R. Grayson and M. E. Stillman. Macaulay 2, a software system for research in algebraic geometry. Available at http://www.math.uiuc.edu/Macaulay2, 2002.

[5] H.-Chr. Graf v. Bothmer and S. Wiedmann. Scripts for finite field experiments. Available at http://www-ifm.math.uni-hannover.de/~bothmer/goettingen.php., 2007.

[6] H.-Chr. Graf v. Bothmer and F. O. Schreyer. A quick and dirty irreducibility test for multivariate polynomials over $\mathbb{F}_q$. *Experimental Mathematics*, 14(4):415–422, 2005.

[7] I. N. Bronstein, K. A. Semendjajew, G. Musiol, and H. Mühlig. *Taschenbuch der Mathematik*. Verlag Harri Deutsch, Thun, expanded edition, 2001. Translated from the 1977 Russian original, With 1 CD-ROM (Windows 95/98/2000/NT, Macintosh and UNIX).

[8] I. R. Shafarevich. *Basic algebraic geometry. 1.* Springer-Verlag, Berlin, second edition, 1994. Varieties in projective space, Translated from the 1988 Russian edition and with notes by Miles Reid.

[9] M. Frommer. Über das Auftreten von Wirbeln und Strudeln (geschlossener und spiraliger Integralkurven) in der Umgebung rationaler Unbestimmtheitsstellen. *Math. Ann.*, 109:395–424, 1934.

[10] D. Schlomiuk. Algebraic particular integrals, integrability and the problem of the center. *Trans. Amer. Math. Soc.*, 338(2):799–841, 1993.

[11] H. Żołądek. The classification of reversible cubic systems with center. *Topol. Methods Nonlinear Anal.*, 4(1):79–136, 1994.

[12] H. Żołądek. Remarks on: "The classification of reversible cubic systems with center" [Topol. Methods Nonlinear Anal. 4 (1994), no. 1, 79–136; MR1321810 (96m:34057)]. *Topol. Methods Nonlinear Anal.*, 8(2):335–342 (1997), 1996.

[13] H.-Chr. Graf v. Bothmer and Martin Cremer. A C++ program for calculating focal values in characteristic p. Available at http://www-ifm.math.uni-hannover.de/~bothmer/surface, 2005.

[14] H.-Chr. Graf v. Bothmer. Experimental results for the Poincaré center problem (including an appendix with Martin Cremer). math.AG/0505547, 2005. (To appear in NoDEA).

[15] D. Eisenbud. *Commutative Algebra with a View Toward Algebraic Geometry.* Graduate Texts in Mathematics 150. Springer, 1995.

[16] F.-O. Schreyer. Small fields in constructive algebraic geometry. In *Moduli of vector bundles (Sanda, 1994; Kyoto, 1994)*, volume 179 of *Lecture Notes in Pure and Appl. Math.*, pages 221–228. Dekker, New York, 1996.

[17] H.-Chr. Graf v. Bothmer, C. Erdenberger, and K. Ludwig. A new family of rational surfaces in $\mathbb{P}^4$. *Journal of Symbolic Computation.*, 29(1):51–60, 2005.

[18] G. Ellingsrud and Ch. Peskine. Sur les surfaces lisses de $\mathbf{P}_4$. *Invent. Math.*, 95(1):1–11, 1989.

[19] W. Decker and F.-O. Schreyer. Non-general type surfaces in $\mathbf{P}^4$: some remarks on bounds and constructions. *J. Symbolic Comput.*, 29(4-5):545–582, 2000. Symbolic computation in algebra, analysis, and geometry (Berkeley, CA, 1998).

[20] R. Hartshorne. *Algebraic Geometry*. Graduate Texts in Math. 52. Springer, Heidelberg, 1977.

[21] D. Cox, J. Little, and D. O'Shea. *Using algebraic geometry*, volume 185 of *Graduate Texts in Mathematics*. Springer-Verlag, New York, 1998.

[22] H.-Chr. Graf v. Bothmer, C. Erdenberger, and K. Ludwig. Macaulay 2 scripts for finding rational surfaces in $\mathbb{P}^4$. Available at http://www-ifm.math.uni-hannover.de/~bothmer/surface. See also the LaTeX-file of this article at http://arXiv.org., 2004.

[23] J. von zur Gathen and J. Gerhard. *Modern computer algebra*. Cambridge University Press, Cambridge, second edition, 2003.

[24] A. Cayley. A momoir on cubic surfaces. *Philos. Trans. Royal Soc.*, CLIX:231–326, 1869.

[25] L. Schläfli. On the distribution of surfaces of the third order into species, in reference to the presence or absence of singular points and the reality of their lines. *Philos. Trans. Royal Soc.*, CLIII:193–241, 1863.

[26] E.-E. Kummer. Über die Flächen vierten grades mit sechzehn singulären Punkten. In *Collected papers*, pages 418–432. Springer-Verlag, Berlin, 1975. Volume II: Function theory, geometry and miscellaneous, Edited and with a foreward by André Weil.

[27] E. G. Togliatti. Una notevole superficie de 5° ordine con soli punti doppi isolati. *Vierteljschr. Naturforsch. Ges. Zürich 85*, 85(Beiblatt (Festschrift Rudolf Fueter)):127–132, 1940.

[28] A. Beauville. Sur le nombre maximum de points doubles d'une surface dans $\mathbf{P}^3$ ($\mu(5) = 31$). In *Journées de Géometrie Algébrique d'Angers, Juillet 1979/Algebraic Geometry, Angers, 1979*, pages 207–215. Sijthoff & Noordhoff, Alphen aan den Rijn, 1980.

[29] W. Barth. Two projective surfaces with many nodes, admitting the symmetries of the icosahedron. *J. Algebraic Geom.*, 5(1):173–186, 1996.

[30] D. B. Jaffe and D. Ruberman. A sextic surface cannot have 66 nodes. *J. Algebraic Geom.*, 6(1):151–168, 1997.

[31] S. V. Chmutov. Examples of projective surfaces with many singularities. *J. Algebraic Geom.*, 1(2):191–196, 1992.

[32] S. Breske, O. Labs, and D. van Straten. Real Line Arrangements and Surfaces with Many Real Nodes. In R. Piene and B. Jüttler, editors, *Geometric Modeling and Algebraic Geometry*, pages 47–54. Springer, 2008.

[33] A. N. Varchenko. Semicontinuity of the spectrum and an upper bound for the number of singular points of the projective hypersurface. *Dokl. Akad. Nauk SSSR*, 270(6):1294–1297, 1983.

[34] Y. Miyaoka. The maximal number of quotient singularities on surfaces with given numerical invariants. *Math. Ann.*, 268(2):159–171, 1984.

[35] O. Labs. A septic with 99 real nodes. *Rend. Sem. Mat. Univ. Padova*, 116:299–313, 2006.

[36] S. Endraß. *Symmetrische Flche mit vielen gewhnlichen Doppelpunkten*. Dissertation, Universität Erlangen, Germany, 1996.

[37] S. Endrass. SURF 1.0.4. Technical report, University of Mainz, University of Saarbrücken, 2003. http://surf.sourceforge.net/.

[38] S. Holzer and O. Labs. SURFEX 0.89. Technical report, University of Mainz, University of Saarbrücken, 2006. www.surfex.AlgebraicSurface.net.

K3 surfaces of Picard rank one which are double covers of the projective plane

Andreas-Stephan Elsenhans and Jörg Jahnel

Universität Göttingen, Mathematisches Institut,
Bunsenstraße 3–5, D-37073 Göttingen, Germany[1]
e-mail: elsenhan@uni-math.gwdg.de, jahnel@uni-math.gwdg.de

Abstract. We construct examples of K3 surfaces over $\mathbb{Q}$ which are of degree 2 and the geometric Picard rank of which is equal to 1. We construct, particularly, examples in the form $w^2 = \det M$ where M is a symmetric (3×3)-matrix of ternary quadratic forms or a symmetric (6×6)-matrix of ternary linear forms. Our method is based on reduction modulo p for $p = 3$ and 5.

1. Introduction

A K3 surface is a simply connected, projective algebraic surface with trivial canonical class. Let $B \subset \mathbb{P}^2$ be a smooth plane curve of degree 6 given by $f_6(x, y, z) = 0$. The equation $w^2 = f_6(x, y, z)$ defines an algebraic surface S in weighted projective space $\mathbb{P}(1, 1, 1, 3)$. We have a double cover $\pi\colon S \to \mathbb{P}^2$ ramified at $\pi^{-1}(B)$. This surface is a K3 surface (of degree two).

Examples 1.1 K3 surfaces embedded into $\mathbb{P}^n$ are automatically of even degree. Small degree cases may be realized as follows: A K3 surface of degree two is a double cover of $\mathbb{P}^2$, ramified in a smooth sextic. K3 surfaces of degree four are smooth quartics in $\mathbb{P}^3$. A K3 surface of degree six is a smooth complete intersection of a quadric and a cubic in $\mathbb{P}^4$. And, finally, K3 surfaces of degree eight are smooth complete intersections of three quadrics in $\mathbb{P}^5$.

The Picard group of a K3 surface is known to be isomorphic to $\mathbb{Z}^n$ where n may range from 1 to 20. It is generally known that a generic K3 surface over $\mathbb{C}$ is of Picard rank one.

Nevertheless, it seems that the first explicit examples of K3 surfaces of geometric Picard rank one have been constructed as late as in 2005 [5]. All these examples are of degree four.

---

[1]The computer part of this work was executed on the Sun Fire V20z Servers of the Gauß Laboratory for Scientific Computing at the Göttingen Mathematisches Institut. Both authors are grateful to Prof. Y. Tschinkel for the permission to use these machines as well as to the system administrators for their support.

Our goal here is to provide explicit examples of K3 surfaces defined over $\mathbb{Q}$ which are of degree two and geometric Picard rank one.

Let $\mathscr{S}$ be a K3 surface over a finite field $\mathbb{F}_q$. We have the first Chern class homomorphism

$$c_1 \colon \operatorname{Pic}(\mathscr{S}) \longrightarrow H^2_{\text{ét}}(\mathscr{S}_{\overline{\mathbb{F}}_q}, \overline{\mathbb{Q}}_\ell(1))$$

into ℓ-adic cohomology. There is a natural operation of the Frobenius on $H^2_{\text{ét}}(\mathscr{S}_{\overline{\mathbb{F}}_q}, \overline{\mathbb{Q}}_\ell(1))$. All eigenvalues are of absolute value 1. The Frobenius operation on the Picard group is compatible with the operation on cohomology.

Every divisor is defined over a finite extension of the ground field. Consequently, on the subspace $\operatorname{Pic}(\mathscr{S}_{\overline{\mathbb{F}}_q}) \otimes_{\mathbb{Z}} \overline{\mathbb{Q}}_\ell \hookrightarrow H^2_{\text{ét}}(\mathscr{S}_{\overline{\mathbb{F}}_q}, \overline{\mathbb{Q}}_\ell(1))$, all eigenvalues are roots of unity. Those correspond to eigenvalues of the Frobenius operation on $H^2_{\text{ét}}(\mathscr{S}_{\overline{\mathbb{F}}_q}, \overline{\mathbb{Q}}_\ell)$ which are of the form $q\zeta$ for ζ a root of unity.

We may therefore bound the rank of the Picard group $\operatorname{Pic}(\mathscr{S}_{\overline{\mathbb{F}}_q})$ from above by counting how many eigenvalues are of this particular form. Bounds from below may be obtained by explicitly constructing divisors. Combining these two bounds it is sometimes possible to calculate $\operatorname{rk}\operatorname{Pic}(\mathscr{S}_{\overline{\mathbb{F}}_q})$.

Our general strategy is to use reduction modulo p. If S is a K3 surface over $\mathbb{Q}$ then there is the inequality

$$\operatorname{rk}\operatorname{Pic}(S_{\overline{\mathbb{Q}}}) \leq \operatorname{rk}\operatorname{Pic}(S_{\overline{\mathbb{F}}_p})$$

which holds for every prime p of good reduction.

Remark 1.2 Consider a complex K3 surface S. Since $H^1(S, \mathscr{O}_S) = 0$, the Picard group of S is discrete and the first Chern class homomorphism

$$c_1 \colon \operatorname{Pic}(S) \to H^2(S, \mathbb{Z}) \subset H^2(S, \mathbb{C})$$

is an injection. For divisors, numerical and homological equivalence are known to coincide [4, Corollary 1]. This shows that $\operatorname{Pic}(S)$ equals the group of divisors modulo numerical equivalence.

2. Geometric constructions of divisors over $\overline{\mathbb{F}}_p$

In order to bound the rank of the Picard group from below, one needs to explicitly construct divisors. Calculating discriminants, it is possible to show that the corresponding divisor classes are linearly independent.

Assumption 2.1 For the algebro-geometric considerations described in this section, we assume that we work over a ground field which is algebraically closed of characteristic $\neq 2$.

Construction 2.2 i) Assume that the branch curve "$f_6 = 0$" has a tritangent line G. The pull-back of G to the K3 surface $\mathscr{S}$ is a divisor splitting into two irreducible components. The corresponding divisor classes are linearly independent.

ii) A second possibility is to use a conic which is tangent to the branch sextic in six points.

Both constructions yield a lower bound of 2 for the rank of the Picard group.

Tritangent. Assume that the line G is tritangent to "$f_6 = 0$". The restriction of f_6 to $G \cong \mathbb{P}^1$ is a section of $\mathscr{O}(6)$, the divisor of which is divisible by 2 in $\mathrm{Div}(G)$. As G is of genus 0, this implies $f_6|_G$ is the square of a section $f \in \Gamma(G, \mathscr{O}(3))$. The form f_6 may, therefore, be written as $f_6 = \widetilde{f}^2 + lq_5$ for l a linear form defining G, $\widetilde{f}$ a cubic form lifting f, and a quintic form q_5.

Consequently, the restriction of π to $\pi^{-1}(G)$ is given by an equation of the form $w^2 = f^2(s,t)$. We, therefore, have $\pi^*(G) = D_1 + D_2$ where D_1 and D_2 are the two irreducible divisors given by $w = \pm f(s,t)$. Both curves are isomorphic to G. In particular, they are projective lines.

The adjunction formula shows $-2 = D_1(D_1 + K) = D_1^2$. Analogously, $D_2^2 = -2$. Finally, we have $G^2 = 1$. It follows that $(D_1 + D_2)^2 = 2$ which yields $D_1 D_2 = 3$. For the discriminant, we find

$$\begin{vmatrix} -2 & 3 \\ 3 & -2 \end{vmatrix} = -5 \neq 0$$

guaranteeing that $\mathrm{rk}\,\mathrm{Pic}(\mathscr{S}) \geq 2$.

Remark 2.3 This argument works without modification if two or all three points of tangency coincide.

Conic tangent in six points. If C is a conic tangent to the branch curve "$f_6 = 0$" in six points then, for the same reasons as above, we have $\pi^*(C) = C_1 + C_2$, where C_1 and C_2 are irreducible divisors. Again, C_1 and C_2 are isomorphic to C and, therefore, of genus 0. This shows $C_1^2 = C_2^2 = -2$. Further, $C^2 = 4$ which implies $(C_1 + C_2)^2 = 8$ and $C_1 C_2 = 6$. The discriminant equals

$$\begin{vmatrix} -2 & 6 \\ 6 & -2 \end{vmatrix} = -32 \neq 0.$$

Thus, $\mathrm{rk}\,\mathrm{Pic}(\mathscr{S}) \geq 2$ in this case as well.

Remark 2.4 Further tritangents or further conics which are tangent in six points lead to even larger Picard groups.

3. Explicit divisors – Practical tests over $\mathbb{F}_q$

A test for tritangents. The property of a line of being a tritangent may easily be written down as an algebraic condition. Therefore, tritangents may be searched for, in practice, by investigating a Gröbner basis.

More precisely, a general line in $\mathbb{P}^2$ can be described by a parametrization

$$g_{a,b} \colon t \mapsto [1 : t : (a + bt)]\,.$$

$g_{a,b}$ is a (possibly degenerate) tritangent of the sextic "$f_6 = 0$" if and only if $f_6 \circ g_{a,b}$ is a perfect square in $\overline{\mathbb{F}}_q[t]$. This means that

$$f_6(g_{a,b}(t)) = (c_0 + c_1 t + c_2 t^2 + c_3 t^3)^2$$

is an equation which encodes the tritangent property of $g_{a,b}$. Comparing coefficients, this yields a system of seven equations in c_0, c_1, c_2, and c_3 which is solvable if and only if $g_{a,b}$ is a tritangent. The latter may be understood as well as a system of equations in a, b, c_0, c_1, c_2, and c_3 encoding the existence of a tritangent of the form above.

Using `Magma`, we compute the length of $\mathbb{F}_q[a, b, c_0, c_1, c_2, c_3]$ modulo the corresponding ideal I. This is twice the number of the tritangents detected.

The remaining one dimensional family of lines may be tested analogously using the parametrizations $g_a \colon t \mapsto [1 : a : t]$ and $g \colon t \mapsto [0 : 1 : t]$.

Remarks 3.1 a) To compute the length of $\mathbb{F}_q[a, b, c_0, c_1, c_2, c_3]/I$, a Gröbner basis of I is needed. The time required to compute such a basis over a finite field is usually a few seconds. From the Gröbner basis, the tritangents may be read off, explicitly.

b) Since the existence of a tritangent is a codimension one condition, one occasionally finds tritangents on randomly chosen examples.

A test for conics tangent in six points. A non-degenerate conic in $\mathbb{P}^2$ allows a parametrization of the form

$$c \colon t \mapsto [(c_0 + c_1 t + c_2 t^2) : (d_0 + d_1 t + d_2 t^2) : (e_0 + e_1 t + e_2 t^2)]\,.$$

With the sextic "$f_6 = 0$", all intersection multiplicities are even if and only if $f_6 \circ c$ is a perfect square in $\overline{\mathbb{F}}_q[t]$. This may easily be checked by factoring $f_6 \circ c$.

For small q, that allows, at least, to search for conics which are defined over $\mathbb{F}_q$ and tangent in six points. To achieve this, we listed all $q^2(q^3 - 1)$ non-degenerate conics over $\mathbb{F}_q$ for $q = 3$ and 5.

Remark 3.2 A analogous general method to find conics defined over $\overline{\mathbb{F}}_q$ does not succeed. The required Gröbner basis computation becomes too large.

4. Upper bounds – The Frobenius operation on l-adic cohomology

The Lefschetz trace formula. The Frobenius operation on $H^2_{\text{ét}}(\mathscr{S}_{\overline{\mathbb{F}}_p}, \overline{\mathbb{Q}}_\ell)$ can be analyzed as follows.

Count the points on $\mathscr{S}$ over $\mathbb{F}_{p^d}$ and apply the Lefschetz trace formula [6] to compute the trace of the Frobenius $\phi_{\mathbb{F}_{p^d}} = \phi^d$. In our situation, this yields

$$\mathrm{Tr}(\phi^d) = \#\mathscr{S}(\mathbb{F}_{p^d}) - p^{2d} - 1\,.$$

We have $\mathrm{Tr}(\phi^d) = \lambda_1^d + \cdots + \lambda_{22}^d =: \sigma_d(\lambda_1, \ldots, \lambda_{22})$ when we denote the eigenvalues of ϕ by $\lambda_1, \ldots, \lambda_{22}$. Newton's identity [8]

$$s_k(\lambda_1, \ldots, \lambda_{22}) = \frac{1}{k} \sum_{r=0}^{k-1} (-1)^{k+r+1} \sigma_{k-r}(\lambda_1, \ldots, \lambda_{22}) s_r(\lambda_1, \ldots, \lambda_{22})$$

shows that, doing this for $d = 1, \ldots, k$, one obtains enough information to determine the coefficient $(-1)^k s_k$ of t^{22-k} of the characteristic polynomial f_p of ϕ.

Observe that we also have the functional equation

$$(*) \qquad\qquad p^{22} f_p(t) = \pm t^{22} f_p(p^2/t)$$

at our disposal. It may be used to convert the coefficient of t^i into the one of t^{22-i}.

Methods for counting points. The number of the points may be determined by

$$\#\mathscr{S}(\mathbb{F}_q) = \sum_{[x:y:z] \in \mathbb{P}^2(\mathbb{F}_q)} \left[1 + \chi\big(f_6(x, y, z)\big) \right] .$$

Here, χ is the quadratic character. The sum is well-defined since $f_6(x, y, z)$ is uniquely determined up to a sixth-power residue. To count the points naively, one would need $q^2 + q + 1$ evaluations of f_6 and χ.

There are several ways to optimize. Here are two possibilities:

i) Symmetry: If f_6 is defined over $\mathbb{F}_p$ then the summands for $[x : y : z]$ and $\phi([x : y : z])$ are equal. This means, over $\mathbb{F}_{p^d}$, we may save a factor of d if, on the affine chart "$x = 1$", we put in for y only values from a fundamental domain of the Frobenius.

ii) Decoupling: Suppose, f_6 contains only monomials of the form $x^i y^{6-i}$ or $x^i z^{6-i}$. Then, on the affine chart "$x = 1$", the form f_6 may be written as a sum of a function in y and a function in z.

In $O(q \log q)$ steps, for each of the two functions, we build up a table stating how many times it adopts each of its values. Again, we may restrict one of the tables to a fundamental domain of the Frobenius. We tabulate the quadratic character, too. After these preparations, less than q^2 additions suffice to determine the number of points.

The advantage of a decoupled situation is, therefore, that an evaluation of a polynomial in $\mathbb{F}_{p^d}$ gets replaced by an addition.

Remark 4.1 Having implemented the point counting in C, these optimizations allow us to determine the number of $\mathbb{F}_{3^{10}}$-rational points on a K3 surface $\mathscr{S}$ within half an hour (without decoupling) on an AMD Opteron processor.

In a decoupled situation, the number of $\mathbb{F}_{5^9}$-rational points may be counted within two hours. In a few cases, we determined the numbers of points over $\mathbb{F}_{5^{10}}$. This took about two days. Without decoupling, the same counts would have taken about one day or 25 days, respectively.

This shows that using the methods above we may effectively compute the traces of $\phi_{\mathbb{F}_{p^d}} = \phi^d$ for $d = 1, \ldots, 9, (10)$.

An upper bound for $\mathrm{rk}\,\mathrm{Pic}(\mathscr{S}_{\overline{\mathbb{F}}_p})$, counting up to $d = 10$.
We know that the characteristic polynomial of the Frobenius f_p has a zero at p since the pull-back of a line in $\mathbb{P}^2$ is a divisor defined over $\mathbb{F}_p$. Suppose, we determined $\mathrm{Tr}(\phi^d)$ for $d = 1, \ldots, 10$. We may achieve an upper bound for $\mathrm{rk}\,\mathrm{Pic}(\mathscr{S}_{\overline{\mathbb{F}}_p})$ as follows.

i) Assume the minus sign in the functional equation $(*)$. Then f_p automatically has coefficient 0 at t^{11}. Therefore, the numbers of points counted suffice in this case to determine f_p, completely.

ii) Assume, on the other hand, that the plus sign is present in $(*)$. In this case, the data collected immediately allow to compute all coefficients of f_p except the one at t^{11}. Use the known zero at p to determine that final coefficient.

iii) Use the numerical test, described below, to decide which sign is actually present.

iv) Factor $f_p(pt)$ into irreducible polynomials. Check which of the factors are cyclotomic polynomials and add their degrees. That sum is an upper bound for $\mathrm{rk}\,\mathrm{Pic}(\mathscr{S}_{\overline{\mathbb{F}}_p})$. If step iii) had failed then one has to work with both candidates for f_p and deal with the maximum.

Verifying $\mathrm{rk}\,\mathrm{Pic}(\mathscr{S}_{\overline{\mathbb{F}}_p}) = 2$ with $d \leq 9$.
Let $\mathscr{S}$ be a K3 surface over $\mathbb{F}_p$ given by Construction 2.2.i) or ii). We know that the rank of the Picard group is at least 2. We suppose that the divisor constructed by pull-back splits already over $\mathbb{F}_p$. This ensures that p is a double zero of f_p. There is the following method to verify $\mathrm{rk}\,\mathrm{Pic}(\mathscr{S}_{\overline{\mathbb{F}}_p}) = 2$.

i) First, assume the minus sign in the functional equation $(*)$. This forces another zero of f_p at $(-p)$. The data collected suffice to determine f_p, completely. The numerical test, described below, may indicate a contradiction.

Otherwise, the verification fails. (In that case, we could still find an upper bound for $\mathrm{rk}\,\mathrm{Pic}(\mathscr{S}_{\overline{\mathbb{F}}_p})$ which is, however, at least equal to 4.)

ii) As we have the plus sign in $(*)$, the data immediately suffice to compute all coefficients of f_p with the exception of those at t^{10}, t^{11}, and t^{12}. The functional equation yields a linear relation for the three remaining coefficients of f_p. From the known double zero at p, one computes another linear condition.

iii) Let n run through all natural numbers such that $\varphi(n) \leq 20$. (The largest such n is 66.) Assume, in addition, that there is another zero of the form $p\zeta_n$. This yields further linear relations. Inspecting this system of linear equations, one either finds a contradiction or determines all three remaining coefficients. In the latter case, the numerical test may indicate a contradiction.

If each value of n is contradictory then $\operatorname{rk}\operatorname{Pic}(\mathscr{S}_{\overline{\mathbb{F}}_p}) = 2$.

Consequently, the equality $\operatorname{rk}\operatorname{Pic}(\mathscr{S}_{\overline{\mathbb{F}}_p}) = 2$ may be effectively provable from $\operatorname{Tr}(\phi^d)$ for $d = 1, \ldots, 9, (10)$.

A numerical test. Given a polynomial f of degree 22, we calculate all its zeroes as floating point numbers. If at least one of them is clearly not of absolute value p then f can not be the characteristic polynomial of the Frobenius for any K3 surface over $\mathbb{F}_p$.

Remarks 4.2 i) This approach will always yield an even number for the upper bound of the Picard rank. Indeed, the bound is

$$\operatorname{rk}\operatorname{Pic}(\mathscr{S}_{\overline{\mathbb{F}}_p}) \leq \dim(H^2_{\text{ét}}(\mathscr{S}_{\overline{\mathbb{F}}_p}, \overline{\mathbb{Q}}_\ell))$$
$$- \#\{\text{ zeroes of } f_p \text{ which are not of the form } \zeta_n p\}.$$

The relevant zeroes come in pairs of complex conjugate numbers. Hence, for a K3 surface the bound is always even.

ii) There is a famous conjecture due to John Tate [7] which implies that the canonical injection $c_1\colon \operatorname{Pic}(\mathscr{S}_{\overline{\mathbb{F}}_p}) \to H^2_{\text{ét}}(\mathscr{S}_{\overline{\mathbb{F}}_p}, \overline{\mathbb{Q}}_\ell(1))$ maps actually onto the sum of all eigenspaces for the eigenvalues which are roots of unity. Together with the conjecture of J.-P. Serre which says that the Frobenius operation on étale cohomology is always semisimple, this would imply that the bound above is actually sharp.

It is a somewhat surprising consequence of the Tate conjecture that the Picard rank of a K3 surface over $\overline{\mathbb{F}}_p$ is always even. For us, this is bad news. The obvious strategy to prove $\operatorname{rk}\operatorname{Pic}(S_{\overline{\mathbb{Q}}}) = 1$ for a K3 surface S over $\mathbb{Q}$ would be to verify $\operatorname{rk}\operatorname{Pic}(S_{\overline{\mathbb{F}}_p}) = 1$ for a suitable place p of good reduction. The Tate conjecture indicates that there is no hope for such an approach.

5. How to prove $\operatorname{rk}\operatorname{Pic}(S_{\overline{\mathbb{Q}}}) = 1$

Using the methods described above we can construct even upper bounds for the Picard rank. On the other hand, we can generate lower bounds by explicitly stating divisors. In optimal situations this may establish an equality $\operatorname{rk}\operatorname{Pic}(\mathscr{S}_{\overline{\mathbb{F}}_p}) = 2$. However, how to reach Picard rank 1 for a surface defined over $\mathbb{Q}$? Here we apply a trick due to R. van Luijk [5, Remark 2].

Fact 5.1 (van Luijk) *Assume that we are given a K3 surface $\mathscr{S}^{(3)}$ over $\mathbb{F}_3$ and a K3 surface $\mathscr{S}^{(5)}$ over $\mathbb{F}_5$ which are both of geometric Picard rank 2. Suppose further that the discriminants of the intersection forms on $\operatorname{Pic}(\mathscr{S}^{(3)}_{\overline{\mathbb{F}}_3})$ and $\operatorname{Pic}(\mathscr{S}^{(5)}_{\overline{\mathbb{F}}_5})$ are essentially different, i.e., their quotient is not a perfect square in $\mathbb{Q}$.*

Then every K3 surface S such that its reduction at 3 is isomorphic to $\mathscr{S}^{(3)}$ and its reduction at 5 is isomorphic to $\mathscr{S}^{(5)}$ is of geometric Picard rank one.

Proof. The reduction maps $\iota_p \colon \mathrm{Pic}(S_{\overline{\mathbb{Q}}}) \to \mathrm{Pic}(S_{\overline{\mathbb{F}}_p}) = \mathrm{Pic}(\mathscr{S}_{\overline{\mathbb{F}}_p}^{(p)})$ are injective [3, Example 20.3.6]. Observe that $\mathrm{Pic}(S_{\overline{\mathbb{Q}}})$ is equal to the group of divisors on $S_{\overline{\mathbb{Q}}}$ modulo numerical equivalence.

This immediately leads to the bound $\mathrm{rk}\,\mathrm{Pic}(S_{\overline{\mathbb{Q}}}) \leq 2$. Assume, by contradiction, that equality holds. Then, the reductions of $\mathrm{Pic}(S_{\overline{\mathbb{Q}}})$ are sublattices of maximal rank in both, $\mathrm{Pic}(S_{\overline{\mathbb{F}}_3}) = \mathrm{Pic}(\mathscr{S}_{\overline{\mathbb{F}}_3}^{(3)})$ and $\mathrm{Pic}(S_{\overline{\mathbb{F}}_5}) = \mathrm{Pic}(\mathscr{S}_{\overline{\mathbb{F}}_5}^{(5)})$.

The intersection product is compatible with reduction. Therefore, the quotients $\mathrm{Disc}\,\mathrm{Pic}(S_{\overline{\mathbb{Q}}})/\mathrm{Disc}\,\mathrm{Pic}(\mathscr{S}_{\overline{\mathbb{F}}_3}^{(3)})$ and $\mathrm{Disc}\,\mathrm{Pic}(S_{\overline{\mathbb{Q}}})/\mathrm{Disc}\,\mathrm{Pic}(\mathscr{S}_{\overline{\mathbb{F}}_5}^{(5)})$ are perfect squares. This is a contradiction to the assumption. $\square$

Remark 5.2 Suppose that $\mathscr{S}^{(3)}$ and $\mathscr{S}^{(5)}$ are K3 surfaces of degree two given by explicit branch sextics in $\mathbb{P}^2$. Then, using the Chinese Remainder Theorem, they can easily be combined to a K3 surface S over $\mathbb{Q}$.

If one of them allows a conic tangent in six points and the other a tritangent then the discriminants of the intersection forms on $\mathrm{Pic}(\mathscr{S}_{\overline{\mathbb{F}}_3}^{(3)})$ and $\mathrm{Pic}(\mathscr{S}_{\overline{\mathbb{F}}_5}^{(5)})$ are essentially different as shown in Section 2.

Remark 5.3 Suppose S is a K3 surface over $\mathbb{Q}$ constructed that way. Then, S cannot be isomorphic, not even over $\overline{\mathbb{Q}}$, to a K3 surface $S' \subset \mathbb{P}^3$ of degree 4. In particular, the explicit examples, which we will describe in the next sections, are different from those of R. van Luijk [5].

Indeed, $\mathrm{Pic}(S_{\overline{\mathbb{Q}}}) = \mathbb{Z} \cdot \langle \mathscr{L} \rangle$ and $\deg S = 2$ mean that the intersection form on $\mathrm{Pic}(S_{\overline{\mathbb{Q}}})$ is given by $\langle \mathscr{L}^{\otimes n}, \mathscr{L}^{\otimes m} \rangle = 2nm$. All self-intersection numbers of invertible sheaves on $S_{\overline{\mathbb{Q}}}$ are of the form $2n^2$ which is always different from 4.

6. An explicit K3 surface of degree two

Examples 6.1 We consider two particular K3 surfaces over finite fields.

i) By $\mathscr{X}^0$, we denote the surface over $\mathbb{F}_3$ given by the equation

$$
\begin{aligned}
w^2 = \;& (y^3 - x^2 y)^2 \\
& + (x^2 + y^2 + z^2)(2x^3 y + x^3 z + 2x^2 yz + x^2 z^2 + 2xy^3 + 2y^4 + z^4) \\
= \;& 2x^5 y + x^5 z + x^4 y^2 + 2x^4 yz + x^4 z^2 + x^3 y^3 + x^3 y^2 z + 2x^3 yz^2 + x^3 z^3 \\
& + 2x^2 y^3 z + x^2 y^2 z^2 + 2x^2 yz^3 + 2x^2 z^4 + 2xy^5 + 2xy^3 z^2 + 2y^4 z^2 + y^2 z^4 + z^6 \,.
\end{aligned}
$$

ii) Further, let $\mathscr{Y}^0$ be the K3 surface over $\mathbb{F}_5$ given by

$$
w^2 = x^5 y + x^4 y^2 + 2x^3 y^3 + x^2 y^4 + xy^5 + 4y^6 + 2x^5 z + 2x^4 z^2 + 4x^3 z^3 + 2xz^5 + 4z^6 \,.
$$

Theorem 6.2 *Let S be a K3 surface over $\mathbb{Q}$ such that its reduction modulo 3 is isomorphic to $\mathscr{X}^0$ and its reduction modulo 5 is isomorphic to $\mathscr{Y}^0$. Then, $\mathrm{rk}\,\mathrm{Pic}(S_{\overline{\mathbb{Q}}}) = 1$.*

Proof. We follow the strategy described in Remark 5.2. For the branch locus of $\mathscr{X}^0$, the conic given by $x^2 + y^2 + z^2 = 0$ is tangent in six points. The branch locus of $\mathcal{Y}_0$ has a tritangent given by $z - 2y = 0$. It meets the branch locus at $[1 : 0 : 0]$, $[1 : 3 : 1]$, and $[0 : 1 : 2]$.

It remains to show that $\operatorname{rk}\operatorname{Pic}(\mathscr{X}^0_{\overline{\mathbb{F}}_3}) \leq 2$ and $\operatorname{rk}\operatorname{Pic}(\mathscr{Y}^0_{\overline{\mathbb{F}}_5}) \leq 2$. To verify these assertions, we used the methods described in Section 4. We counted points over $\mathbb{F}_{3^d}$ and $\mathbb{F}_{5^d}$, respectively, for $d \leq 10$. For $\mathscr{Y}^0$, we could use the faster method since the sextic form on the right hand side is decoupled. $\square$

Corollary 6.3 *Let S be the K3 surface over $\mathbb{Q}$ given by*

$$w^2 = -4x^5y + 7x^5z + x^4y^2 + 5x^4yz + 7x^4z^2 + 7x^3y^3 - 5x^3y^2z + 5x^3yz^2$$
$$+ 4x^3z^3 + 6x^2y^4 + 5x^2y^3z - 5x^2y^2z^2 + 5x^2yz^3 + 5x^2z^4 - 4xy^5$$
$$+ 5xy^3z^2 - 3xz^5 - 6y^6 + 5y^4z^2 - 5y^2z^4 + 4z^6 \,.$$

i) *Then, $\operatorname{rk}\operatorname{Pic}(S_{\overline{\mathbb{Q}}}) = 1$.*

ii) *Further, $S(\mathbb{Q}) \neq \emptyset$. For example, $[2 : 0 : 0 : 1] \in S(\mathbb{Q})$.*

Remarks 6.4 i) For the K3 surface $\mathscr{X}^0$, our calculations show the following.

The numbers of the points defined over $\mathbb{F}_{3^d}$ for $d = 1, \ldots, 10$ are, in this order, 14, 92, 758, 6 752, 59 834, 532 820, 4 796 120, 43 068 728, 387 421 463, and 3 487 077 812. The traces of the Frobenius $\phi_{\mathbb{F}_{p^d}} = \phi^d$ on $H^2_{\text{ét}}(\mathscr{X}^0_{\overline{\mathbb{F}}_3}, \overline{\mathbb{Q}}_\ell)$ are equal to 4, 10, 28, 190, 784, 1 378, 13 150, 22 006, 973, and 293 410.

The sign in the functional equation is positive. For the decomposition of the characteristic polynomial f_p of the Frobenius, we find (after scaling to zeroes of absolute value 1)

$$(t - 1)^2(3t^{20} + 2t^{19} + 2t^{18} + 2t^{17} + t^{16} - 2t^{13} - 2t^{12} - t^{11} - 2t^{10} - t^9$$
$$- 2t^8 - 2t^7 + t^4 + 2t^3 + 2t^2 + 2t + 3)/3$$

with an irreducible polynomial of degree 20. The assumption of the negative sign leads to zeroes the absolute values of which range (without scaling) from 2.598 to 3.464.

ii) For the K3 surface $\mathscr{Y}^0$, our calculations yield the following results.

The numbers of points over $\mathbb{F}_{5^d}$ are, in this order, 41, 751, 15 626, 392 251, 9 759 376, 244 134 376, 6 103 312 501, 152 589 156 251, 3 814 704 296 876, and 95 367 474 609 376. The traces of the Frobenius on $H^2_{\text{ét}}(\mathscr{Y}^0_{\overline{\mathbb{F}}_5}, \overline{\mathbb{Q}}_\ell)$ are 15, 125, 0, 1 625, −6 250, −6 250, −203 125, 1 265 625, 7 031 250, and 42 968 750.

The sign in the functional equation is positive. For the decomposition of the scaled characteristic polynomial of the Frobenius, we find

$$(t - 1)^2(5t^{20} - 5t^{19} - 5t^{18} + 10t^{17} - 2t^{16} - 3t^{15} + 4t^{14} - 2t^{13} - 2t^{12} + t^{11}$$
$$+ 3t^{10} + t^9 - 2t^8 - 2t^7 + 4t^6 - 3t^5 - 2t^4 + 10t^3 - 5t^2 - 5t + 5)/5 \,.$$

The assumption of the negative sign leads to zeroes the absolute values of which range (without scaling) from 3.908 to 6.398.

7. K3 surfaces of degree 2 given by a symmetric (3×3)-determinant

Examples 7.1

i) Let $\mathscr{X}$ be the surface over $\mathbb{F}_3$ given by the equation $w^2 = f_6(x,y,z)$ for

$$
f_6 = \det \begin{pmatrix}
2xy+2y^2+yz & 2x^2+xy+xz+yz+2z^2 & 2x^2+xz+yz+z^2 \\
2x^2+xy+xz+yz+2z^2 & 2x^2+xy & xy+y^2+yz+2z^2 \\
2x^2+xz+yz+z^2 & xy+y^2+yz+2z^2 & 2x^2+2xy+2y^2+2yz
\end{pmatrix}
$$

$$
\begin{aligned}
= \; & 2x^6 + 2x^5y + 2x^5z + 2x^4y^2 + x^4yz + x^3y^3 + x^3yz^2 + x^3z^3 + 2x^2y^4 \\
& + x^2y^3z + 2x^2y^2z^2 + xy^5 + xy^2z^3 + y^6 + y^5z + y^2z^4 + yz^5 + 2z^6 \,.
\end{aligned}
$$

ii) Let $\mathscr{Y}$ be the K3 surface over $\mathbb{F}_5$ given by $w^2 = f_6(x,y,z)$, where

$$
f_6 = \det \begin{pmatrix}
4x^2+4xz+y^2 & 2x^2+3z^2 & 4x^2+2xy+2xz+4y^2+3yz+2z^2 \\
2x^2+3z^2 & 2x^2+4xy+4y^2+yz+3z^2 & 4xy+2xz+y^2+4yz+4z^2 \\
4x^2+2xy+2xz+4y^2+3yz+2z^2 & 4xy+2xz+y^2+4yz+4z^2 & 4x^2+xz+3z^2
\end{pmatrix}
$$

$$
\begin{aligned}
= \; & 4x^6 + 2x^5y + x^5z + x^4y^2 + x^4z^2 + x^3y^3 \\
& + 4x^3z^3 + 2x^2y^4 + 2x^2z^4 + 4xy^5 + xz^5 + 4z^6 \,.
\end{aligned}
$$

Theorem 7.2 *Let S be any K3 surface over $\mathbb{Q}$ such that its reduction modulo 3 is isomorphic to $\mathscr{X}$ and its reduction modulo 5 is isomorphic to $\mathscr{Y}$. Then, $\operatorname{rk}\operatorname{Pic}(S_{\overline{\mathbb{Q}}}) = 1$.*

Proof. Consider the branch locus of $\mathscr{X}$. The conic C, given by $x^2 + xy + 2xz + z^2 = 0$, admits the parametrization

$$
q \colon u \mapsto [u^2 : 2 : (2u^2 + 2u)].
$$

We find

$$
f_6(q(u)) = (u+1)^2(u^5 + u^4 + u^3 + u + 1)^2,
$$

i.e., C is tangent in six points and the corresponding divisor on $\mathscr{X}$ splits already over $\mathbb{F}_3$. The branch sextic of $\mathscr{Y}$ has a degenerate tritangent given by $x = 0$.

    To verify that $\operatorname{rk}\operatorname{Pic}(\mathscr{X}_{\overline{\mathbb{F}_3}}) \leq 2$ and $\operatorname{rk}\operatorname{Pic}(\mathscr{Y}_{\overline{\mathbb{F}_5}}) \leq 2$, again, we used the methods described in Section 4. We counted points over $\mathbb{F}_{3^d}$, respectively $\mathbb{F}_{5^d}$, for $d \leq 10$. Observe that, for $\mathscr{Y}$, we could use the faster method since the sextic form on the right hand side is decoupled. $\qquad\qquad\square$

Corollary 7.3 *Let S be the K3 surface over $\mathbb{Q}$ given by $w^2 = f_6(x,y,z)$ for $f_6 =$*

$$
\det \begin{pmatrix}
-6x^2+5xy-6xz-4339y^2-5yz & 2x^2-5xy-5xz-150y^2-5yz-7z^2 & -x^2-3xy+7xz-6y^2-2yz+7z^2 \\
2x^2-5xy-5xz-150y^2-5yz-7z^2 & 2x^2+4xy-6y^2+6yz+3z^2 & 4xy-3xz+y^2+4yz-z^2 \\
-x^2-3xy+7xz-6y^2-2yz+7z^2 & 4xy-3xz+y^2+4yz-z^2 & -x^2+5xy+6xz+5y^2+5yz+3z^2
\end{pmatrix}
$$

$$
= 14x^6 - 118x^5y - 64x^5z + 8\,021x^4y^2 + 220x^4yz - 114x^4z^2
$$

$$
- 20\,249x^3y^3 - 47\,700x^3y^2z - 635x^3yz^2 + 4x^3z^3
$$

$$
- 64\,753x^2y^4 - 247\,925x^2y^3z + 26\,045x^2y^2z^2 - 2\,745x^2yz^3 - 153x^2z^4
$$

$$- 33\,821xy^5 - 107\,100xy^4z - 463\,245xy^3z^2 - 62\,450xy^2z^3 - 3\,075xyz^4 - 384xz^5$$

$$+ 24\,025y^6 - 77\,345y^5z - 143\,880y^4z^2 - 201\,885y^3z^3 - 39\,455y^2z^4 - 1\,055yz^5 - 196z^6.$$

i) *Then,* $\operatorname{rk}\operatorname{Pic}(S_{\overline{\mathbb{Q}}}) = 1$.

ii) *Further,* $S(\mathbb{Q}) \neq \emptyset$. *For example,* $[155 : 0 : 1 : 0] \in S(\mathbb{Q})$.

Remarks 7.4 i) For the K3 surface $\mathscr{X}$, our calculations show the following.

The numbers of the points defined over $\mathbb{F}_{3^d}$ for $d = 1, \ldots, 10$ are, in this order, 14, 88, 800, 6\,664, 59\,114, 531\,136, 4\,782\,344, 43\,029\,952, 387\,550\,223 and 3\,486\,755\,578. The traces of the Frobenius $\phi_{\mathbb{F}_{p^d}} = \phi^d$ on $H^2_{\text{ét}}(\mathscr{X}_{\overline{\mathbb{F}_3}}, \overline{\mathbb{Q}}_\ell)$ are 5, 7, 71, 103, 65, -305, -625, $-16\,769$, 129\,734, and $-28\,823$.

The decomposition of the scaled characteristic polynomial is

$$(t - 1)^2(3t^{20} + t^{19} + 2t^{18} + t^{16} + t^{15} + 2t^{14} + 2t^{13} + 3t^{12}$$
$$+ 2t^{10} + 3t^8 + 2t^7 + 2t^6 + t^5 + t^4 + 2t^2 + t + 3)/3.$$

It follows that the geometric Picard rank is equal to 2.

ii) For the K3 surface $\mathscr{Y}$, our calculations yield the following results.

The numbers of points over $\mathbb{F}_{5^d}$ are, in this order, 33, 669, 15\,522, 391\,861, 9\,768\,668, 244\,132\,734, 6\,103\,019\,942, 152\,588\,860\,821, 3\,814\,709\,624\,898, and 95\,367\,420\,137\,974. The traces of the Frobenius on $H^2_{\text{ét}}(\mathscr{Y}_{\overline{\mathbb{F}_5}}, \overline{\mathbb{Q}}_\ell)$ are 8, 44, -103, 1\,236, 3\,043, $-7\,891$, $-495\,683$, 970\,196, 12\,359\,273, and $-11\,502\,651$.

The decomposition of the scaled characteristic polynomial is

$$(t - 1)^2(5t^{20} + 2t^{19} + t^{18} + 5t^{17} + 2t^{16} + 2t^{15} + 5t^{14} + 8t^{13} + 4t^{12} + 2t^{11}$$
$$+ 8t^{10} + 2t^9 + 4t^8 + 8t^7 + 5t^6 + 2t^5 + 2t^4 + 5t^3 + t^2 + 2t + 5)/5.$$

Consequently, the geometric Picard rank is equal to 2.

Construction of symmetric (3×3)-matrices with decoupled determinant.
A general ternary sextic has 28 coefficients. It is decoupled if 15 of these vanish. Thus, a randomly chosen sextic form in $\mathbb{F}_q[x, y, z]$ is decoupled with a probability of q^{-15}. This is too low for our purposes.

On the other hand, we can think of decoupling as solving a nonlinear system of 15 equations in 36 variables. One could try to attack this system by a Gröbner base calculation. We use a mixture of both methods. More precisely, we do the following.

Method 7.5 *We construct the matrix M in the particular form*

$$M := \begin{pmatrix} a(x,y,z) & b(x,z) & c_1(x,y,z) \\ b(x,z) & c_2(x,y,z) & c_3(x,y,z) \\ c_1(x,y,z) & c_3(x,y,z) & d(x,z) \end{pmatrix}.$$

i) *We choose the quadratic forms c_1, c_2, c_3, and d, randomly.*

ii) *In a second step, we have to fix the nine coefficients of the quadratic forms a and b. The coefficients of* $\det M$ *at* $x^{6-i-j}y^i z^j$ *for* $i, j > 0$ *are linear functions of the coefficients of* a *and* b. *Observe that the summand* $-b^2 d$ *does not contribute to these critical coefficients.*

Thus, we have to solve a system of 15 *linear equations in nine variables. Naively, such a system is solvable with a probability of* q^{-6}.

If it is not solvable then we go back to the first step.

Remarks 7.6 i) We randomly generated a sample of 30 surfaces over $\mathbb{F}_3$. For each of them, the branch locus was smooth and had passed the two tests described in Section 3, to exclude the existence of a tritangent and to ensure there was exactly one conic over $\mathbb{F}_3$ tangent in six points.

We could establish the equality $\operatorname{rk}\operatorname{Pic}(\mathscr{X}_{\overline{\mathbb{F}}_3}) = 2$ in three of the examples. Example 7.1.i) reproduces one of them.

ii) Using the probabilistic method described above, we generated a sample of 50 surfaces over $\mathbb{F}_5$. We made sure that for each of them the branch sextic was smooth, had exactly one tritangent, and no conic over $\mathbb{F}_5$ tangent in six points. Further, it was decoupled by construction. It took `Magma` approximately one hour to generate that sample.

Having counted points over $\mathbb{F}_{5^d}$ for $d \leq 9$, we were able to establish the equality $\operatorname{rk}\operatorname{Pic}(\mathscr{Y}_{\overline{\mathbb{F}}_5}) = 2$ in two of the examples. For those, we determined, in addition, the numbers of points over $\mathbb{F}_{5^{10}}$. Example 7.1.ii) reproduces one of the two.

8. K3 surface of degree 2 given by a symmetric (6×6)-determinant

Examples 8.1 i) Let $\mathscr{X}'$ be the surface over $\mathbb{F}_3$ given by $w^2 = f_6(x, y, z)$ for

$$f_6 = \det \left[x \begin{pmatrix} 0 & 0 & 0 & 0 & 2 & 2 \\ 0 & 1 & 0 & 1 & 0 & 1 \\ 0 & 0 & 2 & 1 & 1 & 2 \\ 0 & 1 & 1 & 1 & 1 & 1 \\ 2 & 0 & 1 & 1 & 2 & 2 \\ 2 & 1 & 2 & 1 & 2 & 0 \end{pmatrix} + y \begin{pmatrix} 0 & 2 & 1 & 1 & 2 & 1 \\ 2 & 2 & 2 & 2 & 2 & 2 \\ 1 & 2 & 0 & 1 & 1 & 0 \\ 1 & 2 & 1 & 1 & 1 & 1 \\ 2 & 2 & 1 & 1 & 2 & 2 \\ 1 & 2 & 0 & 1 & 2 & 0 \end{pmatrix} + z \begin{pmatrix} 2 & 1 & 1 & 1 & 2 & 0 \\ 1 & 1 & 0 & 1 & 2 & 1 \\ 1 & 0 & 2 & 1 & 1 & 2 \\ 1 & 1 & 1 & 0 & 0 & 1 \\ 2 & 2 & 1 & 0 & 2 & 2 \\ 0 & 1 & 2 & 1 & 2 & 0 \end{pmatrix} \right]$$

$$
\begin{aligned}
= \quad & x^6 + x^5 y + 2x^5 z + 2x^4 y^2 + 2x^4 yz + 2x^2 y^3 z + x^2 z^4 \\
& + 2xy^5 + 2xy^4 z + 2y^6 + 2y^5 z + y^2 z^4 + 2yz^5 \,.
\end{aligned}
$$

ii) Further, let $\mathscr{Y}'$ be the K3 surface over $\mathbb{F}_5$ given by $w^2 = f_6(x, y, z)$ for

$$f_6 = \det \left[x \begin{pmatrix} 3 & 4 & 3 & 4 & 4 & 1 \\ 4 & 3 & 0 & 2 & 1 & 0 \\ 3 & 0 & 4 & 0 & 3 & 0 \\ 4 & 2 & 0 & 2 & 1 & 3 \\ 4 & 1 & 3 & 1 & 0 & 2 \\ 1 & 0 & 0 & 3 & 2 & 1 \end{pmatrix} + y \begin{pmatrix} 0 & 0 & 3 & 3 & 0 & 1 \\ 0 & 2 & 0 & 1 & 3 & 1 \\ 3 & 0 & 5 & 0 & 3 & 5 \\ 3 & 1 & 0 & 3 & 5 & 5 \\ 0 & 3 & 3 & 5 & 0 & 1 \\ 1 & 1 & 5 & 5 & 1 & 3 \end{pmatrix} + z \begin{pmatrix} 2 & 1 & 1 & 1 & 0 & 5 \\ 1 & 0 & 4 & 4 & 4 & 4 \\ 1 & 4 & 2 & 3 & 0 & 2 \\ 1 & 4 & 3 & 1 & 2 & 2 \\ 0 & 4 & 0 & 2 & 3 & 1 \\ 5 & 4 & 2 & 2 & 1 & 0 \end{pmatrix} \right]$$

$$= 2x^6 + x^5 y + 2x^4 y^2 + 3x^4 z^2 + x^3 y^3 + 2x^3 z^3 + x^2 y^4 + 3x^2 z^4 + 2xy^5 + 4z^6 \, .$$

Theorem 8.2 *Let S be any K3 surface over $\mathbb{Q}$ such that its reduction modulo 3 is isomorphic to $\mathscr{X}'$ and its reduction modulo 5 is isomorphic to $\mathscr{Y}'$. Then, $\operatorname{rk}\operatorname{Pic}(S_{\overline{\mathbb{Q}}}) = 1$.*

Proof. Consider the branch locus of $\mathscr{X}'$. For the conic C, given by $xz + y^2 + 2yz + 2z^2 = 0$, there is the parametrization

$$q \colon u \mapsto [(2u^2 + u + 1) : u : 1].$$

We find

$$f_6(q(u)) = (u^2 + u + 2)^2 (u^4 + u + 2)^2,$$

i.e. C admits the property of being tangent in six points and the corresponding divisor on $\mathscr{X}'$ splits already over $\mathbb{F}_3$. The branch sextic of $\mathscr{Y}'$ has a degenerate tritangent given by $x = 0$.

To verify that $\operatorname{rk}\operatorname{Pic}(\mathscr{X}'_{\overline{\mathbb{F}}_3}) \leq 2$ and $\operatorname{rk}\operatorname{Pic}(\mathscr{Y}'_{\overline{\mathbb{F}}_5}) \leq 2$, again, we used the methods described in section 4. We counted points over $\mathbb{F}_{3^d}$ and $\mathbb{F}_{5^d}$, respectively, for $d \leq 10$. Observe, for $\mathscr{Y}'$, we could use the faster method since the sextic form on the right hand side is decoupled. $\qquad\square$

Corollary 8.3 *Let S be the K3 surface over $\mathbb{Q}$ given by $w^2 = f_6(x, y, z)$ for $f_6 =$*

$$\det\left[x\begin{pmatrix} -2382 & -21 & 3 & -6 & -1 & -4 \\ -21 & 28 & 0 & 7 & 6 & -5 \\ 3 & 0 & -1 & -5 & -2 & 5 \\ -6 & 7 & -5 & 7 & 1 & -2 \\ -1 & 6 & -2 & 1 & 5 & 2 \\ -4 & -5 & 5 & -2 & 2 & 6 \end{pmatrix} + y\begin{pmatrix} 0 & 5 & -2 & -2 & 5 & 1 \\ 5 & 2 & 5 & -4 & -7 & -4 \\ -2 & 5 & 0 & -5 & -2 & 0 \\ -2 & -4 & -5 & -2 & -5 & -5 \\ 5 & -7 & -2 & -5 & 5 & -4 \\ 1 & -4 & 0 & -5 & -4 & 3 \end{pmatrix} + z\begin{pmatrix} 2 & 1 & 1 & 1 & 5 & 0 \\ 1 & -5 & -6 & 4 & -1 & 4 \\ 1 & -6 & 2 & -2 & -5 & 2 \\ 1 & 4 & -2 & 6 & -3 & 7 \\ 5 & -1 & -5 & -3 & -7 & -4 \\ 0 & 4 & 2 & 7 & -4 & 0 \end{pmatrix}\right]$$

$$= 76\,139\,167x^6 + 231\,184\,081x^5 y + 210\,075\,725x^5 z$$

$$+ 25\,609\,337x^4 y^2 + 487\,337\,315x^4 yz - 314\,154\,987x^4 z^2$$

$$- 141\,937\,719x^3 y^3 + 283\,035\,180x^3 y^2 z - 434\,149\,815x^3 yz^2 - 5\,367\,468x^3 z^3$$

$$- 175\,763\,034x^2 y^4 + 168\,686\,090x^2 y^3 z$$

$$- 421\,490\,010x^2 y^2 z^2 + 160\,009\,155x^2 yz^3 - 153\,566\,957x^2 z^4$$

$$- 90\,295\,273xy^5 + 175\,779\,575xy^4 z - 285\,747\,180xy^3 z^2$$

$$+ 327\,585\,255xy^2 z^3 - 215\,766\,345xyz^4 + 94\,479\,045xz^5$$

$$+ 133\,220y^6 + 31\,145y^5 z + 380\,715y^4 z^2 - 324\,195y^3 z^3 - 476\,810y^2 z^4$$

$$+ 40\,2845yz^5 - 174\,261z^6 \, .$$

i) *Then, $\operatorname{rk}\operatorname{Pic}(S_{\overline{\mathbb{Q}}}) = 1$.*

ii) *Further, $S(\mathbb{Q}) \neq \emptyset$. For example, $[1286 : 1 : 1 : 1] \in S(\mathbb{Q})$.*

Remarks 8.4 i) For the K3 surface $\mathscr{X}'$, our calculations show the following.

The numbers of the points defined over $\mathbb{F}_{3^d}$ for $d = 1, \ldots, 10$ are, in this order, 12, 90, 783, 6 534, 59 697, 535 329, 4 793 661, 43 079 526, 387 521 091, and 3 487 248 045. The traces of the Frobenius $\phi_{\mathbb{F}_{p^d}} = \phi^d$ on $H^2_{\text{ét}}(\mathscr{X}'_{\overline{\mathbb{F}_3}}, \overline{\mathbb{Q}}_\ell)$ are 3, 9, 54, -27, 648, 3 888, 10 692, 32 805, 100 602, and 463 644.

The decomposition of the scaled characteristic polynomial is

$$(t-1)^2(3t^{20} + 3t^{19} + 3t^{18} + 2t^{17} + 3t^{16} + 2t^{15} - 2t^{13} - 3t^{12} - 4t^{11}$$
$$- 6t^{10} - 4t^9 - 3t^8 - 2t^7 + 2t^5 + 3t^4 + 2t^3 + 3t^2 + 3t + 3)/3 \,.$$

Consequently, the geometric Picard rank is equal to 2.

ii) For the K3 surface $\mathscr{Y}'$, our calculations yield the following results.

The numbers of points over $\mathbb{F}_{5^d}$ are, in this order, 36, 666, 15 711, 391 706, 9 763 601, 244 152 021, 6 103 934 341, 152 589 189 186, 3 814 705 355 181, and 95 367 412 593 451. The traces of the Frobenius on $H^2_{\text{ét}}(\mathscr{Y}'_{\overline{\mathbb{F}_5}}, \overline{\mathbb{Q}}_\ell)$ are 11, 41, 86, 1 081, $-2 024$, 11 396, 418 716, 1 298 561, 8 089 556, and $-19 047 174$.

The decomposition of the scaled characteristic polynomial is

$$(t-1)^2(5t^{20} - t^{19} + t^{18} + 2t^{17} + 3t^{15} + t^{14} - 2t^{13} + t^{12} - t^{11}$$
$$+ 2t^{10} - t^9 + t^8 - 2t^7 + t^6 + 3t^5 + 2t^3 + t^2 - t + 5)/5 \,.$$

It follows that the geometric Picard rank is equal to 2.

Construction of symmetric (6×6)-matrices with decoupled determinant.

Method 8.5 a) We construct a symmetric (6×6)-matrix M_0 the entries of which are linear forms only in y and z. The goal is that its determinant is decoupled, i.e.

$$\det M_0 = ay^6 + bz^6$$

for certain $a, b \in \mathbb{F}_q$, not both vanishing.

This leads to five conditions for the coefficients.

i) We choose all entries in M_0 randomly except for $(M_0)_{11}$.

ii) The determinant is linear in the coefficients of $(M_0)_{11}$. Therefore, we have a system of five linear equations in two variables. Such a system is solvable with a probability of q^{-3} which is enough for our purposes.

If there is no solution then we return to step i).

b) We construct M in the form

$$M := M_0 + xA$$

for A a symmetric matrix with entries in $\mathbb{F}_q$.

i) First, look at the monomials $xy^i z^{5-i}$ for $i = 1, \ldots, 4$, only. Insisting that their coefficients vanish gives a system of four linear equations. In general, its solutions form a 17-dimensional vector space.

ii) For decoupling, six further coefficients have to vanish. We are left with 17 parameters and six non-linear equations.

We choose the parameters randomly and iterate this procedure until a solution is found. Naively, the probability to hit a solution is q^{-6}.

Remarks 8.6 i) We randomly generated a sample of 50 surfaces over $\mathbb{F}_3$. For each of them, the branch sextic was smooth and had passed the two tests described in Section 3, to exclude the existence of a tritangent and to ensure there was exactly one conic over $\mathbb{F}_3$ tangent in six points.

We established $\operatorname{rk}\operatorname{Pic}(\mathscr{X}'_{\overline{\mathbb{F}_3}}) = 2$ in eleven of the examples. Example 8.1.i) is one of them.

ii) Using the probabilistic method described above, we generated a sample of 120 surfaces over $\mathbb{F}_5$. For each of them, the branch sextic was decoupled, by construction. We made sure, in addition, that it was smooth, had exactly one tritangent, and no conic, defined over $\mathbb{F}_5$, which was tangent in six points. It took Magma half a day to generate that sample.

Having counted points over $\mathbb{F}_{5^d}$ for $d \leq 9$, we were able to establish the equality $\operatorname{rk}\operatorname{Pic}(\mathscr{X}'_{\overline{\mathbb{F}_5}}) = 2$ in three of the examples. For those, we determined also the number of points over $\mathbb{F}_{5^{10}}$. Example 8.1.ii) reproduces one of the three.

References

[1]　Beauville, A., *Surfaces algébriques complexes*, Astérisque 54, Société Mathématique de France, Paris 1978

[2]　Elsenhans, A.-S. and Jahnel, J., *The Asymptotics of Points of Bounded Height on Diagonal Cubic and Quartic Threefolds,* Algorithmic number theory, Lecture Notes in Computer Science 4076, Springer, Berlin 2006, 317–332

[3]　Fulton, W., *Intersection theory,* Springer, Berlin 1984

[4]　Lieberman, D. I., *Numerical and homological equivalence of algebraic cycles on Hodge manifolds,* Amer. J. Math. **90** (1968), 366–374

[5]　van Luijk, R., *K3 surfaces with Picard number one and infinitely many rational points,* Algebra & Number Theory **1** (2007), 1–15

[6]　Milne, J. S., *Étale Cohomology,* Princeton University Press, Princeton 1980

[7]　Tate, J., *Conjectures on algebraic cycles in l-adic cohomology,* in: Motives, Proc. Sympos. Pure Math. 55-1, Amer. Math. Soc., Providence 1994, 71–83

[8]　Zeilberger, D., *A combinatorial proof of Newtons's identities,* Discrete Math. **49** (1984), 319

Higher-Dimensional Geometry over Finite Fields
D. Kaledin and Y. Tschinkel (Eds.)
IOS Press, 2008

Beilinson Conjectures in the non-commutative setting

Dmitry Kaledin
e-mail: kaledin@mccme.ru

Abstract. We discuss a p-adic version of Beilinson's conjecture and its relationship with noncommutative geometry.

Introduction

Hodge theory is one of the most important computation tools in modern algebraic geometry, and for many reasons; in these lectures, we will be concerned with only one facet of the story – the properties of the so-called *regulator map*. This actually has a long history which predates both Hodge theory and algebraic geometry and includes, for instance, the well-known Dirichlet Unit Theorem. However, from the modern viewpoint – and we will adopt the modern viewpoint – the regulator map is a gadget which compares *algebraic K-theory* and certain *cohomology groups* of algebraic varieties constructed by means of the Hodge theory. Algebraic K-groups of a variety contain a lot of valuable information, but are notoriously hard to compute; cohomology, on the other hand, is easily computable in most cases. Thus it would be very important to be able to express one through the other. This is what the regulator map does.

Of course, one needs to know that the comparison is exact, so that no information is lost in the process. This is essentially the content of the first of the famous Beilinson conjectures made about twenty years ago ([B], [RSS]).

At the time of writing, there is still no significant progress in *proving* the conjectures. However, we now *understand* them somewhat better. In particular, while Beilinson was working with algebraic varieties defined over $\mathbb{Q}$ and their cohomology with real coefficients, we now have a p-adic version of the story. The goal of these lectures is to give a very brief introduction to a still more recent discovery – it turns out that the p-adic version of the first of Beilinson conjectures can be transfered to the setting of *non-commutative varieties*. We still cannot prove anything; however, since the p-adic conjecture can be now formulated in much larger generality, it becomes more accessible, and a lot of structure used in the original version can be removed as redundant. Hopefully, this will allow someone to concentrate on the essential heart of the problem, and maybe finally solve it.

The paper follows very closely two lectures I gave at a summer school in Goettingen in June 2007. The exposition is very threadbare – we only indicate proofs, with details given elsewhere, and we try to concentrate on the ideas by cutting a lot of technical corners. We follow the most direct path we could find from the definitions, to Beilinson conjectures, to the p-adic analog, to the non-commutative p-adic version. For better or for worse, we choose brevity over completeness at every turn.

Acknowledgements. I would like to thank Yu. Tschinkel for making these lectures possible, and I would like to thank N. Hoffmann for a superb job of taking down the notes and writing them up as a first draft. This research was partially supported by CRDF grant RM1-2694-MO05.

1. Regulator maps and Beilinson conjectures

Let X be a smooth projective algebraic variety over $\mathbb{C}$. Let

$$\mathrm{ch} : K^0(X) \longrightarrow \bigoplus_i H^{2i}(X_{an}, \mathbb{Q})$$

be the Chern character map from the algebraic K-group $K^0(X)$ to the cohomology of the underlying analytic space of X with rational coefficients. For most varieties, the map is of course not even close to being surjective. What can be said about its image? One constraint is well-known: if we denote by

$$H^i(X_{an}, \mathbb{C}) = \bigoplus_{p+q=i} H^{p,q}(X)$$

the Hodge decomposition, then every class $[\mathcal{E}] \in K^0(X)$ satisfies

$$\mathrm{ch}([\mathcal{E}]) \in \bigoplus_i H^{i,i}(X).$$

Definition 1.1. *A pure $\mathbb{Q}$-Hodge structure of weight n consists of*

   (i) *a $\mathbb{Q}$-vector space $V_{\mathbb{Q}}$, and*
   (ii) *a decreasing filtration $F^i V_{\mathbb{C}}$ on its complexification $V_{\mathbb{C}} := V_{\mathbb{Q}} \otimes \mathbb{C}$,*

such that the following two conditions are satisfied:

   (i) $F^i V_{\mathbb{C}} \cap \overline{F^j V_{\mathbb{C}}} = 0$ *whenever $i + j > n$, and*
   (ii) $V_{\mathbb{C}} = \bigoplus_i F^i V_{\mathbb{C}} \cap \overline{F^{n-i} V_{\mathbb{C}}}.$

Here the overline in $\overline{V}$ denotes complex conjugation on $V_{\mathbb{C}}$, i.e. the tensor product $V_{\mathbb{C}} \to V_{\mathbb{C}}$ of the identity on $V_{\mathbb{Q}}$ and the complex conjugation on $\mathbb{C}$.

Example 1.2. The vector space $H^n(X, \mathbb{Q})$ carries a natural pure $\mathbb{Q}$-Hodge structure of weight n, given by the Hodge decomposition:

$$F^i H^n(X, \mathbb{C}) = \bigoplus_{j \geq i} H^{j,n-j}(X).$$

Assume that V is a pure $\mathbb{Q}$-Hodge structure of weight $2i$. Then the space of (i,i)-classes in $V_{\mathbb{Q}}$ is the kernel of the map

$$F^i V_{\mathbb{C}} \oplus V_{\mathbb{Q}} \longrightarrow V_{\mathbb{C}}$$

which is the difference of the two inclusions $F^i V_{\mathbb{C}} \hookrightarrow V_{\mathbb{C}}$ and $V_{\mathbb{Q}} \hookrightarrow V_{\mathbb{C}}$.

Applying this to the vector spaces $V = H^i(X, \mathbb{Q})$, we see that the Chern character maps into the direct sum of these kernels for all $i \geq 0$.

Note that we can turn a pure $\mathbb{Q}$-Hodge structure of weight n into one of weight $n + 2$ by just renumbering the filtration. Thus we can turn the pure $\mathbb{Q}$-Hodge structure $H^{2i}(X, \mathbb{Q})$ into one of any given even weight; we denote by

$$H^{2i}(X, \mathbb{Q}(j))$$

the pure $\mathbb{Q}$-Hodge structure of weight $2i - 2j$ thus obtained. In particular, this produces a pure $\mathbb{Q}$-Hodge structure $H^{2i}(X, \mathbb{Q}(i))$ of weight 0. Altogether, we get a factorisation of the Chern character

$$K^0(X) \longrightarrow \ker\left[F^0 V_{\mathbb{C}} \oplus V_{\mathbb{Q}} \longrightarrow V_{\mathbb{C}}\right] \subseteq V := \bigoplus_i H^{2i}(X, \mathbb{Q}(i)).$$

The famous Hodge conjecture states that this arrow is surjective.

To proceed further, recall that $K^0(X)$ is a part of Quillen's higher K-theory $K^{\bullet}(X)$, which behaves like a cohomology theory (has Mayer-Vietoris sequences, excision etc.) Can we extend the Chern character ch to $K^{\bullet}(X)$? Yes – as shown in [B], based on earlier work by other people, there exists a *regulator map*

$$r : K^{\bullet}(X) \longrightarrow \mathrm{cone}\left(F^0 V_{\mathbb{C}}^{\bullet} \oplus V_{\mathbb{Q}}^{\bullet} \longrightarrow V_{\mathbb{C}}^{\bullet}\right), \qquad V^{\bullet} := \bigoplus_j H^{2j+\bullet}(X, \mathbb{Q}(j)).$$

where, just as one would expect for a cohomology theory, we have replaced the kernel above by a mapping cone. This cone

$$\mathrm{cone}\left(F^0 V_{\mathbb{C}}^{\bullet} \oplus V_{\mathbb{Q}}^{\bullet} \longrightarrow V_{\mathbb{C}}^{\bullet}\right), \qquad V^{\bullet} := \bigoplus_j H^{2j+\bullet}(X, \mathbb{Q}(j))$$

has a name: it is called *Deligne cohomology* and denoted by $H_{\mathcal{D}}^{2j+\bullet}(X, \mathbb{Q}(j))$.

(We note that the Deligne cohomology is usually defined as the hypercohomology $\mathbb{H}^{\bullet}(X, \mathbb{Q}(j))$ of the complex

$$\mathbb{Q}(j) : \quad \mathbb{Q} \longrightarrow \mathcal{O}_X \longrightarrow \Omega_X^1 \longrightarrow \Omega_X^2 \longrightarrow \ldots \longrightarrow \Omega_X^j \longrightarrow 0.$$

In order to compare this definition with the one given above, one notes that the first term $\mathbb{Q}$ in this complex yields $V_{\mathbb{Q}}^{\bullet}$, and that the rest of the complex yields up to quasi-isomorphism the mapping cone of $F^0 V_{\mathbb{C}}^{\bullet} \to V_{\mathbb{C}}^{\bullet}$.)

Replacing the $\mathbb{Q}$-lattices in the Deligne cohomology by $\mathbb{R}$-vector spaces, we obtain a version

$$r : K^{\bullet}(X) \otimes \mathbb{R} \longrightarrow \bigoplus_{j} H_{\mathcal{D}}^{2j+\bullet}(X, \mathbb{R}(j)) \tag{1}$$

of the regulator map above. Roughly speaking, the Beilinson conjecture asserts that this is an isomorphism if X is defined over $\mathbb{Q}$.

More precisely, there is one necessary modification: if a smooth projective variety X is defined over $\mathbb{Q}$, or even over $\mathbb{R} \subset \mathbb{C}$, then it has a real structure — that is, an anti-complex involution $\iota : X \to \overline{X}$. This involution acts on everything in the story above, and in particular, on Deligne cohomology; one twists the action of ι on $H_{\mathcal{D}}^{\bullet}(X, \mathbb{R}(j))$ by $(-1)^j$ and replaces the right-hand side of (1) with the subspace of ι-invariant vectors.

However, even with this modification, the Beilinson conjecture is false for a stupid reason, since the left hand side of (1) is zero in negative degrees, and the right-hand side is not. To kill off the parasitic cohomology classes, one has to replace Deligne cohomology by the so-called Deligne-Beilinson cohomology.

Definition 1.3. *A mixed $\mathbb{R}$-Hodge structure consists of*

(i) *an $\mathbb{R}$-vector space $V_{\mathbb{R}}$,*
(ii) *an increasing filtration $W_{\bullet}V_{\mathbb{R}}$, called the* weight filtration, *and*
(iii) *a decreasing filtration $F^{\bullet}V_{\mathbb{C}}$ on $V_{\mathbb{C}} := V_{\mathbb{R}} \otimes_{\mathbb{R}} \mathbb{C}$, called the* Hodge filtration,

such that the graded piece $\mathrm{gr}_i^W(V)$ with the induced filtration $F^{\bullet}$ is a pure Hodge structure of weight i for all i.

Although the category of filtered vector spaces is not abelian, we have:

Fact 1.4.   (i) *Mixed $\mathbb{R}$-Hodge structures form an abelian category.*
(ii) *This abelian category has homological dimension 1.*

Example 1.5. The mixed $\mathbb{R}$-Hodge structure $\mathbb{R}(j)$ consists of

(i) the vector space $V_{\mathbb{R}} := \mathbb{R}$,
(ii) the weight filtration $W_{-1}V_{\mathbb{R}} = 0$, $W_0 V_{\mathbb{R}} = V_{\mathbb{R}}$, and
(iii) the Hodge filtration $F^{-j}V_{\mathbb{C}} = V_{\mathbb{C}}$, $F^{1-j}V_{\mathbb{C}} = 0$.

For any mixed $\mathbb{R}$-Hodge structure V, one can check that

$$\mathrm{RHom}^{\bullet}(\mathbb{R}(0), V)$$

is (quasi-isomorphic to) the mapping cone of

$$(W_0 V_{\mathbb{C}} \cap F^0 V_{\mathbb{C}}) \oplus W_0 V_{\mathbb{R}} \longrightarrow W_0 V_{\mathbb{C}}.$$

This is the Deligne-Beilinson cohomology – it both gives a conceptual explanation for the Deligne cohomology, and refines it by removing the "parasitic" terms. As shown by Beilinson, the regulator map factors through a map from $K^{\bullet}(X) \otimes \mathbb{R}$ to the Deligne-Beilinson cohomology; what he actually conjectured was that this refined regulator map is an isomorphism onto the subspace of ι-invariant vectors when X is a smooth projective variety over $\mathbb{Q}$.

To finish the section, here are some additional comments.

(i) The Beilinson conjecture comprises both a Hodge-type conjecture which says that the regulator map is surjective, and a generalization of a conjecture by S. Bloch which says that the map is injective. Hodge-type conjecture has a chance of being true even for varieties defined over $\mathbb{R}$, but the injectivity certainly fails unless X is defined over $\mathbb{Q}$.

(ii) There are further conjectures about the determinant of the regulator map in some appropriate basis and its relation to values of L-functions at integral points, but this lies outside the scope of the present paper (and a reader who does not know or does not wish to know what an L-function is may safely read on).

(iii) In the usual definition of the Hodge structure $\mathbb{R}(i)$, one modifies the embedding $V_{\mathbb{R}} \to V_{\mathbb{C}}$ by multiplying it by $(2\pi\sqrt{-1})^i$. This makes no sense in our definition of Hodge structure; the only place where $\sqrt{-1}^i$ actually appears is in the action of the additional complex conjugation ι (this explains the twist by $(-1)^j$ on $H_{\mathcal{D}}^{\bullet}(X, \mathbb{R}(j))$). The multiplier 2π is important for the further Beilinson conjectures on special values; for the purposes of the present paper, it can be ignored.

2. A p-adic version

There is a p-adic version of the above theory, due to Fontaine and Laffaille [FL], Fontaine and Messing [FM], M. Gros [G1,G2].

We work over the ring of Witt vectors $W := W(\overline{\mathbb{F}}_p)$, which is the maximal unramified extension of $\mathbb{Z}_p$. Let $\mathrm{Fr}_W : W \to W$ be the unique lift of the Frobenius automorphism on $\overline{\mathbb{F}}_p$. Given a W-module M, we denote its Frobenius twist by

$$M^{(1)} := M \otimes_{W, \mathrm{Fr}_W} W.$$

Thus a W-linear map $M^{(1)} \to M'$ is the same thing as a Fr_W-semilinear map $M \to M'$.

Definition 2.1 ([FL]). *A filtered Dieudonné module M consists of*

(i) *a finitely generated module M over the ring $W(\overline{\mathbb{F}}_p)$,*
(ii) *a decreasing filtration $F^{\bullet}M$ of M, and*
(iii) *W-linear maps $\varphi_i : F^i M^{(1)} \to M$ for all i,*

such that the following two conditions are satisfied:

(i) *$\varphi_i|_{F^{i+1}M} = p\varphi_{i+1}$ for all i, and*
(ii) *the direct sum $\bigoplus_i \varphi_i : \bigoplus_i F^i M^{(1)} \to M$ is surjective.*

Note that for any i and $j \geq i$, φ_j is determined by φ_i if M is torsion-free. However, it is useful also to include modules M with torsion, and then we need all the φ_i.

Fact 2.2 ([FL]).   (i) *Filtered Dieudonné modules form an abelian category.*
(ii) *This abelian category has homological dimension 1.*

Example 2.3. The filtered Dieudonné module $\mathbb{Z}_p(0)$ consists of

  (i) the free W-module $M := W$,
  (ii) the trivial filtration $F^0 M = M$, $F^1 M = 0$, and
  (iii) the Frobenius map $\varphi_0 := \mathrm{Fr}_W : F^0 M^{(1)} \to M$.

From now on, let X be a smooth projective variety over W; we assume $p > \dim(X)$ (in order to be able to divide by $i!$ when dealing with exterior i-forms on X). Recall that the de Rham cohomology $H^\bullet_{\mathrm{DR}}(X)$ of X is by definition the hypercohomology of its de Rham complex.

Example 2.4. $M := H^\bullet_{\mathrm{DR}}(X)$, together with the filtration $F^\bullet M$ given by the stupid filtration of the de Rham complex, is a filtered Dieudonné module.

About the proof. The main point is to construct the maps $\varphi_i : F^i M^{(1)} \to M$. Let $X^{(1)}$ be the Frobenius twist of X, i.e. the fibered product

$$
\begin{array}{ccc}
X^{(1)} & \longrightarrow & X \\
\downarrow & & \downarrow \\
\mathrm{Spec}\, W(\overline{\mathbb{F}}_p) & \xrightarrow{\ \mathrm{Fr}^*_W\ } & \mathrm{Spec}\, W(\overline{\mathbb{F}}_p).
\end{array}
$$

We make the simplifying assumption that there is a map

$$
\widetilde{\mathrm{Fr}} : X \longrightarrow X^{(1)} \tag{2}
$$

which lifts the (absolute) Frobenius map at the special fiber. Then we get a map

$$
\widetilde{\mathrm{Fr}}^* : \Omega^\bullet_{X^{(1)}} \longrightarrow \Omega^\bullet_X.
$$

The induced map on hypercohomology is our φ_0.

We claim that $\widetilde{\mathrm{Fr}}^* : \Omega^1_{X^{(1)}} \longrightarrow \Omega^1_X$ vanishes modulo p. Indeed, the sheaf Ω^1 is locally generated by exact forms df, and

$$
\widetilde{\mathrm{Fr}}^*(df) = d(\widetilde{\mathrm{Fr}}^*(f)) = d(f^p + pf') = pf^{p-1}df + pdf' \equiv 0 \bmod p
$$

for some function f'. This shows that $\widetilde{\mathrm{Fr}}$ is divisible by p on Ω^1; by multiplicativity, it is then divisible by p^i on Ω^i. Thus we get a map $1/p^i\widetilde{\mathrm{Fr}}$ on the truncated de Rham complex $\Omega^{\geq i}$; the induced map on hypercohomology is our φ_i.

This constructs the filtered Dieudonné module structure on $H^\bullet_{\mathrm{DR}}(X)$ in the case where a lift $\widetilde{\mathrm{Fr}}$ of the absolute Frobenius at the special fiber exists. In general, there are obstructions against such a lift. There is a way of deducing the general case from the special case treated here. However, we can then no longer guarantee that φ_i maps $F^i M^{(1)}$ to $F^i M$; all we can say is that it maps $F^i M^{(1)}$ to M. $\quad\square$

For any filtered Dieudonné module M, it is easy to check that

$$
\mathrm{RHom}(\mathbb{Z}_p(0), M)
$$

is (quasi-isomorphic to) the mapping cone of

$$F^0 M^{(1)} \xrightarrow{\;\mathrm{Id}\,-\varphi_0\;} M.$$

E. g. for $\mathrm{RHom}(\mathbb{Z}_p(0), \mathbb{Z}_p(0))$, we get $W \xrightarrow{\;\mathrm{Id}\,-\,\mathrm{Fr}_W\;} W$, which is just $\mathbb{Z}_p$ in degree 0.

The analogy with Beilinson's definition of the Deligne cohomology leads to:

Definition 2.5 ([FM]). *The* syntomic cohomology *of X is*

$$\mathrm{RHom}(\mathbb{Z}_p(0), H^\bullet_{\mathrm{DR}}(X)).$$

One immediate problem with this definition is that our Dieudonné modules lack the weight filtration – thus what we get is a version of Deligne cohomology, not of Deligne-Beilinson cohomology, and we cannot expect a version of Beilinson conjectures to hold for the same stupid reason as in char 0. At present, it is not known how to cure this. The best we can do is to introduce the following.

Definition 2.6. *Assume that the operations φ^i on the de Rham cohomology groups $H^\bullet_{\mathrm{DR}}(X)$ preserve the Hodge filtration, $\phi^i(F^i) \subset F^i$ (for instance, this is the case when X admits a lifting of the Frobenius, as in (2)). The* reduced syntomic cohomology *of X is the mapping cone of the natural map*

$$F^0 M^{(1)} \xrightarrow{\;\mathrm{Id}\,-\varphi_0\;} F^0 M,$$

where $M = H^\bullet_{\mathrm{DR}}(X)$.

Unfortunately, the assumption needed to define reduced syntomic cohomology is only rarely satisfied; in general, one has to deal with the full syntomic cohomology which contains parasitic classes. Be it as it may, Michel Gros [G1,G2] has constructed a regulator map

$$r : K^\bullet(X) \longrightarrow \text{syntomic cohomology.}$$

He also formulated a precise version of the Beilinson conjectures in this p-adic setting (including those that deal with special values of L-functions).

3. The non-commutative setting

We still work over the ring of Witt vectors $W = W(\overline{\mathbb{F}}_p)$. Let A be a flat W-algebra; all our algebras are associative and unital, but not necessarily commutative. Our goal is to construct

    (i) an analogue of the de Rham cohomology for A,
    (ii) a filtered Dieudonné module structure on it, and
    (iii) a regulator map.

At this point I should add a disclaimer. While my source for the material here is [K1], a large part of it is an independent rediscovery of things discovered by algebraic topogists about 15 years ago — mostly within the theory of the so-called *Topological Cyclic Homology* and *cyclotomic trace* of Bökstedt, Hsiang and Madsen (see [BHM], or a very good exposition in [HM]). However, at the moment I don't completely understand the precise relation to the topological story, and I prefer to completely ignore this in these lectures.

3.1. Non-commutative de Rham cohomology

The main reference for this subsection is Loday's book [L] (which is in particular a reliable source for the many signs involved).

We consider A as a bimodule under left and right multiplication by A, or in other words as a (left) $A \otimes A^{\mathrm{opp}}$-module (where A^{opp} denotes the opposite algebra). This bimodule is in general not flat. E. g. if A is commutative, then this bimodule A corresponds to the structure sheaf of the diagonal as a module over the structure sheaf of $\mathrm{Spec}(A) \times \mathrm{Spec}(A)$.

Definition 3.1. *The* Hochschild homology *of A is*

$$HH_{\bullet}(A) := \mathrm{Tor}^{A \otimes A^{\mathrm{opp}}}_{\bullet}(A, A).$$

The diagonal bimodule A has a standard flat resolution $C_{\bullet}(A) \to A$, namely

$$\ldots \xrightarrow{b'} A \otimes A \otimes A \xrightarrow{b'} A \otimes A \xrightarrow{b'} A, \tag{3}$$

whose differential $b' := \sum_{i=1}^{n}(-1)^i m_i$ involves the multiplication $m : A \otimes A \to A$; m_i is the multiplication at the i-th tensor sign.

Using this flat resolution $C_{\bullet}(A) \to A$ to compute $HH_{\bullet}(A)$, we get the complex

$$\ldots \xrightarrow{b} A \otimes A \otimes A \xrightarrow{b} A \otimes A \xrightarrow{b} A \tag{4}$$

whose differential $b := b' + (-1)^n m_0$ contains the extra summand $m_0 := m_1 \circ \sigma$, where n is the number of A's, and σ is the cyclic permutation

$$\sigma(a_1 \otimes \ldots \otimes a_{n-1} \otimes a_n) := (-1)^{n-1} a_n \otimes a_1 \otimes \ldots \otimes a_{n-1}.$$

The two complexes (3) and (4) can be put together to a periodic bicomplex

$$
\begin{array}{ccccccccc}
\longrightarrow & A & \xrightarrow{\;1\;} & A & \xrightarrow{\;1-\sigma\;} & A & \longrightarrow \\
& \big\uparrow{\scriptstyle b} & & \big\uparrow{\scriptstyle b'} & & \big\uparrow{\scriptstyle b} & \\
\longrightarrow & A\otimes A & \xrightarrow{\;1+\sigma\;} & A\otimes A & \xrightarrow{\;1-\sigma\;} & A\otimes A & \longrightarrow \\
& \big\uparrow{\scriptstyle b} & & \big\uparrow{\scriptstyle b'} & & \big\uparrow{\scriptstyle b} & \\
\longrightarrow & A\otimes A\otimes A & \xrightarrow{\;1+\sigma+\sigma^2\;} & A\otimes A\otimes A & \xrightarrow{\;1-\sigma\;} & A\otimes A\otimes A & \longrightarrow \\
& \big\uparrow{\scriptstyle b} & & \big\uparrow{\scriptstyle b'} & & \big\uparrow{\scriptstyle b} & \\
& \cdots & & \cdots & & \cdots & \\
& \big\uparrow{\scriptstyle b} & & \big\uparrow{\scriptstyle b'} & & \big\uparrow{\scriptstyle b} & \\
\longrightarrow & A^{\otimes n} & \xrightarrow{\;1+\sigma+\cdots+\sigma^{n-1}\;} & A^{\otimes n} & \xrightarrow{\;1-\sigma\;} & A^{\otimes n} & \longrightarrow \\
& \big\uparrow{\scriptstyle b} & & \big\uparrow{\scriptstyle b'} & & \big\uparrow{\scriptstyle b} & \\
\end{array}
$$

$$\tag{5}$$

which we denote by $\mathrm{Per}_{\bullet}(A)$.

Definition 3.2. $HP_{\bullet}(A)$ *is the homology of the total complex of* $\mathrm{Per}_{\bullet}(A)$.

Remark 3.3. Since the bicomplex $\mathrm{Per}_{\bullet}(A)$ is unbounded in one direction, there are two ways of forming its total complex: one that involves infinite direct sums of its entries, and one that involves infinite products. We use products. This is important – for instance, were the base field to have characteristic 0, the total complex understood as a sum would have been acyclic.

Example 3.4. Suppose that A is commutative, that $X := \mathrm{Spec}(A)$ is smooth and that $p > \dim(X)$. Then

$$
HH_i(A) = \Omega^i(X) \qquad \text{and} \qquad HP_i(A) = \bigoplus_j H_{\mathrm{DR}}^{2j+i}(X).
$$

Thus $HP_{\bullet}(A)$ contains less information then $H_{\mathrm{DR}}^{\bullet}(X)$ in the commutative case — we can only recover certain direct sums of $H_{\mathrm{DR}}^n(X)$ from the $HP_i(A)$, not all the $H_{\mathrm{DR}}^n(X)$ themselves. However, as the reader will easily notice, it is exactly these direct sums that are relevant for our story.

3.2. The filtered Dieudonné module structure.

The aim of this subsection is to turn $HP_{\bullet}(A)$ into a filtered Dieudonné module. Here the main problem is to find a non-commutative analogue of the Frobenius.

Even if we reduce everything $\mod p$ and replace A with the $\overline{\mathbb{F}}_p$-algebra A/p, the naive guess does not work: the map $x \mapsto x^p$ is not even additive modulo p. However, we can analyze the difficulty by decomposing it into two maps

$$
A \xrightarrow{\;\varphi\;} A^{\otimes p} \xrightarrow{\;m\;} A, \quad \varphi(a) := a \otimes a \ldots \otimes a, \qquad m(a_1 \otimes \ldots \otimes a_p) := a_1 \cdot \ldots \cdot a_p. \tag{6}
$$

The first map is awful (not additive, etc.), but it is equally awful in the commutative case; it is the multiplication map m which stops being an algebra map in the non-commutative case and creates difficulties.

Fortunately — and this is the main idea — while this map m cannot be made into an algebra map, it can be made to act on Hochschild and cyclic homology.

More precisely, we have two morphisms of complexes

$$\begin{array}{ccccccc}
\xrightarrow{\ b'\ } & A^{\otimes 3} & \xrightarrow{\ b'\ } & A^{\otimes 2} & \xrightarrow{\ b'\ } & A \\[6pt]
& m\uparrow & & m\uparrow & & \uparrow m \\[6pt]
\xrightarrow{\ b'_p\ } & (A^{\otimes 3})^{\otimes p} & \xrightarrow{\ b'_p\ } & (A^{\otimes 2})^{\otimes p} & \xrightarrow{\ b'_p\ } & A^{\otimes p}
\end{array} \qquad (7)$$

and

$$\begin{array}{ccccccc}
\xrightarrow{\ b\ } & A^{\otimes 3} & \xrightarrow{\ b\ } & A^{\otimes 2} & \xrightarrow{\ b\ } & A \\[6pt]
& m\uparrow & & m\uparrow & & \uparrow m \\[6pt]
\xrightarrow{\ b_p\ } & (A^{\otimes 3})^{\otimes p} & \xrightarrow{\ b_p\ } & (A^{\otimes 2})^{\otimes p} & \xrightarrow{\ b_p\ } & A^{\otimes p}
\end{array} \qquad (8)$$

where the differentials $b'_p := \sum_{i=1}(-1)^i m_i^{\otimes p}$ and $b_p := \sum_{i=0}(-1)^i m_i^{\otimes p}$ involve the same multiplications m_i as in (3) and (4) above, but raised to the p-th tensor power, and

$$m : A^{\otimes pn} = \underbrace{A^{\otimes n} \otimes \ldots \otimes A^{\otimes n}}_{p \text{ terms}} \longrightarrow A^{\otimes n}$$

is the identity on the first $n - 1$ tensor factors A, and the multiplication on the remaining $pn - n + 1$ factors. (If for example $n = 2$ and $p = 3$, then

$$m : (A \otimes A) \otimes \underbrace{(A \otimes A) \otimes (A \otimes A)}_{\text{multiply}} \longrightarrow A \otimes A \qquad (9)$$

sends $(a_{11} \otimes a_{12}) \otimes (a_{21} \otimes a_{22}) \otimes (a_{31} \otimes a_{32})$ to $a_{11} \otimes a_{12}a_{21}a_{22}a_{31}a_{32}$.) We can also form a periodic bicomplex $\mathrm{Per}^p_\bullet(A)$, whose vertical differentials are b^p and $(b')^p$, and whose horizontal differentials are the same as in $\mathrm{Per}_\bullet(A)$ (with σ on $A^{\otimes np}$ being the cyclic permutation of order np).

We define the Hodge filtration $F^\bullet$ on the complexes $\mathrm{Per}_\bullet(A)$ and $\mathrm{Per}^p_\bullet(A)$ by the following rule

(i) For the 0-th filtration piece $F^0 HP_\bullet(A)$, we take everything in the bicomplex $\mathrm{Per}_\bullet(A)$ to the right of the column with b.

(ii) For the i-th filtration piece $F^i HP_\bullet(A)$, we shift this to the right by $2i$ columns.

Fact 3.5. *The morphisms of complexes* (7) *and* (8) *are quasi-isomorphisms, and they extend to a quasiisomorphism of filtered complexes* $m : \mathrm{Per}^p_\bullet(A) \to \mathrm{Per}_\bullet(A)$.

About the proof:. The upper rows both in (7) and (8) come from a simplicial abelian group. The first statement is a very general property of simplicial abelian groups, similar to barycentric subdivision. To obtain the second statement, one has to use the notion of *cyclic object* which extends that of a simplicial object, see e. g. [L, Ch. 6]. We say no more and refer the reader to [K2, Lemma 2.2] and [K1, Lemma 1.14] for actual proofs. $\square$

As for the very bad map φ in (6), it turns out that it can be modified quite a bit without changing anything – in particular, it can sometimes be replaced with an actual algebra map. Namely, consider $A^{\otimes p}$ as a representation of $\mathbb{Z}/p\mathbb{Z}$, the generator $\tau \in \mathbb{Z}/p\mathbb{Z}$ acting by cyclic permutation of the tensor factors. The group cohomology of this representation is computed by the periodic complex

$$\longrightarrow A^{\otimes p} \xrightarrow{1+\tau+\cdots+\tau^{p-1}} A^{\otimes p} \xrightarrow{1-\tau} A^{\otimes p} \longrightarrow \,.$$

The above map $\varphi : A \to A^{\otimes p}$, $a \mapsto a \otimes \ldots \otimes a$, induces an isomorphism

$$\varphi : A^{(1)}/p \longrightarrow H_{\mathrm{odd}}(\mathbb{Z}/p\mathbb{Z}, A^{\otimes p}) \tag{10}$$

where $A^{(1)} := A \otimes_{W,\mathrm{Fr}_W} W$ is again the Frobenius twist.

Definition 3.6. *A* quasi-Frobenius map *for A is a $\mathbb{Z}/p\mathbb{Z}$-equivariant algebra homomorphism*

$$\varphi : A^{(1)} \longrightarrow A^{\otimes p}$$

which induces the standard isomorphism (10).

Example 3.7. Let $A = W[G]$ be the group algebra of some (discrete) group G. Then the map $\varphi : A^{(1)} \to A^{\otimes p}$ induced by the diagonal embedding $G \to \underbrace{G \times \ldots \times G}_{p \text{ times}}$ is a quasi-Frobenius map for $A = W[G]$.

If we are given a quasi-Frobenius map φ for A, then we can construct the filtered Dieudonné module structure on $HP_\bullet(A)$ as follows:

(i) The Hodge filtration $F^\bullet$ is as above.

(ii) The required map $\varphi_0 : F^0 HP_\bullet(A^{(1)}) \to HP_\bullet(A)$ is induced by the following morphism of bicomplexes $\varphi_0 : F^0 \operatorname{Per}_\bullet(A) \longrightarrow \operatorname{Per}^p_\bullet(A)$:

  - On F^0/F^1, φ_0 is given by powers of the quasi-Frobenius map φ.
  - On F^i/F^{i+1}, the same times p^i.

(iii) The required maps $\varphi_i : F^i HP_\bullet(A^{(1)}) \to HP_\bullet(A)$ are again obtained by dividing an appropriate restriction of the morphism of bicomplexes φ_0 by p^i.

It is easy to see that this is well-defined. Indeed, the power $\phi^n : A^{\otimes n(1)} \to A^{\otimes pn}$ of the quasi-Frobenius map ϕ commutes with the horizontal differential $1-\sigma$ in the complexes $\operatorname{Per}_\bullet(A)$, $\operatorname{Per}^p_\bullet(A)$ on the nose; to make it send the differential $1 + \sigma + \cdots + \sigma^{n-1}$ to

$$1 + \sigma + \cdots + \sigma^{np-1} = (1 + \sigma + \cdots + \sigma^{n-1})(1 + \tau + \cdots + \tau^{p-1}),$$

we have to multiply it by $(1+\tau+\cdots+\tau^{p-1})$, where τ is the generator of the $\mathbb{Z}/p\mathbb{Z}$-action on $A^{\otimes pn}$. But since $\phi^{\otimes n} : A^{\otimes n} \to A^{\otimes np}$ is $\mathbb{Z}/p\mathbb{Z}$-equivariant with respect to the trivial $\mathbb{Z}/p\mathbb{Z}$-action on the left-hand side, this is equivalent to multiplying by p.

In general, there is no quasi-Frobenius map, but there is a complicated procedure to still obtain such a filtered Dieudonné module structure, cf. [K1], [K3]. It yields the following:

(i) φ_0 exists and is unique up to a quasi-isomorphism if the homological dimension of A is less than $2p$, hom. dim.$(A \otimes A^{\mathrm{opp}}) < 2p$.

(ii) This structure is functorial.

We note that in general, the maps $\varphi^{\bullet}$ do not preserve the filtration $F^{\bullet}$, but if A admits a quasi-Frobenius map, then they do (this is similar to the case of commutative algebraic varieties, where a similar role is palyed by the lifting map (2)). Using the filtered Dieudonné module structure on $HP_{\bullet}(A)$, we can define the *syntomic homology* $HP_{\bullet}^{\mathrm{synt}}(A)$ as the mapping cone of

$$F^0 HP_{\bullet}(A^{(1)}) \xrightarrow{\mathrm{Id}-\varphi_0} HP_{\bullet}(A).$$

If there is a quasi-Frobenius map φ, then we can define *reduced syntomic homology* $\overline{HP}_{\bullet}^{\mathrm{synt}}(A)$, namely as the mapping cone of

$$F^0 HP_{\bullet}(A^{(1)}) \xrightarrow{\mathrm{Id}-\varphi_0} F^0 HP_{\bullet}(A).$$

Of course, the natural embedding $F^0 HP_{\bullet}(A) \to HP_{\bullet}(A)$ induces a natural map $\overline{HP}_{\bullet}^{\mathrm{synt}}(A) \to HP_{\bullet}^{\mathrm{synt}}(A)$.

3.3. The regulator map.

We now turn to the construction of a regulator map. First, let us consider the case $A = W[G]$ for a (discrete) group G. Here we have

$$C_{\bullet}(W[G]) = C_{\bullet}(\widetilde{BG}, W)$$

where the left-hand side is the standard complex which computes $HH_{\bullet}$, and the right-hand side is the chain complex of the simplicial nerve $\widetilde{BG}$ of the groupoid G/G_{ad}, the quotient of the set G modulo the conjugation action of G.

Inside G/G_{ad}, we have the "unity component" $1/G_{\mathrm{ad}} \subset G/G_{\mathrm{ad}}$; its nerve $BG \subset \widetilde{BG}$ is the usual classifying simplicial set of the group G, and we obtain the inclusion $C_{\bullet}(BG) \subset C_{\bullet}(\widetilde{BG})$. One checks easily that $BG \subset \widetilde{BG}$ is preserved by the cyclic permutation σ needed to define the periodic cyclic complex (the scientific formulation is "BG is a cyclic subset in $\widetilde{BG}$", see [L, Ch. 7]). Thus one can define $HP_{\bullet}(BG)$ together with a map $HP_{\bullet}(BG) \to HP_{\bullet}(\widetilde{BG}) = HP_{\bullet}(W[G])$.

Moreover, the quasi-Frobenius map of Example 3.7 preserves $BG \subset \widetilde{BG}$, so that we can define the reduced syntomic homology

$$\overline{HP}_\bullet^{\mathrm{synt}}(BG).$$

Lemma 3.8. *We have* $\overline{HP}_\bullet^{\mathrm{synt}}(BG) \cong H_\bullet(G, \mathbb{Z}_p)$.

Proof. On F^1, φ_0 is divisible by p, so $\mathrm{Id} - \varphi_0$ is invertible. Thus it suffices to consider the cone of

$$F^0/F^1 \xrightarrow{\mathrm{Id} - \varphi_0} F^0/F^1.$$

These two complexes have entries $A^{\otimes n} = W[G^n]$ for various n. Moreover, by definition, the map φ_0 is induced by the identity map on the sets G^n, the components of the simplicial set BG. Indeed, the inclusion $BG \subset \widetilde{BG}$ identifies G^n with the subset of elements

$$\langle g_0, g_1, \ldots, g_n \rangle \in G^{n+1}$$

such that $g_0 \cdot g_1 \cdot \cdots \cdot g_n = 1$, and

$$m(\phi(\langle g_0, g_1, \ldots, g_n \rangle)) = m(\langle g_0, g_1, \ldots, g_n, g_0, g_1, \ldots, g_n, \ldots, g_0, g_1, \ldots, g_n \rangle),$$

with p factors in the right-hand side; plugging this into (9), we obtain

$$m(\phi(\langle g_0, g_1, \ldots, g_n \rangle)) = \langle g_0, g_1, \ldots, g_n (g_0 \cdot g_1 \cdot \cdots \cdot g_n)^{p-1} \rangle = \langle g_0, g_1, \ldots, g_n \rangle.$$

But in the difference $\mathrm{Id} - \varphi_0$, one term is Fr_W-semilinear, whereas the other is W-linear. Thus we obtain many copies of the cone of $\mathrm{Id} - \mathrm{Fr}_W : W \to W$. Replacing each of these copies (quasi-isomorphically) by $\mathbb{Z}_p$, we obtain a complex that computes $H_\bullet(G, \mathbb{Z}_p)$. $\qquad\square$

This Lemma, astonishingly trivial as it may be, is the crucial part in the construction. The rest is a standard and well-known procedure, see e. g. [L]. For any $n \geq 1$, we let $M_n(A)$ be the ring of $(n \times n)$ quadratic matrices over A, and we let $\mathrm{GL}_n(A) \subseteq M_n(A)$ be its group of invertible elements. Then the canonical ring homomorphism

$$W[\mathrm{GL}_n(A)] \longrightarrow M_n(A)$$

and the canonical maps $BG \subset \widetilde{BG}$ etc. yield two maps

$$\overline{HP}_\bullet^{\mathrm{synt}}(B\,\mathrm{GL}_n(A)) \longrightarrow HP_\bullet^{\mathrm{synt}}(W[\mathrm{GL}_n(A)]) \longrightarrow HP_\bullet^{\mathrm{synt}}(M_n(A)).$$

Due to the Morita invariance of $HP^{\mathrm{synt}}_\bullet$, the right-hand side does not depend on n, so that we can pass to the limit with respect to the natural embeddings $GL_n(A) \to GL_{n+1}(A)$ and obtain a map

$$\overline{HP}_\bullet^{\text{synt}}(B\,\mathrm{GL}_\infty(A)) \longrightarrow HP_\bullet^{\text{synt}}(A).$$

This is the desired regulator map. Indeed, according to the Lemma 3.8, its source is

$$\overline{HP}_\bullet^{\text{synt}}(B\,\mathrm{GL}_\infty(A)) \cong H_\bullet(B\,\mathrm{GL}_\infty(A),\mathbb{Z}_p) \cong H_\bullet(B\,\mathrm{GL}_\infty^+(A),\mathbb{Z}_p),$$

Quillen's plus-construction for K-theory, and its target is $HP_\bullet^{\text{synt}}(A)$.

Remark 3.9. To re-iterate: since the weight filtration is missing here, we cannot expect this regulator map to be an isomorphism. In the commutative setting and in char 0, this was healed by the passage from Deligne to Deligne-Beilinson cohomology. We don't know yet how to do this here. A reader who has an idea is kindly requested to contact the author.

References

[B] A. Beilinson, *Higher regulators and values of L-functions*, (in Russian), VINITI Current problems in mathematics, Vol. 24, Moscow, 1984; 181–238.

[BHM] M. Bökstedt, W C Hsiang, and I. Madsen, *The cyclotomic trace and algebraic K-theory of spaces*, Inv. Math. **111** (1993), 465–540.

[FL] J.-M. Fontaine and G. Laffaille, *Construction de représentations p-adiques*, Ann. Sci. cole Norm. Sup. (4) **15** (1982), 547–608

[FM] J.-M. Fontaine and W. Messing, *p-adic periods and p-adic étale cohomology*, in *Current Trends in arithmetical algebraic geometry (Arcata, 1985)*, Contemp. Math. **67**, AMS, Providence, RI, 1987; 179–207.

[G1] M. Gros, *Régulateurs syntomiques et valeurs de fonctions L p-adiques, I*, Invent. Math. **99** (1990), 293–320.

[G2] M. Gros, *Régulateurs syntomiques et valeurs de fonctions L p-adiques, II*, Invent. Math. **115** (1994), 61–79.

[HM] L. Hesselholt and I. Madsen, *On the K-theory of finite algebras over Witt vectors of perfect fields*, Topology **36** (1997), 29–101.

[K1] D. Kaledin, *Non-commutative Hodge-to-de Rham degeneration via the method of Deligne-Illusie*, arXiv:math/0611623.

[K2] D. Kaledin, *Cartier isomorphism and Hodge Theory in the non-commutative case*, arXiv:0708.1574.

[K3] D. Kaledin, *Non-commutative syntomic cohomology and the regulator map*, in preparation.

[L] J.-L. Loday, *Cyclic homology*, Grundlehren der Mathematischen Wissenschaften, **301**, Springer-Verlag, Berlin, 1998.

[RSS] M. Rapoport, N. Schappacher, and P. Schneider, ed., *Beilinson's conjectures on special values of L-functions*, Perspectives in Mathematics, **4**, Academic Press, Inc., Boston, MA, 1988.

Higher-Dimensional Geometry over Finite Fields
D. Kaledin and Y. Tschinkel (Eds.)
IOS Press, 2008

Looking for rational curves on cubic hypersurfaces

János KOLLÁR

Princeton University, Princeton NJ 08544-1000
e-mail: kollar@math.princeton.edu

notes by Ulrich DERENTHAL

Abstract. We study rational points and rational curves on varieties over finite fields. The main new result is the construction of rational curves passing through a given collection of points on smooth cubic hypersurfaces over finite fields.

1. Introduction

The aim of these lectures is to study rational points and rational curves on varieties, mainly over finite fields $\mathbb{F}_q$. We concentrate on hypersurfaces X^n of degree $\leq n + 1$ in $\mathbb{P}^{n+1}$, especially on cubic hypersurfaces.

The theorem of Chevalley–Warning (cf. Esnault's lectures) guarantees rational points on low degree hypersurfaces over finite fields. That is, if $X \subset \mathbb{P}^{n+1}$ is a hypersurface of degree $\leq n + 1$, then $X(\mathbb{F}_q) \neq \emptyset$.

In particular, every cubic hypersurface of dimension ≥ 2 defined over a finite field contains a rational point, but we would like to say more.

- Which cubic hypersurfaces contain more than one rational point?
- Which cubic hypersurfaces contain rational curves?
- Which cubic hypersurfaces contain many rational curves?

Note that there can be rational curves on X even if X has a unique $\mathbb{F}_q$-point. Indeed, $f : \mathbb{P}^1 \to X$ could map all $q + 1$ points of $\mathbb{P}^1(\mathbb{F}_q)$ to the same point in $X(\mathbb{F}_q)$, even if f is not constant.

So what does it mean for a variety to contain many rational curves? As an example, let us look at $\mathbb{CP}^2$. We know that through any 2 points there is a line, through any 5 points there is a conic, and so on. So we might say that a variety X_K contains many rational curves if through any number of points $p_1, \ldots, p_n \in X(K)$ there is a rational curve defined over K.

However, we are in trouble over finite fields. A smooth rational curve over $\mathbb{F}_q$ has only $q + 1$ points, so it can never pass through more than $q + 1$ points in $X(\mathbb{F}_q)$. Thus, for cubic hypersurfaces, the following result, proved in Section 9, appears to be optimal:

Theorem 1.1. *Let $X \subset \mathbb{P}^{n+1}$ be a smooth cubic hypersurface over $\mathbb{F}_q$. Assume that $n \geq 2$ and $q \geq 8$. Then every map of sets $\phi : \mathbb{P}^1(\mathbb{F}_q) \to X(\mathbb{F}_q)$ can be extended to a map of $\mathbb{F}_q$-varieties $\Phi : \mathbb{P}^1 \to X$.*

In fact, one could think of stronger versions as well. A good way to formulate what it means for X to contain many (rational and nonrational) curves is the following:

Conjecture 1.2. [KS03] $X \subset \mathbb{P}^{n+1}$ be a smooth hypersurface of degree $\leq n+1$ defined over a finite field $\mathbb{F}_q$. Let C be a smooth projective curve and $Z \subset C$ a zero-dimensional subscheme. Then any morphism $\phi : Z \to X$ can be extended to C. That is, there is a morphism $\Phi : C \to X$ such that $\Phi|_Z = \phi$.

More generally, this should hold for any separably rationally connected variety X, see [KS03]. We define this notion in Section 4.

The aim of these notes is to explore these and related questions, especially for cubic hypersurfaces. The emphasis will be on presenting a variety of methods, and we end up outlining the proof of two special cases of the Conjecture.

Theorem 1.3. *Conjecture (1.2) holds in the following two cases.*

1. *[KS03] For arbitrary X, when q is sufficiently large (depending on $\dim X, g(C)$ and $\deg Z$), and*
2. *for cubic hypersurfaces when $q \geq 8$ and Z contains only odd degree points.*

As a warm-up, let us prove the case when $X = \mathbb{P}^n$. This is essentially due to Lagrange. The case of quadrics is already quite a bit harder, see (4.7).

Example 1.4 (Polynomial interpolation). Over $\mathbb{F}_q$, let C be a smooth projective curve, $Z \subset C$ a zero-dimensional subscheme and $\phi : Z \to \mathbb{P}^n$ a given map.

Fix a line bundle L on C such that $\deg L \geq |Z| + 2g(C) - 1$ and choose an isomorphism $\mathcal{O}_Z \cong L|_Z$. Then ϕ can be given by $n + 1$ sections $\phi_i \in H^0(Z, L|_Z)$. From the exact sequence

$$0 \to L(-Z) \to L \to L|_Z \to 0$$

we see that $H^0(C, L) \twoheadrightarrow H^0(Z, L|_Z)$. Thus each ϕ_i lifts to $\Phi_i \in H^0(C, L)$ giving the required extension $\Phi : C \to \mathbb{P}^n$.

1.5 (The plan of the lectures). In Section 2, we study hypersurfaces with a unique point. This is mostly for entertainment, though special examples are frequently useful.

Then we prove that a smooth cubic hypersurface containing a K-point is unirational over K. That is, there is a dominant map $g : \mathbb{P}^n \dashrightarrow X$. This of course gives plenty of rational curves on X as images of rational curves on $\mathbb{P}^n$. Note however, that in general, $g : \mathbb{P}^n(K) \dashrightarrow X(K)$ is not onto. (In fact, one expects the image to be very small, see [Man86, Sec.VI.6].) Thus unirationality does not guarantee that there is a rational curve through every K-point.

As a generalization of unirationality, the notion of separably rationally connected varieties is introduced in Section 4. This is the right class to study the existence of many rational curves. Spaces parametrizing all rational curves on a variety are constructed in Section 5 and their deformation theory is studied in Section 6.

The easy case of Conjecture 1.2 is when Z is a single K-point. Here a complete answer to the analogous question is known over $\mathbb{R}$ or $\mathbb{Q}_p$. Over $\mathbb{F}_q$, the Lang-Weil estimates give a positive answer for q large enough; this is reviewed in Section 7.

The first really hard case of (1.2) is when $C = \mathbb{P}^1$ and $Z = \{0, \infty\}$. The geometric question is: given X with $p, p' \in X(\mathbb{F}_q)$, is there a rational curve defined over $\mathbb{F}_q$ passing through p, p'? We see in Section 8 that this is much harder than the 1-point case since it is related to Lefschetz-type results on the fundamental groups of open subvarieties. We use this connection to settle the case for q large enough and p, p' in general position.

Finally, in Section 9 we use the previous result and the "third intersection point map" to prove Theorem 1.1.

Remark 1.6. The first indication that the 2-point case of (1.2) is harder than the 1-point case is the different behavior over $\mathbb{R}$. Consider the cubic surface S defined by the affine equation $y^2 + z^2 = x^3 - x$. Then $S(\mathbb{R})$ has two components (a compact and an infinite part).

- If p, p' lie in different components, there is no rational curve over $\mathbb{R}$ through p, p', since $\mathbb{R}\mathbb{P}^1$ is connected.
- If p, p' lie in the same component, there is no topological obstruction. In fact, in this case an $\mathbb{R}$-rational curve through p, p' always exists, see [Kol99, 1.7].

2. Hypersurfaces with a unique point

The first question has been answered by Swinnerton-Dyer. We state it in a seemingly much sharpened form.

Proposition 2.1. *Let X be a smooth cubic hypersurface of dimension ≥ 2 defined over a field K with a unique K-point. Then $\dim X = 2$, $K = \mathbb{F}_2$ and X is unique up to projective equivalence.*

Proof. We show in the next section that X is unirational. Hence, if K is infinite, then X has infinitely many K-points. So this is really a question about finite fields.

If $\dim X \geq 3$ then $|X(K)| \geq q + 1$ by (2.3). Let us show next that there is no such surface over $\mathbb{F}_q$ for $q \geq 3$.

Assume to the contrary that S contains exactly one rational point $x \in S(\mathbb{F}_q)$. There are q^3 hyperplanes in $\mathbb{P}^3$ over $\mathbb{F}_q$ not passing through x.

The intersection of each hyperplane with S is a curve of degree 3, which is either an irreducible cubic curve, or the union of a line and a conic, or the union of three lines. In the first case, the cubic curve contains a rational point (if C contains a singular point, this points is defined over $\mathbb{F}_q$; if C is smooth, the Weil conjectures show that $|\#C(\mathbb{F}_q) - (q + 1)| \leq 2\sqrt{q}$, so $\#C(\mathbb{F}_q) = 0$ is impossible);

in the second case, the line is rational and therefore contains rational points; in the third case, at least one line is rational unless all three lines are conjugate.

By assumption, for each hyperplane H not passing through x, the intersection $S \cap H$ does not contain rational points. Therefore, $S \cap H$ must be a union of three lines that are not defined over $\mathbb{F}_q$, but conjugate over $\overline{\mathbb{F}}_q$.

This gives $3q^3$ lines on S. However, over an algebraically closed field, a cubic surface contains exactly 27 lines. For $q \geq 3$, we have arrived at a contradiction.

Finally we construct the surface over $\mathbb{F}_2$, without showing uniqueness. That needs a little more case analysis, see [KSC04, 1.39].

We may assume that the rational point is the origin of an affine space on which S is given by the equation

$$f(x, y, z) = z + Q(x, y, z) + C(x, y, z),$$

with Q (resp. C) homogeneous of degree 2 (resp. 3). If C vanishes in (x, y, z), then S has a rational point $(x : y : z)$ on the hyperplane $\mathbb{P}^2$ at infinity. Since C must not vanish in $(1, 0, 0)$, the cubic form C must contain the term x^3, and similarly y^3, z^3. By considering $(1, 1, 0)$, we see that it must also contain x^2y or xy^2, so without loss of generality, we may assume that it contains x^2y, and for similar reasons, we add the terms y^2z, z^2x. To ensure that C does not vanish at $(1, 1, 1)$, we add the term xyz, giving

$$C(x, y, z) = x^3 + y^3 + z^3 + x^2y + y^2z + z^2x + xyz.$$

Outside the hyperplane at infinity, we distinguish two cases: We see by considering a tangent plane $(z = 0)$ that it must intersect S in three conjugate lines. We conclude that $Q(x, y, z) = z(ax + by + cz)$ for certain $a, b, c \in \mathbb{F}_2$. For $z \neq 0$, we must ensure that f does not vanish at the four points $(0, 0, 1)$, $(1, 0, 1)$, $(0, 1, 1)$, $(1, 1, 1)$, resulting in certain restrictions for a, b, c. These are satisfied by $a = b = 0$ and $c = 1$. Therefore, $f(x, y, z) = z + z^2 + C(x, y, z)$ with C as above defines a cubic surface over $\mathbb{F}_2$ that has exactly one $\mathbb{F}_2$-rational point.

There are various ways to check that the cubic is smooth. The direct computations are messy by hand but easy on a computer. Alternatively, one can note that S does contain $3 \cdot 2^3 + 3 = 27$ lines and singular cubics always have fewer than 27. $\qquad\square$

Remark 2.2. A variation of the argument in the proof shows, without using the theorem of Chevalley–Warning, that a cubic surface S defined over $\mathbb{F}_q$ must contain at least one rational point:

If S does not contain a rational point, the intersection of S with any of the $q^3 + q^2 + q + 1$ hyperplanes in $\mathbb{P}^3$ consists of three conjugate lines, giving $3(q^3 + q^2 + q + 1) \geq 45 > 27$ lines, a contradiction.

Exercise 2.3. Using Chevalley–Warning, show that for a hypersurface $X \subset \mathbb{P}^{n+1}$ of degree $n + 1 - r$, the number of $\mathbb{F}_q$-rational points is at least $|\mathbb{P}^r(\mathbb{F}_q)| = q^r + \cdots + q + 1$.

Question 2.4. Find more examples of hypersurfaces $X \subset \mathbb{P}^{n+1}$ of degree at most $n + 1$ with $\#X(\mathbb{F}_q) = 1$.

Example 2.5 (H.-C. Graf v. Bothmer). We construct hypersurfaces $X \subset \mathbb{P}^{n+1}$ over $\mathbb{F}_2$ containing exactly one rational point.

We start by constructing an affine equation. Note that the polynomial $f := x_0 \cdots x_{n+1}$ vanishes in every $\mathbf{x} \in \mathbb{F}_2^{n+2}$ except $(1, \ldots, 1)$, while $g := (x_0 - 1) \cdots (x_{n+1} - 1)$ vanishes in every point except $(0, \ldots, 0)$. Therefore, the polynomial $h := f + g + 1$ vanishes only in $(0, \ldots, 0)$ and $(1, \ldots, 1)$.

The only monomial of degree at least $n+2$ occurring in f and g is $x_0 \cdots x_{n+1}$, while the constant term 1 occurs in g but not in f. Therefore, h is a polynomial of degree $n + 1$ without constant term. We construct the homogeneous polynomial H of degree $n + 1$ from h by replacing each monomial $x_{i_1} x_{i_2} \cdots x_{i_r}$ of degree $r \in \{1, \ldots, n+1\}$ of h with $i_1 < \cdots < i_r$ by $x_{i_1}^{k+1} x_{i_2} \cdots x_{i_r}$ of degree $n+1$ (where $k = n + 1 - r$). Since $a^k = a$ for any $k \geq 1$ and $a \in \mathbb{F}_2$, we have $h(\mathbf{x}) = H(\mathbf{x})$ for any $\mathbf{x} \in \mathbb{F}_2^{n+2}$; the homogeneous polynomial H vanishes exactly in $(0, \ldots, 0)$ and $(1, \ldots, 1)$.

Therefore, H defines a degree $n+1$ hypersurface $\mathbb{P}^{n+1}$ containing exactly one $\mathbb{F}_2$-rational point $(1 : \cdots : 1)$.

Using a computer, we can check for $n = 2, 3, 4$ that H defines a smooth hypersurface of dimension n. Note that for $n = 2$, the resulting cubic surface is isomorphic to the one constructed in Proposition 2.1. For $n \geq 5$, it is unknown whether H defines a smooth variety.

Example 2.6. Let α_1 be a generator or $\mathbb{F}_{q^m}/\mathbb{F}_q$ with conjugates α_i. It is easy to see that

$$X(\alpha) := \left(\prod_i (x_1 + \alpha_i x_2 + \cdots + \alpha_i^{m-1} x_m) = 0 \right) \subset \mathbb{P}^m$$

has a unique $\mathbb{F}_q$-point at $(1 : 0 : \cdots : 0)$. $X(\alpha)$ has degree m, it is irreducible over $\mathbb{F}_q$ but over $\mathbb{F}_{q^m}$ it is the union of m planes.

Assume now that $q \leq m - 1$. Note that $x_i^q x_j - x_i x_j^q$ is identically zero on $\mathbb{P}^m(\mathbb{F}_q)$. Let H be any homogeneous degree m element of the ideal generated by all the $x_i^q x_j - x_i x_j^q$. Then H is also identically zero on $\mathbb{P}^m(\mathbb{F}_q)$, thus

$$X(\alpha, H) := \left(\prod_i (x_1 + \alpha_i x_2 + \cdots + \alpha_i^{m-1} x_m) = H \right) \subset \mathbb{P}^m$$

also has a unique $\mathbb{F}_q$-point at $(1 : 0 : \cdots : 0)$.

By computer it is again possible to find further examples of smooth hypersurfaces with a unique point, but the computations seem exceedingly lengthy for $m \geq 6$.

Remark 2.7. Let $X \subset \mathbb{P}^{n+1}$ be a smooth hypersurface of degree d. Then the primitive middle Betti number is

$$\frac{(d-1)^{n+2} + (-1)^n}{d} + 1 + (-1)^n \leq d^{n+1}.$$

Thus by the Deligne-Weil estimates

$$\left| \#X(\mathbb{F}_q) - \#\mathbb{P}^n(\mathbb{F}_q) \right| \le d^{n+1} q^{n/2}.$$

Thus we get that for $d = n + 1$, there are more then $\#\mathbb{P}^{n-1}(\mathbb{F}_q)$ points in $X(\mathbb{F}_q)$ as soon as $q \ge (n + 1)^{2 + \frac{2}{n}}$.

3. Unirationality

Definition 3.1. A variety X of dimension n defined over a field K is called *unirational* if there is a dominant map $\phi : \mathbb{P}^n \dashrightarrow X$, also defined over K.

Exercise 3.2. [Kol02, 2.3] Assume that there is a dominant map $\phi : \mathbb{P}^N \dashrightarrow X$ for some N. Show that X is unirational.

The following result was proved by Segre [Seg43] in the case $n = 2$, by Manin [Man86] for arbitrary n and general X when K is not a finite field with "too few" elements, and in full generality by Kollár [Kol02].

Theorem 3.3. *Let K be an arbitrary field and $X \subset \mathbb{P}^{n+1}$ a smooth cubic hypersurface ($n \ge 2$). Then the following are equivalent:*

1. *X is unirational over K.*
2. *$X(K) \ne \emptyset$.*

Proof. Let us start with the easy direction: (1) $\Rightarrow$ (2). The proof of the other direction will occupy the rest of the section.

If K is infinite, then K-rational points in $\mathbb{P}^n$ are Zariski-dense, so ϕ is defined on most of them, giving K-rational points of X as their image.

If K is a finite field, ϕ might not be defined on any K-rational point. Here, the result is a special case of the following.

Lemma 3.4 (Nishimura). *Given a smooth variety Y defined over K with $Y(K) \ne \emptyset$ and a rational map $\phi : Y \dashrightarrow Z$ with Z proper, we have $Z(K) \ne \emptyset$.*

Proof (after E. Szabó). We proceed by induction on the dimension of Y. If $\dim Y = 0$, the result is clear. If $\dim Y = d > 0$, we extend $\phi : Y \dashrightarrow Z$ to $\phi' : Y' \dashrightarrow Z$, where Y' is the blow-up of Y in $p \in Y(K)$. Since a rational map is defined outside a closed subset of codimension at least 2, we can restrict ϕ' to the exceptional divisor, which is isomorphic to $\mathbb{P}^{d-1}$. This restriction is a map satisfying the induction hypothesis. Therefore, $X(K) \ne \emptyset$. $\square$

3.5 (Third intersection point map). Let $C \subset \mathbb{P}^2$ be a smooth cubic curve. For $p, p' \in C$ the line $\langle p, p' \rangle$ through them intersects C in a unique third point, denote it by $\phi(p, p')$. The resulting morphism $\phi : C \times C \to C$ is, up to a choice of the origin and a sign, the group law on the elliptic curve C.

For an arbitrary cubic hypersurface X defined over a field K, we can construct the analogous rational map $\phi : X \times X \dashrightarrow X$ as follows. If $p \ne p'$ and if the line $\langle p, p' \rangle$ does not lie completely in X, it intersects X in a unique third point $\phi(p, p')$. If $X_{\bar{K}}$ is irreducible, this defines ϕ on an open subset of $X \times X$.

It is very tempting to believe that out of ϕ one can get an (at least birational) group law on X. This is, unfortunately, not at all the case. The book [Man86] gives a detailed exploration of this direction.

We use ϕ to obtain a dominant map from a projective space to X, relying on two basic ideas:

- Assume that $Y_1, Y_2 \subset X$ are rational subvarieties such that $\dim Y_1 + \dim Y_2 \geq \dim X$. Then, if Y_1, Y_2 are in "general position," the restriction of ϕ gives a dominant map $Y_1 \times Y_2 \dashrightarrow X$. Thus X is unirational since $Y_1 \times Y_2$ is birational to a projective space.
- How can we find rational subvarieties of X? Pick a rational point $p \in X(K)$ and let Y_p be the intersection of X with the tangent hyperplane T_p of X in p. Note that Y_p is a cubic hypersurface of dimension $n-1$ with a singularity at p.

  If p is in "general position," then Y_p is irreducible and not a cone. Thus $\pi : Y_p \dashrightarrow \mathbb{P}^{n-1}$, the projection from p, is birational and so Y_p is rational.

From this we conclude that if $X(K)$ has at least 2 points in "general position," then X is unirational. In order to prove unirationality, one needs to understand the precise meaning of the above "general position" restrictions, and then figure out what to do if there are no points in "general position." This is especially a problem over finite fields.

Example 3.6. [Hir81] Check that over $\mathbb{F}_2, \mathbb{F}_4$ and $\mathbb{F}_{16}$ all points of $(x_0^3 + x_1^3 + x_2^3 + x_3^3 = 0)$ lie on a line. In particular, the curves Y_p are reducible whenever p is over $\mathbb{F}_{16}$. Thus there are no points in "general position."

3.7 (End of the proof of (3.3)). In order to prove (3.3), we describe 3 constructions, working in increasing generality.

(3.7.1) Pick $p \in X(K)$. If Y_p is irreducible and not a cone, then Y_p is birational over K to $\mathbb{P}^{n-1}$. This gives more K-rational points on Y_p. We pick $p' \in Y_p(K) \subset X(K)$, and if we are lucky again, $Y_{p'}$ is also birational over K to $\mathbb{P}^{n-1}$. This results in

$$\Phi_{1,p,p'} : \mathbb{P}^{2n-2} \xrightarrow{\ bir\ } Y_p \times Y_{p'} \dashrightarrow X,$$

where the first map is birational and the second map is dominant.

This construction works when K is infinite and Y_p is irreducible and not a cone.

(3.7.2) Over K, it might be impossible to find $p \in X(K)$ such that Y_p is irreducible. Here we try to give ourselves a little more room by passing to a quadratic field extension and then coming back to K using the third intersection point map ϕ.

Given $p \in X(K)$, a line through p can intersect X in two conjugate points s, s' defined over a quadratic field extension K'/K. If $Y_s, Y_{s'}$ (the intersections of X with the tangent hyperplanes in s resp. s') are birational to $\mathbb{P}^{n-1}$ over K', consider the map

$$\Phi_{1,s,s'} : Y_s \times Y_{s'} \dashrightarrow X.$$

So far $\Phi_{1,s,s'}$ is defined over K'.

Note however that $Y_s, Y_{s'}$ are conjugates of each other by the Galois involution of K'/K. Furthermore, if $z \in Y_s$ and $\bar{z} \in Y_{s'}$ is its conjugate then the line $\phi(z, \bar{z})$ is defined over K. Indeed, the Galois action interchanges $z, \bar{z}$ hence the line $\langle z, \bar{z} \rangle$ is Galois invariant, hence the third intersection point $\Phi_{1,s,s'}(z, \bar{z})$ is defined over K.

That is, the involution $(z_1, z_2) \mapsto (\bar{z}_2, \bar{z}_1)$ makes $Y_s \times Y_{s'}$ into a K-variety and $\Phi_{1,s,s'}$ then becomes a K-morphism. Thus we obtain a dominant map

$$\Phi_{2,p,L} : \mathfrak{R}_{K'/K} \mathbb{P}^{n-1} \dashrightarrow X$$

where $\mathfrak{R}_{K'/K} \mathbb{P}^{n-1}$ is the Weil restriction of $\mathbb{P}^{n-1}$ (cf. Example 3.8).

This construction works when K is infinite, even if Y_p is reducible or a cone. However, over a finite field, it may be impossible to find a suitable line L.

As a last try, if none of the lines work, let's work with all lines together!

(3.7.3) Consider the *universal line* through p instead of choosing a specific line. That is, we are working with all lines at once. To see what this means, choose an affine equation such that p is at the origin:

$$L(x_1, \ldots, x_{n+1}) + Q(x_1, \ldots, x_{n+1}) + C(x_1, \ldots, x_{n+1}) = 0,$$

where L is linear, Q is quadratic and C is cubic. The universal line is given by $(m_1 t, \ldots, m_n t, t)$ where the m_i are algebraically independent over K and the quadratic formula gives the points s, s' at

$$t = \frac{-Q(m_1, \ldots, m_n, 1) \pm \sqrt{D(m_1, \ldots, m_n, 1)}}{2C(m_1, \ldots, m_{n-1}, 1)},$$

where $D = Q^2 - 4LC$ is the discriminant.

Instead of working with just one pair $Y_s, Y_{s'}$, we work with the universal family of them defined over the field

$$K\left(m_1, \ldots, m_n, \sqrt{D(m_1, \ldots, m_n, 1)}\right)$$

It does not matter any longer that Y_s may be reducible for every $m_1, \ldots, m_n \in K$ since we are working with all the Y_s together and the generic Y_s is irreducible and not a cone.

Thus we get a map

$$\Phi_{3,p} : \mathbb{P}^n \times \mathbb{P}^{n-1} \times \mathbb{P}^{n-1} \xdashrightarrow{bir} \mathfrak{R}_{K\left(m_1,\ldots,m_n,\sqrt{D}\right)/K\left(m_1,\ldots,m_n\right)} \mathbb{P}^{n-1} \dashrightarrow X.$$

The last step is the following observation. Unirationality of X_K changes if we extend K. However, once we have a K-map $\mathbb{P}^{3n-2} \dashrightarrow X$, its dominance can be checked after any field extension. Since $\Phi_{3,p}$ incorporates all $\Phi_{2,p,L}$, we see that the K-map $\Phi_{3,p}$ is dominant if $\Phi_{2,p,L}$ is dominant for some $\bar{K}$-line L.

Thus we can check dominance over the algebraic closure of K, where the techniques of the previous cases work.

There are a few remaining points to settle (mainly that Y_p is irreducible and not a cone for general $p \in X(\bar{K})$ and that $\Phi_{1,p,p'}$ is dominant for general $p, p' \in X(\bar{K})$). These are left to the reader. For more details, see [Kol02, Section 2].

Example 3.8. We give an explicit example of the construction of the Weil restriction. The aim of Weil restriction is to start with a finite field extension L/K and an L-variety X and construct in a natural way a K-variety $\mathfrak{R}_{L/K}X$ such that $X(L) = (\mathfrak{R}_{L/K}X)(K)$.

As a good example, assume that the characteristic is $\neq 2$ and let $L = K(\sqrt{a})$ be a quadratic field extension with $G := \mathrm{Gal}(L/K) = \{\mathrm{id}, \sigma\}$. Let X be an L-variety and X^σ its conjugate over K.

Then $X \times X^\sigma$ is an L-variety. We can define a G-action on it by

$$\sigma : (x_1, x_2^\sigma) \mapsto (x_2, x_1^\sigma).$$

This makes $X \times X^\sigma$ into the K-variety $\mathfrak{R}_{L/K}X$.

We explicitly construct $\mathfrak{R}_{L/K}\mathbb{P}^1$, which is all one needs for the surface case of (3.3).

Take the product of two copies of $\mathbb{P}^1$ with the G-action

$$((s_1 : t_1), (s_2^\sigma : t_2^\sigma)) \mapsto ((s_2 : t_2), (s_1^\sigma : t_1^\sigma)).$$

Sections of $\mathcal{O}(1,1)$ invariant under G are

$$u_1 := s_1 s_2^\sigma, \quad u_2 := t_1 t_2^\sigma, \quad u_3 := s_1 t_2^\sigma + s_2^\sigma t_1, \quad u_4 := \frac{1}{\sqrt{a}}(s_1 t_2^\sigma - s_2^\sigma t_1).$$

These sections satisfy $u_3^2 - a u_4^2 = 4 u_1 u_2$, and in fact, this equation defines $\mathfrak{R}_{L/K}\mathbb{P}^1$ as a subvariety of $\mathbb{P}^3$ over K.

Thus $\mathfrak{R}_{L/K}\mathbb{P}^1$ is a quadric surface with K-points (e.g., $(1 : 0 : 0 : 0)$), hence rational over K.

Let us check that $(\mathfrak{R}_{L/K}\mathbb{P}^1)(K) = \mathbb{P}^1(L)$. Explicitly, one direction of this correspondence is as follows. Given $(x_1 + \sqrt{a}x_2 : y_1 + \sqrt{a}y_2) \in \mathbb{P}^1(L)$ with $x_1, x_2, y_1, y_2 \in K$, we get the G-invariant point

$$\big((x_1 + \sqrt{a}x_2 : y_1 + \sqrt{a}y_2), (x_1 - \sqrt{a}x_2 : y_1 - \sqrt{a}y_2)\big) \in ((X \times X^\sigma)(L))^G.$$

From this, we compute

$$u_1 = x_1^2 - a x_2^2, \quad u_2 = y_1^2 - a y_2^2, \quad u_3 = 2(x_1 y_1 - a x_2 y_2), \quad u_4 = 2(x_2 y_1 - x_1 y_2).$$

Then $(u_1 : u_2 : u_3 : u_4) \in \mathfrak{R}_{L/K}\mathbb{P}^1(K)$.

For the precise definitions and for more information, see [BLR90, Section 7.6] or [Kol02, Definition 2.1].

In order to illustrate the level of our ignorance about unirationality, let me mention the following problem.

Question 3.9. Over any field K, find an example of a smooth hypersurface $X \subset \mathbb{P}^{n+1}$ with $\deg X \leq n+1$ and $X(K) \neq \emptyset$ that is not unirational. So far, no such X is known.

The following are 2 further incarnations of the third intersection point map.

Exercise 3.10. Let X^n be an irreducible cubic hypersurface. Show that $S^2 X$ is birational to $X \times \mathbb{P}^n$, where $S^2 X$ denotes the symmetric square of X, that is, $X \times X$ modulo the involution $(x, x') \mapsto (x', x)$.

Exercise 3.11. Let X^n be an irreducible cubic hypersurface defined over K and L/K any quadratic extension. Show that there is a map $\mathfrak{R}_{L/K} X \dashrightarrow X$.

4. Separably rationally connected varieties

Before we start looking for rational curves on varieties over finite fields, we should contemplate which varieties contain plenty of rational curves over an algebraically closed field. There are various possible ways of defining what we mean by lots of rational curves, here are some of them.

4.1. Let X be a smooth projective variety over an algebraically closed field $\overline{K}$. Consider the following conditions:

1. For any given $x, x' \in X$, there is $f : \mathbb{P}^1 \to X$ such that $f(0) = x, f(\infty) = x'$.
2. For any given $x_1, \ldots, x_m \in X$, there is $f : \mathbb{P}^1 \to X$ such that $\{x_1, \ldots, x_m\} \subset f(\mathbb{P}^1)$.
3. Let $Z \subset \mathbb{P}^1$ be a zero-dimensional subscheme and $f_Z : Z \to X$ a morphism. Then f_Z can be extended to $f : \mathbb{P}^1 \to X$.
4. Conditions (3) holds, and furthermore $f^* T_X(-Z)$ is ample. (That is, $f^* T_X$ is a sum of line bundles each of degree at least $|Z| + 1$.)
5. There is $f : \mathbb{P}^1 \to X$ such that $f^* T_X$ is ample.

Theorem 4.2. *[KMM92b], [Kol96, Sec.4.3] Notation as above.*

1. *If $\overline{K}$ is an uncountable field of characteristic 0 then the conditions 4.1.1–4.1.5 are equivalent.*
2. *For any $\overline{K}$, condition 4.1.5 implies the others.*

Definition 4.3. Let X be a smooth projective variety over a field K. We say that X is *separably rationally connected* or *SRC* if the conditions 4.1.1–4.1.5 hold for $X_{\overline{K}}$.

Remark 4.4. There are 2 reasons why the conditions 4.1.1–4.1.5 are not always equivalent.

First, in positive characteristic, there are inseparably unirational varieties. These also satisfy the conditions 4.1.1–4.1.2, but usually not 4.1.5. For instance, if X is an inseparably unirational surface of general type, then 4.1.5 fails. Such examples are given by (resolutions of) a hypersurface of the form $z^p = f(x, y)$ for $\deg f \gg 1$.

Second, over countable fields, it could happen that (4.1.1) holds but X has only countably many rational curves. In particular, the degree of the required f depends on x, x'. These examples are not easy to find, see [BT05] for some over $\overline{\mathbb{F}}_q$. It is not known if this can happen over $\overline{\mathbb{Q}}$ or not.

Over countable fields of characteristic 0, we must require the existence of $f : \mathbb{P}^1 \to X$ of *bounded degree* in these conditions in order to obtain equivalence with 4.1.5.

Example 4.5. Let $S \subset \mathbb{P}^3$ be a cubic surface. Over the algebraic closure, S is the blow-up of $\mathbb{P}^2$ in six points. Considering f mapping $\mathbb{P}^1$ to a line in $\mathbb{P}^2$ not passing through any of the six points, we see that S is separably rationally connected.

More generally, any rational surface is separably rationally connected.

It is not quite trivial to see that for any normal cubic surface S that is not a cone, there is a morphism to the smooth locus $f : \mathbb{P}^1 \to S^{ns}$ such that f^*T_S is ample.

Any normal cubic hypersurface is also separably rationally connected, except cones over cubic curves. To see this, take repeated general hyperplane sections until we get a normal cubic surface $S \subset X$ which is not a cone. The normal bundle of S in X is ample, hence the $f : \mathbb{P}^1 \to S^{ns}$ found earlier also works for X.

In characteristic 0, any smooth hypersurface $X \subset \mathbb{P}^{n+1}$ of degree $\leq n+1$ is SRC; see [Kol96, Sec.V.2] for references and various stronger versions. Probably every normal hypersurface is also SRC, except for cones.

In positive characteristic the situation is more complicated. A general hypersurface of degree $\leq n+1$ is SRC, but it is not known that every smooth hypersurface of degree $\leq n+1$ is SRC. There are some mildly singular hypersurfaces which are not SRC, see [Kol96, Sec.V.5].

4.6 (Effective bounds for hypersurfaces). Let $X \subset \mathbb{P}^{n+1}$ be a smooth SRC hypersurface over $\bar{K}$. Then (4.2) implies that there are rational curves through any point or any 2 points. Here we consider effective bounds for the degrees of such curves.

First, if $\deg X < n+1$ then through every point there are lines. For a general point, the general line is also free (cf. (5.2)).

If $\deg X = n+1$ then there are no lines through a general point, but usually there are conics. However, on a cubic surface there are no irreducible conics through an Eckart point p. (See (7.5) for the definition and details.)

My guess is that in all cases there are free twisted cubics through any point, but this may be difficult to check. I don't know any reasonable effective upper bound.

For 2 general points $x, x' \in X$, there is an irreducible rational curve of degree $\leq n(n+1)/(n+2-\deg X)$ by [KMM92a]. The optimal result should be closer to $n+1$, but this is not known. Very little is known about non-general points.

Next we show that (1.2) depends only on the birational class of X. The proof also shows that (4.1.3) is also a birational property.

Proposition 4.7. *Let K be a field and X, X' smooth projective K-varieties which are birational to each other. Then, if (1.2) holds for X, it also holds for X'.*

Proof. Assume for notational simplicity that K is perfect. Fix embeddings $X \subset \mathbb{P}^N$ and $X' \subset \mathbb{P}^M$ and represent the birational maps $\phi : X \dashrightarrow X'$ and $\phi^{-1} : X' \dashrightarrow X$ with polynomial coordinate functions.

Given $f'_Z : Z \to X'$, we construct a thickening $Z \subset Z_t \subset C$ and $f_{Z_t} : Z_t \to X$ such that if $f : C \to X$ extends f_{Z_t} then $f' := \phi \circ f$ extends f'_Z.

Pick a point $p \in Z$ and let $\hat{\mathcal{O}}_p \cong L[[t]]$ be its complete local ring where $L = K(p)$. Then the corresponding component of Z is $\mathrm{Spec}_K L[[t]]/(t^m)$ for some $m \geq 1$.

By choosing suitable local coordinates at $f'_Z(p) \in X'$, we can define its completion by equations

$$y_{n+i} = G_i(y_1, \ldots, y_n) \quad \text{where } G_i \in L[[y_1, \ldots, y_n]].$$

Thus f'_Z is given by its coordinate functions

$$\bar{y}_1(t), \ldots, \bar{y}_n(t), \cdots \in L[[t]]/(t^m).$$

The polynomials $\bar{y}_i$ for $i = 1, \ldots, n$ can be lifted to $y_i(t) \in L[[t]]$ arbitrarily. These then determine liftings $y_{n+i}(t) = G_i(y_1(t), \ldots, y_n(t))$ giving a map $F' : \mathrm{Spec}_K L[[t]] \to X'$. In particular, we can choose a lifting such that ϕ^{-1} is a local isomorphism at the image of the generic point of $\mathrm{Spec}_K L[[t]]$. Thus $\phi^{-1} \circ F'$ and $\phi \circ \phi^{-1} \circ F'$ are both defined and $\phi \circ \phi^{-1} \circ F' = F'$.

Using the polynomial representations for ϕ, ϕ^{-1}, write

$$\phi^{-1} \circ (1, y_1(t), \ldots, y_n(t), \ldots) = (x_0(t), \ldots, x_n(t), \ldots), \quad \text{and}$$
$$\phi \circ (x_0(t), \ldots, x_n(t), \ldots) = (z_0(t), \ldots, z_n(t), \ldots).$$

Note that $(z_0(t), \ldots, z_n(t), \ldots)$ and $(1, y_1(t), \ldots, y_n(t), \ldots)$ give the same map $F' : \mathrm{Spec}_K L[[t]] \to X'$, but we map to projective space. Thus all we can say is that

$$y_i(t) = z_i(t)/z_0(t) \quad \forall\, i \geq 1.$$

Assume now that we have $(x_0^*(t), \ldots, x_n^*(t), \ldots)$ and the corresponding

$$\phi \circ (x_0^*(t), \ldots, x_n^*(t), \ldots) = (z_0^*(t), \ldots, z_n^*(t), \ldots).$$

such that

$$x_i(t) \equiv x_i^*(t) \mod t^s \quad \forall\, i.$$

Then also

$$z_i(t) \equiv z_i^*(t) \mod t^s \quad \forall\, i.$$

In particular, if $s > r := \mathrm{mult}_0 z_0(t)$, then $\mathrm{mult}_0 z_0^*(t) = \mathrm{mult}_0 z_0(t)$ and so

$$y_i^*(t) := \frac{z_i^*(t)}{z_0^*(t)} \equiv y_i(t) \mod \left(t^{s-r}\right).$$

That is, if $F^* : \mathrm{Spec}_K L[[t]] \to X$ agrees with $\phi^{-1} \circ F'$ up to order $s = r + m$ then $\phi \circ F^*$ agrees with F' up to order $m = s - r$.

We apply this to every point in Z to obtain the thickening $Z \subset Z_t \subset C$ and $f_{Z_t} : Z_t \to X$ as required. $\qquad\qquad\square$

5. Spaces of rational curves

Assume that X is defined over a non-closed field K and is separably rationally connected. Then X contains lots of rational curves over $\overline{K}$, but what about rational curves over K? We are particularly interested in the cases when K is one of $\mathbb{F}_q$, $\mathbb{Q}_p$ or $\mathbb{R}$.

5.1 (Spaces of rational curves). Let X be any variety. Subvarieties or subschemes of X come in families, parametrized by the Chow variety or the Hilbert scheme. For rational curves in X, the easiest to describe is the space of maps $\mathrm{Hom}(\mathbb{P}^1, X)$.

Pick an embedding $X \subset \mathbb{P}^N$ and let F_i be homogeneous equations of X.

Any map $f : \mathbb{P}^1 \to \mathbb{P}^N$ of fixed degree d is given by $N + 1$ homogeneous polynomials $(f_0(s,t), \ldots, f_N(s,t))$ of degree d in two variables s, t (up to scaling of these polynomials). Using the coefficients of $f_0, \ldots, f_N$, we can regard f as a point in $\mathbb{P}^{(N+1)(d+1)-1}$.

We have $f(\mathbb{P}^1) \subset X$ if and only if the polynomials $F_i(f_0(s,t), \ldots, f_N(s,t))$ are identically zero. Each F_i gives $d \cdot \deg F_i + 1$ equations of degree $\deg F_i$ in the coefficients of $f_0, \ldots, f_N$.

If $f_0, \ldots, f_N$ have a common zero, then we get only a lower degree map. We do not count these in $\mathrm{Hom}_d(\mathbb{P}^1, X)$. By contrast we allow the possibility that $f \in \mathrm{Hom}_d(\mathbb{P}^1, X)$ is not an embedding but a degree e map onto a degree d/e rational curve in X. These maps clearly cause some trouble but, as it turns out, it would be technically very inconvenient to exclude them from the beginning.

Thus $\mathrm{Hom}_d(\mathbb{P}^1, X)$ is an open subset of a subvariety of $\mathbb{P}^{(N+1)(d+1)-1}$ defined by equations of degree $\leq \max_i \{\deg F_i\}$.

$\mathrm{Hom}(\mathbb{P}^1, X)$ is the disjoint union of the $\mathrm{Hom}_d(\mathbb{P}^1, X)$ for $d = 1, 2, \ldots$.

Therefore, finding a rational curve $f : \mathbb{P}^1 \to X$ defined over K is equivalent to finding K-points on $\mathrm{Hom}(\mathbb{P}^1, X)$.

In a similar manner one can treat the space $\mathrm{Hom}(\mathbb{P}^1, X, 0 \mapsto x)$ of those maps $f : \mathbb{P}^1 \to X$ that satisfy $f(0) = x$ or $\mathrm{Hom}(\mathbb{P}^1, X, 0 \mapsto x, \infty \mapsto x')$, those maps $f : \mathbb{P}^1 \to X$ that satisfy $f(0) = x$ and $f(\infty) = x'$.

5.2 (Free and very free maps). In general, the local structure of the spaces $\mathrm{Hom}(\mathbb{P}^1, X)$ can be very complicated, but everything works nicely in certain important cases.

We say that $f : \mathbb{P}^1 \to X$ is *free* if $f^* T_X$ is semi-positive, that is a direct sum of line bundles of degree ≥ 0. We see in (6.4) that if f is free then $\mathrm{Hom}(\mathbb{P}^1, X)$ and $\mathrm{Hom}(\mathbb{P}^1, X, 0 \mapsto f(0))$ are both smooth at $[f]$.

We say that $f : \mathbb{P}^1 \to X$ is *very free* if $f^* T_X$ is positive or ample, that is, a direct sum of line bundles of degree ≥ 1. This implies that $\mathrm{Hom}(\mathbb{P}^1, X, 0 \mapsto f(0), \infty \mapsto f(\infty))$ is also smooth at $[f]$.

Remark 5.3. Over a nonclosed field K there can be smooth projective curves C such that $C_{\overline{K}} \cong \mathbb{P}^1_{\overline{K}}$ but $C(K) = \emptyset$, thus C is not birational to $\mathbb{P}^1_K$. When we work with $\operatorname{Hom}(\mathbb{P}^1, X)$, we definitely miss these curves. There are various ways to remedy this problem, but for us this is not important.

Over a finite field K, every rational curve is in fact birational to $\mathbb{P}^1_K$, thus we do not miss anything.

To get a feeling for these spaces, let us see what we can say about the irreducible components of $\operatorname{Hom}_d(\mathbb{P}^1, X)$ for cubic surfaces.

Example 5.4. Let $S \subset \mathbb{P}^3$ be a cubic surface defined over a non-closed field K. Consider $\operatorname{Hom}_d(\mathbb{P}^1, S)$ for low values of d.

- For $d = 1$, over $\overline{K}$, there are 27 lines on S, so $\operatorname{Hom}_1(\mathbb{P}^1, S)$ has 27 components which may be permuted by the action of the Galois group $G = \operatorname{Gal}(\overline{K}/K)$.
- For $d = 2$, over $\overline{K}$, there are 27 one-dimensional families of conics, each obtained by intersecting S with the pencil of planes containing a line on S. These 27 families again may have a non-trivial action of G.
- For $d = 3$, over $\overline{K}$, there are 72 two-dimensional families of twisted cubics on S (corresponding to the 72 ways to map S to $\mathbb{P}^2$ by contracting 6 skew lines; the twisted cubics are preimages of lines in $\mathbb{P}^2$ not going through any of the six blown-up points). Again there is no reason to assume that any of these 72 families is fixed by G.

  However, there is exactly one two-dimensional family of plane rational cubic curves on S, obtained by intersecting S with planes tangent to the points on S outside the 27 lines. This family is defined over K and is geometrically irreducible.

All this is not very surprising. A curve C on S determines a line bundle $\mathcal{O}_S(C) \in \operatorname{Pic}(S) \cong \mathbb{Z}^7$, hence we see many different families in a given degree because there are many different line bundles of a given degree. It turns out that, for cubic surfaces, once we fix not just the degree but also the line bundle $L = \mathcal{O}_S(C)$, the resulting spaces $\operatorname{Hom}_L(\mathbb{P}^1, X)$ are irreducible.

This, however, is a very special property of cubic surfaces and even for smooth hypersurfaces X it is very difficult to understand the irreducible components of $\operatorname{Hom}(\mathbb{P}^1, X)$. See [HRS04,HS05,HRS05,dJS04] for several examples.

Thus, in principle, we reduced the question of finding rational curves defined over K to finding K-points of the scheme $\operatorname{Hom}(\mathbb{P}^1, X)$. The problem is that $\operatorname{Hom}(\mathbb{P}^1, X)$ is usually much more complicated than X.

5.5 (Plan to find rational curves). We try to find rational curves defined over a field K in 2 steps.

1. For any field K, we will be able to write down reducible curves C and morphisms $f : C \to X$ defined over K and show that $f : C \to X$ can be naturally viewed as a smooth point $[f]$ in a suitable compactification of $\operatorname{Hom}(\mathbb{P}^1, X)$ or $\operatorname{Hom}(\mathbb{P}^1, X, 0 \mapsto x)$.

2. Then we argue that for certain fields K, a smooth K-point in a compactification of a variety U leads to a K-point inside U.

   There are 2 main cases where this works.

   (a) (Fields with an analytic inverse function theorem)
       These include $\mathbb{R}, \mathbb{Q}_p$ or the quotient field of any local, complete
       Dedekind domain, see [GR71]. For such fields, any smooth point in
       $\bar{U}(K)$ has an analytic neighborhood biholomorphic to $0 \in K^n$. This
       neighborhood has nontrivial intersection with any nonempty Zariski
       open set, hence with U.

   (b) (Sufficiently large finite fields)
       This method relies on the Lang-Weil estimates. Roughly speaking these
       say that a variety U over $\mathbb{F}_q$ has points if $q \gg 1$, where the bound on q
       depends on U. We want to apply this to $U = \mathrm{Hom}_d(\mathbb{P}^1, X)$. We know
       bounds on its embedding dimension and on the degrees of the defining
       equations, but very little else. Thus we need a form of the Lang-Weil
       estimates where the bound for q depends only on these invariants.

We put more detail on these steps in the next sections, but first let us see an
example.

Example 5.6. Let us see what we get in a first computation trying to find a degree
3 rational curve through a point p on a cubic surface S over $\mathbb{F}_q$.

The intersection of S with the tangent plane at p usually gives a rational
curve C_p which is singular at p. If we normalize to get $n : \mathbb{P}^1 \to C$, about half the
time, $n^{-1}(p)$ is a conjugate pair of points in $\mathbb{F}_{q^2}$. This is not what we want.

So we have to look for planes $H \subset \mathbb{P}^3$ that pass through p and are tangent
to S at some other point. How to count these?

Projecting S from p maps to $\mathbb{P}^2$ and the branch curve $B \subset \mathbb{P}^2$ has degree
4. Moreover, B is smooth if p is not on any line. The planes we are looking for
correspond to the tangent lines of B.

By the Weil estimates, a degree 4 smooth plane curve has at least $q + 1 - 6\sqrt{q}$
points. For $q > 33$ this guarantees a point in $B(\mathbb{F}_q)$ and so we get a plane H
through p which is tangent to S at some point.

However, this is not always enough. First, we do not want the tangency to be
at p. Second, for any line $L \subset S$, the plane spanned by p and L intersects S in L
and a residual conic. These correspond to double tangents of B. The 28 double
tangents correspond to 56 points on B. Thus we can guarantee an irreducible
degree 3 rational curve only if we find an $\mathbb{F}_q$-point on B which different from these
56 points. This needs $q > 121$ for the Weil estimates to work. This is getting quite
large!

Of course, a line is a problem only if it is defined over $\mathbb{F}_q$ and then the
corresponding residual conic is a rational curve over $\mathbb{F}_q$ passing through p, unless
the residual conic is a pair of lines. In fact, if we look for rational curves of degree
≤ 3, then $q > 33$ works.

I do not know what the best bound for q is. In any case, we see that even
this simple case leads either to large bounds or to case analysis.

The current methods work reasonably well when $q \gg 1$, but, even for cubic
hypersurfaces, the bounds are usually so huge that I do not even write them down.

Then we see by another method that for cubics we can handle small values of q. The price we pay is that the degrees of the rational curves found end up very large.

It would be nice to figure out a reasonably sharp answer at least for cubics, Just to start the problem, let me say that I do not know the answer to the following.

Question 5.7. Let X be a smooth cubic hypersurface over $\mathbb{F}_q$ and $p, p' \in X(\mathbb{F}_q)$ two points. Is there a degree ≤ 9 rational curve defined over $\mathbb{F}_q$ passing through p and p'?

Exercise 5.8. Let $S \subset \mathbb{P}^3$ be the smooth cubic surface constructed in (2.1). Show that S does not contain any rational curve of degree ≤ 8 defined over $\mathbb{F}_2$.

Hints. First prove that the Picard group of S is generated by the hyperplane sections. Thus any curve on S has degree divisible by 3.

A degree 3 rational curve would be a plane cubic, these all have at least 2 points over $\mathbb{F}_2$.

Next show that any rational curve defined over $\mathbb{F}_2$ must have multiplicity 3 or more at the unique $p \in S(\mathbb{F}_2)$. A degree 6 rational curve would be a complete intersection of S with a quadric Q. Show that S and Q have a common tangent plane at p and then prove that $S \cap Q$ has only a double point if Q is irreducible.

6. Deformation of combs

Example 6.1. Let S be a smooth cubic surface over $\mathbb{R}$ and p a real point of S. Our aim is to find a rational curve defined over $\mathbb{R}$ passing through p. It is easy to find such a rational curve C defined over $\mathbb{C}$. Its conjugate $\bar{C}$ then also passes through p. Together, they define a curve $C + \bar{C} \subset S$ which is defined over $\mathbb{R}$. So far this is not very useful.

We can view $C + \bar{C}$ as the image of a map $\phi_0 : Q_0 \to S$ where $Q_0 \subset \mathbb{P}^2$ is defined by $x^2 + y^2 = 0$. Next we would like to construct a perturbation $\phi_\varepsilon : Q_\varepsilon \to S$ of this curve and of this map. It is easy to perturb Q_0 to get "honest" rational curves over $\mathbb{R}$, for instance $Q_\varepsilon := (x^2 + y^2 = \varepsilon z^2)$.

The key question is, can we extend ϕ_0 to ϕ_ε? Such questions are handled by deformation theory, originated by Kodaira and Spencer. A complete treatment of the case we need is in [Kol96] and [AK03] is a good introduction.

The final answer is that if $H^1(Q_0, \phi_0^* T_S) = 0$, then ϕ_ε exists for $|\varepsilon| \ll 1$. This allows us to obtain a real rational curve on S, and with a little care we can arrange for it to pass through p.

In general, the above method gives the following result:

Corollary 6.2. *[Kol99] Given $X_{\mathbb{R}}$ such that $X_{\mathbb{C}}$ is rationally connected, there is a real rational curve through any $p \in X(\mathbb{R})$.*

We would like to apply a similar strategy to X_K such that $X_{\overline{K}}$ is separably rationally connected. For a given $x \in X(K)$, we find a curve $g_1 : \mathbb{P}^1 \to X$ defined over $\overline{K}$ such that $g_1(0) = x$, with conjugates $g_2, \ldots, g_m$. Then $C :=$

$g_1(\mathbb{P}^1) + \cdots + g_m(\mathbb{P}^1)$ is defined over K. Because of the singularity of C in x, it is harder to find a smooth deformation of C. It turns out that there is a very simple way to overcome this problem: we need to add a whole new $\mathbb{P}^1$ at the point x and look at maps of curves to X which may not be finite.

Definition 6.3. Let X be a variety over a field K. An *m-pointed stable curve of genus 0 over* X is an object $(C, p_1, \ldots, p_m, f)$ where

1. C is a proper connected curve with $p_a(C) = 0$ defined over K having only nodes,
2. $p_1, \ldots, p_m$ are distinct smooth points in $C(K)$,
3. $f : C \to X$ is a K-morphism, and
4. C has only finitely many automorphisms that commute with f and fix $p_1, \ldots, p_m$. Equivalently, there is no irreducible component $C_i \subset C_{\bar{K}}$ such that f maps C_i to a point and C_i contains at most 2 special points (that is, nodes of C or $p_1, \ldots, p_m$).

Note that if $f : C \to X$ is finite, then $(C, p_1, \ldots, p_m, f)$ is a stable curve of genus 0 over X, even if $(C, p_1, \ldots, p_m)$ is not a stable m-pointed genus 0 curve in the usual sense [FP97].

We have shown how to parametrize all maps $\mathbb{P}^1 \to X$ by the points of a scheme $\mathrm{Hom}(\mathbb{P}^1, X)$. Similarly, the methods of [KM94] and [Ale96] show that one can parametrize all m-pointed genus 0 stable curves of degree d with a single scheme $\overline{M}_{0,m}(X, d)$. For a map $(C, p_1, \ldots, p_m, f)$, the corresponding point in $\overline{M}_{0,m}(X, d)$ is denoted by $[C, p_1, \ldots, p_m, f]$.

Given K-points $x_1, \ldots, x_m \in X(K)$, the family of those maps $f : C \to X$ that satisfy $f(p_i) = x_i$ for $i = 1, \ldots, m$ forms a closed scheme

$$\overline{M}_{0,m}(X, p_i \mapsto x_i) \subset \overline{M}_{0,m}(X).$$

See [AK03, sec.8] for more detailed proofs.

The deformation theory that we need can be conveniently compacted into one statement. The result basically says that the deformations used in (6.1) exist for any reducible rational curve.

Theorem 6.4. *(cf. [Kol96, Sec.II.7] or [AK03]) Let $f : (p_1, \ldots, p_m \in C) \to X$ be an m-pointed genus 0 stable curve. Assume that X is smooth and*

$$H^1(C, f^*T_X(-p_1 - \cdots - p_m)) = 0.$$

Then:

1. *There is a unique irreducible component*

$$\mathrm{Comp}(C, p_1, \ldots, p_m, f) \subset \overline{M}_{0,m}(X, p_i \mapsto f(p_i))$$

   which contains $[C, p_1, \ldots, p_m, f]$.
2. *$[C, p_1, \ldots, p_m, f]$ is a smooth point of $\mathrm{Comp}(C, p_1, \ldots, p_m, f)$. In particular, if $f : (p_1, \ldots, p_m \in C) \to X$ is defined over K then $\mathrm{Comp}(C, p_1, \ldots, p_m, f)$ is geometrically irreducible.*

3. *There is a dense open subset*

$$\text{Smoothing}(C, p_1, \ldots, p_m, f) \subset \text{Comp}(C, p_1, \ldots, p_m, f)$$

which parametrizes free maps of smooth rational curves, that is

$$\text{Smoothing}(C, p_1, \ldots, p_m, f) \subset \text{Hom}^{\text{free}}(\mathbb{P}^1, X, p_i \mapsto f(p_i))$$

(We cheat a little in (6.4.2). In general $[C, p_1, \ldots, p_m, f]$ is smooth only in the stack sense; this is all one needs. Moreover, in all our applications $[C, p, f]$ will be a smooth point.)

The required vanishing is usually easy to check using the following.

Exercise 6.5. Let $C = C_1 + \cdots + C_m$ be a reduced, proper curve with arithmetic genus 0 and $p \in C$ a smooth point. Let $C_1, \cdots, C_m$ be its irreducible components over $\bar{K}$. Let E be a vector bundle on C and assume that $H^1(C_i, E|_{C_i}(-1)) = 0$ for every i. Then $H^1(C, E(-p)) = 0$.

In particular, if $f : C \to X$ is a morphism to a smooth variety and if each $f|_{C_i}$ is free then $H^1(C, f^*T_X(-p)) = 0$.

Definition 6.6 (Combs). A *comb* assembled from a curve B (the handle) and m curves C_i (the teeth) attached at the distinct points $b_1, \ldots, b_m \in B$ and $c_i \in C_i$ is a curve obtained from the disjoint union of B and of the C_i by identifying the points $b_i \in B$ and $c_i \in C_i$. In these notes we only deal with the case when B and the C_i are smooth, rational.

A comb can be pictured as below:

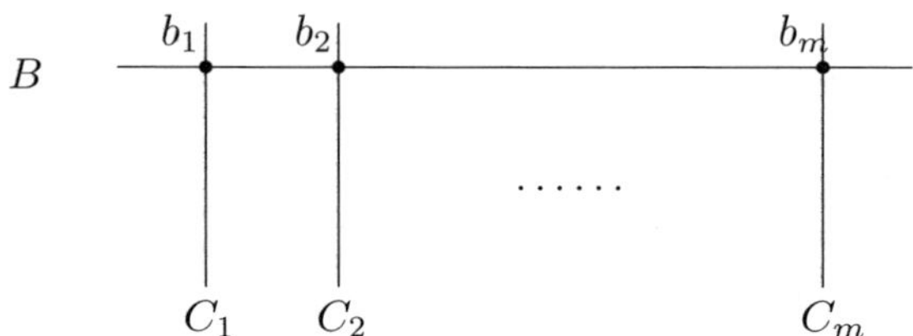

Comb with m-teeth

Assume now that we have a Galois extension L/K and $g_i : (0 \in \mathbb{P}^1) \to (x \in X)$, a conjugation invariant set of maps defined over L.

We can view this collection as just one map as follows. The maps $[g_i] \in \text{Hom}(\mathbb{P}^1, X)$ form a 0-dimensional reduced K-scheme Z. Then the g_i glue together to a single map

$$G : Z \times (0 \in \mathbb{P}^1) - (Z \subset \mathbb{P}^1_Z) \to (x \subset X).$$

Let $j : Z \hookrightarrow \mathbb{P}^1_K$ be an embedding. We can then assemble a comb with handle $\mathbb{P}^1_K$ and teeth $\mathbb{P}^1_Z$. Let us denote it by

$$\mathrm{Comb}(g_1, \ldots, g_m).$$

(The role of j is suppressed, it will not be important for us.)

If K is infinite, an embedding $j : Z \hookrightarrow \mathbb{P}^1_K$ always exists. If K is finite, then Z may have too many points, but an embedding exists whenever Z is irreducible over K.

Indeed, in this case $Z = \mathrm{Spec}_K K(a)$ for some $a \in \bar{K}$. Thus $K[t] \to K(a)$ gives an embedding $Z \hookrightarrow \mathbb{A}^1_K$.

Everything is now ready to obtain rational curves through 1 point.

Corollary 6.7. *[Kol99] Given a separably rationally connected variety X defined over a local field $K = \mathbb{Q}_p$ or $K = \mathbb{F}_q((t))$, there is a rational curve defined over K through any $x \in X(K)$.*

Proof. Given $x \in X(K)$, pick a free curve $g_1 : (0 \in \mathbb{P}^1) \to (x \in X)$ over $\bar{K}$ with conjugates $g_2, \ldots, g_m$. As in (6.6), assemble a K-comb

$$f : (0 \in \mathrm{Comb}(g_1, \ldots, g_m)) \to (x \in X).$$

Using (6.4), we obtain

$$\mathrm{Smoothing}(C, 0, f) \subset \mathrm{Hom}^{\mathrm{free}}(\mathbb{P}^1, X, 0 \mapsto f(p))$$

and by (6.4.2) we see that (5.5.2.a) applies. Hence we get K-points in $\mathrm{Smoothing}(p \in C, f)$, as required. $\qquad\square$

The finite field case, corresponding to (5.5.2.b), is treated in the next section.

7. The Lang-Weil estimates

Theorem 7.1. *[LW54] Over $\mathbb{F}_q$, let $U \subset \mathbb{P}^N$ be the difference of two subvarieties defined by several equations of degree at most D. If $U_0 \subset U$ is a geometrically irreducible component, then*

$$\left| \#U_0(\mathbb{F}_q) - q^{\dim U_0} \right| \leq C(N, D) \cdot q^{\dim U_0 - \frac{1}{2}},$$

where the constant $C(N, D)$ depends only on N and D.

Notes on the proof. The original form of the estimate in [LW54] assumes that U_0 is projective and it uses $\deg U_0$ instead of D. These are, however, minor changes.

First, if $V \subset \mathbb{P}^N$ is an irreducible component of W which is defined by equations of degree at most D, then it is also an irreducible component of some $W' \supset W$ which is defined by $N - \dim V$ equations of degree at most D. Thus, by Bézout's theorem, $\deg V \leq D^{N - \dim V}$.

Thus we have a bound required for $\#\bar{U}_0(\mathbb{F}_q)$ and we need an upper bound for the complement $\#(\bar{U}_0 \setminus U_0)(\mathbb{F}_q)$. We assumed that $\bar{U}_0 \setminus U_0$ is also defined by equations of degree at most D. A slight problem is, however, that it may have components which are geometrically reducible. Fortunately, an upper bound for $\#V(\mathbb{F}_q)$ is easy to get. $\square$

Exercise 7.2. Let $V \subset \mathbb{P}^N$ be a closed, reduced subscheme of pure dimension r and degree d. Show that if $q \geq d$ then V does not contain $\mathbb{P}^N(\mathbb{F}_q)$. Use this to show that there is a projection $\pi : V \to \mathbb{P}^r$ defined over $\mathbb{F}_q$ which is finite of degree d. Conclude from this that

$$\#V(\mathbb{F}_q) \leq d \cdot \#\mathbb{P}^r(\mathbb{F}_q) = d \cdot (q^r + \cdots + q + 1).$$

7.3 (Application to $\mathrm{Hom}_d(\mathbb{P}^1, X)$). We are looking for rational curves of degree d on a hypersurface $X \subset \mathbb{P}^{n+1}$ of degree m. We saw in (5.1) that $\mathrm{Hom}_d(\mathbb{P}^1, X)$ lies in $\mathbb{P}^{(n+2)(d+1)-1}$ (hence we can take $N = (n+2)(d+1) - 1$) and its closure is defined by equations of degree m.

The complement of $\mathrm{Hom}_d(\mathbb{P}^1, X)$ in its closure consists of those $(f_0, \ldots, f_N)$ with a common zero. One can get explicit equations for this locus as follows. Pick indeterminates λ_i, μ_j. Then $f_0, \ldots, f_N$ have a common zero iff the resultant

$$\mathrm{Res}(\textstyle\sum_i \lambda_i f_i, \sum_j \mu_j f_j)$$

is identically zero as a polynomial in the λ_i, μ_j. This gives equations of degree $2d$ in the coefficients of the f_i. Thus we can choose $D = \max\{m, 2d\}$.

Finally, where do we find a geometrically irreducible component of the space $\mathrm{Hom}_d(\mathbb{P}^1, X)$? Here again a smooth point $[f]$ in a suitable compactification of $\mathrm{Hom}(\mathbb{P}^1, X)$ gives the answer by (7.4). Similar considerations show that our methods also apply to $\mathrm{Hom}_d(\mathbb{P}^1, X, 0 \mapsto p)$.

Exercise 7.4. Let W be a K-variety and $p \in W$ a smooth point. Then there is a unique K-irreducible component $W_p \subset K$ which contains p and W_p is also geometrically irreducible if either p is K-point or K is algebraically closed in $K(p)$.

As a first application, let us consider cubic surfaces.

Example 7.5 (Cubic surfaces). Consider a cubic surface $S \subset \mathbb{P}^3$, defined over $K = \mathbb{F}_q$. We would like to use these results to get a rational curve through any $p \in S(\mathbb{F}_q)$.

We need to start with some free rational curves over $\bar{K}$.

The first such possibility is to use conics. If $L \subset S$ is a line, then the plane spanned by p and L intersects S in L plus a residual conic C_L. C is a smooth and free conic, unless p lies on a line.

In general, we get 27 conics and we conclude that if q is large enough, then through every point $p \in S(\mathbb{F}_q)$ which is not on a line, there is rational curve of degree $2 \cdot 27 = 54$, defined over $\mathbb{F}_q$.

If p lies on 1 (resp. 2) lines, then we get only 16 (resp. 8) smooth conics, and so we get even lower degree rational curves.

However, when p lies on 3 lines (these are called Eckart points) then there is no smooth conic through p.

Let us next try twisted cubics. As we saw in (5.4), we get twisted cubics from a morphism $S \to \mathbb{P}^2$ as the birational transforms of lines not passing through any of the 6 blown up points. Thus we get a 2-dimensional family of twisted cubics whenever p is not on one the 6 lines contracted by $S \to \mathbb{P}^2$.

If p lies on 0 (resp. 1, 2, 3) lines, we get 72 (resp. $72 - 16$, $72 - 2 \cdot 16$, $72 - 3 \cdot 16$) such families.

Hence we obtain that for every $p \in S(\mathbb{F}_q)$, the space $\mathrm{Hom}_d(\mathbb{P}^1, X, 0 \mapsto p)$ has a geometrically irreducible component for some $d \leq 3 \cdot 72 = 216$.

As in (7.3), we conclude that if q is large enough, then through every point $p \in S(\mathbb{F}_q)$, there is rational curve of degree at most 216, defined over $\mathbb{F}_q$.

Example 7.6 (Cubic hypersurfaces). Consider a smooth cubic hypersurface $X^n \subset \mathbb{P}^{n+1}$, defined over $K = \mathbb{F}_q$ and let $p \in X(\mathbb{F}_q)$ be a point.

If p lies on a smooth cubic surface section $S \subset X$, then we can assemble a K-comb of degree ≤ 216 and, as before, we can use it to get rational curves through p.

Over a finite field, however, there is no guarantee that X has any smooth cubic surface sections. What can we do then?

We can use a generic cubic surface section through p. This is then defined over a field extension $L = K(y_1, \ldots, y_s)$ where the y_i are algebraically independent over K. By the previous considerations we can assemble an L-comb and conclude that $\mathrm{Hom}_d(\mathbb{P}^1, X, 0 \mapsto p)$ has a smooth L-point for some $d \leq 3 \cdot 72 = 216$.

By (7.4), this implies that it also has a geometrically irreducible component, and we can then finish as before.

It is now clear that the methods of this section together with (4.6) imply the following:

Theorem 7.7. *Let $X \subset \mathbb{P}^{n+1}$ be a smooth SRC hypersurface of degree $m \leq n + 1$ defined over a finite field $\mathbb{F}_q$. Then there is a $C(n)$ such that if $q > C(n)$ then through every point in $X(\mathbb{F}_q)$ there is a rational curve defined over $\mathbb{F}_q$.* $\square$

Exercise 7.8. Prove the following consequence of (7.1):

Let $f : U \to W$ be a dominant morphism over $\mathbb{F}_q$. Assume that W and the generic fiber of f are both geometrically irreducible. Then there is a dense open set W^0 such that $f\big(U(\mathbb{F}_{q^m})\big) \supset W^0(\mathbb{F}_{q^m})$ for $m \gg 1$.

8. Rational curves through two points and Lefschetz-type theorems

8.1 (How not to find rational curves through two points). Let us see what happens if we try to follow the method of (6.7) for 2 points. Assume that over $\overline{K}$ we have a rational curve C_1 through p, p'. Then C_1 is already defined over a finite Galois extension K' of K. As before, consider its conjugates of $C_2, \ldots, C_m$ under $G := \mathrm{Gal}(K'/K)$, and attach copies $C_1', \ldots, C_m'$ to *two* copies of $\mathbb{P}^1$, one over p and one over p'. This results in a curve Y_0 which is defined over K and may be deformed to a smooth curve Y_ε, still passing through p, p'.

The problem is that although all the $\overline{K}$-irreducible components of Y_0 are rational, it has arithmetic genus $m - 1$, hence the smooth curve Y_ε has genus $m - 1$.

Note that finding curves of higher genus through p, p' is not very interesting. Such a curve can easily be obtained by taking the intersection of X with hyperplanes through p, p'.

In fact, no other choice of Y_0 would work, as shown by the next exercise.

Exercise 8.2. Let C be a reduced, proper, connected curve of arithmetic genus 0 defined over K. Let $p \neq p' \in C(K)$ be 2 points. Then there is a closed sub-curve $p, p' \in C' \subset C$ such that C' is connected and every K-irreducible component of C' is isomorphic to $\mathbb{P}^1_K$.

In this section we first connect the existence of rational curves through two points with Lefschetz-type results about the fundamental groups of open subsets of X and then use this connection to find such rational curves in certain cases.

Definition 8.3. Let K be a field, X a normal, projective variety and

$$
\begin{array}{ccc}
C_U & \xrightarrow{\phi} & X \\
\pi \downarrow\uparrow s & & \\
U & &
\end{array}
\tag{8.3.1}
$$

a smooth family of reduced, proper, connected curves mapping to X with a section s. For $x \in X$, set $U_{s \to x} := s^{-1}\phi^{-1}(x)$, parametrizing those maps that send the marked point to x, and

$$
\begin{array}{ccc}
C_{U_{s \to x}} & \xrightarrow{\phi_x} & X \\
\pi_x \downarrow\uparrow s_x & & \\
U_{s \to x} & &
\end{array}
\tag{8.3.2}
$$

the corresponding family.

We say that the family (8.3.1) satisfies the *Lefschetz condition* if, for general $x \in X(\bar{K})$, the map ϕ_x is dominant with geometrically irreducible generic fiber.

Sometimes it is more convenient to give just

$$
U \xleftarrow{\pi} C_U \xrightarrow{\phi} S,
\tag{8.3.3}
$$

without specifying the section $s : U \to C_U$. In this case, we consider the family obtained from the universal section. That is,

$$
\begin{array}{ccc}
C_U \times_U C_U & \xrightarrow{\phi_1} & X \\
\pi_2 \downarrow\uparrow s_1 & & \\
C_U & &
\end{array}
\tag{8.3.4}
$$

where $\pi_2(c, c') = c'$, $\phi_1(c, c') = \phi(c)$ and $s_1(c) = (c, c)$.

If $x \in X$ then $U_x = \phi^{-1}(x) = s_1^{-1}\phi_1^{-1}(x)$ is the set of triples $(C, c, \phi|_C)$ where C is a fiber of π and c a point of C such that $\phi(c) = x$.

Similarly, if $x, x' \in X$ then $U_{x,x'} := \phi^{-1}(x) \times_U \phi^{-1}(x')$ is the set of all $(C, c, c', \phi|_C)$ where C is a fiber of π and c, c' points of C such that $\phi(c) = x$ and $\phi(c') = x'$. Informally (and somewhat imprecisely) $U_{x,x'}$ is the family of curves in U that pass through both x and x'.

Thus the family (8.3.4) satisfies the Lefschetz condition iff $(C_U)_{x,x'}$ is geometrically irreducible for general $x, x' \in X(\bar{K})$.

Exercise 8.4 (Stein factorization). Let $g : U \to V$ be a morphism between irreducible and normal varieties. Then g can be factored as

$$g : U \xrightarrow{c} W \xrightarrow{h} V$$

where W is normal, h is finite and generically étale and there is an open and dense subset W^0 such that $c^{-1}(w)$ is geometrically irreducible for every $w \in W^0$.

Thus $g : U \to V$ is dominant with geometrically irreducible generic fiber iff g can not be factored through a nontrivial finite and generically étale map $W \to V$.

The Appendix explains how the Lefschetz condition connects with the Lefschetz theorems on fundamental groups of hyperplane sections. For now let us prove that a family satisfying the Lefschetz condition leads to rational curves through 2 points.

Example 8.5. Let $S \subset \mathbb{P}^3$ be a smooth cubic surface. Let $U \leftarrow C_U \to S$ be the family of rational hyperplane sections. Note that $C_U \to S$ is dominant with geometrically irreducible generic fiber. Furthermore, for general $p \in S(\bar{K})$, the map ϕ_p is dominant, generically finite and has degree 12.

On the other hand, let U be an irreducible family of twisted cubics on S. Then U satisfies the Lefschetz condition. As discussed in (5.4), U corresponds to the family of lines in $\mathbb{P}^2$ not passing through the 6 blown-up points. Thus U_x consists of lines in $\mathbb{P}^2$ through x, hence $\phi_x : C_{U_x} \to S$ is birational. Thus it cannot factor through a nontrivial finite cover.

Theorem 8.6. *Let X be a smooth projective variety over $\mathbb{F}_q$. Let $U \subset \mathrm{Hom}^{\mathrm{free}}(\mathbb{P}^1, X)$ be a geometrically irreducible smooth subset, closed under $\mathrm{Aut}(\mathbb{P}^1)$. Assume that $U \xleftarrow{\pi} U \times \mathbb{P}^1 \xrightarrow{\phi} X$ satisfies the Lefschetz condition.*

Then there is an open subset $Y^0 \subset X \times X$ such that for $m \gg 1$ and $(x, x') \in Y^0(\mathbb{F}_{q^m})$ there is a point $u \in U(\mathbb{F}_{q^m})$ giving a rational curve

$$\phi_u : \mathbb{P}^1 \to X \quad \text{such that} \quad \phi_u(0) = x, \ \phi_u(\infty) = x'.$$

Proof. Set $s(u) = (u, 0)$ and consider the map

$$\Phi_2 := (\phi \circ s \circ \pi, \phi) : U \times \mathbb{P}^1 \to X \times X.$$

Note that on $U_{s \to x} \times \mathbb{P}^1$ this is just ϕ_x followed by the injection $X \cong \{x\} \times X \hookrightarrow X \times X$.

If the generic fiber of Φ_2 is geometrically irreducible, then by (8.4) and (7.8), there is an open subset $Y^0 \subset X \times X$ such that for $m \gg 1$ and for every $(x, x') \in$

$Y^0(\mathbb{F}_{q^m})$ there is a $(u,p) \in U(\mathbb{F}_{q^m}) \times \mathbb{P}^1(\mathbb{F}_{q^m})$ such that $\Phi_2(u,p) = (x,x')$. This means that $\phi_u(0) = x$ and $\phi_u(p) = x'$. A suitable automorphism γ of $\mathbb{P}^1$ sends $(0,\infty)$ to $(0,p)$. Thus $\phi_u \circ \gamma$ is the required rational curve.

If the generic fiber of Φ_2 is geometrically reducible, then Φ_2 factors through a nontrivial finite cover $W \to X \times X$. For general $x \in X$, the restriction $\mathrm{red}W_x \to \{x\} \times X$ is nontrivial and $U_{s \to x} \to \mathrm{red}W_x$ is dominant. This is impossible by the Lefschetz condition. $\qquad\square$

Next we discuss how to construct families that satisfy the Lefschetz condition.

Lemma 8.7. *Let $U \xleftarrow{\pi} C_U \xrightarrow{\phi} X$ be a smooth family of reduced, proper, generically irreducible curves over $\bar{K}$ such that U_x is irreducible for general $x \in X$. Let $W \subset U$ be a locally closed smooth subset and $W \times \mathbb{P}^1 \cong D_W \subset C_W$ a subfamily. Let $U^0 \subset U$ be an open dense subset. If*

$$W \xleftarrow{\pi} D_W \xrightarrow{\phi} X$$

satisfies the Lefschetz condition, then so does

$$U^0 \xleftarrow{\pi} C_{U^0} \xrightarrow{\phi} X.$$

Proof. Assume that contrary. Then there is a nontrivial finite and generically étale map $Z \to X$ such that the restriction $\phi|_{C_{U^0_x}} : C_{U^0_x} \to X$ factors through Z. Since U_x is irreducible, so is Z.

Let $g_C : C_U \times_X Z \to C_U$ be the projection. By assumption, there is a rational section $s : C_{U^0_x} \to C_U \times_X Z$. Let $B \subset C_U \times_X Z$ be the closure of its image. Then $g_C|_B : B \to C_U$ is finite and an isomorphism over $C_{U^0_x}$. Thus $g_C|_B : B \to C_U$ is an isomorphism at every point where C_U is smooth (or normal). In particular, s restricts to a rational section $s_W : D_W \dashrightarrow C_U \times_X Z$.

Repeating the previous argument, we see that s_W is an everywhere defined section, hence $\phi|_{D_W}$ factors through Z, a contradiction. $\qquad\square$

Corollary 8.8. *Let X be a smooth projective variety over a perfect field K. If there is a $\bar{K}$-family of free curves*

$$U_1 \xleftarrow{\pi_1} U_1 \times \mathbb{P}^1 \xrightarrow{\phi_1} X$$

satisfying the Lefschetz condition then there is a K-family of free curves

$$U \xleftarrow{\pi} U \times \mathbb{P}^1 \xrightarrow{\phi} X$$

satisfying the Lefschetz condition.

Proof. As usual, the first family is defined over a finite Galois extension; let $U_1, \ldots, U_m$ be its conjugates.

Consider the family of all $\bar{K}$-combs

$$\mathrm{Comb}(U) := \big\{\mathrm{Comb}(\phi_{1,u_1}, \ldots, \phi_{m,u_m})\big\}$$

where $u_i \in U_i$ and $\phi_1(u_1, 0) = \cdots = \phi_m(u_m, 0)$ with 0 a marked point on the handle. (We do not assume that the u_i are conjugates of each other.) Each comb is defined by choosing $u_1, \ldots, u_m$ as above and m distinct points in $\mathbb{P}^1 \setminus \{0\}$.

Thus $\mathrm{Comb}(U) \subset \overline{M}_{0,1}(X)$ is defined over K. Furthermore, for each $x \in X$, $\mathrm{Comb}(U)_x \subset \overline{M}_{0,1}(X, 0 \mapsto x)$ is isomorphic to an open subset of

$$(\mathbb{P}^1)^m \times U_{1,x} \times \cdots \times U_{m,x},$$

hence irreducible.

By (6.4), there is a unique irreducible component $\mathrm{Smoothing}(U) \subset \overline{M}_{0,1}(X)$ containing $\mathrm{Comb}(U)$ and $\mathrm{Smoothing}(U)$ is defined over K.

We can now apply (8.7) with $W := \mathrm{Comb}(U)$ and $D_W \to W$ the first tooth of the corresponding comb. This shows that $\mathrm{Smoothing}(U)$ satisfies the Lefschetz condition. $\qquad\qquad\qquad\qquad\qquad\qquad\qquad\qquad\qquad\qquad\qquad\qquad\square$

Example 8.9 (Cubic hypersurfaces). We have already seen in (8.5) how to get a family of rational curves on a smooth cubic surface S that satisfies the Lefschetz condition:

For general $p \in S$, there are 72 one-parameter families of twisted cubics $C_1, \ldots, C_{72}$ through p. Assemble these into a 1-pointed comb and smooth them to get a family $U(S)$ of degree 216 rational curves. (In fact, the family of degree 216 rational curves on S that are linearly equivalent to $\mathcal{O}_S(72)$ is irreducible, and so equals $U(S)$, but we do not need this.)

Let us go now to a higher dimensional cubic $X \subset \mathbb{P}^{n+1}$. Let G denote the Grassmannian of 3-dimensional linear subspaces in $\mathbb{P}^{n+1}$. Over G we have $\mathbf{S} \to G$, the universal family of cubic surface sections of X. For any fiber $S = L^3 \cap X$ we can take $U(S)$. These form a family of rational curves $\mathbf{U}(\mathbf{S})$ on X and we obtain

$$\mathbf{U}(\mathbf{S}) \xleftarrow{\pi} \mathbf{U}(\mathbf{S}) \times \mathbb{P}^1 \xrightarrow{\phi} X.$$

We claim that it satisfies the Lefschetz condition. Indeed, given $x, x' \in X$, the family of curves in $\mathbf{U}(\mathbf{S})$ that pass through x, x' equals

$$\mathbf{U}(\mathbf{S})_{x,x'} = \bigcup_{x,x' \in L^3} \big(U(L^3 \cap X)\big)_{x,x'}.$$

The set of all such L^3-s is parametrized by the Grassmannian of lines in $\mathbb{P}^{n-1}$, hence geometrically irreducible. The general $L^3 \cap X$ is a smooth cubic surface, hence we already know that the corresponding $U(L^3 \cap X)_{x,x'}$ is irreducible. Thus $\mathbf{U}(\mathbf{S})_{x,x'}$ is irreducible.

Although we did not use it for cubics, let us note the following.

Theorem 8.10. *[Kol00, Kol03] Let X be a smooth, projective SRC variety over a field K. Then there is a family of rational curves defined over K*

$$U \xleftarrow{\pi} U \times \mathbb{P}^1 \xrightarrow{\phi} X$$

that satisfies the Lefschetz condition.

8.11 (Going from 2 points to many points). It turns out that going from curves passing through 2 general points to curves passing through m arbitrary points does not require new ideas.

Let us see first how to find a curve through 2 arbitrary points $x, x' \in X$.

We have seen in Section 5 how to produce very free curves in $\mathrm{Hom}(\mathbb{P}^1, X, 0 \mapsto x)$ and in $\mathrm{Hom}(\mathbb{P}^1, X, 0 \mapsto x')$. If $m \gg 1$ then we can find $\psi \in \mathrm{Hom}(\mathbb{P}^1, X, 0 \mapsto x)$ and $\psi' \in \mathrm{Hom}(\mathbb{P}^1, X, 0 \mapsto x')$ such that (8.6) produces a rational curve $\phi : \mathbb{P}^1 \to X$ passing through $\psi(\infty)$ and $\psi'(\infty)$.

We can view this as a length 3 chain

$$(\psi, \phi, \psi') : \mathbb{P}^1 \vee \mathbb{P}^1 \vee \mathbb{P}^1 \to X$$

through x, x'. Using (6.4), we get a family of free rational curves through x, x' and, again for $m \gg 1$ a single free curve through x, x'.

How to go from 2 points to m points $x_1, \ldots, x_m$? For each $i > 1$ we already have very free curves a $g_i : \mathbb{P}^1 \to X$ such that $g_i(0) = x_1$ and $g_i(\infty) = x_i$. We can assemble a comb with $(m-1)$ teeth $\phi : \mathrm{Comb}(g_2, \ldots, g_m) \to X$.

By (6.4), we can smooth it in

$$\overline{M}_{0,m}(X, p_1 \mapsto x_1, \cdots, p_m \mapsto x_m)$$

to get such rational curves.

Appendix. The Lefschetz condition and fundamental groups

The classical Lefschetz theorem says that if X is a smooth, projective variety over $\mathbb{C}$ and $j : C \hookrightarrow X$ is a smooth curve obtained by intersecting X with hypersurfaces, then the natural map

$$j_* : \pi_1(C) \to \pi_1(X) \quad \text{is onto.}$$

Later this was extended to X quasi-projective. Here j_* need not be onto for every curve section C, but j_* is onto for general curve sections. In particular we get the following. (See [GM88] for a general discussion and further results.)

Theorem 8.12. *Let X^n be a smooth, projective variety over $\mathbb{C}$ and $|H|$ a very ample linear system. Then, for every open subset $X^0 \subset X$ and general $H_1, \ldots, H_{n-1} \in |H|$,*

$$\pi_1\big(X^0 \cap H_1 \cap \cdots \cap H_{n-1}\big) \to \pi_1(X) \quad \text{is onto.}$$

It should be stressed that the notion of "general" depends on X^0.

If X is a hypersurface of degree ≥ 3 then the genus of the curves $H_1 \cap \cdots \cap H_{n-1}$ is at least 1. We would like to get a similar result where $\{H_1 \cap \cdots \cap H_{n-1}\}$ is replaced by some family of rational curves.

The following argument shows that if a family of curves satisfies the Lefschetz condition, then (8.12) also holds for that family.

Pick a family of curves $U \xleftarrow{\pi} C_U \xrightarrow{\phi} X$ with a section $s : U \to C_U$ that satisfies the Lefschetz condition.

Given a generically étale $g : Z \to X$, there is an open $X^0 \subset X$ such that $Z^0 := g^{-1}(X^0) \to X^0$ is finite and étale.

Pick a general point $p \in X^0$. There is an open subset $U_p^0 \subset U_p$ such that $\phi_p^{-1}(X^0) \to U_p$ is topologically a locally trivial fiber bundle over $C_{U_p}^0 \to U_p^0$ with typical fiber $C_u^0 = C_u \cap \phi^{-1}(X^0)$ where $u \in U$ is a general point.

Thus there is a right split exact sequence

$$\pi_1(C_u^0) \to \pi_1(C_{U_p}^0) \leftrightarrows \pi_1(U_p^0) \to 1,$$

where the splitting is given by the section s. Since $s(U_p)$ gets mapped to the point p by ϕ, $\pi_1(C_{U_p}^0)$ gets killed in $\pi_1(X^0)$. Hence

$$\mathrm{im}[\pi_1(C_u^0) \to \pi_1(X^0)] = \mathrm{im}[\pi_1(C_{U_p}^0) \to \pi_1(X^0)].$$

Since $C_{U_p} \to X$ is dominant, $\mathrm{im}[\pi_1(C_{U_p}^0) \to \pi_1(X^0)]$ has finite index in $\pi_1(X^0)$. We are done if the image is $\pi_1(X^0)$. Otherwise the image corresponds to a nontrivial covering $Z^0 \to X^0$ and ϕ_p factors through Z^0. This, however, contradicts the Lefschetz condition. $\qquad\square$

A more detailed consideration of the above argument shows that (8.12) is equivalent to the following weaker Lefschetz-type conditions:

1. The generic fiber of $C_U \to X$ is geometrically irreducible, and
2. for general $x \in X$, U_x is geometrically irreducible and $C_{U_x} \to X$ is dominant.

In positive characteristic the above argument has a problem with the claim that something is "topologically a locally trivial fiber bundle" and indeed the two versions are not quite equivalent. In any case, the purely algebraic version of (8.3) works better for us.

9. Descending from $\mathbb{F}_{q^2}$ to $\mathbb{F}_q$

Our methods so far constructed rational curves on hypersurfaces over $\mathbb{F}_q$ for $q \gg 1$. Even for cubics, the resulting bounds on q are huge. The aim of this section is to use the third intersection point map to construct rational curves on cubic hypersurfaces over $\mathbb{F}_q$ from rational curves on cubic hypersurfaces over $\mathbb{F}_{q^2}$. The end result is a proof of (1.1). The price we pay is that the degrees of the rational curves become larger as q gets smaller.

9.1 (Descent method). Let X be a cubic hypersurface, C a smooth curve and $\phi : C(\mathbb{F}_q) \to X(\mathbb{F}_q)$ a map of sets.

Assume that for each $p \in C(\mathbb{F}_q)$ there is a line L_p through $\phi(p)$ which intersects X in two further points $s(p), s'(p)$. These points are in $\mathbb{F}_{q^2}$ and we assume that none of them is in $\mathbb{F}_q$, hence $s(p), s'(p)$ are conjugate over $\mathbb{F}_q$. This gives a lifting of ϕ to $\phi_2 : C(\mathbb{F}_q) \to X(\mathbb{F}_{q^2})$ where $\phi_2(p) = s(p)$. (This involves a choice for each p but this does not matter.)

Assume that over $\mathbb{F}_{q^2}$ there is an extension of ϕ_2 to $\Phi_2 : C \to X$. If $\bar{\Phi}_2$ denotes the conjugate map, then $\bar{\Phi}_2(p) = s'(p)$.

Applying the third intersection point map (3.5) to the Weil restriction (3.8) we get an $\mathbb{F}_q$-map

$$h : \mathfrak{R}_{\mathbb{F}_{q^2}/\mathbb{F}_q} C \to X.$$

Since C is defined over $\mathbb{F}_q$, the Weil restriction has a diagonal

$$j : C \hookrightarrow \mathfrak{R}_{\mathbb{F}_{q^2}/\mathbb{F}_q} C$$

and $\Phi := h \circ j : C \to X$ is the required lifting of ϕ.

Thus, in order to prove (1.1), we need to show that

1. (1.1) holds for $q \gg 1$, and
2. for every $x \in X(\mathbb{F}_q)$ there is a line L as required.

Remark 9.2. In trying to use the above method over an arbitrary field K, a significant problem is that for each point p we get a degree 2 field extension $K(s(p))/K$ but we can use these only if they are all the same. A finite field has a unique extension of any given degree, hence the extensions $K(s(p))/K$ are automatically the same.

There are a few other fields with a unique degree 2 extension, for instance $\mathbb{R}, \overline{\mathbb{Q}}((t))$ or $\overline{\mathbb{F}}_p((t))$ for $p \neq 2$.

If we have only 1 point p, then the method works over any field K. This is another illustration that the 1 point case is much easier.

In the finite field case, the method can also deal with odd degree points of C but not with even degree points.

9.3 (Proof of (9.1.1)). We could just refer to (8.11) or to [KS03, Thm.2], but I rather explain how to prove the 2 point case using (8.6) and the above descent method.

Fix $c, c' \in C(\mathbb{F}_q)$. By (8.6), there is an open subset $Y^0 \subset X \times X$ such that the following holds

(∗) If $\mathbb{F}_{q^m} \supset \mathbb{F}_q$ is large enough then for every $(x, x') \in Y^0(\mathbb{F}_{q^m})$ there is an $\mathbb{F}_{q^m}$-map $\Psi : C \to X$ such that $\Psi(c) = x$ and $\Psi(c') = x'$.

Assume now that we have any $x, x' \in X(\mathbb{F}_{q^m})$. If we can choose the lines L through x and L' through x' such that $(s(x), s(x')) \subset Y^0$, then the descent method produces the required extension $\Psi : C \to X$ over $\mathbb{F}_{q^m}$.

By the Lang-Weil estimates, $Y^0(\mathbb{F}_{q^m})$ has about q^{2nm} points. If, for a line L through x, one of the other two points of $X \cap L$ is in $\mathbb{F}_{q^m}$ then so is the other point.

Thus we have about $\frac{1}{2}q^{nm}$ lines where $s(x), s'(x)$ are in $X(\mathbb{F}_{q^m})$. Accounting for the lines tangent to X gives a contribution $O(q^{(n-1)m})$. Thus about $\frac{1}{4}$ of all line pairs (L, L') work for us.

The proof of (9.1.2) is an elaboration of the above line and point counting argument.

Lemma 9.4. *Let $X \subset \mathbb{P}^{n+1}$ be a normal cubic hypersurface and $p \in X(\mathbb{F}_q)$ a smooth point. Assume that $n \geq 1$ and $q \geq 8$. Then*

1. *either there is a line defined over $\mathbb{F}_q$ through p but not contained in X that intersects X in two further smooth points $s, s' \in X(\mathbb{F}_{q^2}) \setminus X(\mathbb{F}_q)$,*
2. *or projecting X from p gives an inseparable degree 2 map $X \dashrightarrow \mathbb{P}^n$. In this case $q = 2^m$ and X is singular.*

Proof. Start with the case $n = 1$. Thus $C := X$ is plane cubic which we allow to be reducible.

Consider first the case when $C = L \cup Q$ a line through p and a smooth conic Q. There are $q + 1$ $\mathbb{F}_q$-lines through p, one is L and at most 2 of them are tangent to Q, unless projecting Q from p is purely inseparable. If all the remaining $q - 2$ lines intersect Q in two $\mathbb{F}_q$-points, then Q has $2 + 2(q - 2) = 2q - 2$ points in $\mathbb{F}_q$. This is impossible for $q > 3$. In all other reducible cases, C contains a line not passing through p. (Since C is smooth at p, C can not consist of 3 lines passing through p.)

Assume next that C is irreducible and smooth. If projection from p is separable, then at most 4 lines through p are tangent to C away from p and one is tangent at p. If all the remaining $q - 4$ lines intersect C in two $\mathbb{F}_q$-points, then C has $5 + 2(q - 4) = 2q - 3$ points in $\mathbb{F}_q$. For $q \geq 8$ this contradicts the Hasse-Weil estimate $\#C(\mathbb{F}_q) \leq q + 1 + 2\sqrt{q}$. The singular case works out even better.

Now to the general case. Assume that in affine coordinates p is the origin and write the equation as

$$L(x_1, \ldots, x_{n+1}) + Q(x_1, \ldots, x_{n+1}) + C(x_1, \ldots, x_{n+1}) = 0.$$

Let us show first that there is a line defined over $\mathbb{F}_q$ through p but not contained in X that intersects X in two further smooth points s, s'.

If the characteristic is 2, then projection from p is inseparable iff $Q \equiv 0$. If Q is not identically zero, then for $q \geq 3$ there are $a_1, \ldots, a_{n+1} \in \mathbb{F}_q$ such that $(L \cdot Q)(a_1, \ldots, a_{n+1}) \neq 0$. The corresponding line intersects X in 2 further distinct points, both necessarily smooth.

If the characteristic is $\neq 2$, then the line corresponding to $a_1, \ldots, a_{n+1} \in \mathbb{F}_q$ has a double intersection iff the discriminant $Q^2 - 4LC$ vanishes. Note that $Q^2 - 4LC$ vanishes identically only if X is reducible. Thus, for $q \geq 5$ there are $a_1, \ldots, a_{n+1} \in \mathbb{F}_q$ such that $(L \cdot (Q^2 - 4LC))(a_1, \ldots, a_{n+1}) \neq 0$. As before, the corresponding line intersects X in 2 further distinct points, both necessarily smooth.

It is possible that for this line $s, s' \in X(\mathbb{F}_{q^2}) \setminus X(\mathbb{F}_q)$ and we are done. If not then $s, s' \in X(\mathbb{F}_q)$. We can choose the line to be $(x_1 = \cdots = x_n = 0)$ and write

$s = (0, \ldots, 0, s_{n+1})$ and $s' = (0, \ldots, 0, s'_{n+1})$. Our aim now is to intersect X with the planes

$$P(a_1, \ldots, a_n) := \langle (0, \ldots, 0, 1), (a_1, \ldots, a_n, 0) \rangle$$

for various $a_1, \ldots, a_n \in \mathbb{F}_q$ and show that for one of them the intersection does not contain a line not passing through p. Then the curve case discussed above finishes the proof.

Set $x'_{n+1} = x_{n+1} - s_{n+1}$. At s the equation of X is

$$L_s(x_1, \ldots, x'_{n+1}) + Q_s(x_1, \ldots, x'_{n+1}) + C_s(x_1, \ldots, x'_{n+1}) = 0.$$

Since X is irreducible, L_s does not divide either Q_s or C_s. L_s contains x'_{n+1} with nonzero coefficient since the vertical line has intersection number 1 with X. We can use L_s to eliminate x'_{n+1} from Q_s and C_s. As we saw, one of these is nonzero, let it be $B_s(x_1, \ldots, x_n)$. Similarly, at s' we get $B'_s(x_1, \ldots, x_n)$.

If $X \cap P(a_1, \ldots, a_n)$ contains a line through s (resp. s') then $B_s(a_1, \ldots, a_n) = 0$ (resp. $B'_s(a_1, \ldots, a_n) = 0$). Thus we have the required $(a_1, \ldots, a_n)$, unless $B_s \cdot B'_s$ is identically zero on $\mathbb{P}^{n-1}(\mathbb{F}_q)$. This happens only for $q \leq 5$. $\qquad \square$

Exercise 9.5. Let $H(x_1, \ldots, x_n)$ be a homogeneous polynomial of degree d. If H vanishes on $\mathbb{F}_q^n$ and $q \geq d$ then H is identically zero.

Exercise 9.6. Set $F(x_0, \ldots, x_m) = \sum_{i \neq j} x_i^{2^n} x_j$. Show that F vanishes on $\mathbb{P}^m(\mathbb{F}_{2^n})$ and for m odd it defines a smooth hypersurface.

References

[AK03]   Carolina Araujo and János Kollár. Rational curves on varieties. In *Higher dimensional varieties and rational points (Budapest, 2001)*, volume 12 of *Bolyai Soc. Math. Stud.*, pages 13–68. Springer, Berlin, 2003.

[Ale96]   Valery Alexeev. Moduli spaces $M_{g,n}(W)$ for surfaces. In *Higher-dimensional complex varieties (Trento, 1994)*, pages 1–22. de Gruyter, Berlin, 1996.

[BLR90]   Siegfried Bosch, Werner Lütkebohmert, and Michel Raynaud. *Néron models*, volume 21 of *Ergebnisse der Mathematik und ihrer Grenzgebiete (3) [Results in Mathematics and Related Areas (3)]*. Springer-Verlag, Berlin, 1990.

[BT05]   Fedor Bogomolov and Yuri Tschinkel. Rational curves and points on $K3$ surfaces. *Amer. J. Math.*, 127(4):825–835, 2005.

[dJS04]   Aise Johan de Jong and Jason Starr. Cubic fourfolds and spaces of rational curves. *Illinois J. Math.*, 48(2):415–450, 2004.

[FP97]   William Fulton and Rahul Pandharipande. Notes on stable maps and quantum cohomology. In *Algebraic geometry—Santa Cruz 1995*, volume 62 of *Proc. Sympos. Pure Math.*, pages 45–96. Amer. Math. Soc., Providence, RI, 1997.

[GM88]   Mark Goresky and Robert MacPherson. *Stratified Morse theory*, volume 14 of *Ergebnisse der Mathematik und ihrer Grenzgebiete (3) [Results in Mathematics and Related Areas (3)]*. Springer-Verlag, Berlin, 1988.

[GR71]   Hans Grauert and Reinhold Remmert. *Analytische Stellenalgebren*. Springer-Verlag, Berlin, 1971. Unter Mitarbeit von Oswald Riemenschneider, Die Grundlehren der mathematischen Wissenschaften, Band 176.

[Hir81]   James W. P. Hirschfeld. Cubic surfaces whose points all lie on their 27 lines. In *Finite geometries and designs (Proc. Conf., Chelwood Gate, 1980)*, volume 49 of *London Math. Soc. Lecture Note Ser.*, pages 169–171. Cambridge Univ. Press, Cambridge, 1981.

[HRS04]   Joe Harris, Mike Roth, and Jason Starr. Rational curves on hypersurfaces of low degree. *J. Reine Angew. Math.*, 571:73–106, 2004.

[HRS05]   Joe Harris, Mike Roth, and Jason Starr. Curves of small degree on cubic threefolds. *Rocky Mountain J. Math.*, 35(3):761–817, 2005.

[HS05]   Joe Harris and Jason Starr. Rational curves on hypersurfaces of low degree. II. *Compos. Math.*, 141(1):35–92, 2005.

[KM94]   Maxim Kontsevich and Yuri I. Manin. Gromov-Witten classes, quantum cohomology, and enumerative geometry. *Comm. Math. Phys.*, 164(3):525–562, 1994.

[KMM92a]   János Kollár, Yoichi Miyaoka, and Shigefumi Mori. Rational curves on Fano varieties. In *Classification of irregular varieties (Trento, 1990)*, volume 1515 of *Lecture Notes in Math.*, pages 100–105. Springer, Berlin, 1992.

[KMM92b]   János Kollár, Yoichi Miyaoka, and Shigefumi Mori. Rationally connected varieties. *J. Algebraic Geom.*, 1(3):429–448, 1992.

[Kol96]   János Kollár. *Rational curves on algebraic varieties*, volume 32 of *Ergebnisse der Mathematik und ihrer Grenzgebiete. 3. Folge. A Series of Modern Surveys in Mathematics [Results in Mathematics and Related Areas. 3rd Series. A Series of Modern Surveys in Mathematics]*. Springer-Verlag, Berlin, 1996.

[Kol99]   János Kollár. Rationally connected varieties over local fields. *Ann. of Math. (2)*, 150(1):357–367, 1999.

[Kol00]   János Kollár. Fundamental groups of rationally connected varieties. *Michigan Math. J.*, 48:359–368, 2000. Dedicated to William Fulton on the occasion of his 60th birthday.

[Kol02]   János Kollár. Unirationality of cubic hypersurfaces. *J. Inst. Math. Jussieu*, 1(3):467–476, 2002.

[Kol03]   János Kollár. Rationally connected varieties and fundamental groups. In *Higher dimensional varieties and rational points (Budapest, 2001)*, volume 12 of *Bolyai Soc. Math. Stud.*, pages 69–92. Springer, Berlin, 2003.

[KS03]   János Kollár and Endre Szabó. Rationally connected varieties over finite fields. *Duke Math. J.*, 120(2):251–267, 2003.

[KSC04]   János Kollár, Karen E. Smith, and Alessio Corti. *Rational and nearly rational varieties*, volume 92 of *Cambridge Studies in Advanced Mathematics*. Cambridge University Press, Cambridge, 2004.

[LW54]   Serge Lang and André Weil. Number of points of varieties in finite fields. *Amer. J. Math.*, 76:819–827, 1954.

[Man86]   Yuri I. Manin. *Cubic forms*, volume 4 of *North-Holland Mathematical Library*. North-Holland Publishing Co., Amsterdam, second edition, 1986. Algebra, geometry, arithmetic, Translated from the Russian by M. Hazewinkel.

[Seg43]   Beniamino Segre. A note on arithmetical properties of cubic surfaces. *J. London Math. Soc*, 18:24–31, 1943.

Abelian varieties over finite fields

Frans Oort

Mathematisch Instituut, P.O. Box. 80.010, NL - 3508 TA Utrecht
The Netherlands
e-mail: oort@math.uu.nl

Abstract. A. Weil proved that the geometric Frobenius $\pi = F^a$ of an abelian variety over a finite field with $q = p^a$ elements has absolute value $\sqrt{q}$ for every embedding. T. Honda and J. Tate showed that $A \mapsto \pi_A$ gives a bijection between the set of isogeny classes of simple abelian varieties over $\mathbf{F}_q$ and the set of conjugacy classes of q-Weil numbers.

Higher-dimensional varieties over finite fields,
Summer school in Göttingen, June 2007

Introduction

We could try to classify *isomorphism classes of abelian varieties*. The theory of moduli spaces of polarized abelian varieties answers this question completely. This is a geometric theory. However in this general, abstract theory it is often not easy to exhibit explicit examples, to construct abelian varieties with required properties.

A coarser classification is that of studying *isogeny classes of abelian varieties*. A wonderful and powerful theorem, the Honda-Tate theory, gives

a complete classification of isogeny classes of abelian varieties over a finite field,

see Theorem 1.2.

The basic idea starts with a theorem by A. Weil, a proof for the Weil conjecture for an abelian variety A over a finite field $K = \mathbb{F}_q$, see 3.2:

the geometric Frobenius π_A of A/K is an algebraic integer
which for every embedding $\psi : \mathbb{Q}(\pi_A) \to \mathbb{C}$ has absolute value $\mid \psi(\pi_A) \mid = \sqrt{q}$.

For an abelian variety A over $K = \mathbb{F}_q$ the assignment $A \mapsto \pi_A$ associates to A its geometric Frobenius π_A; the isogeny class of A gives the conjugacy class of the algebraic integer π_A, and

conversely an algebraic integer which is a Weil q-number
determines an isogeny class, as T. Honda and J. Tate showed.

Geometric objects are constructed and classified up to isogeny by a simple alge-braic invariant. This arithmetic theory gives access to a lot of wonderful theorems. In these notes we describe this theory, we give some examples, applications and some open questions.

Instead of reading these notes it is much better to read the wonderful and clear [73]. Some proofs have been worked out in more detail in [74].

In §§ 1 ∼ 15 material discussed in the course is described. In the appendices §§ 16 ∼ 22 we have gathered some information we need for statements and proofs of the main result. I hope all relevant notions and information needed for under-standing the main arguments of these notes can be found in the appendices.

Material discussed below will be contained eventually in [GM]. That book by G. van der Geer and B. Moonen can be used as a reference for all material we need, and for all results we discuss. However, as a final version of this book is not yet available, we also give other references. In referring to [GM] we will usually not be precise as the final numbering can be different from the one available now.

Further recommended reading:
Abelian varieties: [47], [35], [15] Chapter V.
Honda-Tate theory: [73], [29], [74].
Abelian varieties over finite fields: [72], [75], [77], [64].
Group schemes: [62], [49].
Endomorphism rings and endomorphism algebras: [68], [24], [72], [75], [54].
CM-liftings: [56], [11].

Contents:
$$\begin{array}{ll}
\S\S\ 1-13: & \text{material for this course,} \\
\S\S\ 14,\ 15: & \text{examples and exercises,} \\
\S\S\ 16-21: & \text{appendices giving definitions and background,} \\
\S\ \ 22: & \text{questions and open problems.}
\end{array}$$

Some notation. In definitions and proofs below we need various fields, in various disguises. We use K, L, M, P, k, $\mathbb{F}_q$, $\overline{\mathbb{F}_p} = \mathbb{F}$, $\mathbb{P}$, m.

We write K for an arbitrary field, usually the base field, in some cases of arbitrary characteristic, however most of the times a finite field. We write k for an algebraically closed field. We write g for the dimension of an abelian variety, unless otherwise stated. We write p for a prime number. We write ℓ for a prime number, which usually is different from the characteristic of the base field, respectively invertible in the sheaf of local rings of the base scheme. We write $\mathbb{F} = \overline{\mathbb{F}_p}$. We use the notation M for a field, sometimes a field of definition for an abelian variety in characteristic zero.

We will use L as notation for a field, usually the center of an endomorphism algebra; we will see that in our cases this will be a totally real field or a CM-field.

We write P for a CM-field, usually of degree $2g$ over $\mathbb{Q}$. We write $\mathbb{P}$ for a prime field: either $\mathbb{P} = \mathbb{Q}$ or $\mathbb{P} = \mathbb{F}_p$.

A discrete valuation on a base field usually will be denoted by v, whereas a discrete valuation on a CM-field usually will be denoted by w. If w divides p, the normalization chosen will be given by $w(p) = 1$.

For a field M we denote by Σ_M the set of discrete valuations (finite places) of M. If moreover M is of characteristic zero, we denote by $\Sigma_M^{(p)}$ the set of discrete valuations with residue characteristic equal to p.

We write $\lim_{\leftarrow i}$ for the notion of "projective limit" or "inverse limit".

We write $\mathrm{colim}_{i \to}$ for the notion of "inductive limit" or "direct limit".

1. Main topic/survey

1.1. Definition. Let p be a prime number, $n \in \mathbb{Z}_{>0}$; write $q = p^n$. A Weil q-number π is an *algebraic integer* such that for every embedding $\psi : \mathbb{Q}(\pi) \to \mathbb{C}$ we have

$$| \psi(\pi) | \;\; = \;\; \sqrt{q}.$$

We say that π and π' are *conjugated* if there exists an isomorphism $\mathbb{Q}(\pi) \cong \mathbb{Q}(\pi')$ mapping π to π'.

Notation: $\pi \;\; \sim \;\; \pi'$.

Equivalently: *the minimum polynomials of π and π' over $\mathbb{Q}$ are equal.* We write $W(q)$ for the set conjugacy classes of Weil q-numbers.

In this definition $| \cdot |$ denotes the *complex absolute value* given by $| a + b\sqrt{-1} | = \sqrt{a^2 + b^2}$ for $a, b \in \mathbb{R}$. We will show that for any Weil q-number π there exists an element $\bar{\pi} = \rho(\pi) \in \mathbb{Q}(\pi)$ such that for any $\psi : \mathbb{Q}(\pi) \to \mathbb{C}$ the number $\psi(\bar{\pi})$ is the complex conjugate of $\psi(\pi)$; moreover we show that $\pi \cdot \bar{\pi} = q$.

As Weil proved, we will see that the geometric Frobenius π_A, see 3.1, of a simple abelian variety over the finite field $\mathbb{F}_q$ is a Weil q-number, see Theorem 3.2. We will see that

$$A \sim B \;\;\; \Rightarrow \;\;\; \pi_A \sim \pi_B,$$

i.e. abelian varieties defined over the same finite field K isogenous over K define conjugated Weil numbers. We will write

$$\{ \text{ simple abelian variety over } \;\; K \} / \sim_K \;\; =: \;\; \mathcal{M}(K, s)$$

for the set of isogeny classes of simple abelian varieties over K.

1.2. Theorem (Honda, Serre and Tate). *Fix a finite field $K = \mathbb{F}_q$. The assignment $A \mapsto \pi_A$ induces a bijection*

$$\boxed{\{\text{simple abelian variety over } \;\; K\} / \sim_K = \mathcal{M}(K, s) \;\; \xrightarrow{\;\sim\;} \;\; W(q), \;\;\;\; A \mapsto \pi_A}$$

from the set of K-isogeny classes of K-simple abelian varieties defined over K and the set $W(q)$ of conjugacy classes of Weil q-numbers.
See [73].

The fact

- that the map is defined follows by Weil,
- the map is injective by Tate, and
- surjective by Honda and Tate.

This map will be denoted by

$$\mathcal{W} : \mathcal{M}(K,s) \quad \longrightarrow \quad W(q).$$

This theorem will be the main topic of these talks. We encounter various notions and results, which will be exposed below (sometimes in greater generality than strictly necessary to understand this beautiful theorem).

1.3. Definition. We say that a Weil q-number π is *effective* if there exists an abelian variety A simple over $\mathbb{F}_q$ such that $\pi \sim \pi_A$. I.e. π is effective if it is in the image of the map $\mathcal{W} : A \mapsto \pi_A / \sim$.

We indicate the steps in a proof of 1.2, which will be elaborated below. Write $K = \mathbb{F}_q$, with $q = p^n$.

ONE (Weil)   *For a simple abelian variety A over a finite field $K = \mathbb{F}_q$ the Weil conjecture implies that π_A is a Weil q-number, see* Section 3, especially Theorem 3.2. *Hence the map*

$$\{\text{simple abelian variety over} \quad K\} \quad \longrightarrow \quad W(K), \qquad A \mapsto \pi_A$$

is well-defined.

TWO (Tate)   *For simple abelian varieties A, B defined over a finite field we have:*

$$A \sim B \quad \Longleftrightarrow \quad \pi_A \sim \pi_B.$$

See 5.3. Note that $A \sim B$ only makes sense if A and B are defined over the same field. Note that $\pi_A \sim \pi_B$ implies that A and B are defined over the same finite field. This shows that the map $\mathcal{W} : \mathcal{M}(\mathbb{F}_q, s) \to W(q)$ is *well-defined and injective.* See Sections 4, 5, especially Theorem 5.3.

THREE (Honda)   *Suppose given $\pi \in W(q)$. There exists a finite extension $K = \mathbb{F}_q \subset K' := \mathbb{F}_{q^N}$ and an abelian variety B' over K' with $\pi^N = \pi_{B'}$.*
See [29], Theorem 1. This step says that for every Weil q-number there *exists* $N \in \mathbb{Z}_{>0}$ such that π^N is effective. See Section 10 , especially Theorem 10.4.

FOUR (Tate) *If $\pi \in W(q)$ and there exists $N \in \mathbb{Z}_{>0}$ such that π^N is effective, then π is effective.* See Section 10, especially 10.5 - 10.9.

This result by Honda plus the last step shows that $(A \bmod \sim) \mapsto (\pi_A \bmod \sim)$ is *surjective*.

These four steps together show that the map

$$\boxed{\mathcal{W}: \quad \{\text{simple abelian variety over } K\}/\sim_K \;=\; \mathcal{M}(K,s) \;\xrightarrow{\;\sim\;}\; W(q)}$$

is bijective, thus proving the main theorem of Honda-Tate theory.

In 1966/1967 Serre wrote a letter to Tate in which he explained a proof of the Manin conjecture; see Section 11. That method proved the surjectivity result proved by Honda. Therefore, sometimes the theory discussed here can be called the Honda-Serre-Tate theory. As Serre's proof was never published we can also use the terminology Honda-Tate theory.

We will see several examples. Here are three special cases, which we mention now in order to convey the flavor of the aspects we will encounter.

1.4. Motivation/Some examples. See 15.5. Consider the following examples.
(1) Choose $q = p^n$, and choose $i \in \mathbb{Z}_{>0}$. Let $\pi := \zeta_i \cdot \sqrt{q}$, where ζ_i is a primitive i-th root of unity.
(2) Choose coprime positive integers $d > c > 0$, and choose p. Let π be a zero of

$$T^2 + p^c T + p^{d+c}.$$

(3) Let $\beta := \sqrt{2 + \sqrt{3}}$, and $q = p^n$. Let π be a zero of

$$T^2 - \beta T + q.$$

In all these cases we see that π is a Weil q-number. *How can we see that these numbers are the Weil number belonging to an isogeny class of an abelian variety simple over $\mathbb{F}_q$?* Using Theorem 1.2 this follows; however these examples might illustrate that this theorem is non-trivial. If such an isogeny class exists *what is the dimension of these abelian varieties? How can we compute this dimension? What are the p-adic properties of such an abelian variety?* See 5.4, 5.5.

2. Weil numbers and CM-fields

2.1. Definition. A field L is said to be a CM-field if

- L is a finite extension of $\mathbb{Q}$ (i.e. L is a number field),
- there is a subfield $L_0 \subset L$ such that $L_0/\mathbb{Q}$ is totally real, i.e. every $\psi_0 : L_0 \to \mathbb{C}$ gives $\psi_0(L_0) \subset \mathbb{R}$, and

- L/L_0 is quadratic totally imaginary, i.e. $[L : L_0] = 2$ and for every $\psi : L \to \mathbb{C}$ we have $\psi(L) \not\subset \mathbb{R}$.

Remark. The quadratic extension L/L_0 gives an involution $\rho \in \mathrm{Aut}(L/L_0)$. For every embedding $\psi : L \to \mathbb{C}$ this involution on a CM-field L corresponds with the restriction of complex conjugation on $\mathbb{C}$ to $\psi(L)$.

2.2. Proposition. *Let π be a Weil q-number.*

($\mathbb{R}$) *Either for at least one $\psi : \mathbb{Q}(\pi) \to \mathbb{C}$ we have $\psi(\pi) \in \mathbb{R}$; in this case we have:*
    ($\mathbb{R}$e) *n is even, $\sqrt{q} \in \mathbb{Q}$, and $\pi = +p^{n/2}$, or $\pi = -p^{n/2}$, or*
    ($\mathbb{R}$o) *n is odd, $\sqrt{q} \in \mathbb{Q}(\sqrt{p})$, and $\psi(\pi) = \pm p^{n/2}$.*
In particular in case ($\mathbb{R}$) we have $\psi(\pi) \in \mathbb{R}$ for every ψ.

($\mathbb{C}$) *Or for every $\psi : \mathbb{Q}(\pi) \to \mathbb{C}$ we have $\psi(\pi) \not\in \mathbb{R}$* (equivalently: for at least one ψ we have $\psi(\pi) \not\in \mathbb{R}$). *In case ($\mathbb{C}$) the field $\mathbb{Q}(\pi)$ is a CM-field.*

See 15.9, where we explain these cases in the Honda-Tate theory.

Proof. The claims in ($\mathbb{R}$) follow from the fact that $\pm p^{n/2}$ are precisely those real numbers with absolute value, taken in $\mathbb{C}$, are equal to $\sqrt{q}$.

If at least one embedding ψ gives $\psi(\pi) \not\in \mathbb{R}$, then we are not in case ($\mathbb{R}$), hence all embeddings have this property. Then

$$\psi(\pi) \cdot \overline{\psi(\pi)} = q.$$

Write $\beta := \pi + \frac{q}{\pi}$. Then for every ψ we have

$$\overline{\psi(\beta)} = \overline{\psi(\pi)} + (q/\overline{\psi(\pi)}) = \frac{q}{\psi(\pi)} + \psi(\pi) = \psi(\beta).$$

Hence $L_0 := \mathbb{Q}(\beta)$ is totally real. For any Weil q-number π with $\psi(\pi) \not\in \mathbb{R}$ we have

$$\beta := \pi + \frac{q}{\pi}, \qquad (T - \psi(\pi))(T - \overline{\psi(\pi)}\,) = T^2 - \beta T + q \in \mathbb{Q}(\beta)[T].$$

In this case $\psi(\pi) \not\in \mathbb{R}$ for every ψ, and $L_0 := \mathbb{Q}(\beta)$ is totally real and L/L_0 is totally complex. Hence L is a CM-field. $\qquad\square$

2.3. Remark. We see a characterization of Weil q-numbers:

$$\beta := \pi + \frac{q}{\pi} \quad \text{is a totally real integer,}$$

and either $\pi = \sqrt{q} \in \mathbb{R}$ or π is a zero of

$$T^2 - \beta \cdot T + q, \quad \text{with} \quad |\psi(\beta)| < 2\sqrt{q} \quad \text{for any} \quad \psi : \mathbb{Q}(\beta) \to \mathbb{R}.$$

Using this it is easy to construct Weil q-numbers, see Section 15 for some examples.

3. The Weil conjecture for abelian varieties over a finite field

3.1. The geometric Frobenius. For a scheme $A \to S$ over a base $S \to \mathrm{Spec}(\mathbb{F}_p)$ in characteristic p there is the relative Frobenius

$$F_{A/S} : A \longrightarrow A^{(p)};$$

see 21.2. If moreover A/S is a group scheme this is a homomorphism. If $S = \mathrm{Spec}(\mathbb{F}_{p^n})$ there is a canonical identification $A^{(p^n)} \cong_S A$, and we define:

$$\pi_A := \left(A \ \xrightarrow{F_{A/S}} \ A^{(p)} \ \xrightarrow{F_{A^{(p)}/S}} \ A^{(p^2)} \ \longrightarrow \ \cdots \ \longrightarrow \ A^{(p^n)} = A \right).$$

This endomorphism is called *the geometric Frobenius* of $A/\mathbb{F}_{p^n}$. Sometimes we will write (in abused notation) " $\pi_A = F^n$ ".

3.2. Theorem (Weil). *Let A be a simple abelian variety over $K = \mathbb{F}_q$; consider the endomorphism $\pi_A \in \mathrm{End}(A)$, the geometric Frobenius of $A/\mathbb{F}_q$. The algebraic number π_A is a Weil q-number, i.e. it is an algebraic integer and for every embedding $\psi : \mathbb{Q}(\pi_A) \to \mathbb{C}$ we have*

$$| \psi(\pi) | \ = \ \sqrt{q}.$$

See [78], page 70; [79], page 138; [47], Theorem 4 on page 206. Using the following two propositions we give a proof of this theorem.

3.3. Proposition. *For any polarized abelian variety A over a field the Rosati-involution $\dagger : D \to D := \mathrm{End}^0(A)$ is positive definite bilinear form on D, i.e. for any non-zero $x \in D$ we have $\mathrm{Tr}(x \cdot x^\dagger) > 0$.* $\qquad\square$
See [47], Th. 1 on page 192, see [15], Th. 17.3 on page 138. For the notation D and for the notion of the Rosati involution defined by a polarization, see Section 16, in particular 16.3 and 16.5.

3.4. Proposition. *For a simple abelian variety A over $K = \mathbb{F}_q$ we have*

$$\pi_A \cdot (\pi_A)^\dagger \ = \ q.$$

Here $\dagger : D \to D := \mathrm{End}^0(A)$ is the Rosati-involution.

One proof can be found in [47], formula (i) on page 206; also see [15], Coroll. 19.2 on page 144.

Another proof of 3.4 can be found in 5.21, 7.34 and Section 15 of [GM]. To this end we study Verschiebung, see 21.3, defined for commutative flat group schemes over a base in characteristic p. The (relative) Frobenius and the Verschiebung homomorphism for abelian varieties are related by two properties:

$$\text{for any abelian variety } B \text{ we have } \left(B \xrightarrow{\ F\ } B^{(p)} \xrightarrow{\ V\ } B \right) = p,$$

also $V \cdot F = p \cdot \mathbf{1}_{B^{(p)}}$, and

$$\left(F_{B/S} : B \to B^{(p)} \right)^t \;=\; \left(V_{B^t/S} : (B^{(p)})^t \to B^t \right);$$

see 21.10. For the definition of the dual abelian scheme, and for the notation A^t see 16.2. From this we see that

$$\pi_{A^t} \cdot (\pi_A)^t = \left(F_{(A^t)(p^{n-1})} \cdots F_{A^t} \right) \left(F_{A^{(p^{n-1})}} \cdots F_A \right)^t =$$

$$= F_{(A^t)(p^{n-1})} \left(\cdots \left(F_{(A^t)(p)} \left(F_{A^t} \cdot V_{A^t} \right) V_{(A^t)(p)} \right) \cdots \right) V_{(A^t)(p^{n-1})} = p^n = q.$$

In abused notation we could write: $\pi_{A^t} \cdot (\pi_A)^t = F^n \cdot (F^n)^t = F^n \cdot V^n = p^n$. $\quad\square$3.4

3.5. We give a proof of 3.2 using 3.4 and 3.3. We use that $L = \mathbb{Q}(\pi_a)$ is the center of D, see 5.4 (1). Hence $\dagger$ on D induces an involution on L. Hence $\dagger$ induces an involution $\dagger_{\mathbb{R}}$ on $L \otimes_{\mathbb{Q}} \mathbb{R}$. This algebra is a finite product of copies of $\mathbb{R}$ and of $\mathbb{C}$. Using 3.3 we conclude that the involution $\dagger_{\mathbb{R}}$ is a positive definite $\mathbb{R}$-linear involution on this product. We see that this implies that $\dagger_{\mathbb{R}}$ is the identity on every real factor, stabilizes every complex factor, and is the complex conjugation on those factors. Conclusion:

$$\forall x \in L, \quad \forall \, \psi : L \to \mathbb{C} \quad \Rightarrow \quad \psi(x^\dagger) = \overline{\psi(x)}.$$

Hence

$$q = \psi(q) = \psi\left(\pi_A \cdot (\pi_A)^\dagger \right) = \psi(\pi_A) \cdot \overline{\psi(\pi_A)}.$$

Hence

$$| \psi(\pi_A) | \;=\; \sqrt{q}.$$

$$\square 3.2$$

3.6. Definition/Notation. Let A be a *simple* abelian variety over $K = \mathbb{F}_q$. We have seen that $\pi_A \in \text{End}(A) =: D$. As A is simple, D is a division algebra, and $\mathbb{Q}(\pi_A) \subset D$ is a number field (a finite extension of $\mathbb{Q}$). We have seen that π_A is a Weil q-number. *We will say that π_A is the Weil q-number attached to the simple abelian variety A.*

3.7. Simple and absolutely simple. We give an example of an abelian variety A over a field which is K-simple, such that for some extension $K' \supset K$ the abelian variety $A \otimes K'$ is not simple, i.e. A is not absolutely simple.

Choose $q = p^n$. Let $i \in \mathbb{Z}_{>0}$, and let $\zeta = \zeta_i$ be a primitive i-th root of unity. Define $\pi = \zeta \cdot \sqrt{q}$. Clearly π is a Weil q-number. Using Th. 1.2, we know there exists an abelian variety A over K, which is simple such that $\pi_A \sim \pi$. Assume $i > 2$; note that for any N which is a multiple of $2i$ we have $\mathbb{Q} = \mathbb{Q}(\pi^N) \subsetneq \mathbb{Q}(\pi)$. We will see: in this case $g := \dim(A) > 1$, and $A \otimes \mathbb{F}_{q^N} \sim (E \otimes \mathbb{F}_{q^N})^g$ where E is a supersingular curve defined over $\mathbb{F}_p$. Hence in this case A is not K-simple.

3.8. Remark/Definition. We say that an abelian variety A over a field K is *isotypic* if there exists an abelian variety B simple over K and an isogeny $A \sim B^\mu$ for some $\mu \in \mathbb{Z}_{>0}$; in this case we will define $\pi_A := \pi_B$; note that $f_A - (f_B)^\mu$ (for the definition of f_A see 16.8).

We have just seen that the property " A is simple " can get lost under a field extension. However

if A is isotypic over K and $\mathbb{F}_q = K \subset K'$ is an extension then $A \otimes K'$ is isotypic;

see 10.8.

Moreover, if K is a finite field and $[K' : K] = N$ then $(\pi_A)^N = \pi_{A \otimes K'}$,

i.e. the formation $A \mapsto \pi_A$ commutes under base extension with exponentiation as explained.

4. Abelian varieties with CM

4.1. smCM We say that an abelian variety X over a field K *admits sufficiently many complex multiplications over K*, abbreviated by "smCM over K", if $\operatorname{End}^0(X)$ contains a commutative semi-simple subalgebra of rank $2 \cdot \dim(X)$ over $\mathbb{Q}$.

Equivalently: for every simple abelian variety Y over K which admits a non-zero homomorphism to X the algebra $\operatorname{End}^0(Y)$ contains a field of degree $2 \cdot \dim(Y)$ over $\mathbb{Q}$.

If no confusion is possible we say "A admits smCM" omitting "over K". However we should be careful; it is possible that A, defined over K, does not admit smCM, but that there exists a field extension $K \subset K'$ such that $A \otimes_K K'$ admits smCM (over K').

Equivalently. Suppose $A \sim \Pi B_i$, where each of the B_i is simple. We say that A admits smCM, if every $\operatorname{End}^0(B_i)$ contains a CM-subfield of degree $2 \cdot \dim(B_i)$ over $\mathbb{Q}$.

For other characterizations, see [18], page 63 and [44], page 347.

4.2. Note that if a simple abelian variety A of dimension g over a field *of characteristic zero* admits smCM then its endomorphism algebra $L = \operatorname{End}^0(X)$ is a field, in fact a CM-field of degree $2g$ over $\mathbb{Q}$; see 5.9. We will use he notion "CM-type" in the case of an abelian variety A over $\mathbb{C}$ which admits smCM, and where the type is given, i.e. the action of the endomorphism algebra on the tangent space $\mathbf{t}_{A,0} \cong \mathbb{C}^g$ is part of the data, see 13.1. See 13.12: we do use CM-types in characteristic zero, but we do not define (and we do not use) such a notion over fields of positive characteristic.

Note that there exist (many) abelian varieties A admitting smCM defined over a field of positive characteristic, such that $\operatorname{End}^0(A)$ is not a field.

We could use the terminology "A has complex multiplication" to denote the cases with $\mathrm{End}(A) \supsetneq \mathbb{Z}$. However this could be misleading, and in these notes we will not use this terminology.

It can be proved that if a simple abelian variety A admits smCM in the sense defined above, then $D = \mathrm{End}^0(A)$ contains a CM-field of degree $2 \cdot \dim(A)$ over $\mathbb{Q}$. Note that a field E with $E \subset \mathrm{End}^0(A)$ and $[E : \mathbb{Q}] = 2 \cdot \dim(A)$ however need not be a CM-field; see 15.7.

Terminology. Let $\varphi \in \mathrm{End}^0(A)$. Then $d\varphi$ is a K-linear endomorphism of the tangent space of A at $0 \in A$. See 16.9. If the base field is $K = \mathbb{C}$, this is just multiplication by a complex matrix x. Suppose $A(\mathbb{C}) \cong \mathbb{C}^g / \Lambda$ where Λ is a lattice in $\mathbb{C}^g$. For $\varphi \in \mathrm{End}^0(A)$ the linear map $d\varphi$ leaves $\Lambda \subset \mathbb{C}^g$ invariant. Conversely any complex linear map complex linear map $x : \mathbb{C}^g \to \mathbb{C}^g$ leaving invariant Λ defines an endomorphism φ of A with $x = d\varphi$.

Consider $g = 1$, i.e. A is an elliptic curve and $\varphi \in \mathrm{End}(A)$. If $\varphi \notin \mathbb{Z}$ then $x \in \mathbb{C}$ and $x \notin \mathbb{R}$. Therefore an endomorphism of an elliptic curve over $\mathbb{C}$ which is not in $\mathbb{Z}$ can be called "a complex multiplication". Later this terminology was extended to all abelian varieties.

Warning. Sometimes the terminology "an abelian variety with CM" is used, when one wants to say "admitting smCM"; we will not adopt that confusing terminology. An elliptic curve E has $\mathrm{End}(E) \supsetneq \mathbb{Z}$ if and only if it admits smCM. However it is easy to give an abelian variety A which "admits CM", meaning that $\mathrm{End}(A) \supsetneq \mathbb{Z}$, such that A does not admit smCM. However we will use the terminology "a CM-abelian variety" for an abelian variety which admits smCM.

It can happen that an abelian variety A over a field K does not admit smCM, and that $A \otimes K'$ does admit smCM.

4.3. Exercise. *Show that for any elliptic curve E defined over $\mathbb{Q}$ we have* $\mathrm{End}(E) = \mathbb{Z}$.

Show there exists an abelian surface A over $\mathbb{Q}$ with $\mathbb{Z} \neq \mathrm{End}(A) = \mathrm{End}(A \otimes \overline{\mathbb{Q}})$.

Show there exists an abelian variety A over a field k such that $\mathbb{Z} \subsetneq \mathrm{End}(A)$ and such that A does not admit smCM.
See 15.6, 18.10.

4.4. Remark. An abelian variety over a field of characteristic zero which admits smCM is defined over a number field. See [69], Proposition 26 on page 109. Also see [51].

We will see that a theorem of Tate, see Theorem 5.4 implies that *an abelian variety defined over a finite field does admit* smCM. By Grothendieck we know that an abelian variety which admits smCM up to isogeny is defined over a finite field, see 4.5.

4.5. Remark. The converse of Tate's result 5.4 (2) is almost true; see 5.7.

It is easy to give an example of an abelian variety, over a field of characteristic p, with smCM which is not defined over a finite field. E. g. see 5.8.

4.6. Lemma. *Let K be a field, and let A be an abelian variety simple over K which admits* smCM. *Choose a* CM-*field P with $[P : \mathbb{Q}] = 2 \cdot \dim(A)$ inside* $\mathrm{End}^0(A)$. (This is possible by Lemma 10.1.) *Then there exists a K-isogeny $A \sim_K B$ such that $\mathcal{O}_P \hookrightarrow \mathrm{End}(B)$, where $\mathcal{O}_P$ is the ring of integers of P.* $\square$See [80], page 308.

In [80] we also find: if A in positive characteristic admits smCM by a CM-field L, and the ring of integers $\mathcal{O}_L$ is contained in $\mathrm{End}(A)$ then A can be defined over a finite field, see [80], Th. 1.3. This gives a new proof of Theorem 4.5, see [80], Th. 1.4.

4.7. Definition CM-type. Let P be a CM-field of degree $2g$. Let C be an algebraically closed field *of characteristic zero*. The set $\mathrm{Hom}(P, C)$ has $2g$ elements. For any $\varphi : P \to C$ the homomorphism $\varphi \cdot \rho$ is different from φ. A subset $\Phi \subset \mathrm{Hom}(P, C)$ is called a CM-type for P if $\mathrm{Hom}(P, C) = \Phi \coprod \rho(\Phi)$. Equivalently: For every $\varphi : P \to C$ either $\varphi \in \Phi$ or $\varphi \cdot \rho \in \Phi$.

4.8. Let A be an abelian variety simple over $\mathbb{C}$ which admits smCM. Let $P = \mathrm{End}^0(A)$. This is a CM-field of degree $2 \cdot \dim(A)$. The action of P on the tangent space $\mathbf{t}_{A,0}$ splits as a direct sum of one-dimensional representations (as P is commutative and $\mathbb{C}$ is algebraically closed of characteristic zero). Hence this representation is given by $\Phi = \{\varphi_1, \cdots, \varphi_g\}$. One shows this is a CM-type (i.e. these homomorphisms $\varphi_i : P \to C$ are mutually different and either $\varphi \in \Phi$ or $\varphi \cdot \rho \in \Phi$). For the converse construction see 19.6.

5. Tate: The structure of $\mathrm{End}^0(A)$: abelian varieties over finite fields.

Main references: [72], [73]. Also see the second printing of [47], especially Appendix 1 by C. P. Ramanujam.

5.1. For a simple abelian variety over a field K the algebra $\mathrm{End}^0(A)$ is a division algebra. By the classification of Albert, see 18.2, we know the structure theorem of such algebras 18.4. Moreover, as Albert, Shimura and Gerritzen showed, for any algebra D in the list by Albert, and for any characteristic, there is an abelian variety having D as endomorphism algebra. However over a finite field not all types do appear, there are restrictions; see 2.2, 15.9.

For an element $\beta \in \overline{\mathbb{Q}}$ we write $\mathrm{Irr}_\mathbb{Q}(\beta) = \mathrm{Irr}(\beta) \in \mathbb{Q}[T]$ for the irreducible, monic polynomial having β as zero, the *minimum polynomial* of β.

5.2. Tate described properties of the endomorphism algebra of a simple abelian variety over $K = \mathbb{F}_q$, with $q = p^n$. We write π_A for the geometric Frobenius of A, and $f_A = f_{A,\pi_A}$ for the characteristic polynomial of π_A. We write Write $\mathrm{Irr}_\mathbb{Q}(\pi_A) = \mathrm{Irr}(\pi_A) \in \mathbb{Z}[T]$ for the minimum polynomial of π_A over $\mathbb{Q}$. For the definition of a characteristic polynomial of an endomorphism, see 16.8.

The following theorems are due to Tate; these results (and much more) can be found: [72], Theorem 1 on page 139, [72], Theorem 2 on page 140 and [73], Th. 1 on page 96, [47], Appendix 1.

5.3. Theorem (Tate). *Let A be an abelian variety over the finite field $K = \mathbb{F}_q$. The characteristic polynomial $f_{A,\pi_A} = f_A \in \mathbb{Z}[T]$ of $\pi_A \in \mathrm{End}(A)$ is of degree $2 \cdot \dim(A)$, the constant term equals $q^{\dim(A)}$ and $f_A(\pi_A) = 0$.*

If an abelian variety A is K-simple then f_A is a power of the minimum polynomial $\mathrm{Irr}(\pi_A) \in \mathbb{Z}[T]$.

Let A and B be abelian variety over $K = \mathbb{F}_q$. Then:

A is K-isogenous to an abelian subvariety of B iff f_A divides f_B.

In particular

$$A \sim_K B \quad \Longleftrightarrow \quad f_A = f_B.$$

Remark. Note that for an abelian variety A over a finite field the characteristic polynomial f_A of $\pi_A \in \mathrm{End}(A)$ is a power of an irreducible polynomial then A is isotypic (not necessarily simple); it seems that a statement in [74] in Th. 1.1 of "The theorem of Honda and Tate" needs a small correction on this point.

For an abelian variety A over a field the endomorphism algebra $\mathrm{End}^0(A)$ is a semi-simple ring. If moreover A is K-simple, then $D = \mathrm{End}^0(A)$ is a division ring (hence a simple ring).

5.4. Theorem (Tate). *Suppose A is a simple abelian variety over the finite field $K = \mathbb{F}_q$.*
(1) *The center of $D := \mathrm{End}^0(A)$ equals $L := \mathbb{Q}(\pi_A)$.*
(2) *Moreover*

$$2g \quad = \quad [L : \mathbb{Q}] \cdot \sqrt{[D : L]},$$

where g is the dimension of A. Hence: every abelian variety over a finite field admits smCM. See 4.1. We have:

$$f_A \quad = \quad (\mathrm{Irr}(\pi_A))^{\sqrt{[D:L]}}.$$

(3)

$$\mathbb{Q} \quad \subset \quad L := \mathbb{Q}(\pi_A) \quad \subset \quad D = \mathrm{End}^0(A).$$

The central simple algebra D/L

- *does not split at every real place of L,*
- *does split at every finite place not above p.*

- *For a discrete valuation w of L with $w \mid p$ the invariant of D/L is given by*

$$\operatorname{inv}_w(D/L) = \frac{w(\pi_A)}{w(q)} \cdot [L_w : \mathbb{Q}_p] \quad \operatorname{mod} \ \mathbb{Z},$$

where L_w is the local field obtained from L by completing at w. Moreover

$$\operatorname{inv}_w(D/L) + \operatorname{inv}_{\overline{w}}(D/L) = 0 \quad \operatorname{mod} \ \mathbb{Z},$$

where $\overline{w} = \rho(w)$ is the complex conjugate of w.

5.5. Corollary/Notation. Using Brauer theory, see Section 17, and using this theorem by Tate we see that the structure of D follows once $\pi = \pi_A$ is given. In particular the dimension g of A follows from π. *We will say that D is the algebra determined by the Weil number π.*

For a given Weil q-number the division algebra with invariants as described by the theorem will be denoted by $D = \mathcal{D}(\pi)$. We write $e(\pi) = [\mathbb{Q}(\pi) : \mathbb{Q}]$, and $r(\pi)^2 = [\mathcal{D}(\pi) : \mathbb{Q}(\pi)]$ and $g(\pi) = e(\pi) \cdot r(\pi)/2$.

Note that $g(\pi) \in \mathbb{Z}$. Indeed, in case $(\mathbb{R}e)$ we have $e = 1, r = 2$. In all other cases we have that e is even. See 15.9.

5.6. Corollary. *Let A be an abelian variety over a finite field. Then A admits* smCM.

It suffices to show this in case A is simple. A splitting field of the central simple algebra $\mathbb{Q}(\pi_A) = L \subset D = \operatorname{End}^0(A)$ is a field of degree $2g$, where $g = \dim(A)$. $\square$

Note that this splitting field in general need not be, but can can be chosen to be a CM-field, see 10.1.

The converse of this corollary is almost true.

5.7. Theorem (Grothendieck). *Let K be a field with prime field $\mathbb{P}$. Let A be an abelian variety over K which admits* smCM *(over K). Write $k = \overline{K}$. There exists an isogeny $B \sim A \otimes_K k$ such that B is defined over a finite extension of $\mathbb{P}$.*

See [51], [80], Th. 1.4. Note that if $\operatorname{char}(K) = 0$ any abelian variety with smCM is defined over a finite extension of $\mathbb{P} = \mathbb{Q}$, i.e. over a number field, see [69], Prop. 26 on page 109. However in positive characteristic there are examples where this is not the case.

5.8. An easy example. *There exists a non-finite field K, and an abelian variety A over K which admits* smCM, *such that A cannot be defined over a finite field. In this case there does not exist a CM-lift of A to characterisitc zero.*

Indeed, choose any abelian variety B over a finite field K' such that $(\alpha_p \times \alpha_p) = N \subset B$. One can take for B the product of two supersingular elliptic curves. More generally one can take any abelian variety C over $\mathbb{F} = \overline{\mathbb{F}_p}$ with $f(C) \leq g - 2$; there exists a finite field K' and an abelian variety B/K' having the property required above such that $B_{\mathbb{F}}$ is in the $\mathbb{F}$-isogeny class of C. Choose $K = K'(t)$. Let $(1,t) : \alpha_p \to N_K$. Define $A = B_K/(1,t)(\alpha_p)$. Show that A cannot be defined over a finite field. Observe that B admits smCM by [72], see [73], Th. 1 (2);

hence A admits smCM. A CM-lifting of A is defined over a number field, by [69], Prop. 26 on page 109; this would show that A can be defined over a finite field, a contradiction.

We will see that the idea of the example above is the basis for a proof of Th. 12.4.

5.9. Remark/Exercise. Let A be an abelian variety of dimension g simple over a field K. Write $D = \mathrm{End}^0(A)$.
(1) If $\mathrm{char}(K) = 0$ and A admits smCM then D is a field.
(2) If K is finite and the p-rank $f = f(A)$ satisfies $f \geq g - 1$, "A is ordinary or A is almost ordinary", then D is commutative; e.g. see [54], Proposition 3.14.
(3) There are many examples where K is finite, $f(A) < g - 1$, and D is not commutative.
(4) There are many examples of a simple abelian variety over a field k, with either $\mathrm{char}(k) = 0$ or $\mathrm{char}(k) = p$ and A ordinary such that D is not commutative; see 18.4

5.10. Lemma. *Let M be a field, and $\pi \in M^{\mathrm{sep}}$ be a separable algebraic element over M. Let $N \in \mathbb{Z}_{>0}$. Let M'/M be the Galois closure over $M(\pi)/M$. Let $\{\gamma_1, \cdots, \gamma_e\}$ be the set of conjugates of π in M'. Then:*

$$M(\pi^N) \subsetneq M(\pi) \quad \Longleftrightarrow \quad \exists z,\ i,\ j:\ 1 \neq z \in M',\quad 1 \leq i < j \leq e,\quad z^N = 1,$$
$$\gamma_j/\gamma_i = z.$$

I thank Yuri Zarhin for drawing my attention to this fact.
Proof. Note that $\#(\{\gamma_1, \cdots, \gamma_e\}) = [M(\pi) : M]$. As $[M(\pi^N) : M]$ equals the number of mutually different elements in $\{\gamma_1^N, \cdots, \gamma_e^N\}$ the result follows. $\qquad\square$

5.11. Proposition. *Let A be an abelian variety simple over a finite field K. Let $N \in \mathbb{Z}_{>0}$ and $[K' : K] = N$. Then*

$$\mathrm{End}(A) \subsetneq \mathrm{End}(A \otimes K') \quad \Longleftrightarrow \quad M(\pi^N) \subsetneq M(\pi).$$

Note that the last condition is described in the previous lemma.
Proof. Note that $\mathrm{End}(A \otimes K')/\mathrm{End}(A)$ is torsion free. Hence $\mathrm{End}(A) \subsetneq \mathrm{End}(A \otimes K')$ iff $\mathrm{End}^0(A) \subsetneq \mathrm{End}^0(A \otimes K')$. Hence this proposition is a corollary of 5.4. $\square$

5.12. Remark. there are two "reasons" (or a combination of both) explaining $\mathrm{End}(A) \subsetneq \mathrm{End}(A \otimes K')$.

It can happen that (although A is k-simple) $A \otimes K'$ is not K'-simple.

It can happen that $A \otimes K'$ is K'-simple but that under $K \subset K'$ the endomorphism ring gets bigger.
Both cases do appear. For some examples see 15.15, 15.16, 15.19.

6. Injectivity

6.1. Exercise/Construction. Let K be a field, and let A and B be abelian varieties over K. Assume there exists an isogeny $\varphi : A \to B$. Choose an integer $N > 0$ which annihilates (the finite group scheme which is) $\mathrm{Ker}(\varphi)$. Show there exists an isogeny $\psi : B \to A$ such that $\psi \cdot \varphi = N \cdot 1_A$. Construct

$$\Phi : \quad \mathrm{End}^0(A) \quad \longrightarrow \quad \mathrm{End}^0(B), \quad \Phi(x) := \frac{1}{N} \cdot \varphi \cdot x \cdot \psi.$$

(1) *Show that Φ is a homomorphism. Construct Ψ by $\Psi(y) = \psi \cdot y \cdot \varphi / N$. Show $\Psi \cdot \Phi = Id$ and $\Phi \cdot \Psi = Id$. Conclude that*

$$\Phi : \quad \mathrm{End}^0(A) \quad \xrightarrow{\sim} \quad \mathrm{End}^0(B)$$

is an isomorphism.
(2) *Show that Φ is independent of the choice of ψ and N.*
(3) *Show that $\varphi \cdot \psi = N \cdot 1_B$.*
Remark. Take $A = B$, and an isogeny $\varphi \in \mathrm{End}(A)$. We have constructed the inverse φ^{-1} in $\mathrm{End}^0(A)$.

6.2. Exercise. Let $A \sim B$ be a K-isogeny of simple abelian varieties over a finite field $K = \mathbb{F}_q$; using the construction 6.1 this isogeny gives an isomorphism $\mathbb{Q}(\pi_A) \cong \mathbb{Q}(\pi_B)$. *Show that this maps π_A tot π_B.*

6.3. By Theorem 3.2 by Weil we see that for a simple abelian variety A over $K = \mathbb{F}_q$ indeed π_A is a Weil q-number. If A and B are K-isogenous, π_A and π_B are conjugated. Hence

$$\boxed{\mathcal{W} : \{\text{simple abelian variety over } K\}/\sim_K \quad \longrightarrow \quad W(q), \qquad A \mapsto \pi_A,}$$

is well-defined.

We have seen in 5.3 (2) that Tate showed that A and B are K-isogenous if and only if $f_A = f_B$. Hence this map $\mathcal{W}$ is *injective*.

7. Abelian varieties with good reduction

References: [48], [12], [67], [63], [6], [53], [13].
This section mostly contains references to known (non-trivial) results.

7.1. Definition. Let A be an abelian variety over a field K. Let v be a discrete valuation of K. We say that A has *good reduction* at v if there exists an abelian scheme $\mathcal{A} \to \mathrm{Spec}(\mathcal{O}_v)$ with generic fiber $\mathcal{A} \otimes K \simeq A$.

We say that A has *potentially good reduction* at v if there exist a finite extension $K \subset K'$, a discrete valuation v' over v such that $A' := A \otimes K'$ has good reduction at v'.

7.2. The Néron minimal model. Let Let A be an abelian variety over a field K. Let v be a discrete valuation of K. Consider the category of smooth morphisms $Y \to \mathrm{Spec}(\mathcal{O}_v) = S$ and the contravariant functor on this category given by

$$Y/S \quad \mapsto \quad \mathrm{Hom}_K(Y \times_S \mathrm{Spec}(K), A).$$

We say that $\mathcal{A} \to S$ is the *Néron minimal model*, abbreviation: Nmm, of A at v if it represents this functor.

7.3. Theorem (Néron). *For every A/K and every v the Néron minimal model of A at v exists.* $\qquad\square$
See [48]; see [15], Section VIII.

7.4. Theorem (Chevalley). *Let G be a group variety over a perfect field m.* (That is: this is an algebraic group scheme $G \to \mathrm{Spec}(m)$ which is connected, and geometrically reduced.) *There exists a filtration by subgroup varieties $G_1 \subset G_2 \subset G$ over m such that G_1 is a torus* (i.e. $G_1 \otimes \overline{m}$ is isomorphic with a product of copies of $\mathbb{G}_m$), *G_2/G_1 is affine, unipotent and G/G_2 is an abelian variety.*
See [12]; see [13], Th. 1.1 on page 3. $\qquad\square$

7.5. Definition. Let A be an abelian variety over a field K. Let v be a discrete valuation of K. Let $A^0_{k_v}$ be the connected component containing 0 of the special fiber of the Néron minimal model $\mathcal{A}$. We say that A has *stable reduction* at v if in the Chevalley decomposition of $A^0_{k_v}$ the unipotent part is equal to zero. We say A has *potentially stable reduction* at $v \in \Sigma_K$ if there exist a finite extension $K \subset K'$, a discrete valuation v' over v such that $A' := A \otimes K'$ has stable reduction at v'.

7.6. We refer to the literature, especially to [63], for the notions of ℓ-adic representations, algebraic monodromy, and the fact that for an abelian variety the ℓ-adic monodromy at a discrete valuation of the base field is quasi-unipotent.

As a corollaries of these ideas on can prove:

7.7. Theorem (The Néron-Ogg-Shafarevich criterion). *Suppose A has stable, respectively good reduction at v and $B \sim_K A$. Then B has stable, respectively good reduction at v.* $\qquad\square$

7.8. Theorem (Grothendieck). *Every A/K has potentially stable reduction at every $v \in \Sigma_K$.* $\qquad\square$

7.9. Corollary. *Let A be an abelian variety over a field K which admits smCM. At every $v \in \Sigma_K$ the abelian variety A has potentially good reduction.*
Sketch of a proof. After extending of the base field and choosing v again we can assume that A has stable reduction at v, where the residue class field of v is perfect. Up to isogeny we can write $A \sim \prod B_i$, with every B_i simple. By the Néron-Ogg-Shafarevich criterion we conclude every B_i has stable reduction. Hence it suffice to show: if A is K-simple + has stable reduction at v + admits smCM then A has good reduction at v.

Let $\mathcal{A}$ be its Nmm, and let $G = A^0_{k_v}$ be the connected component of the special fiber of $\mathcal{A} \to \mathrm{Spec}(\mathcal{O}_v)$. By properties of the Nmm we conclude that $\mathrm{End}^0(A) \subset$

End(G). Consider the Chevalley decomposition in this case $G_1 = G_2 \subset G$. Let μ be the dimension of G_1. We obtain homomorphisms

$$\mathrm{End}^0(A) \to \mathrm{End}(G_1), \qquad \mathrm{End}^0(A) \to \mathrm{End}(G/G_1).$$

If $\mu = \dim(G_1) > 0$ it follows that $\mathrm{End}^0(A) \to \mathrm{End}(G_1) \subset \mathrm{Mat}(\mu, \mathbb{Z})$; it follows that this homomorphism is injective; given the fact that A admits smCM we derive a contradiction. Hence $\mu = 0$. Alternative argument: if $\mu > 0$, the dimension of $B = G/G_1$ is strictly smaller than $\dim(A)$ and the fact that A has smCM shows there does not exist a homomorphism $\mathrm{End}^0(A) \to \mathrm{End}^0(B)$. This contradiction shows $\mu = 0$, and hence A admits good reduction at v. $\qquad\square$

7.10. Remark. Also see 15.10. Let R be a normal integral domain, $\mathcal{A} \to S = \mathrm{Spec}(R)$ an abelian scheme, and $R \to K$ a homomorphism to a field K. Write $A_K = \mathcal{A} \otimes_R K$. We obtain a homomorphism

$$\mathrm{End}(\mathcal{A}) \quad \longrightarrow \quad \mathrm{End}(A_K).$$

This homomorphism is injective.

In general this homomorphism is *not surjective*.

If R is normal and K is the field of fractions of R the homomorphism is surjective (hence bijective).

If ℓ is a prime not equal to the characteristic of K, the additive factor group $\mathrm{End}(A_K)/\mathrm{End}(\mathcal{A})$ has no ℓ-torsion.

There are many examples where $R \to R/I = K$ gives a factor group $\mathrm{End}(A_K)/\mathrm{End}(\mathcal{A})$ which does have p-torsion, where $p = \mathrm{char}(K)$.

8. p-divisible groups

Also see Section 20.

8.1. For an abelian variety A over a base S and a prime number ℓ *which is invertible in the structure sheaf* on S one defines the ℓ-Tate module $T_\ell(A) := \lim_{\leftarrow i} A[\ell^i]$. This is a pro-group scheme. It can also be viewed as a local system with fiber $\mathbb{Z}_\ell$ under the fundamental group of S.

For an arbitrary prime number (not necessarily invertible on the base) we choose another strategy:

8.2. Definition. Let S be a scheme. Let $h \in \mathbb{Z}_{\geq 0}$. A p-divisible group, of height h, over S is an inductive system of finite flat group schemes $G_i \to S$, $i \in \mathbb{Z}_{\geq 0}$, such that:

- the rank of $G_i \to S$ equals $p^{h \cdot i}$;
- p^i annihilates G_i;
- there are inclusions $G_i \hookrightarrow G_{i+1}$ such that
- $G_{i+1}[p^i] = G_i$.
- Consequently $G_{i+j}/G_i = G_j$.

We will write $G = \text{colim}_{i \to} G_i$; this limit considered in the category of inductive systems of finite group schemes. A p-divisible group is also called a Barsotti-Tate group.

Examples. (1) For any abelian scheme $A \to S$ (over any base), and any integer $n \in \mathbb{Z}_{>0}$ the group scheme $A[n] \to S$ is a finite flat group scheme of rank n^{2g} over S, where $g = \dim(A)$; see [47], proposition on page 64, see [15], V, Theorem 8.2 on page 115. Hence

$$\{A[p^i] \mid i \in \mathbb{Z}_{\geq 0}\}$$

is a p-divisible group of height $2g$. This will be denoted by $A[p^\infty]$. This notation should be understood symbolically: there is no morphism "$\times\infty$" and hence, strictly speaking, no "kernel" $A[p^\infty]$.

(2) Consider $\mathbb{G}_{m,S} \to S$, the multiplicative group over any base scheme S. Then

$$\mathbb{G}_{m,S}[p^i] =: G_i = \mu_{p^i,S}, \quad \text{and this defines} \quad \mathbb{G}_{m,S}[p^\infty] \to S,$$

a p-divisible group over S of height one.

(3) Consider $\mathbb{Q}_p/\mathbb{Z}_p$, which is a profinite group, which can be given by $\text{colim}_{i \to}(\mathbb{Z}/p^i)$. By considering the constant group schemes $\underline{(\mathbb{Z}/p^i)}_S$ we obtain a p-divisible group $\underline{(\mathbb{Q}_p/\mathbb{Z}_p)}_S$.

8.3. The Serre dual of a p-divisible group. Let $G = \{G_i \mid i \in \mathbb{Z}_{\geq 0}\}/S$ be a p-divisible group over some base scheme S. The surjections $G_{i+1} \twoheadrightarrow G_{i+1}/G_1 = G_i$ define by Cartier duality inclusions $(G_i)^D \hookrightarrow (G_{i+1})^D$; see 16.5. This defines a p-divisible group

$$G^t := \{(G_i)^D \mid i\} \to S,$$

which is called the Serre dual of $G \to S$.

Note that $G \mapsto G^t$ is a duality for p-divisible groups, which is is defined by purely algebraic methods. We see a duality $A \mapsto A^t$ for abelian schemes, see 16.2, which is a (non-trivial) geometric theory. Notation is chosen in this way, because the duality theorem connects these two operation in a natural way: $A^t[p^\infty] = A[p^\infty]^t$, see 16.6; note that this fact is more involved than this simple notation suggests.

8.4. Exercise. Show that $(\mathbb{G}_{m,S}[p^\infty])^t = \underline{\mathbb{Q}_p/\mathbb{Z}_p}_S$.

9. Newton polygons

For a p-divisible group X or an abelian variety A over a field of characteristic p the Newton polygon $\zeta = \mathcal{N}(X)$, respectively $\xi = \mathcal{N}(A) := \mathcal{N}(A[p^\infty])$ is defined, see Section 21. In this section we give an easier definition in case we work with an abelian variety over a finite field, and we show that this coincides with the more general definition as recorded in Section 21.

9.1. Notation. Let $K = \mathbb{F}_q$ be a finite field, $q = p^n$ and let A be an abelian variety over K of dimension g. We have defined the geometric Frobenius $\pi = \pi_A \in \operatorname{End}(A)$; this endomorphism has a characteristic polynomial $f_A \in \mathbb{Z}[T]$, see 16.8; this is a monic polynomial of degree $2g$.

Suppose that A is simple. The algebraic integer π_A is a zero of its minimum polynomial $\operatorname{Irr}(\pi) \in \mathbb{Z}[T]$; this is a monic polynomial, and its degree equals $e = [\mathbb{Q}(\pi) : \mathbb{Q}]$. In this case $f_A = (\operatorname{Irr}(\pi))^r$, where r^2 is the degree of $D = \operatorname{End}^0(A)$ over its centre $L = \mathbb{Q}(\pi)$.

Suppose $f_A = \sum_j b_j T^{2g-j}$. We define $\xi = \xi(A)$ as a *lower convex hull*, written as lch():

$$\xi(A) = \operatorname{lch}\left(\{(j, v_p(b_j)/n) \mid o \leq j \leq 2g\}\right).$$

This is the Newton polygon of f_A compressed by the factor n. Note that if A is simple with $\operatorname{Irr}(\pi_A) = \sum_i c_i T^{e-i}$ then $\xi(A) = \operatorname{lch}(\{(r\cdot i, r\cdot v_p(c_i)/n) \mid 0 \leq i \leq e\})$.

9.2. Theorem. *Let A be an abelian variety isotypic over a finite field $K = \mathbb{F}_q$, with $q = p^n$. As above we write $\pi = \pi_A$, the geometric Frobenius of A, and $L = \mathbb{Q}(\pi)$ with $[L : \mathbb{Q}] = e$ and $D = \operatorname{End}^0(A)$ with $[D : L] = r^2$ and $\dim(A) = g = er/2$. Let $X = A[p^\infty]$. Consider the set $\Sigma_L^{(p)}$ of discrete valuations of L dividing the rational prime number p. Let $L \subset P \subset D$, where P is a CM-field of degree $2g$ (existence assured by 10.1. If necessary we replace A be a K-isogenous abelian variety (again called A) such that $\mathcal{O}_P \subset \operatorname{End}(A)$, see 4.6. Then also $\mathcal{O}_L \subset \operatorname{End}(A)$.*
(1) *The decomposition*

$$D \otimes \mathbb{Q}_p = \prod_{w \in \Sigma_L^{(p)}} D_w, \qquad \mathcal{O}_L = \prod \mathcal{O}_{L_w},$$

gives a decomposition $X = \prod_w X_w$.
(2) *The height of X_w equals $[L_w : \mathbb{Q}_p]\cdot r$.*
(3) *The p-divisible group X_w is isoclinic of slope γ_w equal to $w(\pi_A)/w(q)$; note that $q = p^n$.*
(4) *Let $\overline{w}$ be the discrete valuation of L obtained from w by complex conjugation on the CM-field L; then $\gamma_w + \gamma_{\overline{w}} = 1$.*
See [77]. We will give a proof of one of the details.
Proof. (3) Fix $w \in \Sigma_L^{(p)}$, and write $Y = X_w$. Write $w(\pi_A)/n = d/h$ with $\gcd(d, h) = 1$. The kernel of

$$Y \xrightarrow{\ F\ } Y^{(p)} \xrightarrow{\ F\ } \cdots \xrightarrow{\ F\ } Y^{(p^{nh})}$$

will be denoted by $Y[F^{nh}]$
Claim. $Y[F^{nh}] = Y[p^{nd}]$.
The action of π on Y is given by F^n. We see that $w(F^{nh}/p^{nd}) = 0$. This proves that this quotient (in $\mathcal{O}_L$) acts by a unit on Y, which proves the claim. $\square$

By the Dieudonné-Manin theory we know that $Y \otimes \mathbb{F} \sim \prod G_{d_i, c_i} \otimes \mathbb{F}$. We know that $G_{d_i, c_i}[F^{c_i + d_i}] = G_{d_i, c_i}[p^{d_i}]$. By the claim this proves that in this decomposition only factors $(d_i, c_i) = (d, h - d)$ do appear, see 21.22. This proves proves that Y is isoclinic of slope equal to d/h. $\qquad\qquad\square(3)$

9.3. Corollary. *The polygon $\xi(A)$ constructed in* 9.1 *for an abelian variety A over a finite field equals the Newton polygon $\mathcal{N}(A)$, as defined in Section 21.*

9.4. Remark. Let A be an abelian variety over a finite field K. By the Dieudonné-Manin theory we know that $A[p^\infty] = X$ has the property that there exists a p-divisible group Y over $\mathbb{F}_p$ such that $X \otimes_K \mathbb{F} \sim Y \otimes_{\mathbb{F}_p} \mathbb{F}$. Hence $\xi(A) = \mathcal{N}(A) = \mathcal{N}(Y)$ as we have seen above. We could try to prove the corollary above by comparing the minimum polynomial of π_A and the same of Y over some common finite field. However in general one cannot compute f_A from the characteristic polynomial of $Y/\mathbb{F}_p$, as is shown by examples below.

9.5. (1) Let E be a supersingular elliptic curve over a finite field $K = \mathbb{F}_q$; see 21.8. We will see, 14.6, that there exists a root of unity ζ_i such that $\pi_E \sim \zeta_i \sqrt{q}$. Hence $\pi' := \pi_{E \otimes K'} = q^i$, with $K' = \mathbb{F}_{q'}$, where $q' = q^{2i} = p^{2ni}$. We can choose $Y/\mathbb{F}_q$ with $F_Y = \pm\sqrt{p}$ and $Y \otimes \mathbb{F} \cong E[p^\infty] \otimes \mathbb{F}$. Note the curious fact that in this case for a finite exension we have equality: $(F_Y)^{2ni} = \pi'$.
(2) Let E be an ordinary elliptic curve over a finite field $K = \mathbb{F}_q$, with $f_E \in \mathbb{Z}[T]$ the characteristic polynomial of π_E. For $Y = G_{(1,0)} + G_{(0,1)}$ we have $E[p^\infty] \otimes_K \mathbb{F} \cong Y \otimes_{\mathbb{F}_p} \mathbb{F}$. However, for every finite field $K' \supset K$ the p-divisible groups $E[p^\infty] \otimes_K K'$ and $Y \otimes_{\mathbb{F}_p} K'$ are *not isomorphic*. In this case the minimum polynomial of the geometric Frobenius of $E \otimes K'$ is different from the same of $Y \otimes K'$, although $\mathcal{N}(E) = \mathcal{N}(Y)$.

9.6. The Shimura-Taniyama formula. Suppose given an abelian variety A of CM-type (P, Φ) over a number field M having *good reduction at a discrete valuation* $v \in \Sigma_M$. Can we compute from these data the slopes of the geometric Frobenius π_0 of the reduction A_0/K_v over the residue class field of v ? The formula of Shimura and Taniyama precisely gives us this information.

Let $\mathcal{A}$ be the Nmm of A at v. We have

$$P = \operatorname{End}^0(A) = \operatorname{End}^0(\mathcal{A}) \hookrightarrow \operatorname{End}^0(A_0).$$

Let ℓ be a prime different form the characteristic of K_v. We see that $P \otimes \mathbb{Q}_\ell \subset \operatorname{End}^0(A) \otimes \mathbb{Q}_\ell$. As $P : \mathbb{Q}] = 2 \cdot \dim(A)$ it follows that $P \subset \operatorname{End}^0(A)$ is its own centralizer; hence $L := \mathbb{Q}(\pi_{A_0}) \subset P$. Moreover $\pi := \pi_{A_0}$ is integral over $\mathbb{Z}$; hence $\pi \in \mathcal{O}_P$.

Let C be an algebraically closed field containing $\mathbb{Q}_p$. We have

$$H := \operatorname{Hom}(P, C), \quad H_w = \operatorname{Hom}(P_w, C), \quad H = \coprod_{w \in \Sigma_P^{(p)}} H_w.$$

We define $\Phi_w := \Phi \cap H_w$. Write $K_v = \mathbb{F}_q$. With these notations we have:

9.7. Theorem (the Shimura-Taniyama formula).

$$\forall w \in \Sigma_P, \quad w \mid p, \quad \frac{w(\pi)}{w(q)} = \frac{\#(\Phi_w)}{\#(H_w)}.$$

See [69], §13; see [40], Corollary 2.3.

Tate gave a proof based on "CM-theory for p-divisible groups". See [73],Lemma 5; see [74], Shimura-Taniyama formula by B. Conrad, Theorem 2.1. $\qquad\square$

See 13.12 for a further discussion.

10. Surjectivity

In this section we prove surjectivity of the map $\mathcal{W} : \mathcal{M}(K, s) \to W(q)$, hence finishing a proof for Theorem 1.2. We indicate the structure of the proof by subdividing it into the various steps.

Step (1) Proving $\mathcal{W}$ is surjective means showing every Weil number is effective, see 1.3. We start with a choice $q = p^n$, and with the choice of a Weil q-number π. In case $\pi \in \mathbb{R}$ we know effectivity. From now on we suppose that π is non-real.

Step (2) A Weil q-number π determines a number field $\mathbb{Q}(\pi) = L$ and a division algebra $D = \mathcal{D}(\pi)$; see 5.5. In the case considered π is non-real and L is a CM-field.

Step (3) We *choose* a CM field $P \subset D$ of degree $2g$ over $\mathbb{Q}$, which is possible by the following lemma.

10.1. Lemma. *Suppose given a CM-field L and a central division algebra $L \subset D$. There exists $L \subset P \subset D$ where P is a CM-field splitting D/L. See [73], Lemme 2 on page 100.* $\qquad\square$

See Exercise 15.7

Step (4) Given π and $L \subset P \subset D = \mathcal{D}(\pi)$ as above we will *choose* a CM-type Φ for P such that

$$\forall w \in \Sigma_L^{(p)}, \quad w \mid p, \quad \frac{w(\pi)}{w(q)} = \frac{\#(\Phi_w)}{\#(H_w)}.$$

Here $\Sigma_L^{(p)}$ is the set of finite places of L dividing p. We have a decomposition $L \otimes \mathbb{Q}_p = \prod L_w$; hence a decomposition

$$H := \mathrm{Hom}(L, \overline{\mathbb{Q}_p}) = \coprod \mathrm{Hom}(L_w, \overline{\mathbb{Q}_p});$$
$$\text{write} \quad H_w = \mathrm{Hom}(L_w, \overline{\mathbb{Q}_p}); \qquad \Phi = \coprod \Phi_w.$$

The set $\Phi \subset H$ defines the sets $\Phi_w \subset H_w$; conversely $\{\Phi_w \mid w \in \Sigma_L^{(p)}\}$ determines Φ.

Claim. *The involution $\varphi \mapsto \varphi \cdot \rho$ has no fixed points on $H := \mathrm{Hom}(L, \overline{\mathbb{Q}_p})$.*

Proof. Embeddings $\overline{\mathbb{Q}} \hookrightarrow \mathbb{C}$ and $\overline{\mathbb{Q}} \hookrightarrow \mathbb{Q}_p$ give an identification $H = \mathrm{Hom}(L, \mathbb{C})$, compatible with $-\cdot\rho$. We know that ρ on L is complex conjugation on every embedding $L \hookrightarrow \mathbb{C}$. Hence, if we would have $\varphi = \varphi\cdot\rho$ we conclude that $\varphi(L) \subset \mathbb{R}$. However a CM-field is totally complex. This contradiction shows that $-\cdot\rho$ has no fixed point on $\mathrm{Hom}(L, \mathbb{C}) = H = \mathrm{Hom}(L, \overline{\mathbb{Q}_p})$. $\qquad\square$

Construction. Notation will be chosen in relation with 9.2. For every $w \in \Sigma_L^{(p)}$ we define:

- $\beta_w = w(\pi)/w(q)$;
- $h_w = [L_w : \mathbb{Q}_p]\cdot r$, where $r = r_\pi = \sqrt{[\mathcal{D}(\pi) : \mathbb{Q}(\pi)]}$;
- $d_w := h_w\cdot\beta_w$.

Note that complex conjugation induces (for every embedding) an involution $\rho : P \to P$, which restricts to an involution $\rho : L \to L$ which is also complex conjugation on L. We see that $\rho(w) = w$ or $\rho(w) \neq w$.

If $\rho(w) = w$ we conclude that $\beta_w = 1/2$. In this case we choose for $\Phi_w \subset H_w$ any subset such that $\#(\Phi_w) = \#(H_w)/2$ and $\Phi_w \cap \Phi_w\cdot\rho = \emptyset$; this is possible as $-\cdot\rho$ has no fixed point on H.

If $\rho(w) \neq w$ we make a choice $\Phi_w \subset H_w$ such that $\#(\Phi_w) = d_w$, and we define $\Phi_{\rho(w)} = H_{\rho(w)} - \Phi_w\cdot\rho$; this ends a choice for the pair $\{w, \rho(w)\}$. This ends the construction.

Step (5)　Given the CM-type (P, Φ) as above, in particular $\Phi_w \cap \Phi_w\cdot\rho = \emptyset$ and $\#(\Phi_w) = d_w$ for every w, we construct B over M as follows.

10.2. *We choose a number field M, an abelian variety B defined over M, and $v \in \Sigma_M^{(p)}$ with residue class field $K_v := \mathcal{O}_v/m_v \supset \mathbb{F}_q$ such that $\mathrm{End}^0(B) = P$, with Φ as CM-type, and such that B has good reduction at v.*

Notation. Write $[K_v : \mathbb{F}_q] = m$; write B_v for the abelian variety defined over K_v obtained by reduction of B at v.

Proof. By 19.6 we construct an abelian variety B' over $\mathbb{C}$ of CM-type (P, Φ). By [69], Proposition 26 on page 109 we know that B'' can be defined over a number field. We can choose a finite extension so that all complex multiplications are defined over that field. By 7.9 we know that an abelian variety with smCM has potentially good reduction; hence we can choose a finite extension of the base field and achieve good reduction everywhere. We choose a discrete valuation dividing p. Conclusion: after a finite extension we can achieve that B is an abelian variety defined over a number field M, with $B \otimes_M \mathbb{C} \cong B'$, and $v \in \Sigma_M^{(p)}$ such that all properties mentioned above are satisfied.

10.3. Lemma. *Let E be a number field, i.e. $[E : \mathbb{Q}] < \infty$. A root of unity $\zeta \in E$ has the properties:*
(i) for every $\psi : E \to \mathbb{C}$ we have $|\zeta| = 1$,
(ii) for every finite prime w we have $w(\zeta) = 0$.
Conversely an element $\zeta \in E$ satisfying (i) and (ii) is a root of unity. $\qquad\square$
See [28], page 402 (page 520 in the second printing).

Step (6)   *Suppose given π, and (P, Φ) and B/M as constructed above. There exist $s \in \mathbb{Z}_{>0}$ and an s-root of unity ζ_s such that*

$$\pi^m = \zeta_s \cdot \pi_{B_v}.$$

This implies that

$$\pi^{ms} = \pi_{B_v}^s = \pi_{B_v \otimes \mathbb{F}_{q^{ms}}}.$$

Hence π^N is effective with $N := ms$.

Proof. We have $\pi \in \mathcal{O}_L \subset P$. Also we have $\pi_{B_v} \in \mathcal{O}_P$. Let $\zeta := \pi^m / \pi_{B_v}$, where $[K_v : \mathbb{F}_q] = m$. As π^m and π_{B_v} are Weil $\#(K_v)$-numbers condition (i) of the previous lemma is satisfied. For every prime not above p these numbers are units, hence condition (ii) is satisfied for primes of P not dividing p. For every $w \in \Sigma_P^{(p)}$ we can apply the Shimura-Taniyama formula, see 9.7, to π_{B_v}; for the restriction of w to L we can apply 9.2 (3) to π; these shows that $w(\zeta) = 1$ for every $w \in \Sigma_P^{(p)}$. Hence the conditions mentioned in the previous lemma are satisfied. By the lemma $\zeta \in \mathcal{O}_P$ is a root of unity, say $\zeta = \zeta_s$. Hence π^N is effective for $N := ms$. This means that $\pi^N = \pi^{ms} = \pi_{B_v \otimes \mathbb{F}_{q^{ms}}}$ is effective. $\square$

The arguments in this section up to here in fact prove the following fundamental theorem.

10.4. Theorem (Honda). *Let $K = \mathbb{F}_q$. Let A_0 be an abelian variety, defined and simple over K. Let $L \subset \mathrm{End}^0(A_0)$ be a CM-field of degree $2g$ over $\mathbb{Q}$. There exists a finite extension $K \subset K'$, an abelian variety B_0 over K' and a K'-isogeny $A_0 \otimes_K K' \sim B_0$ such that B_0/K' satisfies* (CML) *by L.* $\square$
See [29], Th. 1 on page 86, see [73], Th. 2 on page 102. For the notion (CML) see 12.2.

10.5. The Weil restriction functor. Suppose given a finite extension $K \subset K'$ of fields (we could consider much more general situations, but we will not do that); write $S = \mathrm{Spec}(K)$ and $S' = \mathrm{Spec}(K')$. We have the base change functor

$$\mathrm{Sch}_{/S} \quad \to \quad \mathrm{Sch}_{S'}, \qquad T \mapsto T_{S'} := T \times_S S'.$$

The *right adjoint functor* to the base change functor is denoted by

$$\Pi = \Pi_{S'/S} = \Pi_{K'/K} \; : \; \mathrm{Sch}_{S'} \quad \to \quad \mathrm{Sch}_{/S},$$
$$\mathrm{Mor}_S(T, \Pi_{S'/S}(Z)) \cong \mathrm{Mor}_{S'}(T_{S'}, Z).$$

In this situation, with K'/K separable, Weil showed that $\Pi_{S'/S}(Z)$ exists. In fact, consider $\times_{S'}^{[K':K]} = Z \times_{S'} \cdots \times_{S'} Z$, the self-product of $[K' : K]$ copies. It can be shown that $\times_{S'}^{[K':K]}$ can be descended to K in such a way that it solves this problem. Note that $\Pi_{S'/S}(Z) \times_S S' = \times_{S'}^{[K':K]} Z$. For a more general situation, see [25], Exp. 195, page 195-13. Also see [74], Nick Ramsey - CM seminar talk, Section 2.

10.6. Lemma. *Let B' be an abelian variety over a finite field K'. Let $K \subset K'$, with $[K' : K] = N$. Write*

$$B := \Pi_{K'/K}\, B'; \quad \text{then} \quad f_B(T) = f_{B'}(T^N).$$

$\square$

See [73], page 100.

We make a little detour. From [14], 3.19 we cite:

10.7. Theorem (Chow). *Let K'/K be an extension such that K is separably closed in K'. (For example K' is finite and purely inseparable over K.) Let A and B be abelian varieties over K. Then*

$$\mathrm{Hom}(A, B) \quad \xrightarrow{\sim} \quad \mathrm{Hom}(A \otimes K', B \otimes K')$$

is an isomorphism. In particular, if A is K-simple, then $A \otimes K'$ is K'-simple. $\square$

10.8. Claim.

For an isotypic abelian variety A over a field K, and an extension $K \subset K'$, we have that $A \otimes K'$ is isotypic.

Proof. It suffices to this this in case A is K-simple. It suffices to show this in case K'/K is finite. Moreover, by the previous result it suffices to show this in case K'/K is separable.

Let $K \subset K'$ be a separable extension, $[K' : K] = N$. Write $\Pi = \Pi_{\mathrm{Spec}(K')/\mathrm{Spec}(K)}$. For any abelian variety A over K we have $\Pi(A \otimes_K K') \cong A^N$, and for any C over K' we have $\Pi(C) \otimes_K K' \cong C^N$, as can be seen by the construction; e.g. see the original proof by Weil, or see [74], Nick Ramsey - CM seminar talk, Section 2; see 10.5. If there is an isogeny $A \otimes_K K' \sim C_1 \times C_2$, with non-zero C_1 and C_2 we have $\Pi(C_1 \times C_2) \sim A^N$. Hence we can choose positive integers e and f with $\Pi(C_1) \sim A^e$ and $\Pi(C_2) \sim A^f$. Hence

$$\Pi(C_1) \otimes K' \cong (C_1)^N \sim (A \otimes_K K')^{eN}, \quad (C_2)^N \sim (A \otimes_K K')^{fN};$$

hence $\mathrm{Hom}(C_1, C_2) \neq 0$. We conclude: if A is simple, any two isogeny factors of $A \otimes_K K'$ are isogenous. $\square$

By Step 6 and by Lemma 10.6 we conclude:

10.9. Corollary (Tate). *Let π be a Weil q-number and $N \in \mathbb{Z}_{>0}$ such that π^N is effective. Then π is effective.*
See [73], Lemme 1 on page 100. $\square$

Remark. The abelian variety B_v as constructed above is isotypic and hence π_{B_v} is well-defined. It might be that the B_v thus obtained is not simple. Moreover $A := \Pi_{K'/K}(B_v)$ is isotypic with $\pi_A \sim \pi$.

Step (7)   End of the proof. By the theorem by Honda we know that there exists $N \in \mathbb{Z}_{>0}$ such that π^N is effective. We conclude that π is effective. Hence we have proved that $\mathcal{W} : \mathcal{M}(K, s) \to W(q)$ is surjective. $\qquad\qquad$ □Theorem 1.2

Warning (again). For a K-simple abelian variety A over $K = \mathbb{F}_q$ in general it can happen that for a (finite) extension $K \subset K'$ the abelian variety $A \otimes K'$ is not K'-simple.

10.10. Exercise. *Notation and assumptions as above; in particular $K = \mathbb{F}_q$ is a finite field, $[K' : K] = N$. Write $A' = A \otimes K'$. Write $\pi' = \pi_A^N$.*
  Show that $\mathrm{End}(A) = \mathrm{End}(A')$ iff $\mathbb{Q}(\pi_A) = \mathbb{Q}(\pi')$.
  Show that $\mathbb{Q}(\pi_A) = \mathbb{Q}(\pi')$ for every $N \in \mathbb{Z}_{>0}$ implies that A is absolutely simple (i.e. $A \otimes \mathbb{F}$ is simple).
  Construct K, A, K' such that $\mathbb{Q}(\pi_A) \neq \mathbb{Q}(\pi_{A'})$ and A' is K'-simple.

11. A conjecture by Manin

We recall an important corollary from the Honda-Tate theory. This result was observed and proved independently by Honda and by Serre.

11.1. Definition. Let ξ be a Newton polygon. Suppose it consists of slopes $1 \geq \beta_1 \geq \cdots \geq \beta_h \geq 0$. We say that ξ is *symmetric* if $h = 2g$ is even, and for every $1 \leq i \leq h$ we have $\beta_i = 1 - \beta_{h+1-i}$.

11.2. Proposition. *Let A be an abelian variety in positive characteristic, and let $\xi = \mathcal{N}(A)$ be its Newton polygon. Then ξ is symmetric.*
Over a finite field this was proved by Manin, see [39], page 74; in that proof the functional equation of the zeta-function for an abelian variety over a finite field is used. The general case (an abelian variety over an arbitrary field of positive characteristic) follows from [49], Theorem 19.1; see 21.23.

11.3. Exercise. Give a **proof** of this proposition in case we work over a finite field. Suggestion: use 9.2.

Does the converse hold? I.e.:

11.4. Conjecture (Manin, see [39], Conjecture 2 on page 76).
Suppose given a prime number p and a symmetric Newton polygon ξ. Then there exists an abelian variety A over a field of characteristic p with $\mathcal{N}(A) = \xi$.

Actually if such an abelian variety does exist, then there exists an abelian variety with this Newton polygon over a finite field. This follows by a result of Grothendieck and Katz about Newton polygon strata being Zariski closed in $\mathcal{A}_g \otimes \mathbb{F}_p$; see [32], Th. 2.3.1 on page 143.

11.5. Proof of the Manin Conjecture (Serre, Honda), see [73], page 98. We recall that Newton polygons can be described by a sum of ordered pairs (d, c). A symmetric Newton polygon can be written as

$$\xi = f\cdot((1,0)+(0,1)) + s\cdot(1,1) + \sum_i \;((d_i,c_i)+(c_i,d_i))\,,$$

with $f \geq 0$, $\;s \geq 0$ and moreover $d_i > c_i > 0$ being coprime integers.

Note that $\mathcal{N}(A) \cup \mathcal{N}(B) = \mathcal{N}(A \times B)$; here we write $\mathcal{N}(A) \cup \mathcal{N}(B)$ for the Newton polygon obtained by taking all slopes in $\mathcal{N}(A)$ and in $\mathcal{N}(B)$, and arranging them in non-decreasing order.

We know that for an ordinary elliptic curve E we have $\mathcal{N}(E) = (1,0)+(0,1)$, and for a supersingular elliptic curve we have $\mathcal{N}(E) = (1,1)$, and both types exist. Hence the Manin Conjecture has been settled if we can handle the case

$$(d,c)+(c,d) \text{ with } \gcd(d,c)=1 \text{ and } d>c>0.$$

For such integers we consider a zero π of the polynomial

$$U := T^2 + p^c\cdot T + p^n, \qquad n = d+c, \;\; q = p^n.$$

Clearly $(p^c)^2 - 4\cdot p^n < 0$, and we see that π is an imaginary quadratic Weil q-number. Note that

$$(T^2 + p^c\cdot T + p^n)/p^{2c} \;\;=\;\; (\frac{T}{p^c})^2 + (\frac{T}{p^c}) + p^{d-c}.$$

As $d > c$, we see that $L = \mathbb{Q}(\pi)/\mathbb{Q}$ is an imaginary quadratic extension in which p splits. Moreover, using 5.4 (3), the Newton polygon of U tells us the p-adic values of zeros of U; this shows that the invariants of D/L are c/n and d/n. This proves that $[D : L] = n^2$. Using Theorem 1.2 we have proved the existence of an abelian variety A over $\mathbb{F}_q$ with $\pi = \pi_A$, hence $\mathrm{End}^0(A) = D$. In particular the dimension of A equals $n = d+c$. Using 9.2 (3) we compute the Frobenius slopes: we conclude that $\mathcal{N}(A) = (d,c)+(c,d)$. Hence, using the theorem by Honda and Tate, see 1.2, the Manin conjecture is proved. $\qquad\square$

11.6. Exercise. *Let $g > 2$ be a prime number and let A be an abelian variety simple over a finite field K of dimension g. Show that either $\mathrm{End}^0(A)$ is a field, or $\mathrm{End}^0(A)$ is of* Type(1,g), *i.e. a division algebra of rank g^2 central over an imaginary quadratic field. Show that for any odd prime number in every characteristic both types of endomorphism algebras do appear.* See [54], 3.13.

11.7. Exercise. *Fix a prime number p, fix coprime positive integers $d > c > 0$. Consider all division algebras D such that there exists an abelian variety A of dimension $g := d + c$ over some finite field of characteristic p such that $[\mathrm{End}^0(A) : \mathbb{Q}] = 2g^2$ and $\mathcal{N}(A) = (d,c)+(c,d)$. Show that this gives a infinite set of isomorphism classes of such algebras.*

11.8. We have seen a proof of the Manin conjecture using the Honda-Tate theory. For a reference to a different proof see 21.25.

12. CM-liftings of abelian varieties

References: [56], [11].

12.1. Definition. Let A_0 be an abelian variety over a field $K \supset \mathbb{F}_p$. We say A/R is a *lifting of A_0 to characteristic zero* if R is an integral domain of characteristic zero, with a ring homomorphism $R \to K$, and $A \to \mathrm{Spec}(R)$ is an abelian scheme such that $A \otimes_R K = A_0$.

12.2. Definition. Suppose A_0 be an abelian variety over a field $K \supset \mathbb{F}_p$ such that A_0 admits smCM. We say A is a CM-*lifting of A_0 to characteristic zero* if A/R is a lifting of A_0, and if moreover A/R admits smCM. If this is the case we say that A_0/K satisfies (CML). Moreover, if $L \subset \mathrm{End}^0(A_0)$ is a CM- field of degree $2g$ over $\mathbb{Q}$ and $\mathrm{End}^0(A) = L$ we say that A_0/K satisfies (CML) by L.

We say that A_0/K satisfies (CMLN), if A_0 admits a CM-*lifting to a normal characteristic zero domain*.

Note that in these cases $\mathrm{End}^0(A_M) = \mathrm{End}^0(A) \hookrightarrow \mathrm{End}^0(A_0)$ need not be bijective.

12.3. As Honda proved, [29], Th. 1 on page 86, see [73], Th. 2, see 10.4, for an abelian variety A over a finite field, after a finite field extension, and after an isogeny we obtain an abelian vareity $B_0 \sim A \otimes K'$ which admits a CM-lifting to characteristic zero.

Question 1. *Is an isogeny necessary ?*

Question 2. *Is a field extension necessary ?*

12.4. Theorem I. *For any $g \geq 3$ and for any $0 \leq f \leq g-2$ there exists an abelian variety A_0 over $\mathbb{F} = \overline{\mathbb{F}_p}$, with $\dim(A) = g$ and of p-rank $f(A) = f$, such that A_0 does not admit a CM-lifting to characteristic zero.*
See [56], Th. B on page 131. Compare 5.8.

We indicate the essence of the proof; for details, see [56].
(1) Suppose given a prime number p, and a symmetric Newton polygon ξ which is non-supersingular with $f(\xi) \leq g - 2$. Using [36] choose an abelian variety C over $\mathbb{F} = \overline{\mathbb{F}_p}$ with $\mathcal{N}(C) = \xi$ such that $\mathrm{End}^0(C)$ is a *field*.
(2) Choose an abelian variety B over a finite field K such that $B \otimes \mathbb{F} \sim C$, such that $a(B) = 2$ and such that for every $\alpha_p \subset B$ we have $a(B/\alpha_p) \leq 2$. For a definition of the a-number, see 21.7. Fix an isomorphism $(\alpha_p \times \alpha_p)_K \xrightarrow{\sim} B[F, V] \subset B$. Important observation. Suppose $t \in \mathbb{F}$; suppose $B_{\mathbb{F}}/((1,t)(\alpha_p) =: A_t$ can be defined over K', with $K \subset K' \subset \mathbb{F}$. Then $t \in K'$.
(3) We study all quotients of the form $B_{\mathbb{F}}/((1,t)(\alpha_p) = A_t$ and see which one can be CM-lifted to characteristic zero. Because $\mathrm{End}^0(B)$ is a field, we can classify all such CM-liftings over $\mathbb{C}$, and arrive at:
(4) There exist $K \subset K' \subset \Gamma \subset \mathbb{F}$ such that $[K' : K] < \infty$, moreover Γ/K' is a pro-p-extension, and if $t \notin \Gamma$ then A_t does not a CM-lift to characteristic zero. Note that $\Gamma \subsetneq \mathbb{F}$, and hence the theorem is proved. $\qquad\square$

Conclusion. An isogeny is necessary. *In general, an abelian variety defined over a finite field does not admit a CM-lifting to characteristic zero.*

12.5. Definition. Let $K = \mathbb{F}_q$. Let A_0 be an abelian variety, defined over K. We say that A_0/K satisfies (CMLI), *can be* CM-*lifted after an isogeny,* if there exist $A_0 \sim B_0$ such that B_0 satisfies (CML). We say A_0/K satisfies (CMLNI), if moreover if B_0 can be chosen satisfying (CMLN).

12.6. At present it is an open problem whether any abelian variety defined over a finite field satisfies (CMLI), see 22.2

12.7. Theorem IIs / Example. (Failure of CMLNI.) (B. Conrad) *Let $\pi = p\zeta_5$. This is a Weil p^2-number. Suppose $p \equiv 2, 3$ (mod 5). Note that this implies that p is inert in $\mathbb{Q}(\zeta_5)/\mathbb{Q}$. Let A be any abelian variety over $\mathbb{F}_{p^2}$ in the isogeny class corresponding to this Weil number by the Honda-Tate theory, see 1.2. Note that $\dim(A) = 2$ and $\mathrm{End}^0(A) \cong L = \mathbb{Q}(\zeta_5)$ and A is supersingular. The abelian variety $A/\mathbb{F}_{p^2}$ does not satisfy CMLN up to isogeny.*
A proof, taken from [11], will be given in Section 13.

12.8. Remark. The previous example can be generalized. Let ℓ be a prime number such that $L = \mathbb{Q}(\zeta_\ell)$ contains no proper CM field (e.g. ℓ is a Fermat prime). Let p be a rational prime, such that the residue class field of L above p has degree more than 2. Let $\pi = p\zeta_\ell$ and proceed as above. Note that also in this example we obtain a supersingular abelian variety.

12.9. Theorem IIns / Example. (Failure of CMLNI.) (Chai) *Let p be a rational prime number such that $p \equiv 2, 3$ (mod 5), i.e. p is inert in $\mathbb{Q}(\zeta_5)/\mathbb{Q}$. Suppose $K/\mathbb{Q}$ is imaginary quadratic, such that p is split in $K/\mathbb{Q}$ with an element $\beta \in O_K$ such that $O_K \cdot \beta$ is one of the primes above p in O_K (to ensure existence of β, assume for example K to be chosen in such a way that the class number of K is equal to 1). Let L/K be an extension of degree 5 generated by $\pi := \sqrt[5]{p^2\beta}$. We see that π is a Weil p-number. Let A be any abelian variety over $\mathbb{F}_p$ in the isogeny class corresponding to this Weil number by the Honda-Tate theory, see 1.2. Note that $\dim(A) = 5$, the Newton polygon of A has slopes equal to 2/5 respectively 3/5, and $\mathrm{End}^0(A)$ is a field of degree 10 over $\mathbb{Q}$. The abelian variety $A/\mathbb{F}_p$ does not satisfy (CMLN) up to isogeny.*
A proof, taken from [11], will be suggested in Section 13.

Conclusion. A field extension is necessary. *In general, an abelian variety defined over a finite field does not satisfy (CMLNI).*

13. The residual reflex condition ensures (CMLNI)

13.1. The reflex field. See [69] Section 8 (the dual of a CM-type), [34], I.5. Let P be a CM-field, and let $\rho \in \mathrm{Aut}(P)$ be the involution on P which is complex conjugation under every complex embedding $P \hookrightarrow \mathbb{C}$.

Let (P, Φ) be a CM-type. The *reflex field* L' defined by (L, Φ) is the finite extension of $\mathbb{Q}$ generated by all traces:

$$L' := \mathbb{Q}(\sum_{\varphi \in \Phi} \varphi(x) \mid x \in L).$$

If $L/\mathbb{Q}$ is Galois we have $L' \subset L$. It is known that L' is a CM-field.

Suppose B is an abelian variety, simple over $\mathbb{C}$, with smCM by $P = \text{End}^0(B)$. The representation of P on the tangent space of B defines a CM-type. It follows that any field of definition for B contains L'; see [69], 8.5, Prop. 30; see [34], 3.2 Th. 1.1. Conversely for every such CM-type and every field M containing L' there exists an abelian variety B over M having smCM by L with CM-type Φ.

13.2. Remark. Suppose P is a CM-field, and let Φ be a CM-type for P. Let w' be a discrete valuation of the reflex field P'; write $K_{w'}$ for its residue class field. Suppose B is an abelian variety defined over a number field M such that B/M admits smCM of type (P, Φ). In particular $[P : \mathbb{Q}] = 2\dim(B)$. Then $M \supset L'$; see 13.1 for references. Let v be a discrete valuation of M extending w'. Suppose B has good reduction at v. Let B_v/K_v be the reduction of B at v.

The residual reflex condition. Then K_v contains $K_{w'}$.

(A remark on notation. We use to write w for a discrete valuation of a CM-filed, and v for a discrete valuation of a base field.)

13.3. Proof of 12.7. We see that $\text{End}^0(A) = \mathbb{Q}(\zeta_5) = L$. Note that $L/\mathbb{Q}$ is Galois; hence $L' \subset L$; moreover $L'/\mathbb{Q}$ is a CM-field; hence $L' = L$; this equality can also be checked directly using the possible CM-types for $L = \mathbb{Q}(\zeta_5)$. Suppose there would exist up to isogeny over $K = \mathbb{F}_{p^2}$ a CM-lifting B/M to a field of characteristic zero. We see that the residue class field $K' = K_v$ of M contains the residue class field $K_{w'}$ of L'. As p is inert in $L = L'$ it follows that $K \supset K_{w'} = \mathbb{F}_{p^4}$. This contradicts the fact that A is defined over $\mathbb{F}_{p^2}$. $\qquad\qquad \square$12.7

A proof of 12.9 can be given along the same lines, by showing that $K_{w'} \supset \mathbb{F}_{p^2}$.

13.4. Given a CM-type (P, Φ) and a discrete valuation w' of the reflex field P' we obtain $K_{w'} \supset \mathbb{F}_p$. We see that in order that A_0/K with $K = \mathbb{F}_q$ does allow a lifting with CM-type (P, Φ) it is necessary that it satisfies the *residual reflex condition*: $K_{w'} \subset K$. Moreover note that the triple (P, Φ, w') determines the Newton polygon of B_v (notation as above): see [73], page 107, Th. 3, see 9.7. The triple (P, Φ, w') will be called a *p-adic CM-type*, where p is the residue characteristic of w'. The following theorem says that the *residual reflex condition is sufficient for ensuring* (NLCM) *up to isogeny*.

13.5. Theorem III. *Let A_0/K be an abelian variety of dimension g simple over a finite field $K \supset \mathbb{F}_p$. Let $L \subset \text{End}^0(A_0)$ be a CM -field of degree $2 \cdot g$ over $\mathbb{Q}$. Suppose there exists a p-adic CM-type (L, Φ, w') such that it gives the Newton polygon of A_0 and such that $K_{w'} \subset K$. Then A_0 satisfies (CMNL) up to isogeny.* We expect more details will appear in [11].

13.6. In order to be able to lift an abelian variety from characteristic p to characteristic zero, and to have a good candidate in characteristic zero whose reduction modulo p gives the required Weil number we have to realize that in general an endomorphism algebra in positive characteristic does not appear for that dimension as an endomorphism algebra in characteristic zero. However "less structure" will do:

13.7. Exercise *. Let E be an elliptic curve over a field $K \supset \mathbb{F}_p$. Let $X = E[p^\infty]$ be its p-divisible group. Show:

(1) For every $\beta \in End(X)$ the pair (X, β) can be lifted to characteristic zero.

For every $b \in \mathrm{End}(E)$ the pair (E, b) can be llifted to characteristic zero.
This was proved in [20]. See [55], Section 14, in particular 14.7.

13.8. Remark/Exercise * (Lubin and Tate). *There exists an elliptic curve E over a local field M such that E has good reduction, such that* $\mathrm{End}(E) = \mathbb{Z}$ *and* $\mathrm{End}(E[p^\infty]) \supsetneq \mathbb{Z}_p$. *(We could say: E does not have CM, but $E[p^\infty]$ does have CM.)* See [38], 3.5.

13.9. Remark. We have seen that the Tate conjecture holds for abelian varieties over a base field of finite type over the prime field; see 20.5. By the previous exercise we see that an analogue of the Tate conjecture for abelian varieties does not hold over a local field.

Grothendieck formulated his "anabelian conjecture" for hyperbolic curves; see [27], Section 3. Maybe his motivation was partly the Tate conjecture, partly the description of algebraic curves defined over $\overline{\mathbb{Q}}$ by Bielyi. Grothendieck stressed the fact that the base field should be a number field. This "anabelian" conjecture by Grothendieck generalizes the Neukirch-Uchida theorem for number fields to curves over number fields. Various forms of this conjecture for curves have been proved (Nakamura, Tamagawa, Moichizuki).

It came as a big surprise that this anabelian conjecture for curves actually is true over local fields, as Mochizuchi showed, see [41]. The ℓ-adic representation for abelian varieties is in an *abelian* group: H^1-ℓ-adic or $\pi_1(A)$. It turned out that for curves the representation in the *non-abelian group* $\pi_1(C)$ gives much more information. This is an essential tool in Mochizuki's result.

13.10. We keep notation as in 12.7: $\pi = p\zeta_5$ with $p \equiv 2, 3 \pmod 5$. Write $L = \mathbb{Q}(\zeta_5)$ where $\zeta = \zeta_5$ and write $\mathcal{O} = \mathcal{O}_L$ for the ring of integers of L. We choose A over $\mathbb{F}_p$ such that $\pi_A \sim \pi$ and $\mathcal{O}_L = \mathrm{End}(A)$, see 4.6. We consider

$$\rho_{0,F} : \mathcal{O} \longrightarrow \mathrm{End}(A[F]) = \mathrm{End}(\mathbb{D}(A[F])) = \mathrm{Mat}(2, \mathbb{F}_{p^2}),$$

and

$$\rho_{0,p} : \mathcal{O} \longrightarrow \mathrm{End}(A[p]) = \mathrm{End}(\mathbb{D}(A[p])) = \mathrm{Mat}(4, \mathbb{F}_{p^2}).$$

Let $u \in \mathbb{F}_{p^4}$ be a primitive 5-th root of unity.

Claim. (1) *The set of eigenvalues of $\rho_{0,F}(\zeta)$ is either $\{u, u^4\}$ or $\{u^2, u^3\}$.*
(2) *The set of eigenvalues of $\rho_{0,p}(\zeta)$ is $\{u, u^2, u^3, u^4\}$.*
(3) *The abelian variety A over $\mathbb{F}_{p^2}$ defined above does not admit* (CML).
Proof. Clearly the eigenvalues considered are a power of u. As the trace of $\rho_{0,p}(\zeta)$ is in $\mathbb{F}_{p^2}$ this shows **(1)**.

Consider $\mathcal{O} \otimes_{\mathbb{Z}} W_\infty(\mathbb{F}_{p^2})$. This ring is isomorphic with a product $\Lambda_1 \times \Lambda_2$ according to the two irreducible factors $[(T-\zeta)(T-\zeta^4)]$ respectively $[(T-\zeta^2)(T-\zeta^3)]$ of $\mathrm{Irr}_{\mathbb{Q}}(\pi) = (T^5 - 1)/(T - 1) \in W_\infty(\mathbb{F}_{p^2})[T]$. The action of $\Lambda_1 \times \Lambda_2$ on the additive group $\mathbb{D}(A[p]$ gives a splitting into $\mathbb{D}(A[F])$ and $\mathrm{Ker}(\mathbb{D}(A[p] \to \mathbb{D}(A[F])$. This proves **(2)**.

Suppose there would exist a CM-lifting of A. Then there would be a normal CM-lifting of $B_0 := A \otimes \mathbb{F}$. I.e. the would exist: a normal integral local domain R with residue class field $\mathbb{F}$ and field of fractions M, an abelian scheme $\mathcal{B} \to \mathrm{Spec}(R)$ such that $\mathcal{B} \otimes \mathbb{F} \cong B_0$, and such that $\Gamma := \mathrm{End}(\mathcal{B})$ is an order in $\mathcal{O}$; hence the field of fractions of Γ is L. Write $B = \mathcal{B} \otimes M$ for the generic fiber. Let $z \in W_\infty(\mathbb{F})$ be a primitive root 5-th of unity such that $z \bmod p = u$. Consider $T = \mathbf{t}_{\mathcal{B},0}$ the tangent bundle of $\mathcal{B} \to S := \mathrm{Spec}(R)$ along the zero section. We obtain an action $\Gamma \to \mathrm{End}(T/S)$. Note that $\mathcal{B} \to S$ admits smCM, hence $\mathcal{B} \otimes \mathbb{C}$ has a CM-type. Hence on the generic fiber $T \otimes_R M$ the action of $\zeta \in L$ is either with eigenvalues $\{z, z^2\}$, or $\{z, z^3\}$ or $\{z^4, z^2\}$ or $\{z^4, z^3\}$. This action also can be computed as follows. Consider the p-divisible group $\mathcal{B}[p^\infty]$, with action $\rho : \Gamma \to \mathrm{End}(\mathcal{B}[p^\infty])$. The action on the generic fiber $\rho_\eta : \Gamma \to \mathrm{End}(B[p^\infty])$ extends to $\rho_\eta : L \to \mathrm{End}(B[p^\infty])$. Hence we see that the action of ζ on $T_\eta := T \otimes M$ has eigenvalues as given by the CM-type.

As Γ acts via $\rho : \Gamma \to \mathcal{B}$ we obtain an action

$$\rho_{p^\infty} : \Gamma \to \mathrm{End}(\mathcal{B}[p^\infty]).$$

The closed fiber

$$\rho_{0,p^\infty} : \Gamma \to \mathrm{End}(B_0[p^\infty])$$

of this action extends to the original $\mathcal{O} \to \mathrm{End}(B_0[p^\infty])$.

We conclude that on the one hand ζ acts on $B_0[F]$ by eigenvalues either $\{u, u^4\}$ or $\{u^2, u^3\}$, on the other hand by one of the four possibilities given by a CM-type. This is a contradiction. This proves **(3)**. $\qquad\square$

13.11. This complements 12.4. We expect that for every prime number p there exists an example with $f = 0$ and $g = 2$ of an abelian variety over a finite field which does not admit a CM-lifting.

13.12. (0) For an abelian variety A over a field M of characteristic *zero* with smCM an embedding $M \subset \mathbb{C}$ we obtain a CM-type. Of course, an isogeny does not change the CM-type. Is there an analogue in positive characteristic?

(p) In the example just discussed we see that in the isogeny class of $A \otimes \mathbb{F}$ the action of ζ on the tangent space of different members of the isogeny class can have different "types". An isogeny may change the "CM-type" in positive characteristic. In a more general situation than the one just considered it also not so clear what to expect for a reasonable definition of a "CM-type".

However in 9.2 we see a description of a notion which is intrinsic in the isogeny class of an abelian variety with smCM: not the action on the tangent space, but the action on the p-divisible group does split the p-divisible group into isogeny factors; this splitting is stable under isogenies.

We see the general strategy: in characteristic zero it often suffices to study the tangent space of an abelian variety, whereas in positive characteristic the whole p-divisible group is the right concept to study "infinitesimal properties".

The Shimura-Taniyama formula and the contents of this section are the study of these two aspects, and the way they fit together under reduction modulo p and under lifting to characteristic zero.

14. Elliptic curves

14.1. Reminder. Let E be an eliptic curve over a field $K \supset \mathbb{F}_p$. We say that E is *supersingular* if $E[p](k) = 0$, for an algebraically closed field $k \supset K$. In 21.20 and 21.21 we discuss the definition of an abelian variety being supersingular. We mention that any supersingular abelian variety has p-rank equal to zero; however the converse is not true: for any $g \geq 3$ there exist abelian varieties of p-rank equal to zero of that dimension which are not supersingular.

We say that an abelian variety A of dimension g over a field $K \supset \mathbb{F}_p$ is *ordinary* if its p-rank equals g, i.e. $A[p](k) \cong (\mathbb{Z}/p)^g$. Note that

an elliptic curve E is ordinary iff $E[p](k) \neq 0$, i.e. iff E is not supersingular.

14.2. Exercise. Let E be an elliptic curve over $K \supset \mathbb{F}_p$.
(1) *Show that*

$$\mathrm{Ker}(E \xrightarrow{F_E} E^{(p)} \xrightarrow{F_{E^{(p)}}} E^{(p^2)}) = E[p].$$

(2) *Show that $j(E) \in \mathbb{F}_{p^2}$.*
(3) *Show that E can be defined over $\mathbb{F}_{p^2}$.*
For the notion of "can be defined over K", see 15.1.
(4) (Warning)
Give an example of an elliptic curve E over a field $K \supset \mathbb{F}_p$ with $F_{E^{(p)}} \cdot F_{E^{(p)}} = p$ and give an example with $F_{E^{(p)}} \cdot F_{E^{(p)}} \neq p$.

14.3. Remark. As Deuring showed, for any elliptic curve E we have $(j(E) \in K) \Rightarrow$ (E can be defined over K). An obvious generalization for abelian varieties of dimension $g > 1$ does not hold; in general it is difficult to determine a field of definition for A, even if a field of definition for its moduli point is given.
In fact, as in formulas given by Tate, see [71] page 52, we see that for $j \in K$ an elliptic curve over K with that j invariant exists:

-    $\mathrm{char}(K) \neq 3, \quad j = 0: \quad Y^2 + Y = X^3;$

- $\mathrm{char}(K) \neq 2, \quad j = 1728: \quad Y^2 = X^3 + X;$
- $j \neq 0, \quad j \neq 1728 \quad :$

$$Y^2 + XY = X^3 - \frac{36}{j - 1728}X - \frac{1}{j - 1728}.$$

Deuring showed that the endomorphism algebra of a supersingular elliptic curve over $\mathbb{F} = \overline{\mathbb{F}_p}$ is the quaternion algebra $\mathbb{Q}_{p,\infty}$; this is the division algebra, of degree 4, central over $\mathbb{Q}$ unramified outside $\{p, \infty\}$ and ramified at p and at ∞. This was an inspiration for Tate to prove his structure theorems for endomorphism algebras of abelian varieties defined over a finite field, and, as Tate already remarked, it reproved Deuring's result.

14.4. Endomorphism algebras of elliptic curves. *Let E be an elliptic curve over a finite field $K = \mathbb{F}_q$. We write $\mathbb{Q}_{p,\infty}$ for the quaternion algebra central over $\mathbb{Q}$, ramified exactly at the places ∞ and p. One of the following three (mutually exclusive) cases holds:*

(1)   (2.1.s) $\boxed{E \text{ is ordinary; } e = 2, \quad d = 1 \text{ and } \mathrm{End}^0(E) = L = \mathbb{Q}(\pi_E)}$

*is an imaginary quadratic field in which p **splits**. Conversely if $\mathrm{End}^0(E) = L$ is a quadratic field in which p splits, E is ordinary. In this case, for every field extension $K \subset K'$ we have $\mathrm{End}(E) = \mathrm{End}(E \otimes K')$.*

(2)   (1.2) $\boxed{E \text{ is supersingular, } e = 1, \quad d = 2 \text{ and } \mathrm{End}^0(E) \cong \mathbb{Q}_{p,\infty}.}$

This is the case if and only if $\pi_E \in \mathbb{Q}$. For every field extension $K \subset K'$ we have $\mathrm{End}^0(E) = \mathrm{End}^0(E \otimes K')$.

(3)   (2.1.ns) $\boxed{E \text{ is supersingular, } e = 2, \quad d = 1 \text{ and } \mathrm{End}^0(E) = L \supsetneq \mathbb{Q}.}$

*In this case $L/\mathbb{Q}$ is an imaginary quadratic field in which p does **not split**. There exists an integer N such that $\pi_E^N \in \mathbb{Q}$. In that case $\mathrm{End}^0(E \otimes K') \cong \mathbb{Q}_{p,\infty}$ for any field K' containing $\mathbb{F}_{q^N}$.*

If E is supersingular over a finite field either (2.1.ns) or (1.2) holds.

A proof can be given using 14.6. Here we indicate a proof independent of that classification of all elliptic curves over a finite field, but using 5.4 and 1.2.

Proof. By 5.4 we know that for an elliptic curve E over a finite field we have $L := \mathbb{Q}(\pi_E)$ and $D = \mathrm{End}^0(E)$ and

$$[L : \mathbb{Q}] \cdot \sqrt{[D : L]} = ed = 2g = 2.$$

Hence $e = 2, d = 1$ or $e = 1, d = 2$. We obtain three cases:

(2.1.s) $[L : \mathbb{Q}] = e = 2$ and $D = L$, hence $d = 1$, and p is split in $L/\mathbb{Q}$.

(2.1.ns) $[L : \mathbb{Q}] = e = 2$ and $D = L$, hence $d = 1$, and p is not split in $L/\mathbb{Q}$.

(1.2) $L = \mathbb{Q}, \quad [D : \mathbb{Q}] = 4$; in this case $e = 1, \quad d = 2$ and $D \cong \mathbb{Q}_{p,\infty}$.

Moreover we have seen that either $\pi_E \in \mathbb{R}$, and we are in case (1.2), note that $\dim(E) = 1$, or $\pi_E \notin \mathbb{R}$ and $D = L := \mathbb{Q}(\pi_E) = \mathbb{Q}$ and $L/\mathbb{Q}$ is an imaginary quadratic field.

For a p-divisible group X write $\mathrm{End}^0(X) = \mathrm{End}(X) \otimes_{\mathbb{Z}_p} \mathbb{Q}_p$. We have the natural maps

$$\mathrm{End}(E) \hookrightarrow \mathrm{End}(E) \otimes \mathbb{Z}_p \hookrightarrow \mathrm{End}(E[p^\infty]) \hookrightarrow \mathrm{End}^0(E[p^\infty]) \hookrightarrow \mathrm{End}^0((E \otimes \mathbb{F})[p^\infty]).$$

Indeed the ℓ-adic map $\mathrm{End}(A) \otimes \mathbb{Z}_\ell \hookrightarrow \mathrm{End}(T_\ell(E))$ is injective, as was proved by Weil, see 18.1. The same arguments of that proof are valid for the injectivity of $\mathrm{End}(A) \otimes \mathbb{Z}_p \hookrightarrow \mathrm{End}(A[p^\infty])$ for any abelian variety over any field, see 20.7, see [77], Theorem 5 on page 56. Hence

$$\mathrm{End}^0(E) \hookrightarrow \mathrm{End}^0(E) \otimes \mathbb{Q}_p \hookrightarrow \mathrm{End}^0(E[p^\infty]).$$

Claim (One) (2.1.ns) or (1.2) $\implies$ E is supersingular.

Proof. *Suppose* (2.1.ns) *or* (1.2), *suppose that E is ordinary, and arrive at a contradiction.*
If E is ordinary we have

$$E[p^\infty] \otimes \overline{K} \quad \cong \quad \mu_{p^\infty} \times \underline{\mathbb{Q}_p/\mathbb{Z}_p}.$$

Moreover

$$\mathrm{End}^0(\mu_{p^\infty}) \quad = \quad \mathbb{Z}_p, \qquad \mathrm{End}^0(\underline{\mathbb{Q}_p/\mathbb{Z}_p}) \quad = \quad \mathbb{Z}_p$$

(over any base field). In case (2.1.ns) we see that $D_p = \mathrm{End}^0(E) \otimes \mathbb{Q}_p$ is a quadratic extension of $\mathbb{Q}_p$. In case (1.2) we see that $D_p = \mathrm{End}^0(E) \otimes \mathbb{Q}_p$ is a quaternion algebra over $\mathbb{Q}_p$. In both cases we obtain

$$\mathrm{End}(E) \to \mathrm{End}^0(E) \otimes \mathbb{Q}_p \to \mathrm{End}^0(\overline{E}[p^\infty] \otimes \overline{K}) = \mathrm{End}^0(\mu_{p^\infty} \times \underline{\mathbb{Q}_p/\mathbb{Z}_p}) = \mathbb{Q}_p \times \mathbb{Q}_p.$$

As $(D_p \to \mathbb{Q}_p) = 0$ we conclude that $(\mathrm{End}(E) \to \mathrm{End}(E[p^\infty])) = 0$; this is a contradiction with the fact that the map $\mathbb{Z} \hookrightarrow \mathrm{End}(E) \to \mathrm{End}(E[p^\infty])$ is non-zero. Hence Claim (One) has been proved. $\qquad\square$

Claim (Two) (2.1.s) $\implies$ E is ordinary.

Proof. *Suppose* (2.1.s), *suppose that E that E is supersingular, and arrive at a contradiction.*
Note that $E'[p^\infty]$ is a simple p-divisble group for any supersingular curve E' over any field. Hence $\mathrm{End}^0(E[p^\infty])$ is a division algebra. Suppose that we are in case (2.1.s). Then $\mathbb{Q}(\pi_E) \otimes \mathbb{Q}_p \cong \mathbb{Q}_p \times \mathbb{Q}_p$. This shows that if this were true we obtain an injective map

$$\mathbb{Q}(\pi_E) \otimes \mathbb{Q}_p \cong \mathbb{Q}_p \times \mathbb{Q}_p \hookrightarrow \mathrm{End}^0(E) \otimes \mathbb{Q}_p \hookrightarrow \mathrm{End}^0(E[p^\infty])$$

from $\mathbb{Q}_p \cong \mathbb{Q}_p$ into a division algebra; this is a contradiction. This proves Claim (Two). $\square$

By Claim (One) and Claim (Two) it follows that

$$E \text{ is ordinary} \iff (2.1.\mathrm{s}), \qquad E \text{ is supersingular} \iff ((2.1.\mathrm{ns}) \text{ or } (1.2)).$$

Claim (Three) If E is supersingular then for some $N \in \mathbb{Z}_{>0}$ we have $\pi_E^N \in \mathbb{Q}$.

Proof. If we are in case (1.2) we know $\pi_E \in \mathbb{Q}$.

Suppose we are in case (2.1.ns), and write $L = \mathbb{Q}(\pi_E)$. Write $\pi = \pi_E$ and consider $\zeta = \pi^2/q \in L$.

- Note that ζ has absolute value equal to one for every complex embedding (by the Weil conjecture), see 3.2.
- Note that for any discrete valuation v' of L not dividing p the element ζ is a unit at v'. Indeed π factors p^n, so π is a unit at w.
- As we are in case (2.1.ns) there is precisely one prime v above p.

The product formula $\Pi_w \mid \zeta \mid_w = 1$, the product running over all places of L, in the number field L (see [28], second printing, §20, absolute values suitably normalized) shows that ζ is also a unit at v. By 10.3 we conclude that ζ is a root of unity. This proves Claim (Three). $\square$

We finish the proof. If E is ordinary, $\mathrm{End}^0(E \otimes M)$ is not of degree four over $\mathbb{Q}$, hence $\mathrm{End}^0(E) = \mathrm{End}^0(E \otimes K')$ for any ordinary eliptic curve over a finite field K, and any extension $K \subset K'$.

If we are in case (1.2) clearly we have $\mathrm{End}^0(E) = \mathrm{End}^0(E \otimes K')$ for any extension $K \subset K'$.

If we are in case (2.1.ns) we have seen in Claim (Three) that for some $N \in \mathbb{Z}_{>0}$ we have $\pi_E^N \in \mathbb{Q}$. Hence for every $K \subset \mathbb{F}_{q^N} \subset K'$ we have

$$\mathrm{End}^0(E) = L = \mathbb{Q}(\pi_E) \subsetneq \mathrm{End}^0(E \otimes K') \cong \mathbb{Q}_{p,\infty}.$$

$$\square 14.4$$

14.5. Exercise. Let A be an elliptic curve over a local field in mixed characteristic zero/p, such that $\mathrm{End}(A) \supsetneq \mathbb{Z}$. Let $L = \mathrm{End}^0(A)$. Note that $E/\mathbb{Q}$ is an imaginary quadratic extension. Suppose A has good reduction A_0 modulo the prime above p. Show:

If p is ramified or if p is inert in $\mathbb{Q} \subset E$ then A_0 is supersingular.

If p is is split in $\mathbb{Q} \subset E$ then A_0 is ordinary.

(Note that in the case studied $\mathrm{End}(A) \hookrightarrow \mathrm{End}(A_0)$; you may use this.)

14.6. Classification of isogeny classes of all elliptic curves over finite fields.
See [75], Th. 4.1 on page 536.

Let E be an elliptic curve over a finite field $K = \mathbb{F}_q$, with $q = p^n$, and $\pi = \pi_E$. Then $\mid \pi \mid = \sqrt{q}$ (for every embedding into $\mathbb{C}$). Hence $\pi + \overline{\pi} =: \beta \in \mathbb{Z}$ has the property $\mid \beta \mid \leq 2\sqrt{q}$. For every E over a finite field $\pi = \pi_E$ is a zero of

$$U := T^2 - \beta{\cdot}T + q, \qquad \beta^2 \leq 4q.$$

The Newton polygon of E equals the Newton polygon of U with the vertical axis compressed by n, see 9.3. Hence:

$$(p \text{ does not divide } \beta) \quad \Longleftrightarrow \quad (E \text{ is ordinary}),$$

and

$$(v_p(\beta) > 0) \; \Leftrightarrow \; (E \text{ is supersingular}) \; \Leftrightarrow \; (v_p(\beta) \geq n/2) \; \Leftrightarrow \; (q \text{ divides } \beta^2);$$

$$(E \text{ is supersingular}) \quad \Longleftrightarrow \quad \beta^2 \in \{0, q, 2q, 3q, 4q\}.$$

We write

$$D = \mathrm{End}^0(E), \quad L = \mathbb{Q}(\pi), \quad e = [L : \mathbb{Q}], \quad \sqrt{[D : \mathbb{Q}]} = d.$$

Note that $ed = 2$. Hence $L = \mathbb{Q}$ iff $D \cong \mathbb{Q}_{p,\infty}$. If $L/\mathbb{Q}$ is quadratic, then L is imaginary. Note that if L is quadratic over $\mathbb{Q}$ then E is supersingular iff p is non-split in $L/\mathbb{Q}$.

We have the following possibilities. Moreover,
using 1.2 *we see that these cases do all occur for an elliptic curve over some finite field.*

(1) $\boxed{p \text{ does not divide } \beta}$,
   E is *ordinary*, $L = \mathbb{Q}(\pi_E)$ is imaginary quadratic over $\mathbb{Q}$, and p is split in $L/\mathbb{Q}$; no restrictions on p, no restrictions on n.

In all cases below p divides β (and E is *supersingular*). We write either $q = p^{2j}$ or $q = p^{2j+1}$.
For supersingular E we have that q divides β^2. As moreover $0 \leq \beta^2 \leq 4q$ we conclude

$$\beta^2 \in \{0, q, 2q, 3q, 4q\}.$$

(2) $\beta^2 = 4q$ $\boxed{\beta = \mp 2\sqrt{q} = \mp 2p^j, \quad n = 2j \text{ is even}}$.
   Here $\pi = \pm p^j = \pm\sqrt{q} \in \mathbb{Q}$, and $L = \mathbb{Q}, \quad D \cong \mathbb{Q}_{p,\infty}$.

In all cases below (E is *supersingular* and) $\pi_E \notin \mathbb{Q}$; hence

$$\mathbb{Q} \subsetneq L = D \ncong \mathbb{Q}_{p,\infty} \quad \text{and} \quad L \subsetneq \operatorname{End}(E \otimes \mathbb{F}) \cong \mathbb{Q}_{p,\infty}$$

(3) $\beta^2 = 3q$ $\boxed{p = 3, \quad \beta = \pm 3^{j+1}}$, $\quad q = 3^{2j+1}$.
　　Here $p = 3, \quad n = 2j + 1$ is odd, and
$\pi \sim \zeta_3 \sqrt{-q}$ or $\pi \sim \zeta_6 \sqrt{-q}$: $\quad \pi \sim \zeta_{12}\sqrt{q}, \quad L = \mathbb{Q}(\sqrt{-3})$.

(4) $\beta^2 = 2q$ $\boxed{p = 2, \quad \beta = \pm 2^{j+1}}$, $\quad q = 2^{2j+1}$.
　　Here $p = 2, \quad n = 2j + 1$ is odd, and $\pi \sim \zeta_8 \sqrt{q}$; $\quad L = \mathbb{Q}(\sqrt{-1})$.

(5) $\beta^2 = q$ $\boxed{\beta = \pm\sqrt{q} = \pm p^j, \quad p \not\equiv 1 \pmod 3}$, $\quad n = 2j$ is even, and $L = \mathbb{Q}(\sqrt{-3})$.
　　Here $\pi \sim \zeta_6 \sqrt{q}$, respectively $\pi \sim \zeta_3 \sqrt{q}$.

If we are not in one of the cases above we have $\beta = 0$.

(6) $\boxed{\beta = 0, \quad n \text{ is odd}}$, $\quad \pi \sim \pm\sqrt{-q}, \quad$ no restrictions on p; $\quad L = \mathbb{Q}(\sqrt{-p})$.

(7) $\boxed{\beta = 0, \quad n \text{ is even}, \quad p \not\equiv 1 \pmod 4}$, $\quad \pi \sim \pm p^j \sqrt{-1}, \quad q = p^{2j}$; $\quad L = \mathbb{Q}(\sqrt{-1})$.

In particular we see:

> *if E is supersingular over a finite field, then $\pi_E \sim \zeta_r \sqrt{q}$ with*
> $r \in \{1, 2, 3, 4, 6, 8, 12\}$.

Proof. Let E be an elliptic curve over $\mathbb{F}_q$. We have seen restrictions on β. If p does not divide $\beta \in \mathbb{Z}$, we see that $\beta^2 - 4q < 0$, and (1) is clear.
If we are not in case (1) then p divides β and E is supersingular and $\beta^2 \in \{0, q, 2q, 3q, 4q\}$.
　　If $\beta^2 = 4q$, we are in Case (2); this is clear; also see 15.9.
　　If $\beta^2 = 3q$, we obtain $p = 3$ and we are in case (3)
　　If $\beta^2 = 2q$, we obtain $p = 2$ and we are in case (4).
　　If $\beta^2 = q$ we obtain $L - \mathbb{Q}(\zeta_3)$; because p is non-split in $L/\mathbb{Q}$ we obtain $p \not\equiv 1 \pmod 3$ in this case; this proves (5).
　　If $\beta = 0$ and n odd, we have $L = \mathbb{Q}(\pi) = \mathbb{Q}(\sqrt{-p})$. We are in case (6), no restrictions on p.

If $\beta = 0$ and n is even, we have $L = \mathbb{Q}(\pi) = \mathbb{Q}(\sqrt{-1})$. As p is non-split in $L/\mathbb{Q}$ we see that $p \not\equiv 1 \pmod 4$. We are in case (7).

This ends the proof of the classification of all isogeny classes of elliptic curves over a finite field as given in [75], pp. 536/7. □14.6

15. Some examples and exercises

15.1. Definition / Remark. Let A be an abelian variety over a field K and let $K_0 \subset K$. We say that *A can be defined over K_0* if there exists a field extension $K \subset K'$ and and abelian variety B_0 over K_0 such that $B_0 \otimes_{K_0} K' \cong A \otimes_K K'$. – The following exercise shows that this *does not imply in general that we can choose B_0 over K_0 such that $B_0 \otimes_{K_0} K \cong A$.*

15.2. Exercise. Let p be a prime number, $p \equiv 3 \pmod 4$. Let $\pi := p \cdot \sqrt{-1}$.
(1) *Show that π is a p^2-Weil number. Let A be an abelian variety simple over $K := \mathbb{F}_{p^2}$ such that $\pi_A \sim \pi$. Determine $\dim(A)$. Describe $\operatorname{End}^0(A)$.*
(2) *Show there does not exist an abelian variety B_0 over $K_0 := \mathbb{F}_p$ such that $B_0 \otimes_{K_0} K \cong A$.*
(3) *Show there exists a field extension $K \subset K'$ and and abelian variety B_0 over K_0 such that $B_0 \otimes_{K_0} K' \cong A \otimes_K K'$. I.e. A can be defined over K_0.*

15.3. Exercise. *Give an example of a simple abelian variety A over a field such that $A \otimes \overline{K}$ is not simple.*

15.4. Exercise. *For each of the numbers below show it is a Weil number, determine q, determine the invariants e_0, e, d, g, describe the structure of D, and describe the structure of $\operatorname{End}^0(A \otimes_K K')$ for any field extension $K \subset K'$.*
(1)　$\pi = \sqrt{-p}$,
(2)　$\zeta = \zeta_3 = -\frac{1}{2} + \frac{1}{2}\sqrt{-3}$,　$\pi = \zeta \cdot \sqrt{-p}$,
(3)　π is a zero of $T^2 - \sqrt{2} \cdot T + 8$,

15.5. Exercise. Consider the following examples.
(1) Let $\beta := \sqrt{2 + \sqrt{3}}$, and $q = p^n$. Let π be a zero of

$$T^2 - \beta T + q.$$

(2) Choose coprime positive integers $d > c > 0$, and choose p. Let π be a zero of

$$T^2 + p^c T + p^{d+c}, \quad q = p^{d+c}.$$

See Section 11, in particular 11.5.
(3) Choose $q = p^n$ and $i \in \mathbb{Z}_{>0}$. Let $\pi := \zeta_i \cdot \sqrt{q}$, where ζ_i is a primitive i−th root of unity.

(a) *Show that every of these numbers π indeed is a Weil q-number.*
For each of these let A_π be an abelian variety simple over $\mathbb{F}_q$ having this number

as geometric Frobenius endomorphism.

(b) *Determine* $\dim(A_\pi)$ *and its Newton poygon* $\mathcal{N}(A_\pi)$.

(c) *For every possible choice of* π *determine the smallest* $N \in \mathbb{Z}_{>0}$ *such that for every* $t > 0$ *we have*

$$\mathrm{End}^0(A_\pi \otimes \mathbb{F}_{q^N}) = \mathrm{End}^0(A_\pi \otimes \mathbb{F}_{q^{Nt}}).$$

You might want to use 5.10.

15.6. Exercise. (1) *Let* E *be an elliptic curve over* $\mathbb{F}_p$. *Show that* $\mathrm{End}^0(E)$ *is a field.*

(2) *Give an example of an abelian variety simple over* $\mathbb{F}_p$ *such that* $\mathrm{End}(A)$ *is non-commutative.*

(3) *Let* E *be an elliptic curve over* $\mathbb{Q}$. *Show that* $\mathrm{End}(E) = \mathbb{Z}$.

(4) *Show there exists an abelian variety* A *simple over* $\mathbb{Q}$ *such that* $\mathbb{Z} \neq \mathrm{End}(A)$. Compare 18.10.

15.7. Remark/Exercise. Suppose A is an abelian variety over a field K which admits smCM. Let $D = \mathrm{End}^0(A)$ and let $L' \subset D$ be a subfield of degree $[L' : \mathbb{Q}] = 2g = 2 \cdot \dim(A)$. In this case L' not be a CM field.

Construct A, K, L', *where* A *is an abelian variety over* K, *a finite field, such that* $D = \mathrm{End}^0(A)$ *is of Type IV(1,g), i.e.* A *admits smCM, and* D *is a division algebra central over degree* g^2 *over an imaginary quadratic field* $L = \mathbb{Q}(\pi_A)$, *and* $L \subset L' \subset D$ *is a field which splits* D/L *such that* L' *is not a CM-field.*

15.8. Exercise. Consider the number π constructed in 12.7, respectively 12.9. Prove it is a Weil number and determine $\mathcal{D}(\pi)$, and $g(\pi)$ and the Newton polygon of the isogeny class thus constructed. For notation see 5.5.

15.9. Let π be a Weil q-number. Let $\mathbb{Q} \subset L \subset D$ be the central simple algebra determined by π. We remind the reader that

$$[L : \mathbb{Q}] =: e, \quad [D : L] =: d^2, \quad 2g := e \cdot d. \qquad \text{See Section 18, see 5.5.}$$

For the different types of Albert algebras see 18.2. As we have seen in Proposition 2.2 there are three possibilities in case we work over a finite field:

($\mathbb{R}$e) *Either* $\sqrt{q} \in \mathbb{Q}$, *and* $q = p^n$ *with* n *an* **even** *positive integer.*

$$\boxed{\text{Type III(1)}, \quad g = 1}$$

In this case $\pi = +p^{n/2}$, or $\pi = -p^{n/2}$. Hence $L = L_0 = \mathbb{Q}$. We see that $D/\mathbb{Q}$ has rank 4, with ramification exactly at ∞ and at p. We obtain $g = 1$, we have that $A = E$ is a supersingular elliptic curve, $\mathrm{End}^0(A)$ is of Type III(1), a definite quaternion algebra over $\mathbb{Q}$. This algebra was denoted by Deuring as $\mathbb{Q}_{p,\infty}$. Note that "all endomorphisms of E are defined over K", i.e. for any

$$\forall \quad K \subset K' \quad \text{we have} \quad \mathrm{End}(A) = \mathrm{End}(A \otimes K').$$

($\mathbb{R}$o) *Or* $q = p^n$ *with* n *an* **odd** *positive integer and hence* $\sqrt{q} \notin \mathbb{Q}$.

$$\boxed{\text{Type III}(2), \quad g = 2}$$

In this case $L_0 = L = \mathbb{Q}(\sqrt{p})$, a real quadratic field. We see that D ramifies exactly at the two infinite places with invariants equal to $(n/2) \cdot 2/(2n) = 1/2$. Hence D/L_0 is a definite quaternion algebra over L_0, it is of Type III(2). We conclude $g = 2$. If $K \subset K'$ is an extension of odd degree we have $\mathrm{End}(A) = \mathrm{End}(A \otimes K')$. If $K \subset K'$ is an extension of even degree $A \otimes K'$ is non-simple, it is K'-isogenous with a product of two supersingular elliptic curves, and $\mathrm{End}^0(A \otimes K')$ is a 2×2 matrix algebra over $\mathbb{Q}_{p,\infty}$, and

$$\forall \quad 2 \mid [K' : K] \quad \text{we have} \quad \mathrm{End}(A) \neq \mathrm{End}(A \otimes K').$$

($\mathbb{C}$) *For at least one embedding $\psi : \mathbb{Q}(\pi) \to \mathbb{C}$ we have $\psi(\pi) \notin \mathbb{R}$.*

$$\boxed{\text{IV}(e_0, d), \quad g := e_0 \cdot d}$$

In this case all conjugates of $\psi(\pi)$ are non-real. We can determine $[D : L]$ knowing all $v(\pi)$ by 5.4 (3); here d is the greatest common divisor of all denominators of $[L_v : \mathbb{Q}_p] \cdot v(\pi)/v(q)$, for all $v \mid p$. This determines $2g := e \cdot d$. The endomorphism algebra is of Type IV(e_0, d). For $K = \mathbb{F}_q \subset K' = \mathbb{F}_{q^m}$ we have

$$\mathrm{End}(A) = \mathrm{End}(A \otimes K') \quad \Longleftrightarrow \quad \mathbb{Q}(\pi) = \mathbb{Q}(\pi^m).$$

15.10. Suppose $M \supset R \twoheadrightarrow K$, where R is a normal domain and $M = Q(R)$ the field of fractions, and K a residue field. Suppose $\mathcal{A} \to \mathrm{Spec}(R)$ is an abelian scheme. Then

$$\mathrm{End}(A_M) \xrightarrow{\sim} \mathrm{End}(\mathcal{A}) \hookrightarrow \mathrm{End}(A_K).$$

Exercise. *In case ℓ is a prime number not equal to the characteristic of K, show that $\mathrm{End}(A_K)/\mathrm{End}(A)$ has no ℓ-torsion.*

Exercise. *Give an example where $\mathrm{End}(A_K)/\mathrm{End}(\mathcal{A})$ does have torsion.*

We conclude that we obtain $\mathrm{End}^0(\mathcal{A}) \hookrightarrow \mathrm{End}^0(A_K)$. In general this is not an equality.

Exercise. *Give examples of A over R such that $\mathrm{End}^0(\mathcal{A}) \subsetneqq \mathrm{End}^0(A_K)$.*

15.11. Remark/Exercise. It is interesting to study the behavior of isomorphism classes and of isogeny classes of abelian varieties over finite fields under field extensions. See [75], page 538:

15.11.1 Example. Let $q = p^n$ with n *even*. Consider $\beta_+ = +2\sqrt{q}$, and $\beta_- = -2\sqrt{q}$. The polynomial $P = T^2 - \beta \cdot T + q$ in both cases gives a Weil q-number. The resulting (isogeny classes) E_+, respectively E_- consist of elliptic curves, with $\mathrm{End}^0(E)$ quaternionic over $\mathbb{Q}$, the case of "all endomorphisms are defined over the base field". These isogeny classes do not coincide over $\mathbb{F}_q$:

$$\beta_\pm = \pm 2\sqrt{q}, \qquad E_+ \not\sim_{\mathbb{F}_q} E_-; \quad \text{however} \quad E_+ \otimes K' \sim_{K'} E_- \otimes K'$$

for the quadratic extension $K = \mathbb{F}_q \subset K' := \mathbb{F}_{q^2}$.

Note that in these cases the characteristic polynomial $f_{E_\pm}$ of the geometric Frobenius equals P^2.

Waterhouse writes: "But the extension which identifies these two classes created also a new isogeny class ... It is this sort of non-stable behavior which is overlooked in a treatment like Deuring's which considers only endomorphism rings over $\overline{k}$..." See [75], page 538.

15.11.2 Exercise/Example. *Classify all isogeny classes of elliptic curves, and their endomorphism algebras for every p, for every $q = p^n$. See 14.6.*

15.11.3 Exercise. *Write* $\mathrm{EIsom}(q)$ *for the set of isomorphism classes of elliptic curves over* $\mathbb{F}_q$. *Let* $K = \mathbb{F}_q \subset K' = \mathbb{F}_{q^N}$ *be an extension of finite fields. There is a natural map*

$$\mathrm{EIsom}(q) \quad \longrightarrow \quad \mathrm{EIsom}(q^N) \qquad [E] \mapsto [E \otimes_K K'].$$

Show that this map is not injective, and is not surjective.

15.11.4 Exercise. *Write* $\mathrm{Isog}(q)$ *for the set of isogeny classes of abelian varieties over* $\mathbb{F}_q$. *Show that for* $N \in \mathbb{Z}_{>1}$ *the natural map* $\mathrm{Isog}(q) \to \mathrm{Isog}(q^N)$ *is not injective, and is not surjective.*

15.12. Exercise. *Show that* $h := Y^3 - 6Y^2 + 9T - 1 \in \mathbb{Q}[Y]$ *is irreducible. Let* β *be a zero of* h. *Show that for any* $\psi_0 : \mathbb{Q}(\beta) \to \mathbb{C}$ *we have* $\psi_0(\beta) \in \mathbb{R}$, *i.e.* β *is totally real, and that* $0 < \psi_0(\beta) < 5$, *hence* β *is totally positive. Let* π *be a zero of* $T^2 - \beta{\cdot}T + 3$. *Determine the dimension of* A *simple over* $\mathbb{F}_3$ *such that* $\pi_A = \pi$.

15.13. Exercise. *Let* $L_0 = \mathbb{Q}(\sqrt{2})$. *Choose a rational prime number* p *inert in* $L_0/\mathbb{Q}$. *Let* $\beta := (2 - \sqrt{2}){\cdot}p$. *Let* π *be a zero of the polynomial*

$$g := T^2 - \beta T + p^4 \in L_0[T].$$

(a) *Show that the discriminant of* g *is totally negative.*
(b) *Show that* π *is a* q-*Weil number with* $q = p^4$.
(c) *Let* A *be an abelian variety over* $\mathbb{F}_q$ *with* $\pi_A = \pi$. *Let*

$$\mathbb{Q} \quad \subset \quad L_0 = \mathbb{Q}(\beta) \quad \subset \quad L = \mathbb{Q}(\pi) \quad \subset \quad D := \mathrm{End}^0(A).$$

Determine: $g = \dim(A)$, *the structure of* D *and the Newton polygon* $\mathcal{N}(A)$.

This can be generalized to:

15.14. Exercise. *Let* $g \in \mathbb{Z}_{>0}$. *Let* $e_0, d \in \mathbb{Z}_{>0}$ *with* $e_0{\cdot}d = g$. *Show there exists an abelian variety* A *over* $\mathbb{F} = \overline{\mathbb{F}_\mathrm{p}}$ *with* $D = \mathrm{End}^0(A)$ *of* $\mathrm{Type}(e_0, d)$.

15.15. Exercise. Let $b \in \mathbb{Z}$ and p and $q = p^n$ satisfy $b^2 < 4q^3$ such that 3 divides b but 3^2 does not divide b. Let β be a zero of $f := X^3 - 3qX - b$. Let ρ be a zero of $Y^2 - bY + q^3$. Let π be a zero of $T^2 - \beta T + q$. Let N be the Galois closure of $\mathbb{Q}(\pi)/\mathbb{Q}$. Show:
(a) $f \in \mathbb{Z}[X]$ is irreducible; β is totally real; write $L_0 = \mathbb{Q}(\beta)$;
(b) π is a Weil q-number; ρ is a Weil q^3-number;
(c) $\pi^3 \sim \rho$; there exists an inclusion $\mathbb{Q}(\rho) \subset \mathbb{Q}(\pi) =: L$;
(e)   there exists an element $1 \neq z \in N$ with $z^3 = 1$; such an element is not contained in $\mathbb{Q}(\pi)$;
(f)   for $w' \in \Sigma_{L_0}^{(p)}$ compute $w'(\beta)$.
(g)   Let A be a K-simple abelian variety with $\pi_A = \pi$. Show how to compute $\dim(A)$ once b and q are given. Is A absolutely simple?

15.16. Exercise. Formulate and solve an exercise analogous to the previous one with $f = X^5 - 5qT^3 + 5q^2T - b$.

15.17. Exercise. Let $N \in \mathbb{Z}_{\geq 2}$ be a prime number. Let π be a Weil q-number, and $L = \mathbb{Q}(\pi)$. Suppose $L' := \mathbb{Q}(\pi^N) \subsetneq L = \mathbb{Q}(\pi)$. Show:
(a) If ζ_N is not conjugated to an element in L then $[L : L'] = N$.
(b) If ζ_N is conjugated to an element in L then $[L : L']$ divides $N - 1$.

15.18. Exercise. Let E be an elliptic curve over a field of characteristic $p > 0$, and let $L \subset \mathrm{End}^0(E)$ be a field quadratic over $\mathbb{Q}$. *Show that L is imaginary. Show there exists a CM-lifting of (E, L) to characteristic zero.* See 22.1(4).

15.19. Exercise. Let p be a prime number, and let $P := T^{30} + pT^{15} + p^{15}$. Write $K_n = \mathbb{F}_{p^n}$.
(a) *Show that $P \in \mathbb{Q}[T]$ is irreducible.* Let π be a zero of g. *Show that π is a Weil p-number.* Let A be an abelian variety over $\mathbb{F}_p$ such that $\pi_A \sim \pi$.
(b) *Describe the structure of* $\mathrm{End}(A)$ *and compute* $\dim(A)$.
(c) *Show that*

$$\mathrm{End}(A) \quad \subsetneq \quad \mathrm{End}(A \otimes K_3) \quad \subsetneq \quad \mathrm{End}(A \otimes K_{15}),$$

and describe the structures of these endomorphism algebras. Show that A is absolutely simple.

15.20. Exercise. (See Section 9.) Let m and n be coprime integers, $m > n \geq 0$. Write $h := m + n$. For every $b \in \mathbb{Z}_{>1}$ write

$$g_b := T^2 + p^{2bn}(1 - 2p^{be}) + p^{2bh}, \quad e := h - 2n = m_n.$$

(a) *Show that the discriminant of g_b is negative; conclude that $g_b \in \mathbb{Q}[T]$ is irreducible. Let π_b be a zero of g_b. Show that π_b is a p^{2bh}-Weil number.* Let A_b be an abelian variety with $\pi_{A_b} \sim \pi_b$.
(b) *Describe the structure of* $\mathrm{End}(A_b)$ *and determine the Newton polygon* $\mathcal{N}(A_b)$.
(c) *Show that*

$$\#\left(\{\ell \mid \ell \text{ is a prime number and } \exists b \in \mathbb{Z}_{>0} \text{ such that } \ell \text{ divides } (4p^{be} - 1)\}\right)$$
$$= \infty.$$

[Hint: you might want to use the reminder below.]
(d) *Show that the set* $\{\mathbb{Q}(\pi_b) \mid b \in \mathbb{Z}_{>0}\}/ \cong_{\mathbb{Q}}$ *is an infinite set of isomorphism classes of quadratic fields.*
(e) *Conclude that*

$$\{A_b \otimes \overline{\mathbb{F}_p} \mid b \in \mathbb{Z}_{>1}\}$$

defines an infinite number of $\overline{\mathbb{F}_p}$-*isogeny classes with Newton polygon equal to* $(m, n) + (n, m)$. **(f)** *Show that for any symmetric Newton polygon* $\xi \neq \sigma$ *which is not supersingular, there exists infinitely many isogeny classes of hypersymmetric abelian varieties over* $\mathbb{F}_p$ *having Newton polygon equal to* ξ.

15.21. Reminder. Let S be a set of primes, and $\mathbb{Z}_S$ the ring of rational numbers with denominators using only products of elements of S; write $(\mathbb{Z}_S)^*$ for the multiplicative group of units in this ring. A conjecture by Julia Robinson, later proved as a corollary of a theorem by Siegel and Mahler says:

$$\#\left(\{\lambda \mid \lambda \in (\mathbb{Z}_S)^*, \ \lambda - 1 \in (\mathbb{Z}_S)^*\}\right) < \infty;$$

this is a very special case of: [33], Th. 3.1 in 8.3 on page 194.

16. Appendix 1: Abelian varieties

Basic references: [47], [15], [GM].
    For the notion of abelian variety over a field, abelian scheme over a base scheme, isogenies, and much more we refer to the literature. But let us at least give one definition.

16.1. Definition. Let S be a scheme. We say that $G \to S$ is a *group scheme* over S if $\mathrm{Mor}_S(-, G)$ represents a group functor on the category of schemes over S. A group scheme $A \to S$ is an *abelian scheme* if A/S is smooth and proper with geometrically irreducible fibers. If $S = \mathrm{Spec}(K)$, an abelian scheme over S is called an *abelian variety* defined over K.

From these properties it follows that A/S is a commutative group scheme. However the name does not come from this, but from the fact that certain integrals of differential forms on a Riemann surface where studied by Niels Henrik Abel, and that the values of such integral are in an algebraizable complex torus; hence these objects were called abelian varieties.

Warning. In most recent papers there is a distinction between an abelian variety defined over a field K on the one hand, and $A \otimes_K K'$ over $K' \supsetneq K$ on the other hand. The notation $\mathrm{End}(A)$ stands for "the ring of endomorphisms of A over K". This is the way Grothendieck taught us to choose our notation.

In pre-Grothendieck literature and in some modern papers there is a confusion between on the one hand A/K and "the same" abelian variety over any extension field. In such papers there is a confusion. Often it is not clear what is meant by "a point on A", the notation $\mathrm{End}_K(A)$ can stand for the "endomorphisms defined over K", but then sometimes $\mathrm{End}(A)$ can stand for the "endomorphisms defined over $\overline{K}$".

Please adopt the Grotendieck convention that a scheme $T \to S$ is what it is, and any scheme obtained by base extension $S' \to S$ is denoted by $T \times_S S' = T_{S'}$, etc.

An abelian variety A over a field K, as defined above, is a "complete group variety defined over K" (in pre-Grothendieck terminology). In particular $A \otimes \overline{K}$ is an integral algebraic scheme.

Exercise. *Show that $G \to S$ is a group scheme over S iff there exist morphisms $S \to G$, and $m : G \times G \to A$ and $i : G \to G$ satisfying certain properties encoded in commutative diagrams as given by the group axioms.*

16.2. For an abelian variety over a field K the *dual abelian variety* $A^t = \underline{\mathrm{Pic}}^0_A$ exists. This is an abelian variety of the same dimension as A.

For the definition of a polarization see [47]; [45], 6.2; see [GM]. A divisor D on an abelian variety A defines a homomorphism $\phi_D : A \to A^t$; in case this divisor is ample ϕ_D is an isogeny. For an abelian variety A over a field K an isogeny $\varphi : A \to A^t$ is called a *polarization* if over some over-field of K this homomorphism can be defined by an ample divisor. We say we have a *principal polarization* if $\varphi : A \to A^t$ is an isomorphism.

As every abelian variety admits a polarization we see that A and A^t are isogenous.

16.3. The Rosati involution. Let A be an abelian variety over a field K We write $D = \mathrm{End}^0(A) = \mathrm{End}(A) \otimes_{\mathbb{Z}} \mathbb{Q}$, called the *endomorphism algebra* of A. Let $\varphi : A \to A^t$ be a polarization. We define $\dagger : D \to D$ by $\dagger(x) := \varphi^{-1} \cdot x^t \cdot \varphi$; for the existence of φ^{-1} in D, see 6.1. This map is an *anti-involution* on the algebra D, called the Rosati-involution. If φ is a principal polarization, we have $\dagger : \mathrm{End}(A) \to \mathrm{End}(A)$. See [47], pp.189 - 193. See [15], Chap. V, §17; note however that the subset of $\mathrm{End}^0(A)$ fixed by the Rosati involution need not be a subalgebra.

16.4. Exercise. *Show there exists a polarized abelian variety (A, μ) over a field k such that the Rosati involution $\dagger : \mathrm{End}^0(A) \to \mathrm{End}^0(A)$ does not map $\mathrm{End}(A) \subset \mathrm{End}^0(A)$ into itself.*

16.5. Duality for finite group schemes. For a finite, locally free, *commutative* group scheme $N \to S$ there is a dual group scheme, denoted by N^D, called the *Cartier dual* of N; for $N = \mathrm{Spec}(B) \to \mathrm{Spec}(A) = S$ we take $B^D := \mathrm{Hom}_A(B, A)$, and show that $N^D := \mathrm{Spec}(B^D)$ exists and this is a finite group scheme over S. See [49], I.2.

Equivalent definition. Let $N \to S$ be as above. Consider the functor $T \mapsto \mathrm{Hom}_T(N_T, \mathbb{G}_{m,T})$. By the Cartier-Shatz formula this functor is representable by

$\mathrm{Hom}(-, N^D)$, see [49], Theorem 16.1. Conversely this can be used as defintion of Cartier duality $N \mapsto N^D$.

16.6. Duality theorem. *Let S be a locally noetherian base scheme. Let $\varphi : A \to B$ be an isogeny of abelian schemes over S, with kernel $N = \mathrm{Ker}(\varphi)$. The exact sequence*

$$0 \quad \to \quad N \quad \longrightarrow \quad A \quad \xrightarrow{\varphi} \quad B \quad \to \quad 0$$

gives rise to an exact sequence

$$0 \quad \to \quad N^D \quad \longrightarrow \quad B^t \quad \xrightarrow{\varphi^t} \quad A^t \quad \to \quad 0.$$

$\square$

See [49]. Theorem 19.1. For the definition of N^D, see 16.5.

16.7. Corollary. *Let S be a locally noetherian base scheme and let $A \to S$ be an abelian scheme. There is a natural isomorphism $A^t[p^\infty] = A[p^\infty]^t$.* $\square$

16.8. The characteristic polynomial of an endomorphism. Let A be an abelian variety over a field K of $\dim(A) = g$, and and let $\varphi \subset \mathrm{End}(A)$; then there exists a polynomial $f_{A,\varphi} \in \mathbb{Z}[T]$ of degree $2g$ called *the characteristic polynomial of φ*; it has the defining property that for any $t \in \mathbb{Z}$ we have $f_{A,\varphi}(\varphi - t) = \deg(\varphi - t)$; see [15] page 125.

See 20.1 for the definition of $T_\ell(A)$; for every $\ell \neq \mathrm{char}(K)$ the polynomial $f_{A,\varphi}$ equals the characteristic polynomial of $T_\ell(\varphi) \in \mathrm{End}(T_\ell(A)(\overline{K})) \cong M_{2g}(\mathbb{Z}_\ell)$. This can be used to give a definition of $f_{A,\varphi}$.

If $\varphi \in \mathrm{End}(A)$ and $\psi \in \mathrm{End}(B)$ then $f_{A \times B,(\varphi,\psi)} = f_{A,\varphi} \times f_{B,\psi}$.
If $A = B^\mu$ and B is simple over a finite field, then $f_{A,\pi_A} = (f_{B,\pi_B})^\mu$.

16.9. Exercise. Let K be a field, and A an abelian variety over K of dimension g. *Show there is a natural homomorphism*

$$\mathrm{End}(A) \quad \longrightarrow \quad \mathrm{End}_K(t_A) \cong M_g(K)$$

by $\varphi \mapsto d\varphi$.
    If $\mathrm{char}(K) = 0$, show this map is injective.
    If $\mathrm{char}(K) = p > 0$, show this map is not injective.
    Let E be an elliptic curve over $\mathbb{Q}$. Show that $\mathrm{End}(E) = \mathbb{Z}$. Construct an elliptic curve E over $\mathbb{Q}$ with $\mathrm{End}(E) \subsetneq \mathrm{End}(E \otimes \mathbb{C})$.

Remark. There does exist an abelian variety A over $\mathbb{Q}$ with $\mathbb{Z} \subsetneq \mathrm{End}(A)$. See 15.6.

16.10. Exercise. *Show that over a field of characteristic p, the kernel of $\mathrm{End}(A) \to \mathrm{End}(t_A) \cong M_g(K)$ can be bigger than $\mathrm{End}(A) \cdot p$.*

We say an abelian variety $A \neq 0$ over a field K is *simple*, or K-simple if confusion might occur, if for any abelian subvariety $B \subset A$ we have either $0 = B$ or $B = A$.

16.11. Theorem (Poincaré-Weil). *For any abelian variety $A \neq 0$ over a field K there exist simple abelian varieties B_i over K and an isogeny $A \sim_K \Pi_i \ B_i$.*
See [47], Th. 1 on page 173 for abelian varieties over an algebraically closed field. See [GM] for the general case. $\qquad\qquad\square$

17. Appendix 2: Central simple algebras

Basic references: [7], [61], [8] Chapter 7, [65] Chapter 10. We will not give a full treatment of this theory here.

17.1. A module over a ring is *simple* if it is non-zero, and it has no non-trivial submodules.

A module over a ring is *semisimple* if it is a direct sum of simple submodules.

A ring is called *semisimple* if it is non-zero, and if it is semisimpe as a left module over itself.

A ring is called *simple* if it is semisimple and if there is only one class of simple left ideals.

A finite product of simple rings is semisimple.

The matrix algebra $\mathrm{Mat}(r, D)$ over a division algebra D for $r \in \mathbb{Z}_{>0}$ is simple.

Wedderburn's theorem says that for a central simple algebra (see below) R over a field L there is a central division algebra D over L and an isomorphism $R \cong \mathrm{Mat}(r, D)$ for some $r \in \mathbb{Z}_{\geq 0}$.

Examples of rings which are not semisimple: $\mathbb{Z}$, $K[T]$, $\mathbb{Z}/p^2$.
Examples of rings which are simple: a field, a division algebra (old terminology: "a skew field"), a matrix algebra over a division algebra.

17.2. Exercise. Let $A \neq 0$ be an abelian variety over a field K. (Suggestion, see 16.11, and see 15.9.)
(1) *Show that* $\mathrm{End}^0(A)$ *is a semisimple ring.*
(2) *Prove: if A is simple, then* $\mathrm{End}^0(A)$ *is a division algebra.*
(3) *Prove: if $A \sim B^s$, where B is simple and $s \in \mathbb{Z}_{>0}$, then* $\mathrm{End}^0(A)$ *is a simple ring.*

17.3. Definition. Let L be a field. A *central simple algebra* over L is an L-algebra Γ such that
(1) Γ is finite dimensional over L,
(2) L is the center of Γ,
(3) Γ is a simple ring.
We say that $\Gamma = D$ is a *central division algebra* over L if moreover D is a division algebra.

Suppose a field L is given. Let D and D' be central simple algebras over L. We say that D and D' are similar, notation $D \sim D'$ if there exist $m, m' \in \mathbb{Z}_{\geq 0}$

and an isomorphism $D \otimes_L \mathrm{Mat}(m, L) \cong D' \otimes_L \mathrm{Mat}(m', L)$. The set of 'similarity classes" of central simple algebras over L will be denoted by $\mathrm{Br}(L)$. On this set we define a "multiplication" by $[D_1] \cdot [D_2] := [D_1 \otimes_L D_2]$; this is well defined, and there is an "inverse" $[D] \mapsto [D^{\mathrm{opp}}]$, where D^{opp} is the opposite algebra. As every central simple algebra is a matrix algebra over a central division algebra over L (Wedderburn's Theorem) one can show that under the operations given $\mathrm{Br}(L)$ is a group, called the *Brauer group* of L. See the literature cited for definitions, and properties.

17.4. Facts (Brauer theory).
(1) *For any local field L there is a canonical homomorphism*

$$\mathrm{inv}_L : \mathrm{Br}(L) \to \mathbb{Q}/\mathbb{Z}.$$

(2) *If L is non-archimedean, then* $\mathrm{inv}_L : \mathrm{Br}(L) \xrightarrow{\sim} \mathbb{Q}/\mathbb{Z}$ *is an isomorphism.*
 If $L \cong \mathbb{R}$ then $Br(L) \cong \frac{1}{2}\mathbb{Z}/\mathbb{Z}$.
 If $L \cong \mathbb{C}$ then $\mathrm{Br}(L) = 0$.
(3) *If L is a global field, there is an exact sequence*

$$0 \;\to\; \mathrm{Br}(L) \;\longrightarrow\; \bigoplus_w \mathrm{Br}(L_w) \;\longrightarrow\; \mathbb{Q}/\mathbb{Z} \;\to\; 0.$$

Note the use of this last statement: any central simple algebra over a global field L is uniquely determined by a finite set of non-zero invariants at places of L. We will see that this gives us the possibility to describe endomorphism algebras of (simple) abelian varieties.
(4) *Let $L \subset D$ be simple central division algebra; by (3) we know it is given by a set of invariants $\{\mathrm{inv}_w(D) \mid w \in \Sigma_L\}$, with $\mathrm{inv}_w(D) \in \mathbb{Q}/\mathbb{Z}$. Let r be the least common multiple of the denominators of these rational numbers written as a a quotient with coprime nominator and denominator. Then $[D : L] = r^2$.*

For explicit descriptions of some division algebras see [5]. Note that such explicit methods can be nice to have a feeling for what is going on, but for the general theory you really need Brauer theory.

17.1.

Example. For a (rational) prime number p we consider the invariant $1/2$ at the prime p in $\mathbb{Z}$, i.e. $p \in \Sigma_{\mathbb{Q}}$ and the invariant $1/2$ at the infinite prime of $\mathbb{Q}$. As these invariants add up to zero in $\mathbb{Q}/\mathbb{Z}$ there is a division algebra central over $\mathbb{Q}$ given by these invariants. This is a quaternion algebra, split at all finite places unequal to p. In [20] this algebra is denoted by $\mathbb{Q}_{p,\infty}$. By 5.4 we see that a supersingular elliptic curve E over $\mathbb{F}$ has endomorphism algebra $\mathrm{End}^0(E) \cong \mathbb{Q}_{p,\infty}$

18. Appendix 3: Endomorphism algebras.

Basic references: [68], [47], [35] Chapt. 5, [54].
We will see: *endomorphism algebras* of abelian varieties can be classified. In many

cases we know which algebras do appear. However we will also see that it is difficult in general to describe all orders in these algebras which can appear as the *endomorphism ring* of an abelian variety.

18.1. Proposition (Weil). *Let A, B be abelian varieties over a field K. The group* $\mathrm{Hom}(A, B)$ *is free abelian of finite rank. In fact,*

(1) $\mathrm{rank}\,(\mathrm{Hom}(A, B)) \leq 4 \cdot g_A \cdot g_B;$

(2)   *if the characteristic of K equals zero,* $\mathrm{rank}\,(\mathrm{Hom}(A, B)) \leq 2 \cdot g_A \cdot g_B.$

Let ℓ be a prime different from the characteristic of K. Let $T_\ell(A)$, respectively $T_\ell(B)$ be the Tate-ℓ-groups as defined in 20.1.

(3) *The natural homomorphisms*

$$\mathrm{Hom}(A, B) \hookrightarrow \mathrm{Hom}(A, B) \otimes_{\mathbb{Z}} \mathbb{Z}_\ell \hookrightarrow \mathrm{Hom}(T_\ell(A), T_\ell(B))$$

are injective.

See [47], Th. 3 on page 176. $\square$

We write $\mathrm{End}(A)$ for the endomorphism ring of A and $\mathrm{End}^0(A) = \mathrm{End}(A) \otimes_{\mathbb{Z}} \mathbb{Q}$ for the endomorphism algebra of A. By Wedderburn's theorem every central simple algebra is a matrix algebra over a division algebra. If A is K-simple the algebra $\mathrm{End}^0(A)$ is a division algebra; in that case we write:

$$\mathbb{Q} \quad \subset \quad L_0 \quad \subset \quad L := \mathrm{Centre}(D) \quad \subset \quad D = \mathrm{End}^0(A);$$

here L_0 is a totally real field, and either $L = L_0$ or $[L : L_0] = 2$ and in that case L is a CM-field. In case A is simple $\mathrm{End}^0(A)$ is one of the four types in the Albert classification. We write:

$$[L_0 : \mathbb{Q}] = e_0, \quad [L : \mathbb{Q}] = e, \quad [D : L] = d^2.$$

The Rosati involution $\dagger : D \to D$ is positive definite. A simple division algebra of finite degree over $\mathbb{Q}$ with a positive definite anti-isomorphism which is positive definite is called an Albert algebra. Applications to abelian varieties and the classification has been described by Albert, [1], [2], [3].

18.2. Albert's classification.

Type I(e_0)   Here $L_0 = L = D$ is a totally real field.

Type II(e_0)   Here $d = 2$, $e = e_0$, $\mathrm{inv}_w(D) = 0$ for all infinite w, and D is an indefinite quaternion algebra over the totally real field $L_0 = L$.

Type III(e_0)   Here $d = 2$, $e = e_0$, $\mathrm{inv}_w(D) \neq 0$ for all infinite w, and D is an definite quaternion algebra over the totally real field $L_0 = L$.

Type IV(e_0, d)   Here L is a CM-field, $[F : \mathbb{Q}] = e = 2e_0$, and $[D : L] = d^2$.

18.3. Theorem. *Let A be an abelian variety over a field K. Then* $\mathrm{End}^0(A)$ *is an Albert algebra.* $\square$

See[47], Theorem 2 on page 201.

18.4. As Albert, Shimura and Gerritzen proved: for any prime field $\mathbb{P}$, and every Albert algebra D there exists an algebraically closed field $k \supset \mathbb{P}$ and an abelian variety A over k such that $\mathrm{End}^0(A) \cong D$; see [54], Section 3 for a discussion and references. In case $\mathbb{P} = \mathbb{F}_p$ in all these cases one can choose for A an ordinary abelian variety.

In particular Gerritzen proves the following more precise result. For an Albert algebra D define $t_0(D) \in \{1, 2\}$ by:

$t_0(D) = 1$ if D is of type I, or II, or IV$(e_0, d > 1)$ (D is generated by the †-invariants),

$t_0(D) = 2$ if D is of type III, or IV$(e_0, d = 1)$ (D is not generated by the †-invariants).

18.5. Theorem (Gerritzen). *For a given prime field $\mathbb{P}$, and a given Albert algebra D, choose any integer $t \geq t_0(D) = 1$, and define $g := t \cdot [D : \mathbb{Q}]$; there exists an algebraically closed field k containing $\mathbb{P}$ and an abelian variety A over k such that* $\mathrm{End}^0(A) \cong D$.
See [24], Th. 12. See [54], Th. 3.3.

18.6. A more refined question is to study the *endomorphism ring* of an abelian variety.
Remark. Suppose A is an abelian variety over a finite field. Let π_A be its geometric Frobenius, and $\nu_A = q/\pi_A$ its geometric Verschiebung. We see that $\pi_A, \nu_A \in \mathrm{End}(A)$. Hence the index of $\mathrm{End}(A)$ in a maximal order in $\mathrm{End}^0(A)$ is quite small, in case A is an abelian variety over a finite field. This is in sharp contrast with:

18.7. Exercise. Let L be a field quadratic over $\mathbb{Q}$ with ring of integers $\mathcal{O}_L$. Show that for any order $R \subset L$ there is a number $f \in \mathbb{Z}_{>0}$ such that $\mathcal{O}_L = \mathbb{Z} + f \cdot \mathcal{O}_L$ (and, usually, this number f is called the conductor). *Show that for any imaginary quadratic L and any $f \in \mathbb{Z}_{>0}$ there exists an elliptic curve E over $\mathbb{C}$ such that* $\mathrm{End}(E) \cong \mathbb{Z} + f \cdot \mathcal{O}_L$.
Conclusion. The index of $\mathrm{End}(A)$ in a maximal order in $\mathrm{End}^0(A)$ is in general not bounded when working over $\mathbb{C}$.

18.8. Exercise. Show that there for every integer m and for every algebraically closed field $k \supset \mathbb{F}_p$ not isomorphic to $\mathbb{F}$ there exists a simple abelian surface over k such that $E := \mathrm{End}^0(A)$ and $[\mathcal{O}_E : \mathrm{End}(A)] > m$.

18.9. Remark. For a simple *ordinary* abelian variety A over a finite field the orders in $\mathrm{End}^0(A)$ containing π_A and ν_A are precisely all possible orders appearing as endomorphism ring in the isogeny class of A, see [75], Th. 7.4.

However this may fail for a non-ordinary abelian variety, see [75], page 555/556, where an example is given of an order containing π_A and ν_A, but which does not appear as the endomorphism ring of any abelian variety.

We see difficulties in determining which orders in $\mathrm{End}^0(A)$ can appear as the endomorphism ring of some $B \sim A$.

Much more information on endomorphism rings of abelian varieties over finite fields can be found in [75].

18.10. Exercise. Let A be a simple abelian variety over an algebraically closed field k *which admits* smCM.

(1)   *If the characteristic of k equals zero, $\mathrm{End}^0(A)$ is commutative.*

(2)   *If A is simple and ordinary over $\mathbb{F}$ then $\mathrm{End}^0(A)$ is commutative.*

(3)   However if A is simple and non-ordinary over $\mathbb{F}$ there are many examples showing that $\mathrm{End}^0(A)$ may be non-commutative. *Give examples.*

(4)   *Show there exists an ordinary, simple abelian variety B over an algebraically closed field of positive characteristic such that* $\mathrm{End}(B)$ *is not commutative.* (Hence $k \not\cong \mathbb{F}$, and B does not admit smCM.)

18.11. Exercise. *Let $K \subset K'$ be a an extension of finite field. Let A be an ordinary abelian variety over K such that $A \otimes K'$ is simple. Show that* $\mathrm{End}^0(A) \to \mathrm{End}^0(A \otimes K')$ *is an isomorphism.*

In [75], Theorem 7.2 we read that for simple and ordinary abelian varieties "$\mathrm{End}(A)$ is commutative and unchanged by base change". Some care has to be take in understanding this. The conclusion of the preceding exercise is not correct without te condition "$A \otimes K'$ is simple".

18.12. Exercise. Choose a prime number p, and let π be a zero of the polynomial $T^4 - T^2 + p^2$. Show that π is a Weil p-number; let A be an abelian variety over $\mathbb{F}_p$ (determined up to isogeny) which has π as geometric Frobenius. Show that A is a simple, ordinary abelian surface. Show that $\mathrm{End}^0(A) \to \mathrm{End}^0(A \otimes \mathbb{F}_{p^2})$ is not an isomorphism.

18.13. Remark/Exercise. *Choose $p > 0$, choose a symmetric Newton polygon ξ which is not supersingular. Then there exists a simple abelian variety A over $\mathbb{F}$ with $\mathcal{N}(A) = \xi$ such that* $\mathrm{End}^0(A)$ *is commutative* (hence a field); see [36]. For constructions of other endomorphism algebras of an abelian variety over $\mathbb{F}$ see [9], Th. 5.4.

18.14. Let A be a simple abelian variety over $\mathbb{F}_p$. Suppose that $\psi(\pi_A) \notin \mathbb{R}$. *Show that* $\mathrm{End}(A)$ *is commutative* (hence $\mathrm{End}^0(A)$ is a field) (an easy exercise, or see [75], Th.6.1). In this case every order containing π_A and ν_A in $D = L = \mathrm{End}^0(A)$ is the endomorphism algebra of an abelian variety over $\mathbb{F}_p$.
Exercise. Show *there does exist a simple abelian variety over $\mathbb{F}_p$ such that* $\mathrm{End}^0(A)$ *is not commutative.*

18.15. For abelian varieties over a *finite field* separable isogenies give an equivalence relation, see [75], Th. 5.2.
Exercise. *Show that there exists an abelian variety A over a field $K \supset \mathbb{F}_p$ such that separable isogenies do not give an equivalence relation in the isogeny class of A.*

18.16. Remark. If $K \subset K'$ is an extension of fields, and A is a simple abelian variety over K, then $A' := A \otimes_K K'$ may be K'-simple or non-K'-simple; both cases do appear, and examples are easy to give. The natural map $\mathrm{End}(A) \to$

End(A') is an embedding which may be an equality, but also inequality does appear; examples are easy to give, see 16.9, 15.19.

18.17. Exercise. *Let g be an odd prime number, and let A be a simple abelian variety over a finite field of dimension g. Show:*

- *either* End(A) *is commutative,*
- *or* End$^0(A)$ *is of* Type$(1, g)$*, and $\mathcal{N}(A)$ has exactly two slopes and the p-rank of A is equal to zero.*

See [54], (3.13).

18.18. Existence of endomorphism fields. Let A be an abelian variety which admits smCM over a field K. If $\operatorname{char}(K) = 0$ and A is simple then $D := \operatorname{End}^0(A)$ is a field. However if $\operatorname{char}(K) = p > 0$, the ring End($A$) need not be commutative. For examples see Section 15.

Suppose k is an algebraically closed field of $\operatorname{char}(k) = p$, and let A be a supersingular abelian variety, i.e. $\mathcal{N}(A) = \sigma$, all slopes are equal to $1/2$; then $A \otimes k \sim E^g$, where E is a supersingular elliptic curve. We have $D := \operatorname{End}^0(A) = \operatorname{Mat}(K_{p,\infty}, g)$; in particular D is *not commutative* and for $g > 1$ the abelian variety A is *not simple*. However this turns out to be the only exceptional case in characteristic p where such a general statement holds.

18.19. Theorem (H. W. Lenstra and FO). *Let ξ be a symmetric Newton polygon, and let p be a prime number. Suppose that $\xi \neq \sigma$, i.e. not all slopes in ξ are equal to $1/2$. Then there exists an abelian variety A over $m = \overline{\mathbb{F}_p}$ such that $D = L = \operatorname{End}^0(A)$ is a field. Necessarily A is simple and L is a CM-field of degree $2 \cdot \dim(A)$ over $\mathbb{Q}$.*
See [36].

18.20. Corollary. *For any p and for any $\xi \neq \sigma$ there exists a simple abelian variety A over $\overline{\mathbb{F}_p}$ with $\mathcal{N}(A) = \xi$.*

For more general constructions of endomorphism algebra with given invariants of an abelian variety over a finite field, see [9], Section 5.

19. Appendix 4: Complex tori with smCM

See [69], [47], [35], [60].

19.1. Let A be an abelian variety over $\mathbb{C}$. Write $T := A(\mathbb{C})$. This is a *complex torus*, i.e. a complex Lie group obtained as quotient $\mathbb{C}^g / \Lambda$, where $\mathbb{Z}^{2g} \cong \Lambda \subset \mathbb{C}^g \cong \mathbb{R}^{2g}$ is a discrete subgroup. Indeed, we have an exact sequence

$$ 0 \; \to \mathbb{Z}^{2g} \cong \; \Lambda \; \longrightarrow \; V \cong \mathbb{C}^g \; \xrightarrow{e} \; T = A(\mathbb{C}) \; \to \; 0. $$

There there are at least two different interpretations of the homomorphism e.

One can take the tangent space $V := \mathbf{t}_{A,0}$. This is also the tangent space of the complex Lie groep T. The *exponential map* of commutatieve complex Lie groups gives $e : V \to T$.

One can also consider the topological space T, and construct its *universal covering space* $V := \tilde{T}$. This is a complex Lie group (in a unique way) such that the covering map e is a homomorphism. The kernel is the fundamental group $\pi_1(T,0) = \Lambda \cong \mathbb{Z}^{2g}$.

19.2. The complex torus $T := A(\mathbb{C})$ is algebraizable, i.e. comes from an algebraic variety. If this is the case, the structure of algebraic variety, and the structure of algebraic group giving the complex torus is unique up to isomorphism (note that a complex torus is compact); see [66], corollaire on page 30.

In general a complex torus of dimension at least two need not be algebraizable as is show by the following two examples.

19.3. Example. Choose any abelian variety A over $\mathbb{C}$ of dimension $g > 1$. There exists an analytic family $\mathcal{T} \to \mathcal{M}$, where $\mathcal{M}$ is a unit cube of dimension g^2, such that over that infinitesimal thickening of the origin the restriction of $\mathcal{T} \to \mathcal{M}$ is the formal deformation space $\mathrm{Def}(A)$. Every polarization μ on A gives a regular formal subscheme $S_\mu \subset \mathrm{Def}(A)$ of dimension $g(g+1)/2$. Let $C \to \mathcal{M}$ be a one dimensional regular analytic curve inside $\mathcal{M}$ whose tangent space is not contained in the tangent spaces to S_μ for any μ; such a curve exists because the set of polarizations on A is countable and because $g(g+1)/2 < g^2$ for $g > 1$. One shows that there exists a point $s \in C$ such that $\mathcal{T}_s$ is not algebraizable.

19.4. Example (Zarhin - FO). Choose a division algebra of finite degree over $\mathbb{Q}$ *which is not an Albert algebra*. For example take a field which is not totally real, and which is not a CM-field; e.g. $D = \mathbb{Q}(\sqrt[3]{2})$. By [60], Corollary 2.3 we know there exists a complex torus T with $\mathrm{End}^0(T) \cong D$. If this torus would be algebraizable, $A(\mathbb{C}) \cong T$, then this would imply $\mathrm{End}^0(A) \cong D$ by GAGA, see [66], Proposition 15 on page 29. By Albert's classification this is not possible, see 18.3.

19.5. Let A be an abelian variety over $\mathbb{C}$. Suppose it is simple. Suppose it admits smCM. In that case $\mathrm{End}^0(A) = P$ is a field of degree $2g$ over $\mathbb{Q}$. Moreover P is a CM-field. We obtain a representation $\rho_0 : P \to \mathrm{End}(\mathbf{t}_{A,0}) \cong \mathrm{GL}(g, \mathbb{C})$. As P is commutative and $\mathbb{C}$ is algebraically closed this representation splits a a direct sum of 1-dimensional representations. Each of these is canonically equivalent to giving a homomorphism $P \to \mathbb{C}$. One shows that these g homomorphisms are mutually different, and that no two are complex conjugated. Conclusion: ρ_0 is a CM-type, call it Φ; conversely a CM-type gives such a representation P operating via a diagonal matrix given by the elements of Φ. This process $(A/\mathbb{C}, P) \mapsto (P, \Phi)$ can be reversed, and the construction gives complex tori which are algebraizable.

19.6. Theorem. *Let* (P, Φ) *be a CM-type. There exists an abelian variety A over* $\mathbb{C}$ *with $P \cong \mathrm{End}^0(A)$ such that the representation ρ_0 of P on the tangent space* $\mathbf{t}_{A,0}$ *is given by the CM-type Φ.* $\qquad\square$

See [69], §6. There are many more references possible.

20. Appendix 5: Tate-ℓ and Tate-p conjectures for abelian varieties

Most important reference: [72] Also see [22], [83].

20.1. Notation. Let A be an abelian vareity over a scheme S, let ℓ be a prime number invertible in the sheaf of local rings on S. Write

$$T_\ell(A) = \varprojlim_i A[\ell^i].$$

This is called the Tate-ℓ-group of A/S.

20.2. Let G be a finite flat group scheme over a base scheme S such that the rank of G is prime to every residue characteristic of S, i.e. the rank of G is invertible in the sheaf of local rings on S. Then $G \to S$ is étale; [50].

20.3. Etale finite group schemes as Galois modules. (Any characteristic.) Let K be a field, and let $G = \mathrm{Gal}(K^{\mathrm{sep}}/K)$. The main theorem of Galois theory says that there is an equivalence between the category of algebras étale and finite over K, and the category of finite sets with a continuous G-action. Taking group-objects on both sides we arrive at:

Theorem. *There is an equivalence between the category of étale finite group schemes over K and the category of finite continuous G-modules.*
See [76], 6.4. Note that this equivalence also holds in the case of not necessarily commutative group schemes.

Naturally this can be generalized to: let S be a connected scheme, and let $s \in S(\Omega)$ be a base point, where Ω is an algebraically closed field; let $\pi = \pi_1(S, s)$. *There is an equivalence between the category of étale finite group schemes* (not necessarily commutative) *over S and the category of finite continuous π-systems.*

Exercise. *Write out the main theorem of Galois theory as a theory describing separable field extensions via sets with continuous action by the Galois group. Then formulate and prove the equivalent theorem for étale finite group scheme over an arbitrary base as above.*

Conclusion. The Tate-ℓ-group of an abelian scheme A/S such that ℓ is invertible on S either can be seen as a pro-finite group scheme, or equivalently it can be seen as a projective system of finite modules with a continuous action of the fundamental group of S.

20.4. For an abelian variety A over a field K and a prime number $\ell \neq \mathrm{char}(K)$ the natural map

$$\mathrm{End}(A) \otimes_{\mathbb{Z}} \mathbb{Z}_\ell \;\hookrightarrow\; \mathrm{End}(T_\ell(A)(\overline{K}))$$

is *injective*, as Weil showed; see 18.1.

20.5. Theorem (Tate, Faltings, and many others). *Suppose K is of finite type over its prime field.* (Any characteristic different from ℓ.) *The canonical map*

$$\operatorname{End}(A) \otimes_{\mathbb{Z}} \mathbb{Z}_\ell \quad \xrightarrow{\sim} \quad \operatorname{End}(T_\ell(A)) \cong \operatorname{End}_{G_K}((\mathbb{Z}_\ell)^{2g})$$

is an isomorphism. $\qquad\qquad\qquad\qquad\qquad\qquad\qquad\qquad\qquad\qquad\qquad\square$

This was conjectured by Tate. In 1966 Tate proved this in case K is a finite field, see [72]. The case of function field in characteristic p was proved by Zarhin and by Mori, see [81], [82], [43]; also see [42], pp. 9/10 and VI.5 (pp. 154-161).

The case K is a number field this was open for a long time; it was finally proved by Faltings in 1983, see [21]. For the case of a function field in characteristic zero, see [22], Th. 1 on page 204.

20.6. We like to have a p-adic analogue of 20.5. For this purpose it is convenient to have p-divisible groups instead of Tate-ℓ-groups:
Definition. Let A/S be an abelian scheme, and let p be a prime number (no restriction on p). We write

$$A[p^\infty] = \operatorname{colim}_{i \to} A[p^i],$$

called the p-divisible group (or the Barsotti-Tate group) of A/S.

Remark. Historically a Tate-ℓ-group is defined as a projective system, and the p-divisible group as an inductive system; it turns out that these are the best ways of handling these concepts (but the way in which direction to choose the limit is not very important). We see that the p-divisible group of an abelian variety should be considered as the natural substitute for the Tate-ℓ-group. Note that $A[p^\infty]$ is defined over any base, while $T_\ell(A)$ is only defined when ℓ is invertible on the base scheme.

The notation $A[p^\infty]$ is just symbolic; there is no morphism "p^∞", and there is no kernel of this.

20.7. Exercise. Let A and B be abelian varieties over a field K. In 18.1 we have seen that $\operatorname{Hom}(A, B)$ is of finite rank as $\mathbb{Z}$-module. Let p be a prime number. Using 18.1, show that the natural map

$$\operatorname{Hom}(A, B) \otimes_{\mathbb{Z}} \mathbb{Z}_p \quad \hookrightarrow \quad \operatorname{Hom}((A)[p^\infty], B[p^\infty])$$

is *injective*. Also see [77], theorem 5 on page 56. Also see [83].

20.8. Remark. On could feel the objects $T_\ell(A)$ and $A[p^\infty]$ as *arithmetic objects* in the following sense. If A and B are abelian varieties over a field K which are isomorphic over $\overline{K}$, then they are isomorphic over a finite extension of K; these are geometric objects. Suppose X and Y are p-divisible groups over a field K which are isomorphic over $\overline{K}$ then they need not be isomorphic over any finite extension of K, these are arithmetic objects. The same statement for pro-ℓ-group schemes.

20.9. Theorem (Tate and De Jong). *Let K be a field finitely generated over $\mathbb{F}_p$. Let A and B be abelian varieties over K. The natural map*

$$\mathrm{Hom}(A, B) \otimes \mathbb{Z}_p \xrightarrow{\ \sim\ } \mathrm{Hom}(A[p^\infty], B[p^\infty])$$

is an isomorphism. □

This was proved by Tate in case K is a finite field; a proof was written up in [77]. The case of a function field over a finite field was proved by Johan de Jong, see [30], Th. 2.6. This case follows from the result by Tate and from the following result on extending homomorphisms 20.10.

20.10. Theorem (Tate, De Jong). *Let R be an integrally closed, Noetherian integral domain with field of fractions K.* (Any characteristic.) *Let X, Y be p-divisible group over $\mathrm{Spec}(R)$. Let $\beta_K : X_K \to Y_K$ be a homomorphism. There exists (uniquely) $\beta : X \to Y$ over $\mathrm{Spec}(R)$ extending β_K.*

This was proved by Tate, under the extra assumption that the characteristic of K is zero. For the case $\mathrm{char}(K) = p$, see [30], 1.2 and [31], Th. 2 on page 261. □

21. Appendix 6: Some properties in characteristic p

See [39]. For information on group schemes see [49], [62], [76], [10].

In characteristic zero we have strong tools at our disposal: besides algebraic-geometric theories we can use analytic and topological methods. It seems that we are at a loss in positive characteristic. However the opposite is true. Phenomena, only occurring in positive characteristic provide us with strong tools to study moduli spaces. And, as it turns out again and again, several results in characteristic zero can be derived using reduction modulo p. These tools in positive characteristic will be of great help in this talk.

21.1. A finite group scheme in characteristic zero, of more generally a finite group scheme of rank prime to all residue characteristics, is étale over the base; e.g. see [50]. However if the rank of a finite group scheme is not invertible on the base, it need not be étale.

21.2. The Frobenius morphism. For a scheme T over $\mathbb{F}_p$ (i.e. $p \cdot 1 = 0$ in all fibers of $\mathcal{O}_T$), we define the *absolute Frobenius morphism* $\mathrm{fr} : T \to T$; if $T = \mathrm{Spec}(R)$ this is given by $x \mapsto x^p$ in R.

For a scheme $A \to S$ over $\mathrm{Spec}(\mathbb{F}_p)$ we define $A^{(p)}$ as the fiber product of $A \to S \xleftarrow{\ \mathrm{fr}\ } S$. The morphism $\mathrm{fr} : A \to A$ factors through $A^{(p)}$. This defines $F_{A/S} = F_A : A \to A^{(p)}$, a morphism over S; this is called *the relative Frobenius morphism*. If A is a group scheme over S, the morphism $F_A : A \to A^{(p)}$ is a homomorphism of group schemes. For more details see [62], Exp. VII$_A$.4. The notation $A^{(p/S)}$ is (maybe) more correct.

Example. Suppose $A \subset \mathbb{A}_R^n$ is given as the zero set of a polynomial $\sum_I a_I X^I$ (multi-index notation). Then $A^{(p)}$ is given by $\sum_I a_I^p X^I$, and $A \to A^p$ is given, on coordinates, by raising these to the power p. Note that if a point $(x_1, \cdots, x_n) \in A$ then indeed $(x_1^p, \cdots, x_n^p) \in A^{(p)}$, and $x_i \mapsto x_i^p$ describes $F_A : A \to A^{(p)}$ on points.

Let $S = \mathrm{Spec}(\mathbb{F}_p)$; for any $T \to S$ we have a canonical isomorphism $T \cong T^{(p)}$. In this case $F_{T/S} = \mathrm{fr} : T \to T$.

21.3. Verschiebung. Let A be a *commutative* group scheme flat over a characteristic p base scheme. In [62], Exp. VII$_A$.4 we find the definition of the "relative Verschiebung"

$$V_A : A^{(p)} \to A; \quad \text{we have:} \quad F_A \cdot V_A = [p]_{A^{(p)}}, \quad V_A \cdot F_A = [p]_A.$$

In case A is an abelian variety we see that F_A is surjective, and $\mathrm{Ker}(F_A) \subset A[p]$. In this case we do not need the somewhat tricky construction of [62], Exp. VII$_A$.4, but we can define V_A by $V_A \cdot F_A = [p]_A$ and check that $F_A \cdot V_A = [p]_{A^{(p)}}$.

21.4. Examples of finite group scheme of rank p. Let $k \supset \mathbb{F}_p$ be an algebraically closed field, and let G be a commutative group scheme of rank p over k. Then we are in one of the following three cases:

$G = \mathbb{Z}/p_{\,k}$. This is the scheme $\mathrm{Spec}(k^p)$, with the group structure given by $\mathbb{Z}/p$. Here $V_G = 0$ and F_G is an isomorphism.

$G = \alpha_p$. We write $\alpha_p = \mathbb{G}_{a,\mathbb{F}_p}[F]$ the kernel of the Frobenius morphism on the linear group $\mathbb{G}_{a,\mathbb{F}_p}$. This group scheme is defined over $\mathbb{F}_p$, and we have the habit to write for any scheme $S \to \mathrm{Spec}(\mathbb{F}_p)$ just α_p, although we should write $\alpha_p \times_{\mathrm{Spec}(\mathbb{F}_p)} S$. For any field $K \supset \mathbb{F}_p$ we have $\alpha_{p,K} = \mathrm{Spec}(K[\tau]/(\tau^p))$ and the group structure is given by the comultiplication $\tau \mapsto \tau \otimes 1 + 1 + \tau$ on the algebra $K[\tau]/(\tau^p)$. Here $V_G = 0 = F_G$.

$G = \mu_{p,k}$. We write $\mu_{t,K} = \mathbb{G}_{m,K}[t]$ for any field K and any $t \in \mathbb{Z}_{\geq 1}$. Here $F_G = 0$ and V_G is an isomorphism. Note that the algebras defining $\alpha_{p,\mathbb{F}_p}$ and $\mu_{p,\mathbb{F}_p}$ are isomorphic, but the comultiplications are different.

Any finite commutative group scheme over k of rank a power of p is a successive extension of group schemes of these three types. For an arbitrary field $K \supset \mathbb{F}_p$ the first and the last example can be "twisted" by a Galois action. However if $G \otimes_K k \cong \alpha_{p,k}$ then $G \cong \alpha_{p,K}$.

For duality, and for the notion of "local" and "etale" group scheme see [49].

Commutative group scheme of p-power rank over a perfect base field can be classified with the help of Dieudonné modules, not dicussed here, but see [39], see [19].

21.5. The p-rank. For an variety A over a field $K \supset \mathbb{F}_p$ we define its p-rank $f(A) = f$ as the integer such that $A[p](\overline{K}) \cong (\mathbb{Z}/p)^f$.
    We say A is *ordinary* iff $f(A) = \dim(A) =: g$.

21.6. For a classification of isomorphism classes of ordinary abelian varieties over finite fields (using Serre-Tate canonical lifts, and classical theory) see the wonderful paper [17]. This is a much finer classification than the Honda-Tate theory which studies isogeny classes.

21.7. The a-number. Let G be a group scheme over a field K of characteristic p. We write

$$a(G) = \dim_k(\mathrm{Hom}(\alpha_p, G \otimes k)),$$

where k is an algebraically closed field containing K. For a further discussion, see [10], 5.4 - 5.8

21.8. Examples. If E is an elliptic curve in characteristic p then:

$$E \quad \text{is ordinary} \quad \Leftrightarrow \quad E[p](\overline{K}) \neq 0 \quad \Leftrightarrow \quad \mathrm{Ker}(F : E \to E^{(p)}) \otimes k \cong \mu_p.$$

In this case $E[p] \otimes k \cong \mu_p \times \underline{\mathbb{Z}/p}$.

$$E \quad \text{is supersingular} \quad \Leftrightarrow \quad E[p](K) = 0 \quad \Leftrightarrow \quad E[F] := \mathrm{Ker}(F : E \to E^{(p)}) \cong \alpha_p.$$

In this case $E[p]$ is a non-trivial extension of α_p by α_p.

Warning. For a higher dimensional abelian varieties $A[F]$ and $A[p]$ can be quite complicated.

21.9. Exercise. *Show that the following properties are equivalent:*
(1) *A is ordinary,*
(2) *$\mathrm{Hom}(\alpha_p, A) = 0$,*
(3) *the kernel of $V : A^{(p)} \to A$ is étale,*
(4) *the rank of the group $\mathrm{Hom}(\mu_p, A \otimes \overline{K})$ equals p^g.*
(5) *$\mathrm{Hom}(\mu_p, A \otimes \overline{K}) \cong (\mathbb{Z}/p)^g$.*

21.10. Duality; see [GM], Chapter V. For a finite locally free group scheme $G \to S$ over a base $S \to \mathrm{Spec}(\mathbb{F}_p)$ we study $F_{G/S} : G \to G^{(p)}$. We can apply Cartier-duality, see 16.5.

Fact.

$$\left(F_{G/S} : G \to G^{(p)}\right)^D = \left(V_{G^D} : (G^{(p)})^D = (G^D)^{(p)} \to G^D\right).$$

In the same way Cartier duality gives $(V_G)^D = F_{G^D}$.

Using duality of abelian varieties, in particular see [49], Theorem 19.1, we arrive at:

For an abelian scheme $A \to S$ over a base $S \to \mathrm{Spec}(\mathbb{F}_p)$ we have

$$\left(F_{A/S} : A \to A^{(p)}\right)^t \;=\; \left(V_{A^t} : (A^{(p)})^t = (A^t)^{(p)} \to A^t\right), \quad \text{and} \quad (V_A)^t = F_{A^t}.$$

21.11. Newton polygons. In order to being able to handle the isogeny class of $A[p^\infty]$ we need the notion of Newton polygons.

Suppose given integers $h, d \in \mathbb{Z}_{\geq 0}$; here $h = $ "height", $d = $ "dimension", and in case of abelian varieties we will choose $h = 2g$, and $d = g$. A Newton polygon γ (related to h and d) is a polygon $\gamma \subset \mathbb{Q} \times \mathbb{Q}$ (or, if you wish in $\mathbb{R} \times \mathbb{R}$), such that:

-   γ starts at $(0,0)$ and ends at (h, d);

-   γ is lower convex;

-   any slope β of γ has the property $0 \leq \beta \leq 1$;

-   the breakpoints of γ are in $\mathbb{Z} \times \mathbb{Z}$; hence $\beta \in \mathbb{Q}$.

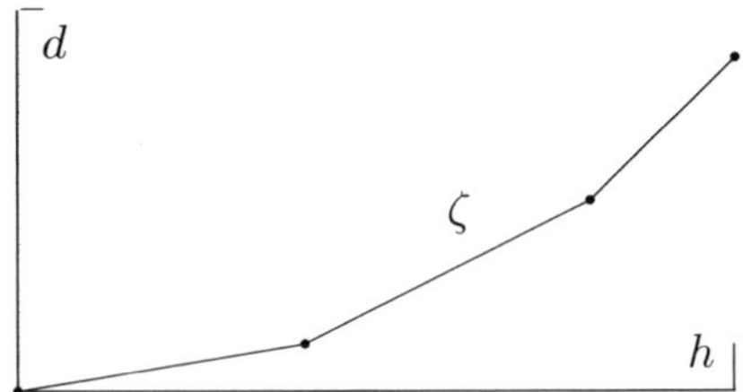

Note that a Newton polygon determines (and is determined by)

$$\beta_1, \cdots, \beta_h \in \mathbb{Q} \text{ with } 0 \leq \beta_h \leq \cdots \leq \beta_1 \leq 1 \quad \leftrightarrow \quad \zeta.$$

Sometimes we will give a Newton polygon by data $\sum_i (d_i, c_i)$; here $d_i, c_i \in \mathbb{Z}_{\geq 0}$, with $\gcd(d_i, c_i) = 1$, and $d_i/(d_i + c_i) \leq d_j/(d_j + c_j)$ for $i \leq j$, and $h = \sum_i (d_i + c_i)$, $d = \sum_i d_i$. From these data we construct the related Newton polygon by choosing the slopes $d_i/(d_i + c_i)$ with multiplicities $h_i = d_i + c_i$. Conversely clearly any Newton polygon can be encoded in a unique way in such a form.

Remark. The Newton polygon of a polynomial. Let $g \in \mathbb{Q}_p[T]$ be a monic polynomial of degree h. We are interested in the p-adic values of its zeroes (in an algebraic closure of $\mathbb{Q}_p$). These can be computed by the Newton polygon of this polynomial. Write $g = \sum_j \gamma_j T^{h-j}$. Plot the pairs $(j, v_p(\gamma_j))$ for $0 \leq j \leq h$. Consider the lower convex hull of $\{(j, v_p(\gamma_j)) \mid j\}$. This is a Newton polygon according to the definition above. *The slopes of the sides of this polygon are precisely the p-adic values of the zeroes of g, ordered in non-decreasing order.*
Exercise. Prove this.
Hint. Write $g = \Pi \, (T - z_i)$, with $z_i \in \overline{\mathbb{Q}_p}$. Write $\beta_i := v_p(z_i) \in \mathbb{Q}_{\geq 0}$. Suppose the order of the $\{z_i\}$ chosen in such a way that

$$0 \leq \beta_h \leq \beta_2 \leq \cdots \leq \beta_{i+1} \leq \beta_i \leq \cdots \leq \beta_1.$$

Let σ_j be the elementary symmetric functions in z_i. Show that:

$$\sigma_j = \gamma_j, \quad v_p(\sigma_j) \geq \beta_h + \cdots + \beta_{h-j+1}, \quad \beta_1 = v_p(\gamma_h).$$

21.12. *A p-divisible group X over a field of characteristic p determines uniquely a Newton polygon.* The general definition can be found in [39]. The isogeny class of a p-divisible group over and algebraically closed field k uniquely determines (and is uniquely determined by) its Newton polygon. We use "the Newton polygon of Frobenius", a notion to be explained below.

21.13. Theorem (Dieudonné and Manin), see [39], "Classification theorem " on page 35 .

$$\{X\}/\sim_k \quad \xrightarrow{\ \sim\ } \quad \{\text{Newton polygon}\}$$

21.14. We sketch the construction of a Newton polygon of a p-divisible group X, or of an abelian variety.

(Incorrect.) Here we indicate what the Newton polygon of a p-divisible group is (in a slightly incorrect way ...). Consider "the Frobenius endomorphism" of X. This has a "characteristic polynomial". This polynomial determines a Newton polygon, which we write as $\mathcal{N}(X)$, the Newton polygon of X. For an abelian variety A we write $\mathcal{N}(A)$ instead of $\mathcal{N}(A[p^\infty])$.

21.15. Exercise. Show that for an abelian variety A over the prime field $\mathbb{F}_p$ this construction is valid, and does give the Newton polygon of A as defined in Section 9.

Although, this "definition" is correct over $\mathbb{F}_p$ as ground field, over any other field $F : X \to X^{(p)}$ is not an endomorphism, and the above "construction" fails.

21.16. Dieudonné-Manin theory. (We only give some definitions and facts.) For coprime integers $d, c \in \mathbb{Z}_{\geq 0}$ one can define a p-divisible group $G_{d,c}$. This is a p-divisible group of dimension d and of height $d+c$. In fact, $G_{1,0} = \mathbb{G}_m[p^\infty]$, and $G_{0,1} = (\mathbb{Q}_p/\mathbb{Z}_p)$. For $d > 0$ and $c > 0$ we have a formal p-divisible group $G_{d,c}$ of dimension d and of height $h = d + c$. We do not give the construction here; see the first two chapters of Manin's thesis [39]; the definition of $G_{d,c}$ is on page 35 of [39]. The p-divisible group $G_{d,c}$ is defined over $\mathbb{F}_p$; we will use the same symbol for this group over any base field or base scheme over $\mathbb{F}_p$, i.e. we write $G_{d,c}$ instead of $G_{d,c} \otimes_{\mathbb{F}_p} K$. Moreover the p-divisible groups $G_{d,c}$ and $G_{c,d}$ over $\mathbb{F}_p$ satisfy $(G_{d,c})^t \cong (G_{c,d})$; here X^t denotes te Serre dual of X, see 8.3.
Remark. With this defintion we have $G_{d,c}[F^{d+c}] = G_{d,c}[p^d]$ and $G_{d,c}[V^{d+c}] = G_{d,c}[p^c]$

21.17. Exercise. Assume the existence of $X = G_{d,c}$ over $\mathbb{F}_p$ as explained above. Let ζ be the Newton polygon of the Frobenius endomorphism of X. Show that ζ

consists of $d + c$ slopes equal to $d/(d + c)$: this polygon is isoclinic (it is a straight line) and it ends at $(d + c, d)$.

Let $K = \mathbb{F}_{p^n}$, and $X = G_{d,c} \otimes_{\mathbb{F}_p} K$. Let $\pi_X \in \mathrm{End}(X)$ be the geometric Frobenius. Then

$$\boxed{v_p(\pi_X) = \frac{d \cdot n}{h}, \quad h := d + c, \quad q = p^n.}$$

21.18. In [39], Chapter II we find:
Theorem. *Let k be an algebraically closed field of characteristic p. Let X be a p-divisible group over k. Then there exists an isogeny*

$$X \quad \sim \quad \prod_i G_{d_i, c_i}.$$

$\qquad\qquad\qquad\qquad\qquad\qquad\qquad\qquad\qquad\qquad\qquad\qquad\qquad\qquad\qquad\quad \square$

See [39], Classification Theorem on page 35.

21.19. Definition of the Newton polygon of a p-divisible group. The isogeny class of $\prod_i G_{d_i, c_i}$ will be encoded in the form of a Newton polygon. The simple p-divisible group $G_{d,c}$ will be represented by $d + c$ slopes equal to $d/(d + c)$. The slopes of $\sum_i G_{d_i, c_i}$ will be ordered in non-decreasing order. For a p-divisible group of dimension d, height h with $h = d + c$ together these slopes form a polygon in $\mathbb{Q} \times \mathbb{Q}$.

For an abelian variety over a field of characteristic p we define $\mathcal{N}(A) := \mathcal{N}(A[p^\infty])$.

Note that for a p-divisible group X over K its Newton polygon only depends on $\mathcal{N}(X \otimes k)$, this only depends on the isogeny factors of $X \otimes k$, and we can choose these isogeny factors in such a way that they are defined over $\mathbb{F}_p$.

Example. Suppose $A[p^\infty] = X \sim G_{d,c} \times G_{c,d}$. Then the Newton polygon $\mathcal{N}(A)$ of A equals $(d, c) + (c, d)$; this has $d + c$ slopes equal to $d/(d + c)$ and $d + c$ slopes equal to $c/(d + c)$.

21.20. Definition. An abelian variety A over a field $K \supset \mathbb{F}_p$ is called *supersingular* if $\mathcal{N}(A)$ is isoclinic with all slopes equal to $1/2$.
Equivalently. An abelian variety A over a field $K \supset \mathbb{F}_p$ is *supersingular* if there exists an isogeny $(A \otimes k)[p^\infty] \sim (G_{1,1})^g$.
Exercise. Show that for an elliptic curve this definition and the one given in 21.8 coincide.

Theorem (Tate, Shioda, Deligne, FO). *An abelian variety A is supersingular iff there exists a supersingular elliptic curve E and an isogeny $A \otimes k \sim E^g \otimes k$.*
See [72], Th. 2 on page 140, see [70], [52], Section 4.

21.21. Definition/Remark/Exercise. (1) Note that the definition of A being supersingular can be given knowing only the p-divisible group $A[p^\infty]$;

$$A \text{ is supersingular} \quad \Longleftrightarrow \quad \mathcal{N}(A) = \sigma,$$

where $\sigma = g(1,1)$ is the Newton polygon having only slopes equal to $1/2$. Equivalently this definition can be given by the property in the theorem just mentioned.

(2) We see that $g > 1$ and $\mathcal{N}(A) = \sigma$ imply that A is not absolutely simple. This is an exceptional case. Indeed, for any symmetric Newton polygon $\xi \neq \sigma$ and any p there exists an absolutely simple abelian variety A in characteristic p with $\mathcal{N}(A) = \xi$; see [36], see 18.13.

(3) Let A be a simple abelian variety over the finite field $\mathbb{F}_q$. Show:

$$A \text{ is supersingular} \quad \Longleftrightarrow \quad \pi_A \sim \zeta \cdot \sqrt{q},$$

where ζ is a root of unity.

21.22. Exercise. Let Y be a p-divisible group over a field K. Suppose $Y \sim \prod_i G_{d_i,c_i}$. Suppose there exist integers $d, h \in \mathbb{Z}_{>0}$ such that $Y[F^h] = Y[p^d]$. Show: *only factors G_{d_i,c_i} do appear with $d_i/(d_i + c_i) = d/h$.*

21.23. Proposition. *For every pair (d, c) of coprime non-negative integers we have $G_{d,c} \cong (G_{c,d})^t$. Let A be an abelian variety over a field $K \supset \mathbb{F}_p$, and $X = A[p^\infty]$. The Newton polygon $\mathcal{N}(A) := \mathcal{N}(X)$ is symmetric, in the sense of 11.1.*
Proof. The first equality follows from the definitions.

By 16.6 we have $A[m]^D = A^t[m]$ for every $m \in \mathbb{Z}_{>0}$. Hence $A[p^\infty]^t - A^t[p^\infty]$; use the definition of the Serre dual X^t; this formula is less trivial than notation suggests. Hence $G_{d,c}$ and $G_{c,d}$ appear with the same multiplicity in the isogeny type of $X = A[p^\infty]$. This proves symmetry of $\mathcal{N}(X)$. ⊔

21.24. Remark. The theory as developed by Dieudonné and Manin gives the Newton polygon of a p-divisible group, and of an abelian variety over an arbitrary field in characteristic p. Note that for an abelian variety an easier construction is possible, which gives the same result, see Section 9, especially 9.3.

21.25. A proof for the Manin Conjecture. We have seen that the Manin Conjecture can be proved using the Honda-Tate theory, see Section 11. In [58], Section 5 we find a proof of that conjecture, using only methods of characteristic p. We sketch that proof (and please see the reference cited for notations and details).

We know that the conjecture holds for $G_{1,1}$: in every characteristic p there exists a supersingular elliptic curve, and $E[p^\infty] \cong G_{1,1}$. Hence every supersingular p-divisible group is algebraizable. We show that for a given $g \geq 1$ there exists an abelian variety A_0 with a principal polarization λ_0 such that A_0 is supersingular, and $a(A_0) = 1$. Methods of [58] show that for a given symmetric Newton polygon ξ, which automatically lies below $\sigma = \mathcal{N}(A_0)$, there exists a formal deformation of $(X_0, \lambda_0) = (A_0, \lambda_0)[p^\infty]$ to (X, λ) with $\mathcal{N}(X) = \xi$. By the Serre-Tate Theorem we know that a formal deformation of an algebraizable p-divisible group is algebraizable; hence there exists (A, λ) with $(X, \lambda) = (A, \lambda)[p^\infty]$; this proves the Manin Conjecture.

22. Some questions

In this section we gather some remarks, questions and open problems.

22.1. Definition; see 12.2. Let B_0 be an abelian variety over a field K of characteristic $p > 0$. We say B is a CM-lift of B_0 if there exists an integral domain R of characteristic zero with a surjective homomorphism $R \twoheadrightarrow K$ with field of fractions $Q(R)$ and an abelian scheme $B \to \mathrm{Spec}(R)$ such that $B \otimes K \cong B_0$ and such that $B \otimes Q(R)$ admits smCM.

Remarks. See Section 12. **(1)** If A_0 admits a CM-lift, then $A_0 \otimes K$ admits smCM. **(2)** By Tate we know that any abelian variety over a finite field admits smCM, [72].
(3) If A_0 is an *ordinary* abelian variety over a finite field K, then by using the canonical Serre-Tate lift we see that A_0 admits a CM-lift.
(4) Deuring has proved that any elliptic curve over a finite field admits a CM-lift; see [20], pp. 259 – 263; for a proof also see [55], Section 14, in particular 14.7.
(5) The previous method can be used to show that any abelian variety of dimension g defined over a finite field of p-rank equal to $g - 1$ admits a CM-lift; use [55], 14.6.
(6) We have seen that for an abelian variety A_0 over a finite field K there exists a finite extension $K \subset K'$, and a K'-isogeny $A_0 \otimes K' \sim B_0$ such that B_0 admits a CM-lift.
Do we really need the finite extension and the isogeny to assure a CM-lift ?
(7) (We need the isogeny.) In [56], Theorem B we find: *suppose $g \geq 3$, and let f be an integer, $0 \leq f \leq g - 2$. Then there exists an abelian variety A_0 over $\mathbb{F} := \overline{\mathbb{F}_p}$ of dimension g with p-rank equal to f such that A_0 does not admit a CM-lift.*

22.2. Question. (Do we need a finite extension?) *Does there exist a finite field K and an abelian variety A_0 over K such that any B_0 over K isogenous over K with A_0 does not admit a CM-lift?*

22.3. In the proof of the Honda-Tate theorem analytic tools are used. Indeed we construct CM abelian varieties over $\mathbb{C}$ in order to prove surjectivity of the map $A \mapsto \pi_A$. As a corollary of the Honda-Tate theory we have seen a proof of the Manin Conjecture. However it turns out that for the Manin Conjecture we now have a purely geometric proof, indeed a proof which only uses characteristic p methods, see [58], Section 5.

22.4. Open Problem. *Does there exist a proof of the Honda-Tate theorem 1.2 only using methods in characteristic p ?*

22.5. Over an algebraically closed field k of characteristic zero for a given g it is exactly known which algebras can appear as the endomorphism algebra of a simple abelian variety over k; see [68], pp. 175/176; also see [47], pp. 202/203; see [35], 5.5.

For any Albert algebra (an algebra of finite dimension over $\mathbb{Q}$, with a positive definite anti-involution, equivalently: a finite product of matrix algebras of algebras in the classification list of Albert), and any characteristic, there exists a simple abelian variety over an algebraically closed field of that characteristic having that endomorphism algebra; see [68], pp. 175/176 and [47] pp. 202/203 for characteristic zero; for arbitrary characteristic see [24]; for a discussion see [54], Theorem 3.3 and Theorem 3.4.

22.6. Open Problem. *Suppose a prime number $p > 0$ given. Determine for every $g \in \mathbb{Z}_{>0}$ the possible endomorphism algebras appearing for that g in characteristic p.*

22.7. Open Problem. *For every characteristic and every $g \in \mathbb{Z}_{>0}$ determine all possible endomorphism rings of an abelian variety over an algebraically closed field in that characteristic.*

22.8. Exercise. *For an abelian variety of dimension g over a field K of characteristic zero we have*

$$m(X) := \frac{2g}{[\mathrm{End}^0(A) : \mathbb{Q}]} \quad \in \quad \mathbb{Z}.$$

Give examples of an abelian variety A in positive characteristic where

$$\frac{2g}{[\mathrm{End}^0(A) : \mathbb{Q}]} \quad \notin \quad \mathbb{Z}.$$

22.9. Expectation. *For every $\gamma \in \mathbb{Q}_{>0}$ and every prime number $p > 0$ there exists a field k in characteristic p, and an abelian variety A over k such that*

$$\frac{2g}{[\mathrm{End}^0(A) : \mathbb{Q}]} \quad = \quad \gamma.$$

See [57], Section 2.

References

[1] A. A. Albert – *On the construction of Riemann matrices, I, II.* Ann. Math. **35** (1934), 1 – 28; **36** (1935), 376 – 394.

[2] A. A. Albert – *A solution of the principal problem in the theory of Riemann matrices.* Ann. Math. **35** (1934), 500 – 515.

[3] A. A. Albert – *Involutorial simple algebras and real Riemann matrices.* Ann. Math. **36** (1935), 886 – 964.

[4] C. Birkenhake & H. Lange – *Complex tori.* Progr. Math. 177, Birkhäuser 1999.

[5] A. Blanchard - *Les corps non commutatifs.* Coll. Sup, Presses Univ. France, 1972.

[6] S. Bosch, W. Lütkebohmert & M. Raynaud – *Néron models.* Ergebn. Math. (3) Vol. 21, Springer – Verlag 1990.

[7] N. Bourbaki – *Algèbre.* Chap.VIII: *modules et anneaux semi-simples.* Hermann, Paris 1985.

[8] J. W. S. Cassels & A. Fröhlich (Editors) – *Algebraic number theory.* Academic Press 1967. Chapter VI: J-P. Serre – *Local class field theory* pp. 129–161.

[9] C.-L. Chai & F. Oort – *Hypersymmetric abelian varieties.* Quarterly Journal of Pure and Applied Mathematics, **2** (Special Issue: In honor of John H. Coates), (2006), 1–27.

[10] C.-L. Chai & F. Oort – *Moduli of abelian varieites and p-divisible groups.* Conference on arithmetic geometry, Göttingen July/August 2006. To appear: Clay Mathematics Proceedings.

[11] C.-L. Chai, B. Conrad & F. Oort - *CM-lifting of abelian varieties.* [In preparation]

[12] C. Chevalley – *Une démonstration d'un théorème sur les groupes algébriques.* Journ. de Math. 39 (1960), 307 – 317.

[13] B. Conrad – *A modern proof of Chevalley's theorem on algebraic groups*. J. Ramanujan Math. Soc. **18** (2002), 1 – 18.

[14] B. Conrad – *Chow's K/k-image and K/k-trace, and the Lang-Néron theorem*. Enseign. Math. (2) **52** (2006), 37–108.

[15] G. Cornell, J. H. Silverman (Editors) – *Arithmetic geometry*. Springer – Verlag 1986.

[16] C. W. Curtis & I. Reiner – *Representation theory of finite groups and associative algebras*. Intersc. Publ.1962.

[17] P. Deligne – *Variétés abéliennes sur un corps fini*. Invent. Math. **8** (1969), 238 – 243.

[18] P. Deligne – *Hodge cycles on abelian varieties*. Hodge cycles, motives and Shimura varieties (Eds P. Deligne et al). Lect. Notes Math. **900**, Springer – Verlag 1982; pp. 9 - 100.

[19] M. Demazure – *Lectures on p-divisible groups*. Lecture Notes Math. 302, Springer – Verlag 1972.

[20] M. Deuring – *Die Typen der Multiplikatorenringe elliptischer Funktionenkörper*. Abh. Math. Sem. Hamburg **14** (1941), 197 – 272.

[21] G. Faltings – *Endlichkeitssätze für abelsche Varietäten über Zahlkörpern*. Invent. Math. **73** (1983), 349 – 366.

[22] G. Faltings & G. Wüstholz – *Rational points*. Seminar Bonn / Wuppertal 1983/84. Asp. Math. E6, Vieweg 1984.

[23] = [GM] G. van der Geer & B. Moonen – *Abelian varieties*. [In preparation] This will be cited as [GM].

[24] L. Gerritzen – *On multiplications of Riemann matrices*. Math. Ann **194** (1971), 109 – 122.

[25] A. Grothendieck – *Fondements de la géométrie algébrique*. Extraits du Séminaire Bourbaki 1957 - 1962. Secr. math., Paris 1962.

[26] A. Grothendieck – *Groupes de Barsotti-Tate et cristaux de Dieudonné*. Sém. Math. Sup. **45**, Presses de l'Univ. de Montreal, 1970.

[27] A. Grothendieck – *Esquisse d'un programme*. Manuscript 56 pp., January 1984. Reproduced in: Geometric Galois actions (Ed. L. Schneps & P. Lochak). Vol. 1: Around Grothendieck's *Esquisse d'un programme*. London Math. Soc. Lect. Note Series 242, Cambridge Univ. Press 1997; pp. 5 – 48 (English translation pp. 243 – 283). http://www.institut.math.jussieu.fr/ leila/grothendieckcircle/EsquisseEng.pdf

[28] H. Hasse - *Zahlentheorie*. Akad. Verlag, Berlin 1949 (first printing, second printing 1963).

[29] T. Honda – *Isogeny classes of abelian varieties over finite fields*. Journ. Math. Soc. Japan **20** (1968), 83 – 95.

[30] A. J. de Jong – *Homomorphisms of Barsotti-Tate groups and crystals in positive characteristics*. Invent. Math. **134** (1998) 301-333, Erratum **138** (1999) 225.

[31] A. J. de Jong – *Barsotti-Tate groups and crystals*. Documenta Mathematica, Extra Volume ICM 1998, II, 259 – 265.

[32] N. M. Katz – *Slope filtration of F–crystals*. Journ. Géom. Alg. Rennes, Vol. I, Astérisque **63** (1979), Soc. Math. France, 113 - 164. are due to Tate

[33] S. Lang – *Fundamentals of diophantine geometry*. Springer – Verlag 1983.

[34] S. Lang – *Complex multiplication*. Grundl. math. Wissensch. 255, Springer – Verlag 1983.

[35] H. Lange & C. Birkenhake - *Complex abelian varieties*. Grundl. math. Wissensch. 302, Springer – Verlag 1992.

[36] H. W. Lenstra jr & F. Oort – *Simple abelian varieties having a prescribed formal isogeny type*. Journ. Pure Appl. Algebra **4** (1974), 47 - 53.

[37] K.-Z. Li & F. Oort – *Moduli of supersingular abelian varieties*. Lecture Notes Math. 1680, Springer - Verlag 1998.

[38] J. Lubin & J. Tate – . *Formal moduli for one-parameter formal Lie groups*. Bull. Soc. Math. France **94** (1966), 49 – 66.

[39] Yu. I. Manin – *The theory of commutative formal groups over fields of finite characteristic*. Usp. Math. **18** (1963), 3-90; Russ. Math. Surveys **18** (1963), 1-80.

[40] J. Milne – it The fundamental theorem of complex multiplication. arXiv:0705.3446v1, 23 May 2007

[41] S. Mochizuki – *The local pro-p anabelian geometry of curves*. Invent. Math. **138** (1999), 319 – 423.

[42] L. Moret-Bailly – *Pinceaux de variétés abéliennes*. Astérisque 129. Soc. Math. France

1985.

[43] S. Mori – *On Tate's conjecture concerning endomorphisms of abelian varieties*. Itl. Sympos. Algebr. Geom. Kyoto 1977 (Ed. M. Nagata). Kinokuniya Book-store 1987, pp. 219 - 230.

[44] D. Mumford – *A note of Shimura's paper "Discontinuous groups and abelian varieties"*. Math. Ann. **181** (1969), 345 - 351.

[45] D. Mumford – *Geometric invariant theory*. Ergebn. Math. Vol. 34, Springer – Verlag 1965 (second version 1982, 1994).

[46] D. Mumford - A note of Shimura's paper "Discontinuous groups and abelian varieties". Math. Ann. **181** (1969), 345-351.

[47] D. Mumford – *Abelian varieties*. Tata Inst. Fund. Research and Oxford Univ. Press 1970 (2nd printing 1974).

[48] A. Néron – *Modèles minimaux des variétés abéliennes sur les corps locaux et globaux*. Publ. Math. IHES **21**, 1964.

[49] F. Oort – *Commutative group schemes*. Lect. Notes Math. 15, Springer - Verlag 1966.

[50] F. Oort – *Algebraic group schemes in characteristic zero are reduced*. Invent. Math. **2** (1966), 79 - 80.

[51] F. Oort – *The isogeny class of a CM-type abelian variety is defined over a finite extension of the prime field*. Journ. Pure Appl. Algebra **3** (1973), 399 - 408.

[52] F. Oort – *Subvarieties of moduli spaces*. Invent. Math. **24** (1974), 95 - 119.

[53] F. Oort – *Good and stable reduction of abelian varieties*. Manuscr. Math. **11** (1974), 171 - 197.

[54] F. Oort – *Endomorphism algebras of abelian varieties*. Algebraic Geometry and Commut. Algebra in honor of M. Nagata (Ed. H. Hijikata et al), Kinokuniya Cy Tokyo, Japan, 1988, Vol II; pp. 469 - 502.

[55] F. Oort – — *Lifting algebraic curves, abelian varieties and their endomorphisms to characteristic zero*. Algebraic Geometry, Bowdoin 1985 (Ed. S. J. Bloch). Proceed. Sympos. Pure Math. **46** Part 2, AMS 1987; pp. 165 -195.

[56] F. Oort – *CM-liftings of abelian varieties*. Journ. Algebraic Geometry **1** (1992), 131 - 146.

[57] F. Oort – *Some questions in algebraic geometry*, preliminary version. Manuscript, June 1995. http://www.math.uu.nl/people/oort/

[58] F. Oort — *Newton polygons and formal groups: conjectures by Manin and Grothendieck*. Ann. Math. **152** (2000), 183 - 206.

[59] F. Oort – *Newton polygon strata in the moduli space of abelian varieties*. In: *Moduli of abelian varieties*. (Ed. C. Faber, G. van der Geer, F. Oort). Progress Math. 195, Birkhäuser Verlag 2001; pp. 417 - 440.

[60] F. Oort & Yu. G. Zarhin - *Endomorphism algebras of complex tori*. Math. Ann. **303** (1995), 11 - 29.

[61] I. Reiner – *Maximal orders*. London Math. Soc. Monographs Vol. 28. Oxford 2003.

[62] M. Demazure & A. Grothendieck – *Schémas en groupes, Séminaire de géométrie algébrique, SGA3*. Vol I: Lect. Notes Math. **151**, Springer – Verlag 1970.

[63] A. Grothendieck – *Séminaire de Géométrie Algébrique, Groupes de monodromie en géométrie algébrique, SGA 7*. Lect. Notes Math. **288**, Springer – Verlag 1972.

[64] R. Schoof – *Nonsingular plane cubic curves over finite fields*. Journal Computat. Theory, Series A, **46** (1987) 183 – 211.

[65] J-P. Serre – *Corps locaux*. Hermann Paris 1962.

[66] J-P. Serre – *Géométrie algébrique et géométrie analytique*. Ann. Inst. Fourier 6) (1956), 1 – 42.

[67] J-P. Serre & J. Tate – *Good reduction of abelian varieties*. Ann. Math. **88** (1968), 492 – 517.

[68] G. Shimura – *On analytic families of polarized abelian varieties and automorphic functions*. Ann. Math. **78** (1963), 149 – 193.

[69] G. Shimura & Y. Taniyama – *Complex multiplication of abelian varieties and its applications to number theory*. Publ. Math. Soc. Japan **6**, Tokyo 1961.

[70] T. Shioda – *Supersingular K3 surfaces*. In: *Algebraic Geometry*, Copenhagen 1978 (Ed. K. Lønsted). Lect. Notes Math. 732, Springer - Verlag (1979), 564 - 591.

[71] J. Silverman – *The arithmetic of elliptic curves*. Grad. Texts Math. 106, Springer – Verlag, 1986.

[72] J. Tate – *Endomorphisms of abelian varieties over finite fields*. Invent. Math. **2** (1966), 134-144.

[73] J. Tate – *Classes d'isogénies de variétés abéliennes sur un corps fini (d'àpres T. Honda)*. Sém. Bourbaki **21** (1968/69), Exp. 352.

[74] 2005-05 VIGRE number theory working group. Organized by Brian Conrad and Chris Skinner. On: http://www.math.Isa.umich.edu/ bdconrad/vigre04.html

[75] W. C. Waterhouse – *Abelian varieties over finite fields*. Ann. Sc. Ec. Norm. Sup. 4.Ser, **2** (1969), 521 – 560).

[76] W. C. Waterhouse – *Introduction to affine group schemes*. Grad. Texts Math. 66, Springer – Verlag, 1979.

[77] W. C. Waterhouse & J. S. Milne – *Abelian varieties over finite fields*. Proc. Sympos. pure math. Vol. XX, 1969 Number Theory Institute (Stony Brook), AMS 1971, pp. 53 – 64.

[78] A. Weil – *Sur les courbes algébriques et les variétés qui s'en déduisent*. Hermann, 1948.

[79] A. Weil – *Variétés abéliennes et courbes algébriques*. Hermann, 1948.

[80] C.-F. Yu – *The isomorphism classes of abelian varieties of CM-type*. Journ. Pure Appl. Algebra **187** (2004), 305 – 319.

[81] J. G. Zarhin – *Isogenies of abelian varieties over fields of finite characteristic*. Math. USSR Sbornik **24** (1974), 451 – 461.

[82] J. G. Zarhin – *A remark on endomorphisms of abelian varieties over function fields of finite characteristic*. Math. USSR Izv. **8** (1974), 477 – 480.

[83] J. G. Zarhin – *Homomorphisms of abelian varieties over finite fields*. Summer school in Göttingen, June 2007. See this volume.

Higher-Dimensional Geometry over Finite Fields
D. Kaledin and Y. Tschinkel (Eds.)
IOS Press, 2008

How to obtain global information
from computations over finite fields

Michael STOLL

*School of Engineering and Science, Jacobs University Bremen, P.O.Box 750561,
28725 Bremen, Germany.*
e-mail: m.stoll@jacobs-university.de

Abstract. This is an extended version of the talk I gave at the summer
school in Göttingen in July 2007. We discuss the Mordell-Weil Sieve and
some applications.

1. The Problem

Let A be an abelian variety over $\mathbb{Q}$ (for simplicity; we could work over an arbitrary number field instead), and let $V \subset A$ be a "transversal" subvariety, i.e., a subvariety that does not contain a translate of a nontrivial subabelian variety of A.

Our goal is to obtain information on $V(\mathbb{Q})$, the set of rational points on V. For example, we would like to prove that $V(\mathbb{Q}) = \emptyset$.

The standard example for this situation is when we have a curve C over $\mathbb{Q}$ of genus $g \geq 2$. If we we know a rational divisor class D of degree 1 on C, then we can use D as a base-point for an embedding $\iota : C \hookrightarrow J$, $P \mapsto [P] - D$. Here $A = J$ is the Jacobian variety of C, and $V = \iota(C) \subset A$.

2. The Idea

Our approach is to combine global and local information in the following way. The *global* input is the knowledge of the Mordell-Weil group $A(\mathbb{Q})$. This means that we need to know explicit generators of this group (which is a finitely generated abelian group, by the Mordell-Weil Theorem). Note that it requires some nontrivial computations and a bit of luck to obtain this information. If A is the Jacobian of a curve of genus 2, it is usually possible to perform the necessary computations successfully. This includes a 2-descent on A as described in [1], a search for rational points on A (see for example [2]), possibly visualization computations to improve the upper bound for the rank obtained by 2-descent (see [3] and [4]) and canonical height computations in order to make sure that one has generators of the full group (see [5] and [6]). The latter part, which is currently only available for genus 2 Jacobians, can be replaced by a computation that checks that the index of the known subgroup is prime to a finite set of primes. Compare the genus 3 example from [7] discussed in Section 7 below.

The *local* input is obtained by looking at the situation over $\mathbb{F}_p$, for a suitable finite set S of primes p. We assume (for now) that p is a prime of good reduction for A and V. We can then compute the finite abelian group $A(\mathbb{F}_p)$ and determine its subset $V(\mathbb{F}_p)$. Denote by

$$\alpha_p : V(\mathbb{F}_p) \hookrightarrow A(\mathbb{F}_p)$$

the inclusion map.

Since we assume we know generators of $A(\mathbb{Q})$, we can also compute the group homomorphism

$$\beta_p : A(\mathbb{Q}) \to A(\mathbb{F}_p)\,.$$

If $P \in A(\mathbb{Q})$ is in $V(\mathbb{Q})$, then $\beta_p(P) \in \alpha_p\big(V(\mathbb{F}_p)\big)$.

Thus we obtain *congruence conditions* on the coefficients of P with respect to our generators of $A(\mathbb{Q})$.

We now combine the information we obtain from all the primes in the set S. Consider the following commutative diagram.

$$
\begin{array}{ccc}
V(\mathbb{Q}) & \hookrightarrow & A(\mathbb{Q}) \\
\downarrow & & \downarrow \; \beta = \prod\limits_{p\in S} \beta_p \\
\prod\limits_{p\in S} V(\mathbb{F}_p) & \xrightarrow{\;\alpha = \prod\limits_{p\in S} \alpha_p\;} & \prod\limits_{p\in S} A(\mathbb{F}_p)
\end{array}
$$

As before, if $P \in A(\mathbb{Q})$ is in $V(\mathbb{Q})$, then $\beta(P) \in \mathrm{im}(\alpha)$.

In particular, if $\mathrm{im}(\alpha) \cap \mathrm{im}(\beta) = \emptyset$, then this *proves* that $V(\mathbb{Q}) = \emptyset$.

This technique is called the *Mordell-Weil Sieve*. It appears first in Scharaschkin's thesis [8]. It was later applied to many genus 2 curves by Flynn [9], and more recently used and improved by Bruin and Stoll [10] in a project whose aim it was to decide for all genus 2 curves $C : y^2 = f(x)$, where f has integral coefficients of absolute value ≤ 3, whether C has rational points or not; see Section 4 below.

3. The Poonen Heuristic

Assuming that indeed $V(\mathbb{Q}) = \emptyset$, what are our chances to prove this fact in the way just described?

The following considerations are due to *Bjorn Poonen* [11].

Let B be some large integer. We will consider all primes $p < B^2$.

For $\rho > 0$, there is a number $\delta_\rho > 0$ such that there are at least $\delta_\rho B^\rho$ B-*smooth* integers $\leq B^\rho$, for B large. (An integer is "B-smooth" if all its prime divisors are $\leq B$.)

We assume that a similar statement is true for the set $\{\#A(\mathbb{F}_p) : p < B^2\}$. More precisely, we make the following

Assumption 1 *Let*

$$S_B = \{p < B^2 : p \text{ is good and } \#A(\mathbb{F}_p) \text{ is } B\text{-smooth}\}.$$

Then

$$\liminf_{B \to \infty} \frac{\#S_B}{\pi(B^2)} > 0.$$

By the Weil bounds, we have

$$\#A(\mathbb{F}_p) \leq (\sqrt{p} + 1)^{2\dim A} \leq B^{2\dim A}(1 + o(1)).$$

If the group orders $\#A(\mathbb{F}_p)$ behave like random integers in this range, then the assumption should be valid, by the result on the density of B-smooth numbers up to B^ρ (taking $\rho = 2\dim A$).

The exponent of $A(\mathbb{F}_p)$ for $p \in S_B$ divides

$$\prod_{q \leq B} q^{\lfloor \log_q \#A(\mathbb{F}_p) \rfloor} \leq B^{2\pi(B)\dim A}(1 + o(1)) \approx e^{2B\dim A}.$$

The inequality comes from $q^{\lfloor \log_q \#A(\mathbb{F}_p) \rfloor} \leq \#A(\mathbb{F}_p) \leq B^{2\dim A}(1 + o(1))$, and for the estimate, we use the Prime Number Theorem $\pi(x) \sim x/\log x$.

Let r be the rank of $A(\mathbb{Q})$. Then the image of $A(\mathbb{Q})$ in $\prod_{p \in S_B} A(\mathbb{F}_p)$ has size at most $c\, e^{2rB\dim A}$ for some constant c. This is because each generator of $A(\mathbb{Q})$ maps to an element of order $\ll e^{2B\dim A}$.

On the other hand, for B large, we have

$$\#\prod_{p \in S_B} A(\mathbb{F}_p) \approx e^{2\delta_B B^2 \dim A},$$

where $\delta_B = \dfrac{\#S_B}{\pi(B^2)} \geq \delta > 0$, by Assumption 1.

We now make the following

Assumption 2 *$V(\mathbb{F}_p)$ behaves like a random subset of $A(\mathbb{F}_p)$ of size $\approx p^{\dim V}$.*

Then $\prod_{p \in S_B} V(\mathbb{F}_p)$ is a random subset of $\prod_{p \in S_B} A(\mathbb{F}_p)$ of size $\approx e^{2\delta_B B^2 \dim V}$. Recall the diagram of maps

$$\begin{array}{ccc} V(\mathbb{Q}) & \lhook\joinrel\longrightarrow & A(\mathbb{Q}) \\ \downarrow & & \downarrow \beta_B \\ \prod_{p \in S_B} V(\mathbb{F}_p) & \xrightarrow{\ \alpha_B\ } & \prod_{p \in S_B} A(\mathbb{F}_p) \end{array}$$

We have seen that we have the following estimates.

$$\# \prod_{p \in S_B} A(\mathbb{F}_p) \approx e^{2\delta_B B^2 \dim A}, \quad \#\mathrm{im}(\alpha_B) \approx e^{2\delta_B B^2 \dim V}, \quad \#\mathrm{im}(\beta_B) \leq c\, e^{2rB \dim A}$$

So the probability that $\mathrm{im}(\alpha) \cap \mathrm{im}(\beta) \neq \emptyset$ is (roughly)

$$\frac{\#\mathrm{im}(\alpha_B) \cdot \#\mathrm{im}(\beta_B)}{\# \prod_{p \in S_B} A(\mathbb{F}_p)} < c\, e^{2(rB \dim A - \delta_B B^2 (\dim A - \dim V))}.$$

(This is in fact the expected size of the intersection, which gives an upper bound for the relevant probability.) Since $\delta_B \geq \delta > 0$, this tends to zero when $B \to \infty$. Thus we obtain the following result.

Proposition 3 *Under Assumptions 1 and 2, the Mordell-Weil Sieve will be successful with probability 1.*

Note that Assumption 2 will not be valid when $V(\mathbb{Q}) \neq \emptyset$, since in this case, $V(\mathbb{F}_p)$ will always contain the images of the global points in $V(\mathbb{Q})$. Of course, the Mordell-Weil Sieve computation cannot succeed in this case. On the other hand, in the absence of global points, there does not seem to be any reason for a non-random behavior of the sets $V(\mathbb{F}_p)$, and so Assumption 2 should make sense in this case. In any case, if we perform the computation and it succeeds, this will prove *unconditionally* that $V(\mathbb{Q}) = \emptyset$; the assumptions are only necessary to convince us that we will succeed eventually.

4. Application: Proving That Curves Do Not Have Rational Points

In a joint project with Nils Bruin [10], we considered all "small" curves of genus 2:

$$C : y^2 = f_6 x^6 + f_5 x^5 + f_4 x^4 + f_3 x^3 + f_2 x^2 + f_1 x + f_0$$

with $f_0, f_1, \ldots, f_6 \in \{-3, -2, -1, 0, 1, 2, 3\}$.

Our goal was to decide whether C has rational points, for *all* such curves C.

Among the $\approx 200\,000$ isomorphism classes of such curves, there were $\approx 1\,500$, for which more straight-forward approaches (searching for rational points, checking for local points, performing a "2-cover descent"; for details see [10]) were unsuccessful.

We were able to determine generators of $J(\mathbb{Q})$ for these curves (this is conditional on the Birch and Swinnerton-Dyer Conjecture in 42 cases). We then applied the *Mordell-Weil Sieve* to these curves and their Jacobians; for *all* of them, we could prove in this way that $C(\mathbb{Q}) = \emptyset$.

5. Practical Considerations and Improvements

In practice, the computation suggested by the heuristic is infeasible. The sets we have to deal with would be much too large.

Instead, we pick a smooth number N and work with

$$
\begin{array}{ccc}
V(\mathbb{Q}) & \xrightarrow{\hspace{3cm}} & \dfrac{A(\mathbb{Q})}{NA(\mathbb{Q})} \\[2ex]
\downarrow & & \downarrow{\scriptstyle \beta^{(N)}} \\[2ex]
\displaystyle\prod_{p\in S} V(\mathbb{F}_p) & \xrightarrow{\hspace{1cm}\alpha^{(N)}\hspace{1cm}} & \displaystyle\prod_{p\in S} \dfrac{A(\mathbb{F}_p)}{NA(\mathbb{F}_p)}
\end{array}
$$

where S is a set of primes such that $A(\mathbb{F}_p)/NA(\mathbb{F}_p)$ is reasonably large (i.e., such that a large part of the exponent divides N)

Instead of computing the subset of $A(\mathbb{Q})/NA(\mathbb{Q})$ of elements that map under $\beta^{(N)}$ into the image of $\alpha^{(N)}$ directly in one go, we build N successively as a product of prime factors, keeping track of the sets $\Sigma(N') = (\beta^{(N')})^{-1}\big(\mathrm{im}(\alpha^{(N')})\big)$ at each step. If we go from N' to $N'q$, we then only have to check all possible lifts to $A(\mathbb{Q})/N'qA(\mathbb{Q})$ of the elements of $\Sigma(N')$. The number of such checks is $q^r \#\Sigma(N')$, and the total complexity will be much less than N^r (which corresponds to the one-step approach) if we can make sure that the sets $\Sigma(N')$ are considerably smaller than $(N')^r$. For more details on our implementation, see [12].

The procedure as decribed so far works well when the rank is at most 2. To go further than this, we need to use more information than just what we can obtain mod p for primes p of good reduction. For the method, this restriction is unnecessary, and we can work more generally with finite quotients of $A(\mathbb{Q}_p)$ in place of $A(\mathbb{F}_p)$. In this way, we can include information at bad primes and "deep" information modulo higher powers of p. For example, the component group of the Néron model of A at a prime p of bad reduction can provide useful information.

These improvements make the Mordell-Weil Sieve practical for a curve sitting in an abelian surface when $r \leq 3$ and maybe even $r = 4$ (but the evidence in this case is too sparse to say something definite).

6. A Variation

Even when V does have rational points, we can use the Mordell-Weil Sieve to rule out rational points on V with certain additional properties.

For example, we can show that there is no $P \in V(\mathbb{Q})$ such that

- P is in a certain *residue class* mod n, or
- P is in a certain *coset* mod $nA(\mathbb{Q})$.

(These two kinds of condition are actually equivalent: via the maps $A(\mathbb{Q}) \to A(\mathbb{Q}_p)/A(\mathbb{Q}_p)_n$ and $C(\mathbb{Q}_p) \to A(\mathbb{Q}_p)/A(\mathbb{Q}_p)_n$, congruence conditions mod p^n can be translated into coset conditions mod $eA(\mathbb{Q})$, where e is the exponent of $A(\mathbb{Q}_p)/A(\mathbb{Q}_p)_n$, and conversely. Here $A(\mathbb{Q}_p)_n$ is the nth kernel of reduction.)

To deal with the first kind of condition, we restrict to the relevant subset of $V(\mathbb{Q}_p)$ for the primes p dividing n.

To deal with the second kind of condition, we use values of N that are multiples of n and restrict to the relevant cosets in $A(\mathbb{Q})/NA(\mathbb{Q})$.

If we can determine an integer n such that no two points in $V(\mathbb{Q})$ are in the same coset mod $nA(\mathbb{Q})$, then this refinement of the Mordell-Weil Sieve allows us (assuming a suitably modified version of the Poonen Heuristic) to *determine* the set $V(\mathbb{Q})$ in the following way. For each coset of $nA(\mathbb{Q})$, we search for points in this coset that are on V, and at the same time, we run the Mordell-Weil Sieve in an attempt to show that no such point exists. One of the two procedures should be successful, and so we will either have shown that there is no point on V in the coset, or else we will have found such a point, and then we know that it must be the only one.

If V is a curve in its Jacobian A, and $r < \dim A$ (which is the genus of the curve), then we can use *Chabauty's method* to obtain such a "separating" integer n. For details, see [12].

7. An Example

Consider the smooth plane quartic curve

$$C : -2x^3 y - 2x^3 z + 6x^2 yz + 3xy^3 - 9xy^2 z + 3xyz^2 - xz^3 + 3y^3 z - yz^3 = 0 \,.$$

It has the known rational points

$$(1:0:0), \quad (0:1:0), \quad (0:0:1), \quad (1:1:1) \,.$$

Any point $P \in C(\mathbb{Q})$ such that

$$P \equiv (0:1:0) \bmod 3 \quad \text{and} \quad P \equiv (1:0:0) \text{ or } (1:1:1) \bmod 2$$

would lead to a primitive integral solution of $x^2 + y^3 = z^7$. Note that the known points do not satisfy this condition.

We want to prove that no rational point on C satisfies the condition.

(This was the last step in the complete solution of $x^2 + y^3 = z^7$, see [7].)

Let J be the Jacobian of C. We can prove that the rank of $J(\mathbb{Q})$ is 3, and we find generators of a subgroup of $J(\mathbb{Q})$ of finite index prime to 14.

We need to use information at the *bad primes* 2 and 3; we will use the component groups of the Néron model at these primes. We find

$$J(\mathbb{Q}_2) \longrightarrow \Phi_2 \cong \frac{\mathbb{Z}}{4\mathbb{Z}} \times \frac{\mathbb{Z}}{4\mathbb{Z}}$$

$$J(\mathbb{Q}_3) \longrightarrow \Phi_3 \cong \frac{\mathbb{Z}}{7\mathbb{Z}}$$

The congruence conditions on $P \in C(\mathbb{Q})$ correspond to subsets of size 3 and 1, respectively.

With the additional information coming from

$$J(\mathbb{F}_{23}) \cong \frac{\mathbb{Z}}{2\mathbb{Z}} \times \frac{\mathbb{Z}}{16\mathbb{Z}} \times \frac{\mathbb{Z}}{16\mathbb{Z}} \times \frac{\mathbb{Z}}{32\mathbb{Z}}$$

$$J(\mathbb{F}_{97}) \cong \frac{\mathbb{Z}}{98\mathbb{Z}} \times \frac{\mathbb{Z}}{98\mathbb{Z}} \times \frac{\mathbb{Z}}{98\mathbb{Z}}$$

$$J(\mathbb{F}_{13}) \longrightarrow \frac{\mathbb{Z}}{14\mathbb{Z}}$$

we get a contradiction. Thus we have shown that no rational points on C exist that satisfy the congruences mod 2 and mod 3.

Since we are working in $J(\mathbb{Q})/NJ(\mathbb{Q})$ with $N = 2^a \cdot 7^b$, it suffices to know that the known points in $J(\mathbb{Q})$ generate a subgroup of index prime to 14. In particular, it is not necessary to know that we actually have generators of $J(\mathbb{Q})$. Since there is no explicit theory of canonical heights available for Jacobians of genus 3 curves, we would not be able to prove that we do have generators. On the other hand, we can verify that the index of the subgroup generated by the points we know is prime to a given prime number q, by considering maps $J(\mathbb{Q}) \to J(\mathbb{F}_p)$ for primes p such that $q \mid \#J(\mathbb{F}_p)$.

8. Another Application

We can use the Mordell-Weil Sieve to show that for every $P \in V(\mathbb{Q})$ there is a known point $Q \in V(\mathbb{Q})$ such that $P - Q$ is in a subgroup of very large index in $A(\mathbb{Q})$. More precisely, if at some stage in the computation, we find that the set $\Sigma(N) \subset A(\mathbb{Q})/NA(\mathbb{Q})$ of elements that are consistent with the local information coincides with the image of the known points in $V(\mathbb{Q})$, then this implies that for any unknown point $P \in V(\mathbb{Q})$, there must be a known point $Q \in V(\mathbb{Q})$ such that $P \in Q + NA(\mathbb{Q})$. Since (by assumption), $P \neq Q$, this implies that $\hat{h}(P) \gg N^2$, and so any unknown point in $V(\mathbb{Q})$ must be extremely large (if N is not very small).

In some cases, we can use Baker's Method to get a (very large) bound on the height of *integral points* on V. We can then combine this with the Mordell-Weil Sieve information to show that we know all the integral points on V. This is ongoing work of Bugeaud, Mignotte, Siksek, Stoll, and Tengely, see [13]. For example, we can determine the set of integral points on the curve

$$C : y^2 - y = x^5 - x \,.$$

The Jacobian J of C has Mordell-Weil rank 3. The Mordell-Weil Sieve computation gave

$$N = 444932978061474820647297268617994065251575448327430679656821404800 0,$$

and after another step based on similar ideas that replaces $NJ(\mathbb{Q})$ be a sublattice of much larger index, this can be used to show that

$$\log x(P) \geq 0.95 \times 10^{2159}$$

for every unknown integral point P on C. This turned out to be much more than sufficient to contradict the upper bound, and so we can conclude that there are no unknown integral points. The complete list of integral points is therefore given by

$$(x, y) = (-1, 0), \ (-1, 1), \ (0, 0), \ (0, 1), \ (1, 0), \ (1, 1), \ (2, -5),$$
$$(2, 6), \ (3, -15), \ (3, 16), \ (30, -4929), \ (30, 4930).$$

References

[1]   M. Stoll, *Implementing 2-descent on Jacobians of hyperelliptic curves*, Acta Arith. **98**, 245–277 (2001).

[2]   N. Bruin and M. Stoll, *Finding Mordell-Weil generators on genus 2 Jacobians*, in preparation.

[3]   N. Bruin, *Visualisation of Sha[2] in Abelian Surfaces*, Math. Comp. **73**, no. 247, 1459–1476 (2004).

[4]   N. Bruin and E.V. Flynn, *Exhibiting Sha[2] on Hyperelliptic Jacobians*, Journal of Number Theory **118**, 266–291 (2006).

[5]   M. Stoll, *On the height constant for curves of genus two*, Acta Arith. **90** (1999), 183–201.

[6]   M. Stoll, *On the height constant for curves of genus two, II*, Acta Arith. **104** (2002), 165–182.

[7]   B. Poonen, E.F. Schaefer, and M. Stoll, *Twists of $X(7)$ and primitive solutions to $x^2 + y^3 = z^7$*, Duke Math. J. **137**, 103–158 (2007).

[8]   V. Scharaschkin, *Local-global problems and the Brauer-Manin obstruction*, Ph.D. thesis, University of Michigan (1999).
      See also *The Brauer-Manin obstruction for curves*, Manuscript (1998).

[9]   E.V. Flynn, *The Hasse Principle and the Brauer-Manin obstruction for curves*, Manuscripta Math. **115**, 437–466 (2004).

[10] N. Bruin and M. Stoll, *Deciding existence of rational points on curves: an experiment*, to appear in Experiment. Math.

[11] B. Poonen, *Heuristics for the Brauer-Manin obstruction for curves*, Experiment. Math. **15**, 415–420 (2006).

[12] N. Bruin and M. Stoll, *The Mordell-Weil sieve: Proving non-existence of rational points on curves*, in preparation.

[13] Y. Bugeaud, M. Mignotte, S. Siksek, M. Stoll, and S. Tengely, *Integral points on hyperelliptic curves*, in preparation.

Geometry of Shimura varieties of Hodge type over finite fields

Adrian VASIU

Department of Mathematical Sciences, Binghamton University,
Binghamton, NY 13902-6000, U.S.A.
e-mail: adrian@math.binghamton.edu

Abstract. We present a general and comprehensive overview of recent
developments in the theory of integral models of Shimura varieties of
Hodge type. The paper covers the following topics: construction of integral models, their possible moduli interpretations, their uniqueness,
their smoothness, their properness, and basic stratifications of their special fibres.

1. Introduction

This paper is an enlarged version of the three lectures we gave in July 2007 during
the summer school *Higher dimensional geometry over finite fields*, June 25 - July
06, 2007, Mathematisches Institut, Georg-August-Universität Göttingen.

The goal of the paper is to provide to non-specialists an *efficient, accessible,
and in depth* introduction to the theory of *integral models* of *Shimura varieties
of Hodge type*. Accordingly, the paper will put a strong accent on defining the
main objects of interest, on listing the main problems, on presenting the main
techniques used in approaching the main problems, and on stating very explicitly
the main results obtained so far. This is not an easy task, as only to be able to list
the main problems one requires a good comprehension of the *language of schemes*,
of *reductive groups*, of *abelian varieties*, of *Hodge cycles* on abelian varieties, of
cohomology theories (including *étale and crystalline* ones), of *deformation theories*,
of *p-divisible groups*, and of *F-crystals*. Whenever possible, proofs are included.

We begin with a motivation for the study of Shimura varieties of Hodge
type. Let X be a connected, smooth, projective variety over $\mathbb{C}$. We recall that
the *albanese variety* of X is an abelian variety $\mathrm{Alb}(X)$ over $\mathbb{C}$ equipped with
a morphism $a_X : X \to \mathrm{Alb}(X)$ that has the following universal property. If
$b_X : X \to B$ is another morphism from X to an abelian variety B over $\mathbb{C}$,
then there exists a unique morphism $c : \mathrm{Alb}(X) \to B$ such that the following
identity $b_X = c \circ a_X$ holds. This universal property determines $\mathrm{Alb}(X)$ uniquely
up to isomorphisms. Not only $\mathrm{Alb}(X)$ is uniquely determined by X, but also the
image $\mathrm{Im}(a_X)$ is uniquely determined by X up to isomorphisms. Thus to X one
associates an abelian variety $\mathrm{Alb}(X)$ over $\mathbb{C}$ as well as a closed subvariety $\mathrm{Im}(a_X)$
of it. If X belongs to a good class $\mathfrak{C}$ of connected, smooth, projective varieties

over $\mathbb{C}$, then deformations of X would naturally give birth to deformations of the closed embedding $\mathrm{Im}(a_X) \hookrightarrow \mathrm{Alb}(X)$. Thus the study of moduli spaces of objects of the class $\mathfrak{C}$ is very much related to the study of moduli spaces of abelian schemes endowed with certain closed subschemes (which naturally give birth to some polarizations). For instance, if X is a curve, then $\mathrm{Alb}(X) = \mathrm{Jac}(X)$ and the morphism a_X is a closed embedding; to this embedding one associates naturally a principal polarization of $\mathrm{Jac}(X)$. This implies that different moduli spaces of geometrically connected, smooth, projective curves are subspaces of different moduli spaces of principally polarized abelian schemes.

For the sake of generality and flexibility, it does not suffice to study moduli spaces of abelian schemes endowed with polarizations and with certain closed subschemes. More precisely, one is naturally led to study moduli spaces of polarized abelian schemes endowed with *families* of *Hodge cycles*. They are called *Shimura varieties of Hodge type* (see Subsection 3.4). The classical *Hodge conjecture* predicts that each Hodge cycle is an *algebraic cycle*. Thus we refer to Subsection 2.5 for a quick introduction to Hodge cycles on abelian schemes over reduced $\mathbb{Q}$–schemes. Subsections 2.1 to 2.5 review basic properties of algebraic groups, of Hodge structures, and of families of tensors.

Shimura varieties can be defined abstractly via few axioms due to Deligne (see Subsection 3). They are in natural bijection to *Shimura pairs* $(G, \mathcal{X})$. Here G is a reductive group over $\mathbb{Q}$ and X is a hermitian symmetric domain whose points form a $G(\mathbb{R})$-conjugacy class of homomorphisms $(\mathbb{C} \setminus \{0\}, \cdot) \to G_{\mathbb{R}}$ of real groups, that are subject to few axioms. Initially one gets a complex Shimura variety $\mathrm{Sh}(G, \mathcal{X})_{\mathbb{C}}$ defined over $\mathbb{C}$ (see Subsection 3.1). The totally discontinuous, locally compact group $G(\mathbb{A}_f)$ acts naturally on $\mathrm{Sh}(G, \mathcal{X})_{\mathbb{C}}$ from the right. Cumulative works of Shimura, Taniyama, Deligne, Borovoi, Milne, etc., have proved that $\mathrm{Sh}(G, \mathcal{X})_{\mathbb{C}}$ has a *canonical model* $\mathrm{Sh}(G, \mathcal{X})$ over a number field $E(G, \mathcal{X})$ which is intrinsically associated to the Shimura pair $(G, \mathcal{X})$ and which is called the *reflex field* of $(G, \mathcal{X})$ (see Subsection 3.2). One calls $\mathrm{Sh}(G, \mathcal{X})$ together with the natural right action of $G(\mathbb{A}_f)$ on it, as the Shimura variety defined by the Shimura pair $(G, \mathcal{X})$. For instance, if $G = \mathbf{GL}_2$ and $\mathcal{X} \overset{\sim}{\to} \mathbb{C} \setminus \mathbb{R}$ is isomorphic to two copies of the upper half-plane, then $\mathrm{Sh}(G, \mathcal{X})$ is the *elliptic modular variety* over $\mathbb{Q}$ and is the projective limit indexed by $N \in \mathbb{N}$ of the *affine modular curves* $Y(N)$.

Let H be a compact, open subgroup of $G(\mathbb{A}_f)$. The quotient scheme $\mathrm{Sh}(G, \mathcal{X})/H$ exists and is a normal, quasi-projective scheme over $E(G, \mathcal{X})$. If v is a prime of $E(G, \mathcal{X})$ of residue field $k(v)$ and if $\mathcal{N}$ is a good integral model of $\mathrm{Sh}(G, \mathcal{X})/H$ over the local ring $O_{(v)}$ of v, then one gets a Shimura variety $\mathcal{N}_{k(v)}$ over the finite field $k(v)$. The classical example of a good integral model is *Mumford moduli scheme* $\mathcal{A}_{r,1}$. Here $r \in \mathbb{N}$, the $\mathbb{Z}$-scheme $\mathcal{A}_{r,1}$ is the *course moduli scheme* of principally polarized abelian scheme of relative dimension r, and the $\mathbb{Q}$–scheme $\mathcal{A}_{r,1,\mathbb{Q}}$ is of the form $\mathrm{Sh}(G, \mathcal{X})/H$ for $(G, \mathcal{X})$ a Shimura pair that defines (see Example 3.1.2) a *Siegel modular variety*.

In this paper, we are mainly interested in Shimura varieties of Hodge type. Roughly speaking, they are those Shimura varieties for which one can naturally choose $\mathcal{N}$ to be a finite scheme over $\mathcal{A}_{r,1,O_{(v)}}$. In this paper we study $\mathcal{N}$ and its special fibre $\mathcal{N}_{k(v)}$. See Subsections 4.1 and 4.2 for some moduli interpretations of $\mathcal{N}$. See Section 5 for different results pertaining to the uniqueness of $\mathcal{N}$. See

Section 6 for basic results that pertain to the smooth locus of $\mathcal{N}$. See Section 7 for the list of cases in which $\mathcal{N}$ is known to be (or it is expected to be) a projective $O_{(v)}$-scheme. Section 8 presents four main stratifications of the (smooth locus of the) special fibre $\mathcal{N}_{k(v)}$ and their basic properties. These four stratifications are defined by (see Subsections 8.3, 8.4, 8.6, and 8.7 respectively):

(a) Newton polygons of p-divisible groups;

(b) isomorphism classes of principally quasi-polarized F-isocrystals with tensors;

(c) inner isomorphism classes of the reductions modulo integral powers of p of principally quasi-polarized F-crystals with tensors;

(d) isomorphism classes of principally quasi-polarized F-crystals with tensors.

The principally quasi-polarized F-crystals with tensors attached naturally to points of the smooth locus of $\mathcal{N}_{k(v)}$ with values in algebraically closed fields are introduced in Subsection 8.1. Generalities on stratifications of reduced schemes over fields are presented in Subsection 8.2. Subsection 8.5 shows that the smooth locus of $\mathcal{N}_{k(v)}$ is a *quasi Shimura p-variety of Hodge type* in the sense of [Va5, Def. 4.2.1]. Subsection 8.5 is used in Subsections 8.6 and 8.7 to define the last two stratifications, called the *level m* and *Traverso stratifications*.

2. A group theoretical review

In this section we review basic properties of algebraic groups, of Hodge structure, of families of tensors, and of Hodge cycles on abelian schemes over reduced $\mathbb{Q}$–schemes. We denote by $\bar{k}$ an algebraic closure of a field k.

We denote by $\mathbb{G}_a$ and $\mathbb{G}_m$ the affine, smooth groups over k with the property that for each commutative k-algebra C, the groups $\mathbb{G}_a(C)$ and $\mathbb{G}_m(C)$ are the additive group of C and the multiplicative group of units of C (respectively). As schemes, we have $\mathbb{G}_a = \mathrm{Spec}(k[x])$ and $\mathbb{G}_m = \mathrm{Spec}(k[x][\frac{1}{x}])$. Thus the dimension of either $\mathbb{G}_a$ or $\mathbb{G}_m$ is 1. For $t \in \mathbb{N}$, let $\boldsymbol{\mu}_t$ be the kernel of the tth-power endomorphism of $\mathbb{G}_m$. An algebraic group scheme over k is called *linear*, if it is isomorphic to a subgroup scheme of $\mathbf{GL}_n$ for some $n \in \mathbb{N}$.

2.1. Algebraic groups

Let G be a smooth group over k which is of finite type. Let G^0 be the identity component of G. We have a short exact sequence

$$(1) \qquad\qquad 0 \to G^0 \to G \to G/G^0 \to 0,$$

where the quotient group G/G^0 is finite and étale. A classical theorem of Chevalley shows that, if k is either perfect or of characteristic 0, then there exists a short exact sequence

$$(2) \qquad\qquad 0 \to L \to G^0 \to A \to 0,$$

where A is an abelian variety over k and where L is a connected, smooth, linear group over k. In what follows we assume that (2) exists. Let L^{u} be the *unipotent radical* of L. It is the maximal connected, smooth, normal subgroup of L which is *unipotent* (i.e., which over $\bar{k}$ has a composition series whose factors are $\mathbb{G}_a$ groups). We have a short exact sequence

$$(3) \qquad\qquad 0 \to L^{\mathrm{u}} \to L \to R \to 0,$$

where $R := L/L^{\mathrm{u}}$ is a *reductive group* over k (i.e., it is a smooth, connected, linear group over k whose unipotent radical is trivial). By the k-*rank* of R we mean the greatest non-negative integer s such that $\mathbb{G}_m^s$ is a subgroup of R. If the k-rank of R is equal to the $\bar{k}$-rank of $R_{\bar{k}}$, then we say that R is *split*.

Let $Z(R)$ be the (scheme-theoretical) center of R. It is a group scheme of *multiplicative type* (i.e., over $\bar{k}$ it is the extension of a finite product of $\boldsymbol{\mu}_t$ group schemes by a torus $\mathbb{G}_m^n$; here $n \in \mathbb{N} \cup \{0\}$ and $t \in \mathbb{N}$). The quotient group $R^{\mathrm{ad}} := R/Z(R)$ is called the *adjoint group* of R; it is a reductive group over k whose (scheme-theoretical) center is trivial. Let R^{der} be the *derived group* of R; it is the minimal, normal subgroup of R with the property that the quotient group $R^{\mathrm{ab}} := R/R^{\mathrm{der}}$ is abelian. The group R^{ab} is a *torus* (i.e., over $\bar{k}$ it is isomorphic to $\mathbb{G}_m^n$). The groups R^{ad} and R^{der} are *semisimple*. We have two short exact sequences

$$(4) \qquad\qquad 0 \to Z(R) \to R \to R^{\mathrm{ad}} \to 0$$

and

$$(5) \qquad\qquad 0 \to R^{\mathrm{der}} \to R \to R^{\mathrm{ab}} \to 0.$$

The short exact sequences (1) to (5) are intrinsically associated to G.

If $G = \mathbf{GL}_n$, then $Z(G)$ and G^{ab} are isomorphic to $\mathbb{G}_m$, $G^{\mathrm{der}} = \mathbf{SL}_n$, and $G^{\mathrm{ad}} = \mathbf{PGL}_n$. If $G = \mathbf{GSp}_{2n}$, then $Z(G)$ and G^{ab} are isomorphic to $\mathbb{G}_m$, $G^{\mathrm{der}} = \mathbf{Sp}_{2n}$, and $G^{\mathrm{ad}} = \mathbf{PGSp}_{2n} = \mathbf{Sp}_{2n}/\boldsymbol{\mu}_2$. If $G = \boldsymbol{SO}_{2n+1}$, then $Z(G)$ and G^{ab} are trivial and therefore from (4) and (5) we get that $G = G^{\mathrm{der}} = G^{\mathrm{ad}}$.

2.1.1. *Examples of semisimple groups over* $\mathbb{Q}$

Let $a, b \in \mathbb{N} \cup \{0\}$ with $a+b > 0$. Let $\mathbf{SU}(a,b)$ be the simply connected semisimple group over $\mathbb{Q}$ whose $\mathbb{Q}$-valued points are the $\mathbb{Q}(i)$-valued points of $\mathbf{SL}_{a+b,\mathbb{Q}}$ that leave invariant the hermitian form $-z_1\bar{z}_1 - \cdots - z_a\bar{z}_a + z_{a+1}\bar{z}_{a+1} + \cdots + z_{a+b}\bar{z}_{a+b}$ over $\mathbb{Q}(i)$. Let $\mathbf{SO}(a,b)$ be the semisimple group over $\mathbb{Q}$ of $a+b$ by $a+b$ matrices of determinant 1 that leave invariant the quadratic form $-x_1^2 - \cdots - x_a^2 + x_{a+1}^2 + \cdots + x_{a+b}^2$ on $\mathbb{Q}^{a+b}$. Let $\mathbf{SO}_a := \mathbf{SO}(0,a)$. Let $\mathbf{SO}^*(2a)$ be the semisimple group over $\mathbb{Q}$ whose group of $\mathbb{Q}$-valued points is the subgroup of $\mathbf{SO}_{2n}(\mathbb{Q}(i))$ that leaves invariant the skew hermitian form $-z_1\bar{z}_{n+1} + z_{n+1}\bar{z}_1 - \cdots - z_n\bar{z}_{2n} + z_{2n}\bar{z}_n$ over $\mathbb{Q}(i)$ (z_i's and x_i's are related here over $\mathbb{Q}(i)$ via $z_i = x_i$).

Definition 1. *By a reductive group scheme $\mathcal{R}$ over a scheme Z, we mean a smooth group scheme over Z which is an affine Z-scheme and whose fibres are reductive groups over fields.*

As above, one defines group schemes $Z(\mathcal{R})$, $\mathcal{R}^{\mathrm{ad}}$, $\mathcal{R}^{\mathrm{der}}$, and $\mathcal{R}^{\mathrm{ab}}$ over Z which are affine Z-schemes. The group scheme $Z(\mathcal{R})$ is of multiplicative type. The group schemes $\mathcal{R}^{\mathrm{ad}}$ and $\mathcal{R}^{\mathrm{der}}$ are semisimple. The group scheme $\mathcal{R}^{\mathrm{ab}}$ is a torus.

2.2. Weil restrictions

Let $i : l \hookrightarrow k$ be a separable finite field extension. Let G be a group scheme over k which is of finite type. Let $\mathrm{Res}_{k/l}$ be the group scheme over l obtained from G through the Weil restriction of scalars. Thus $\mathrm{Res}_{k/l} G$ is defined by the functorial group identification

$$(6) \qquad \mathrm{Hom}(Y, \mathrm{Res}_{k/l} G) = \mathrm{Hom}(Y \times_l k, G),$$

where Y is an arbitrary l-scheme. We have

$$(7) \qquad (\mathrm{Res}_{k/l} G)_{\bar{k}} = \mathrm{Res}_{k \otimes_l \bar{k}/\bar{k}} G_{k \otimes_l \bar{k}} = \prod_{e \in \mathrm{Hom}_l(k,\bar{k})} G \times_{k,e} \bar{k}.$$

From (7) we easily get that:

(*) if G is a reductive (resp. connected, smooth, affine, linear, unipotent, torus, semisimple, or abelian variety) group over k, then $\mathrm{Res}_{k/l} G$ is a reductive (resp. connected, smooth, affine, linear, unipotent, torus, semisimple, or abelian variety) group over l.

If $j : m \hookrightarrow l$ is another separable finite field extension, then we have a canonical and functorial identification

$$\mathrm{Res}_{l/m} \mathrm{Res}_{k/l} G = \mathrm{Res}_{k/m} G$$

as one can easily check starting from formula (6).

If H is a group scheme over l, then we have a natural closed embedding homomorphism

$$(8) \qquad H \hookrightarrow \mathrm{Res}_{k/l} H_k$$

over l which at the level of l-valued points induces the standard monomorphism $H(l) \hookrightarrow H(k) = \mathrm{Res}_{k/l} H_k(l)$.

2.3. Hodge structures

Let $\mathbb{S} := \operatorname{Res}_{\mathbb{C}/\mathbb{R}} \mathbb{G}_m$ be the two dimensional torus over $\mathbb{R}$ whose group of $\mathbb{R}$-valued points is the multiplicative group $(\mathbb{C} \setminus \{0\}, \cdot)$ of $\mathbb{C}$. As schemes, we have $\mathbb{S} = \operatorname{Spec}(\mathbb{R}[x,y][\frac{1}{x^2+y^2}])$. By applying (8) we get that we have a short exact sequence

$$(9) \qquad\qquad 0 \to \mathbb{G}_m \to \mathbb{S} \to \mathbf{SO}_{2,\mathbb{R}} \to 0.$$

The group $\mathbf{SO}_{2,\mathbb{R}}(\mathbb{R})$ is isomorphic to the unit circle and thus to $\mathbb{R}/\mathbb{Z}$. The short exact sequence (9) does not split; this is so as $\mathbb{S}$ is isomorphic to $(\mathbb{G}_m \times_{\mathbb{R}} \mathbf{SO}_{2,\mathbb{R}})/\boldsymbol{\mu}_2$, where $\boldsymbol{\mu}_2$ is embedded diagonally into the product.

We have $\mathbb{S}(\mathbb{R}) = \mathbb{C} \setminus \{0\}$. We identify $\mathbb{S}(\mathbb{C}) = \mathbb{G}_m(\mathbb{C}) \times \mathbb{G}_m(\mathbb{C}) = (\mathbb{C} \setminus \{0\}) \times (\mathbb{C} \setminus \{0\})$ in such a way that the natural monomorphism $\mathbb{S}(\mathbb{R}) \hookrightarrow \mathbb{S}(\mathbb{C})$ induces the map $z \to (z, \overline{z})$, where $z \in \mathbb{C} \setminus \{0\}$.

Let S be a $\mathbb{Z}$-subalgebra of $\mathbb{R}$ (in most applications, we have $S \in \{\mathbb{Z}, \mathbb{Q}, \mathbb{R}\}$). Let V_S be a free S-module of finite rank. Let $V_{\mathbb{R}} := V_S \otimes_S \mathbb{R}$. By a *Hodge S-structure* on V_S we mean a homomorphism

$$(10) \qquad\qquad \rho : \mathbb{S} \to \mathbf{GL}_{V_{\mathbb{R}}}.$$

We have a direct sum decomposition

$$(11) \qquad\qquad V_{\mathbb{R}} \otimes_{\mathbb{R}} \mathbb{C} = \oplus_{(r,s) \in \mathbb{Z}^2} V^{r,s}$$

with the property that $(z_1, z_2) \in \mathbb{S}(\mathbb{C})$ acts via $\rho_{\mathbb{C}}$ on $V^{r,s}$ as the scalar multiplication with $z_1^{-r} z_2^{-s}$. Thus the element $z \in \mathbb{S}(\mathbb{R})$ acts via ρ on $V^{r,s}$ as the scalar multiplication with $z^{-r} \overline{z}^{-s}$. Therefore z acts on $\overline{V^{r,s}}$ as the scalar multiplication with $z^{-s} \overline{z}^{-r}$. This implies that for all $(r,s) \in \mathbb{Z}^2$ we have an identity

$$(12) \qquad\qquad V^{s,r} = \overline{V^{r,s}}.$$

Conversely, each direct sum decomposition (11) that satisfies the identities (12), is uniquely associated to a homomorphism as in (10).

By the *type* of the Hodge S-structure on V_S, we mean any symmetric subset τ of $\mathbb{Z}^2$ with the property that we have a direct sum decomposition

$$V_{\mathbb{R}} \otimes_{\mathbb{R}} \mathbb{C} = \oplus_{(r,s) \in \tau} V^{r,s}.$$

Here symmetric refers to the fact that if $(r,s) \in \tau$, then we also have $(s,r) \in \tau$. If we can choose τ such that the sum $n := r + s$ does not depend on $(r,s) \in \tau$, one says that the Hodge S-structure on V_S has *weight n*.

2.3.1. Polarizations

For $n \in \mathbb{Z}$, let $S(n)$ be the Hodge S-structure on $(2\pi i)^n S$ which has type $(-n, -n)$. Suppose that the Hodge S-structure on V_S has weight n. By a polarization of the Hodge S-structure on V_S we mean a morphism $\psi : V_S \otimes_S V_S \to S(-n)$ of Hodge S-structures such that the bilinear form $(2\pi i)^n \psi(x \otimes \rho(i)y)$ defined for $x, y \in V_{\mathbb{R}}$, is symmetric and positive definite. Here we identify ψ with its scalar extension to $\mathbb{R}$.

2.3.2. Example

Let A be an abelian variety over $\mathbb{C}$. We take $S = \mathbb{Z}$. Let $V_{\mathbb{Z}} = H^1(A^{\mathrm{an}}, \mathbb{Z})$ be the first cohomology group of the analytic manifold $A^{\mathrm{an}} := A(\mathbb{C})$ with coefficients in $\mathbb{Z}$. Then the classical Hodge theory provides us with a direct sum decomposition

$$(13a) \qquad\qquad V_{\mathbb{R}} \otimes_{\mathbb{R}} \mathbb{C} = V^{1,0} \oplus V^{0,1},$$

where $V^{1,0} = H^0(A, \Omega)$ and $V^{0,1} = H^1(A, \mathcal{O}_A)$ (see [Mu, Ch. I, 1]). Here $\mathcal{O}_A$ is the structured ring sheaf on A and Ω is the $\mathcal{O}_A$-module of 1-forms on A. We have $\overline{V^{1,0}} = V^{0,1}$ and therefore (13a) defines a Hodge $\mathbb{Z}$-structure on V_S. Let $F^1(V_{\mathbb{R}} \otimes_{\mathbb{R}} \mathbb{C}) := V^{1,0}$; it is called the *Hodge filtration* of $V_{\mathbb{R}} \otimes_{\mathbb{R}} \mathbb{C}$.

Let $W_{\mathbb{Z}} := \mathrm{Hom}(V_{\mathbb{Z}}, \mathbb{Z}) = H_1(A^{\mathrm{an}}, \mathbb{Z})$. Let $W_{\mathbb{R}} := W_{\mathbb{Z}} \otimes_{\mathbb{Z}} \mathbb{R}$. Taking the dual of (13a), we get a Hodge $\mathbb{Z}$-structure on $W_{\mathbb{Z}}$ of the form

$$(13b) \qquad\qquad W_{\mathbb{R}} \otimes_{\mathbb{R}} \mathbb{C} = W^{-1,0} \oplus W^{0,-1}.$$

One can identify naturally $W^{-1,0} = \mathrm{Hom}(V^{1,0}, \mathbb{C}) = \mathrm{Lie}(A)$. Each $z \in \mathbb{S}(\mathbb{R})$ acts on the complex vector space $\mathrm{Lie}(A)$ as the multiplication with z and this explains the convention on negative power signs used in the paragraph after formula (11). We have canonical identifications

$$A^{\mathrm{an}} = W_{\mathbb{Z}} \backslash \mathrm{Lie}(A) = W_{\mathbb{Z}} \backslash (W_{\mathbb{R}} \otimes_{\mathbb{R}} \mathbb{C})/W^{0,-1}.$$

If λ is a polarization of A, then the non-degenerate form

$$(14) \qquad\qquad \psi : W_{\mathbb{Z}} \otimes_{\mathbb{Z}} W_{\mathbb{Z}} \to \mathbb{Z}(1)$$

defined naturally by λ, is a polarization of the Hodge $\mathbb{Z}$-structure on $W_{\mathbb{Z}}$.

We have $\mathrm{End}(V_{\mathbb{Z}}) = V_{\mathbb{Z}} \otimes_{\mathbb{Z}} W_{\mathbb{Z}} = \mathrm{End}(W_{\mathbb{Z}})$. Due to the identities (13a) and (13b), the Hodge $\mathbb{Z}$-structure on $\mathrm{End}(V_{\mathbb{Z}})$ is of type

$$(15) \qquad\qquad \tau_{\mathrm{ab}} := \{(-1,1), (0,0), (1,-1)\}.$$

Definition 2. *We use the notations of Example 2.3.1. Let $W := W_{\mathbb{Z}} \otimes_{\mathbb{Z}} \mathbb{Q}$. By the Mumford–Tate group of the complex abelian variety A, we mean the smallest subgroup H_A of $\mathbf{GL}_W$ with the property that the homomorphism $x_A : \mathbb{S} \to \mathbf{GL}_{W_{\mathbb{R}}}$ that defines the Hodge $\mathbb{Z}$-structure on $W_{\mathbb{Z}}$, factors through $H_{A,\mathbb{R}}$.*

Proposition 1. *The group H_A is a reductive group over $\mathbb{Q}$.*

Proof: From its very definition, the group H_A is connected. To prove the Proposition it suffices to show that the unipotent radical H_A^{u} of H_A is trivial. Let W_1 be the largest rational subspace of W on which H_A^{u} acts trivially. As H_A^{u} is a normal subgroup of H_A, W_1 is an H_A-module. Thus x_A normalizes $W_1 \otimes_{\mathbb{Q}} \mathbb{R}$ and therefore we have a direct sum decomposition

$$W_1 \otimes_{\mathbb{Q}} \mathbb{C} = [(W_1 \otimes_{\mathbb{Q}} \mathbb{C}) \cap W^{-1,0}] \oplus [(W_1 \otimes_{\mathbb{Q}} \mathbb{C}) \cap W^{0,-1}].$$

Thus $(W_{\mathbb{Z}} \cap W_1) \backslash (W_1 \otimes_{\mathbb{Q}} \mathbb{C}) / [(W_1 \otimes_{\mathbb{Q}} \mathbb{C}) \cap W^{0,-1}]$ is a closed analytic submanifold A_1^{an} of A^{an}. A classical theorem of Serre asserts that A_1^{an} is algebraizable i.e., it is the analytic submanifold associated to an abelian subvariety A_1 of A. The short exact sequence $0 \to A_1 \to A \to A/A_1 \to 0$ splits up to isogenies (i.e., A is isogeneous to $A_1 \times_{\mathbb{C}} A_2$, where $A_2 := A/A_1$). Let $W_2 := H_1(A_2^{\mathrm{an}}, \mathbb{Q})$. We have a direct sum decomposition $W = W_1 \oplus W_2$ whose extension to $\mathbb{R}$ is normalized by x_A. Thus the direct sum decomposition $W = W_1 \oplus W_2$ is normalized by H_A. In particular, W_2 is an H_A^{u}-module.

If $W_1 \neq W$, then the unipotent group H_A^{u} acts trivially on a non-zero subspace of W_2 and this represents a contradiction with the largest property of W_1. Thus $W_1 = W$ i.e., H_A^{u} acts trivially on W. Therefore H_A^{u} is the trivial group. $\square$

2.4. *Tensors*

Let M be a free module of finite rank over a commutative $\mathbb{Z}$-algebra C. Let $M^* := \mathrm{Hom}_C(M, C)$. By the *essential tensor algebra* of $M \oplus M^*$ we mean the C-module

$$\mathcal{T}(M) := \oplus_{s,t \in \mathbb{N} \cup \{0\}} M^{\otimes s} \otimes_C M^{* \otimes t}.$$

Let $F^1(M)$ be a direct summand of M. Let $F^0(M) := M$ and $F^2(M) := 0$. Let $F^1(M^*) := 0$, $F^0(M^*) := \{y \in M^* | y(F^1(M)) = 0\}$, and $F^{-1}(M^*) := M^*$. Let $(F^i(\mathcal{T}(M)))_{i \in \mathbb{Z}}$ be the tensor product filtration of $\mathcal{T}(M)$ defined by the exhaustive, separated filtrations $(F^i(M))_{i \in \{0,1,2\}}$ and $(F^i(M^*))_{i \in \{-1,0,1\}}$ of M and M^* (respectively). We refer to $(F^i(\mathcal{T}(M)))_{i \in \mathbb{Z}}$ as the filtration of $\mathcal{T}(M)$ defined by $F^1(M)$ and to each $F^i(\mathcal{T}(M))$ as the F^i-filtration of $\mathcal{T}(M)$ defined by $F^1(M)$.

We identify naturally $\mathrm{End}(M) = M \otimes_C M^* \subseteq \mathcal{T}(M)$ and $\mathrm{End}(\mathrm{End}(M)) = M^{\otimes 2} \otimes_C M^{* \otimes 2}$. Let $x \in C$ be a non-divisor of 0. A family of tensors of $\mathcal{T}(M[\frac{1}{x}]) = \mathcal{T}(M)[\frac{1}{x}]$ is denoted $(u_\alpha)_{\alpha \in \mathcal{J}}$, with $\mathcal{J}$ as the set of indexes. Let M_1 be another free C-module of finite rank. Let $(u_{1,\alpha})_{\alpha \in \mathcal{J}}$ be a family of tensors of $\mathcal{T}(M_1[\frac{1}{x}])$ indexed also by the set $\mathcal{J}$. By an isomorphism

$$(M, (u_\alpha)_{\alpha \in \mathcal{J}}) \xrightarrow{\sim} (M_1, (u_{1,\alpha})_{\alpha \in \mathcal{J}})$$

we mean a C-linear isomorphism $M \xrightarrow{\sim} M_1$ that extends naturally to a C-linear isomorphism $\mathcal{T}(M[\frac{1}{x}]) \xrightarrow{\sim} \mathcal{T}(M_1[\frac{1}{x}])$ which takes u_α to $u_{1,\alpha}$ for all $\alpha \in \mathcal{J}$. We emphasize that we will denote two tensors or bilinear forms in the same way, provided they are obtained one from another via either a reduction modulo some ideal or a scalar extension.

2.5. *Hodge cycles on abelian schemes*

We will use the terminology of [De3] on Hodge cycles on an abelian scheme B_X over a reduced $\mathbb{Q}$–scheme X. Thus we write each Hodge cycle v on B_X as a pair $(v_{\mathrm{dR}}, v_{\mathrm{ét}})$, where v_{dR} and $v_{\mathrm{ét}}$ are the *de Rham* and the *étale component* of v (respectively). The étale component $v_{\mathrm{ét}}$ at its turn has an l-component $v_{\mathrm{ét}}^l$, for each rational prime l.

In what follows we will be interested only in Hodge cycles on B_X that involve no Tate twists and that are tensors of different essential tensor algebras. Accordingly, if X is the spectrum of a field F, then in applications $v^l_{\acute{e}t}$ will be a suitable $\mathrm{Gal}(\overline{E}/E)$-invariant tensor of $\mathcal{T}(H^1_{\acute{e}t}(B_{\overline{X}}, \mathbb{Q}_l))$, where $\overline{X} := \mathrm{Spec}(\overline{E})$. If $\overline{E}$ is a subfield of $\mathbb{C}$, then we will also use the Betti realization v_B of v. The tensor v_B has the following two properties (that define Hodge cycles on B_X which involve no Tate twist; see [De3, Sect. 2]):

(i) it is a tensor of $\mathcal{T}(H^1((B_X \times_X \mathrm{Spec}(\mathbb{C}))^{\mathrm{an}}, \mathbb{Q}))$ that corresponds to v_{dR} (resp. to $v^l_{\acute{e}t}$) via the canonical isomorphism that relates the Betti cohomology of $(B_X \times_X \mathrm{Spec}(\mathbb{C}))^{\mathrm{an}}$ with $\mathbb{Q}$–coefficients with the de Rham (resp. the $\mathbb{Q}_l$ étale) cohomology of $B_X \times_X \mathrm{Spec}(\mathbb{C})$;

(ii) it is also a tensor of the F^0-filtration of the filtration of $\mathcal{T}(H^1((B_X \times_X \mathrm{Spec}(\mathbb{C}))^{\mathrm{an}}, \mathbb{C}))$ defined by the Hodge filtration $F^1(H^1((B_X \times_X \mathrm{Spec}(\mathbb{C}))^{\mathrm{an}}, \mathbb{C}))$ of $H^1((B_X \times_X \mathrm{Spec}(\mathbb{C}))^{\mathrm{an}}, \mathbb{C})$.

We have the following particular example:

(iii) if $v_B \in \mathrm{End}(H^1((B_X \times_X \mathrm{Spec}(\mathbb{C}))^{\mathrm{an}}, \mathbb{Q}))$, then from Riemann theorem we get that v_B is the Betti realization of a $\mathbb{Q}$–endomorphism of $B_X \times_X \mathrm{Spec}(\mathbb{C})$ and therefore the Hodge cycle $(v_{\mathrm{dR}}, v_{\acute{e}t})$ on B_X is defined uniquely by a $\mathbb{Q}$–endomorphism of B_X.

The class of Hodge cycles is stable under pull backs. In particular, if X is a reduced $\mathbb{Q}$–scheme of finite type, then the pull back of $(v_{\mathrm{dR}}, v_{\acute{e}t})$ via a complex point $\mathrm{Spec}(\mathbb{C}) \to X$, is a Hodge cycle on the complex abelian variety $B_X \times_X \mathrm{Spec}(\mathbb{C})$.

2.5.1. Example

Let A be an abelian variety over $\mathbb{C}$. Let S be an irreducible, closed subvariety of A. Let n be the codimension of S in A. To S one associates classes $[S]_{\mathrm{dR}} \in H^{2n}_{\mathrm{dR}}(A/\mathbb{C})$, $[S]_l \in H^{2n}_{\acute{e}t}(A, \mathbb{Q}_l)(n)$, and $[S]_B \in H^{2n}(A^{\mathrm{an}}, \mathbb{Q})(n)$. If $[S]_{\acute{e}t} := ([S]_l)_{l \text{ a prime}}$, then the pair $([S]_{\mathrm{dR}}, [S]_{\acute{e}t})$ is a Hodge cycle on A which involves Tate twists and whose Betti realization is $[S]_B$. One can identify $H^{2n}_{\acute{e}t}(A, \mathbb{Q}_l)(n)$ with a $\mathbb{Q}_l$–subspace of $H^1_{\acute{e}t}(A, \mathbb{Q}_l)^{\otimes n} \otimes_{\mathbb{Q}_l} [(H^1_{\acute{e}t}(A, \mathbb{Q}_l))^*]^{\otimes n}$ and $H^{2n}_{\mathrm{dR}}(A/\mathbb{C})$ with a $\mathbb{C}$–subspace of $H^1_{\mathrm{dR}}(A/\mathbb{C})^{\otimes n} \otimes_{\mathbb{C}} [(H^1_{\mathrm{dR}}(A/\mathbb{C}))^*]^{\otimes n}$; thus one can naturally view $([S]_{\mathrm{dR}}, [S]_{\acute{e}t})$ as a Hodge cycle on A which involves no Tate twists. The $\mathbb{Q}$–linear combinations of such cycles $([S]_{\mathrm{dR}}, [S]_{\acute{e}t})$ are called algebraic cycles on A.

3. Shimura varieties

In this section we introduce Shimura varieties and their basic properties and main types. All continuous actions are in the sense of [De2, Subsubsect. 2.7.1] and are right actions. Thus if a totally discontinuous, locally compact group Γ acts continuously (from the right) on a scheme Y, then for each compact, open subgroup Δ of Γ the geometric quotient scheme Y/Δ exists and the epimorphism $Y \twoheadrightarrow Y/\Delta$ is pro-finite; moreover, we have an identity $Y = \mathrm{proj.lim._{\Delta}} Y/\Delta$.

3.1. Shimura pairs

A *Shimura pair* $(G, \mathcal{X})$ consists of a reductive group G over $\mathbb{Q}$ and a $G(\mathbb{R})$-conjugacy class $\mathcal{X}$ of homomorphisms $\mathbb{S} \to G_{\mathbb{R}}$ that satisfy Deligne's axioms of [De2, Subsubsect. 2.1.1]:

(i) the Hodge $\mathbb{Q}$–structure on $\mathrm{Lie}(G)$ defined by each element $x \in \mathcal{X}$ is of type $\tau_{\mathrm{ab}} = \{(-1, 1), (0, 0), (1, -1)\}$;

(ii) no simple factor of the adjoint group G^{ad} of G becomes compact over $\mathbb{R}$;

(iii) $\mathrm{Ad}(x(i))$ is a Cartan involution of $\mathrm{Lie}(G_{\mathbb{R}}^{\mathrm{ad}})$, where $\mathrm{Ad} : G_{\mathbb{R}} \to \mathbf{GL}_{\mathrm{Lie}(G_{\mathbb{R}}^{\mathrm{ad}})}$ is the adjoint representation.

Axiom (iii) is equivalent to the fact that the adjoint group $G_{\mathbb{R}}^{\mathrm{ad}}$ has a faithful representation $G_{\mathbb{R}}^{\mathrm{ad}} \hookrightarrow \mathbf{GL}_{V_{\mathbb{R}}}$ with the property that there exists a polarization of the Hodge $\mathbb{R}$-structure on $V_{\mathbb{R}}$ defined naturally by any $x \in \mathcal{X}$ which is fixed by $G_{\mathbb{R}}^{\mathrm{ad}}$. These axioms imply that $\mathcal{X}$ has a natural structure of a hermitian symmetric domain, cf. [De2, Cor. 1.1.17].

For $x \in \mathcal{X}$ we consider the *Hodge cocharacter*

$$\mu_x : \mathbb{G}_m \to G_{\mathbb{C}}$$

defined on complex points by the rule: $z \in \mathbb{G}_m(\mathbb{C})$ is mapped to $x_{\mathbb{C}}(z, 1) \in G_{\mathbb{C}}(\mathbb{C})$.

Let $E(G, \mathcal{X}) \hookrightarrow \mathbb{C}$ be the number subfield of $\mathbb{C}$ that is the field of definition of the $G(\mathbb{C})$-conjugacy class $[\mu_{\mathcal{X}}]$ of the cocharacters μ_x's of $G_{\mathbb{C}}$, cf. [Mi2, p. 163]. More precisely $[\mu_{\mathcal{X}}]$ is defined naturally by a $G(\overline{\mathbb{Q}})$-conjugacy class $[\mu_{\mathcal{X}}^{\overline{\mathbb{Q}}}]$ of cocharacters $\mathbb{G}_m \to G_{\overline{\mathbb{Q}}}$; the Galois group $\mathrm{Gal}(\mathbb{Q})$ acts naturally on the set of such $G(\overline{\mathbb{Q}})$-conjugacy classes and $E(G, \mathcal{X})$ is the number field which is the fixed field of the stabilizer subgroup of $[\mu_{\mathcal{X}}^{\overline{\mathbb{Q}}}]$ in $\mathrm{Gal}(\mathbb{Q})$. One calls $E(G, \mathcal{X})$ the *reflex field* of $(G, \mathcal{X})$.

We define the *complex Shimura space*

$$\mathrm{Sh}(G, \mathcal{X})_{\mathbb{C}} := \mathrm{proj.lim.}_{K \in \sigma(G)} G(\mathbb{Q}) \backslash (\mathcal{X} \times G(\mathbb{A}_f)/K),$$

where $\sigma(G)$ is the set of compact, open subgroups of $G(\mathbb{A}_f)$ endowed with the inclusion relation (see [De1], [De2], and [Mi1] to [Mi4]). Thus $\mathrm{Sh}(G, \mathcal{X})_{\mathbb{C}}(\mathbb{C})$ is a normal complex space on which $G(\mathbb{A}_f)$ acts. We have an identity

$$(16) \qquad \mathrm{Sh}(G, \mathcal{X})_{\mathbb{C}}(\mathbb{C}) = G(\mathbb{Q}) \backslash [\mathcal{X} \times (G(\mathbb{A}_f)/\overline{Z(G)(\mathbb{Q})})],$$

where $\overline{Z(G)(\mathbb{Q})}$ is the topological closure of $Z(G)(\mathbb{Q})$ in $G(\mathbb{A}_f)$ (cf. [De2, Prop. 2.1.10]). Let $x \in \mathcal{X}$ and $a, g \in G(\mathbb{A}_f)$. Let $[x, a] \in \mathrm{Sh}(G, \mathcal{X})_{\mathbb{C}}(\mathbb{C})$ be the point defined naturally by the equivalence class of $(x, a) \in \mathcal{X} \times G(\mathbb{A}_f)$, cf. (16). The $G(\mathbb{A}_f)$-action on $\mathrm{Sh}(G, \mathcal{X})_{\mathbb{C}}(\mathbb{C})$ is defined by the rule $[x, a] \cdot g := [x, ag]$.

For $\ddagger$ a compact subgroup of $G(\mathbb{A}_f)$ let $\mathrm{Sh}_{\ddagger}(G, \mathcal{X})_{\mathbb{C}}(\mathbb{C}) := \mathrm{Sh}(G, \mathcal{X})_{\mathbb{C}}(\mathbb{C})/\ddagger$. Let $K \in \sigma(G)$. We can write $\mathrm{Sh}_K(G, \mathcal{X})_{\mathbb{C}}(\mathbb{C}) = G(\mathbb{Q}) \backslash (\mathcal{X} \times G(\mathbb{A}_f)/K)$ as a disjoint union of normal complex spaces of the form $\Sigma \backslash \mathcal{X}^0$, where $\mathcal{X}^0$ is a connected component of $\mathcal{X}$ and Σ is an *arithmetic subgroup* of $G(\mathbb{Q})$ (i.e., is the intersection of $G(\mathbb{Q})$ with a compact, open subgroup of $G(\mathbb{A}_f)$). A classical result of Baily and Borel allows us to view naturally $\mathrm{Sh}_K(G,\mathcal{X})_{\mathbb{C}}(\mathbb{C}) = G(\mathbb{Q})\backslash(\mathcal{X} \times G(\mathbb{A}_f)/K)$ as the complex space associated to a finite, disjoint union $\mathrm{Sh}_K(G,\mathcal{X})_{\mathbb{C}}$ of normal, quasi-projective, connected varieties over $\mathbb{C}$ (see [BB, Thm. 10.11]). Thus $\mathrm{Sh}_K(G,\mathcal{X})_{\mathbb{C}}$ is a normal, quasi-projective $\mathbb{C}$-scheme and

$$\mathrm{Sh}(G,\mathcal{X})_{\mathbb{C}} := \mathrm{proj.lim.}_{K \in \sigma(G)} \mathrm{Sh}_K(G,\mathcal{X})_{\mathbb{C}}$$

is a normal $\mathbb{C}$-scheme on which $G(\mathbb{A}_f)$ acts. We have a canonical identification $\mathrm{Sh}_K(G,\mathcal{X})_{\mathbb{C}} = \mathrm{Sh}(G,\mathcal{X})_{\mathbb{C}}/K$. If K is small enough, then K acts freely on $\mathrm{Sh}(G,\mathcal{X})_{\mathbb{C}}$ and thus $\mathrm{Sh}_K(G,\mathcal{X})_{\mathbb{C}}$ is in fact a smooth, quasi-projective $\mathbb{C}$-scheme.

3.1.1. Example

Let A be an abelian variety over $\mathbb{C}$. Let H_A be its Mumford–Tate group. Let $x_A : \mathbb{S} \to H_{A,\mathbb{R}}$ be the homomorphism that defines the Hodge $\mathbb{Z}$-structure on $W_A := H_1(A^{\mathrm{an}}, \mathbb{Z})$, cf. Definition 2. Let $\mathcal{X}_A$ be the $H_A(\mathbb{R})$-conjugacy class of x_A. We check that the pair $(H_A, \mathcal{X}_A)$ is a Shimura pair. The fact that the axiom 3.1 (i) holds for $(H_A, \mathcal{X}_A)$ is implied by (15). If H_A^{ad} has a (non-trivial) simple factor $\diamond$ which over $\mathbb{R}$ is compact, then the fact that $\mathcal{X}_A$ is a hermitian symmetric domain implies that the image of x_A in $\diamond_{\mathbb{R}}$ is trivial and this contradicts the smallest property (see Definition 2) of the Mumford–Tate group H_A. Thus the axioms 3.1 (ii) holds for $(H_A, \mathcal{X}_A)$. The fact that the axioms 3.1 (iii) holds is implied by the fact that B has a polarization and thus by the fact that (14) holds. We emphasize that the reflex field $E(H_A, \mathcal{X}_A)$ can be any CM number field.

3.1.2. Example

The most studied Shimura pairs are constructed as follows. Let W be a vector space over $\mathbb{Q}$ of even dimension $2r$. Let ψ be a non-degenerate alternative form on W. Let $\mathcal{S}$ be the set of all monomorphisms $\mathbb{S} \hookrightarrow \mathbf{GSp}(W \otimes_{\mathbb{Q}} \mathbb{R}, \psi)$ that define Hodge $\mathbb{Q}$-structures on W of type $\{(-1,0),(0,-1)\}$ and that have either $2\pi i\psi$ or $-2\pi i\psi$ as polarizations. Thus $\mathcal{S}$ is two copies of the Siegel domain of genus r (the two copies correspond to either $2\pi i\psi$ or $-2\pi i\psi$ being a polarization of the resulting Hodge $\mathbb{Q}$-structures on W). It is easy to see that $\mathcal{S}$ is a $\mathbf{GSp}(W,\psi)(\mathbb{R})$-conjugacy class of homomorphisms $\mathbb{S} \to \mathbf{GSp}(W \otimes_{\mathbb{Q}} \mathbb{R}, \psi)$. One can choose an abelian variety A over $\mathbb{C}$ such that in fact we have $(\mathbf{GSp}(W,\psi), \mathcal{S}) = (H_A, \mathcal{X}_A)$ and therefore $(\mathbf{GSp}(W,\psi), \mathcal{S})$ is a Shimura pair, cf. Example 3.1.1. We call $(\mathbf{GSp}(W,\psi), \mathcal{S})$ a Shimura pair that defines a Siegel modular variety $\mathrm{Sh}(\mathbf{GSp}(W,\psi), \mathcal{S})$ (to be defined in Subsection 3.2 below). As $\mathbf{GSp}(W,\psi)$ is a split group, the $\mathbf{GSp}(W,\psi)(\overline{\mathbb{Q}})$-conjugacy class $[\mu_{\mathcal{X}}^{\overline{\mathbb{Q}}}]$ is defined naturally by a cocharacter of $\mathbf{GSp}(W,\psi)$ and therefore we have $E(\mathbf{GSp}(W,\psi), \mathcal{S}) = \mathbb{Q}$.

3.1.3. Example

Let n be a positive integer. Let $G := \mathbf{SO}(2,n)$; it is the identity component of the group that fixes the quadratic from $-x_1^2 - x_2^2 + x_3^2 + \cdots + x_{n+2}^2$ on $\mathbb{Q}^{n+2}$. The group G has a subgroup $\mathbf{SO}_2 \times_{\mathbb{Q}} \mathbf{SO}_n$ which normalizes the rational vector subspaces of $\mathbb{Q}^{n+2}$ generated by the first two and by the last n vectors of the standard $\mathbb{Q}$–basis for $\mathbb{Q}^{n+2}$. Let $x : \mathbb{S} \to G_{\mathbb{R}}$ be a homomorphism whose image is the subgroup $\mathbf{SO}_{2,\mathbb{R}}$ of $G_{\mathbb{R}}$ and whose kernel is the split torus $\mathbb{G}_m$ of $\mathbb{S}$. Let $\mathcal{X}$ be the $G(\mathbb{R})$-conjugacy class of x. Then the pair $(G, \mathcal{X})$ is a Shimura pair.

The group $G_{\mathbb{Q}(i)}$ is split (i.e., $\mathbb{G}_m^{[\frac{n}{2}]}$ is a subgroup of it) and thus the $G(\overline{\mathbb{Q}})$-conjugacy class $[\mu_{\mathcal{X}}^{\overline{\mathbb{Q}}}]$ is defined naturally by a cocharacter $\mu_0 : \mathbb{G}_m \to G_{\mathbb{Q}(i)}$. We can choose μ_0 such that the non-trivial element of $\mathrm{Gal}(\mathbb{Q}(i)/\mathbb{Q})$ takes μ_0 under Galois conjugation to μ_0^{-1}. It is easy to see that the two cocharacters μ_0 and μ_0^{-1} are $G(\mathbb{Q}(i))$-conjugate. Therefore $E(G, \mathcal{X}) = \mathbb{Q}$.

If $n = 19$, then $(G, \mathcal{X})$ is the Shimura pair associated to moduli spaces of polarized K3 surfaces.

3.1.4. Example

Let T be a torus over $\mathbb{Q}$. Let $x : \mathbb{S} \to T_{\mathbb{R}}$ be an arbitrary homomorphism. Then the pair $(T, \{x\})$ is a Shimura pair. Its reflex field $E := E(T, \{x\})$ is the field of definition of the cocharacter $\mu_x : \mathbb{G}_m \to T_{\mathbb{C}}$. We denote also by $\mu_x : \mathbb{G}_m \to T_E$ the homomorphism whose extension to $\mathbb{C}$ is μ_x.

From the homomorphism $\mu_x : \mathbb{G}_m \to T_E$ we get naturally a new one

$$N_x : \mathrm{Res}_{E/\mathbb{Q}}\,\mathbb{G}_m \xrightarrow{\mathrm{Res}_{E/\mathbb{Q}}(\mu_x)} \mathrm{Res}_{E/\mathbb{Q}}\,T_E \xrightarrow{\mathrm{Norm}\,E/\mathbb{Q}} T.$$

Thus for each commutative $\mathbb{Q}$–algebra C we get a homomorphism $N_x(C) : \mathbb{G}_m(E \otimes_{\mathbb{Q}} C) \to T(C)$.

Let E^{ab} be the maximal abelian extension of E. The reciprocity map

$$r(T, \{x\}) : \mathrm{Gal}(E^{\mathrm{ab}}/E) \to T(\mathbb{A}_f)/\overline{T(\mathbb{Q})}$$

is defined as follows: if $\tau \in \mathrm{Gal}(E^{\mathrm{ab}}/E)$ and if $s \in \mathbb{J}_E$ is an idèle (of E) such that $\mathrm{rec}_E(s) = \tau$, then $r(T, \{x\})(\tau) := N_x(\mathbb{A}_f)(s_f)$, where s_f is the finite part of s. Here the Artin reciprocity map rec_E is such that a uniformizing parameter is mapped to the geometric Frobenius element.

Definition 3. *By a map $f : (G_1, \mathcal{X}_1) \to (G_2, \mathcal{X}_2)$ of Shimura pairs we mean a homomorphism $f : G_1 \to G_2$ of groups over $\mathbb{Q}$ such that for each $x \in \mathcal{X}_1$ we have $f(x) := f_{\mathbb{R}} \circ x \in \mathcal{X}_2$. If $f : G_1 \to G_2$ is a monomorphism, then we say $f : (G_1, \mathcal{X}_1) \to (G_2, \mathcal{X}_2)$ is an injective map. If G_1 is a torus and if $f : (G_1, \mathcal{X}_1) \hookrightarrow (G_2, \mathcal{X}_2)$ is an injective map, then f is called a special pair in $(G_2, \mathcal{X}_2)$.*

3.2. Canonical models

By a *model* of $\mathrm{Sh}(G, \mathcal{X})_{\mathbb{C}}$ over a subfield k of $\mathbb{C}$, we mean a scheme S over k endowed with a continuous right action of $G(\mathbb{A}_f)$ (defined over k), such that there exists a $G(\mathbb{A}_f)$-equivariant isomorphism

$$\mathrm{Sh}(G, \mathcal{X})_{\mathbb{C}} \xrightarrow{\sim} S_{\mathbb{C}}.$$

The *canonical model* of $\mathrm{Sh}(G, \mathcal{X})_{\mathbb{C}}$ (or of $(G, \mathcal{X})$ itself) is the model $\mathrm{Sh}(G, \mathcal{X})$ of $\mathrm{Sh}(G, \mathcal{X})_{\mathbb{C}}$ over $E(G, \mathcal{X})$ which satisfies the following property:

(*) *if $(T, \{x\})$ is a special pair in $(G, \mathcal{X})$, then for each element $a \in G(\mathbb{A}_f)$ the point $[x, a]$ of $\mathrm{Sh}(G, \mathcal{X})(\mathbb{C}) = \mathrm{Sh}(G, \mathcal{X})_{\mathbb{C}}(\mathbb{C})$ is rational over $E(T, \{x\})^{\mathrm{ab}}$ and every element τ of $\mathrm{Gal}(E(T, \{x\})^{\mathrm{ab}}/E(T, \{x\}))$ acts on $[x, a]$ according to the rule*

$$\tau[x, a] = [x, ar(\tau)],$$

where $r := r(T, \{x\})$ is as in Example 3.1.4.

The canonical model of $\mathrm{Sh}(G, \mathcal{X})$ exists and is uniquely determined by the property (*) up to a unique isomorphism (see [De1], [De2], [Mi2], and [Mi4]).

By the *dimension d* of $\mathrm{Sh}(G, \mathcal{X})$ (or $(G, \mathcal{X})$ or $\mathrm{Sh}(G, \mathcal{X})_{\mathbb{C}}$) we mean the dimension of $\mathcal{X}$ as a complex manifold. One computes d as follows. For $x \in \mathcal{X}$, let $\mathrm{Lie}(G_{\mathbb{C}}) = F_x^{-1,0} \oplus F_x^{0,0} \oplus F_x^{0,-1}$ be the Hodge decomposition defined by x. Let K_{∞} be the centralizer of x in $G_{\mathbb{R}}$; it is a reductive group over $\mathbb{R}$ (cf. [Bo, Ch. IV, 13.17, Cor. 2]). We have $\mathrm{Lie}(K_{\infty}) \otimes_{\mathbb{R}} \mathbb{C} = F_x^{0,0}$ and (as analytic real manifolds) $\mathcal{X} = [G(\mathbb{R})]/[K_{\infty}(\mathbb{R})]$. Thus as $\dim_{\mathbb{C}}(F^{-1,0}) = \dim_{\mathbb{C}}(F^{0,-1})$, we get that

(17)
$$d = \frac{1}{2} \dim(G_{\mathbb{R}}/K_{\infty}) = \frac{1}{2} \dim_{\mathbb{C}}(\mathrm{Lie}(G_{\mathbb{C}})/F_x^{0,0}) = \dim_{\mathbb{C}}(F_x^{-1,0}) = \dim_{\mathbb{C}}(F_x^{0,-1}).$$

For $\ddagger$ a compact subgroup of $G(\mathbb{A}_f)$ let $\mathrm{Sh}_{\ddagger}(G, \mathcal{X}) := \mathrm{Sh}(G, \mathcal{X})/\ddagger$. If $K \in \sigma(G)$, then $\mathrm{Sh}_K(G, \mathcal{X})$ is a normal, quasi-projective $E(G, \mathcal{X})$-scheme which is equidimensional of dimension d and whose extension to $\mathbb{C}$ is (canonically identified with) the $\mathbb{C}$-scheme $\mathrm{Sh}_K(G, \mathcal{X})_{\mathbb{C}}$ we have introduced in Subsection 3.1.

If $f : (G_1, \mathcal{X}_1) \to (G_2, \mathcal{X}_2)$ is a map between two Shimura pairs, then $E(G_2, \mathcal{X}_2)$ is a subfield of $E(G_1, \mathcal{X}_1)$ and there exists a unique $G_1(\mathbb{A}_f)$-equivariant morphism (still denoted by f)

(18)
$$f : \mathrm{Sh}(G_1, \mathcal{X}_1) \to \mathrm{Sh}(G_2, \mathcal{X}_2)_{E(G_1, \mathcal{X}_1)}$$

which at the level of complex points is the map $[x, a] \to [f(x), f(a)]$ ([De1, Cor. 5.4]). We get as well a $G(\mathbb{A}_f)$-equivariant morphism (denoted in the same way)

$$f : \mathrm{Sh}(G_1, \mathcal{X}_1) \to \mathrm{Sh}(G_2, \mathcal{X}_2)$$

of $E(G_2, \mathcal{X}_2)$-schemes. If f is an injective map, then based on (16) one gets that (18) is in fact a closed embedding.

3.3. Classification of Shimura pairs

Let $(G, \mathcal{X})$ be a Shimura pair. If $x \in \mathcal{X}$, let $x^{\mathrm{ab}} : \mathbb{S} \to G_{\mathbb{R}}^{\mathrm{ab}}$ and $x^{\mathrm{ad}} : \mathbb{S} \to G_{\mathbb{R}}^{\mathrm{ad}}$ be the homomorphisms defined naturally by $x : \mathbb{S} \to G_{\mathbb{R}}$. The homomorphism x^{ab} does not depend on $x \in \mathcal{X}$ and the Shimura pair $(G^{\mathrm{ab}}, \{x^{\mathrm{ab}}\})$ has dimension 0. Let $\mathcal{X}^{\mathrm{ad}}$ be the $G^{\mathrm{ad}}(\mathbb{R})$-conjugacy class of x^{ad}. The Shimura pairs $(G^{\mathrm{ab}}, \{x^{\mathrm{ab}}\})$ and $(G^{\mathrm{ad}}, \mathcal{X}^{\mathrm{ad}})$ are called the *toric* and the *adjoint* (respectively) Shimura pairs of

$(G, \mathcal{X})$. The centralizer $K_{\infty,\mathrm{ad}}$ of x^{ad} in $G_{\mathbb{R}}^{\mathrm{ad}}$ is a reductive group over $\mathbb{R}$ which is a maximal compact subgroup of $G_{\mathbb{R}}^{\mathrm{ad}}$. The hermitian symmetric domain structure on $\mathcal{X}^{\mathrm{ad}}$ is obtained via the natural identification $\mathcal{X}^{\mathrm{ad}} = [G^{\mathrm{ad}}(\mathbb{R})]/[K_{\infty,\mathrm{ad}}(\mathbb{R})]$. The hermitian symmetric domain $\mathcal{X}$ is a finite union of connected components of $\mathcal{X}^{\mathrm{ad}}$. In particular, we have $\mathcal{X} \subseteq \mathcal{X}^{\mathrm{ad}}$.

We have a product decomposition

$$(19) \qquad (G^{\mathrm{ad}}, \mathcal{X}^{\mathrm{ad}}) = \prod_{i \in I}(G_i, \mathcal{X}_i)$$

into *simple adjoint* Shimura pairs, where each G_i is a simple group over $\mathbb{Q}$. For each $i \in I$ there exists a number field F_i such that we have an isomorphism $G_i \xrightarrow{\sim} \mathrm{Res}_{F_i/\mathbb{Q}} G_i^{F_i}$, where $G_i^{F_i}$ is an absolutely simple adjoint group over F_i (see [Ti, Subsubsect. 3.1.2]). The number field F_i is uniquely determined up to $\mathrm{Gal}(\mathbb{Q})$-conjugation (i.e., up to isomorphism).

Axiom 3.2 (iii) is equivalent to the fact that $G_{\mathbb{R}}^{\mathrm{ad}}$ is an inner form of its compact form $G_{\mathbb{R}}^{\mathrm{ad,c}}$, cf. [De2, p. 255]. Thus $G_{\mathbb{R}}^{\mathrm{ad}}$ is a product of absolutely simple, adjoint groups over $\mathbb{R}$. But for each $i \in I$ we have $G_{i,\mathbb{R}} \xrightarrow{\sim} \mathrm{Res}_{F_i \otimes_{\mathbb{Q}} \mathbb{R}/\mathbb{R}}[G_i^{F_i} \times_{F_i} (F_i \otimes_{\mathbb{Q}} \mathbb{R})]$. From the last two sentences, we get that for each $i \in I$ the $\mathbb{R}$-algebra $F_i \otimes_{\mathbb{Q}} \mathbb{R}$ is isomorphic to a finite number of copies of $\mathbb{R}$. In other words, for each $i \in I$ the number field F_i is totally real.

We have the following conclusions of the last three paragraphs:

(i) Let G be a reductive group over $\mathbb{Q}$. To give a Shimura pair $(G, \mathcal{X})$ is the same thing as to give a Shimura pair $(G^{\mathrm{ab}}, \{x^{\mathrm{ab}}\})$ of dimension 0 (i.e., a homomorphism $x^{\mathrm{ab}} : \mathbb{S} \to G_{\mathbb{R}}^{\mathrm{ab}}$) and an adjoint Shimura pair $(G^{\mathrm{ad}}, \mathcal{X}^{\mathrm{ad}})$, with the properties that for an (any) element $x^{\mathrm{ad}} \in \mathcal{X}^{\mathrm{ad}}$ the homomorphism $(x^{\mathrm{ab}}, x^{\mathrm{ad}})$: $\mathbb{S} \to G_{\mathbb{R}}^{\mathrm{ab}} \times_{\mathbb{R}} G_{\mathbb{R}}^{\mathrm{ad}}$ lifts to a homomorphism $x : \mathbb{S} \to G_{\mathbb{R}}$, where $G_{\mathbb{R}} \to G_{\mathbb{R}}^{\mathrm{ab}} \times_{\mathbb{R}} G_{\mathbb{R}}^{\mathrm{ad}}$ is the standard isogeny. One takes $\mathcal{X}$ to be the $G(\mathbb{R})$-conjugacy class of x. We emphasize that the Shimura pair $(G, \mathcal{X})$ can depend on the choice of $x^{\mathrm{ad}} \in \mathcal{X}^{\mathrm{ad}}$ (though its isomorphism class does not).

(ii) To give an adjoint Shimura pair $(G^{\mathrm{ad}}, \mathcal{X}^{\mathrm{ad}})$ is the same thing as to give a finite set $(G_i, \mathcal{X}_i)$ of simple adjoint Shimura pairs, cf. (19).

(iii) To give a simple adjoint Shimura pair $(G_i, \mathcal{X}_i)$, one has to first give a totally real number field F_i and an absolutely simple, adjoint group $G_i^{F_i}$ over F_i that satisfies the following property:

(*) for each embedding $j : F_i \hookrightarrow \mathbb{R}$, the group $G_i^{F_i} \times_{F_i, j} \mathbb{R}$ is either compact (and then one defines $\mathcal{X}_{i,j}$ to be a set with one element) or is not compact and associated naturally to a connected hermitian symmetric domain $\mathcal{X}_{i,j}$.

The product $\prod_{j \in \mathrm{Hom}(F_i, \mathbb{R})} \mathcal{X}_{i,j}$ is a connected hermitian symmetric domain isomorphic to the connected components of $\mathcal{X}_i$. If $G_{i,\mathbb{C}}^{F_i}$ is of classical Lie type and if $G_i^{F_i} \times_{F_i, j} \mathbb{R}$ is not compact, then $G_i^{F_i} \times_{F_i, j} \mathbb{R}$ is isomorphic to either $\mathbf{SU}(a, b)_{\mathbb{R}}^{\mathrm{ad}}$ with $a, b \geq 1$, or $\mathbf{SO}(2, n)_{\mathbb{R}}^{\mathrm{ad}}$ with $n \geq 1$, or $\mathbf{Sp}_{2n, \mathbb{R}}^{\mathrm{ad}}$ with $n \geq 1$, or $\mathbf{SO}^*(2n)_{\mathbb{R}}^{\mathrm{ad}}$

with $n \geq 4$. The last think one has to give is a family of homomorphisms $x_{i,j} : \mathbb{S}/\mathbb{G}_m \to G_i^{F_i} \times_{F_{i,j}} \mathbb{R}$, where

- $x_{i,j}$ is trivial if $G_i^{F_i} \times_{F_{i,j}} \mathbb{R}$ is compact, and

- $x_{i,j}$ identifies $\mathbb{S}/\mathbb{G}_m = \mathbf{SO}_{2,\mathbb{R}}$ with the identity component of the center of a maximal compact subgroup of $G_i^{F_i} \times_{F_{i,j}} \mathbb{R}$ if $G_i^{F_i} \times_{F_{i,j}} \mathbb{R}$ is not compact.

One takes $\mathcal{X}_i$ to be the $G_i(\mathbb{R})$-conjugacy class of the composite of the natural epimorphism $\mathbb{S} \twoheadrightarrow \mathbb{S}/\mathbb{G}_m$ with $\prod_{j \in \mathrm{Hom}(F_i, \mathbb{R})} x_{i,j} : \mathbb{S}/\mathbb{G}_m \to G_{i,\mathbb{R}}$. Once F_i and $G_i^{F_i}$ are given, there exist a finite number of possibilities for $\mathcal{X}_i$ (they correspond to possible replacements of some of the $x_{i,j}$'s by their inverses).

3.3.1. Shimura types

A Shimura variety $\mathrm{Sh}(G_1, \mathcal{X}_1)$ is called *unitary* if the adjoint group G_1^{ad} is non-trivial and all simple factors of $G_{1,\mathbb{C}}^{\mathrm{ad}}$ are **PGL** groups over $\mathbb{C}$.

Let $(G, \mathcal{X})$ be a simple, adjoint Shimura pair. Let $\mathfrak{L}$ be the Lie type of anyone of the simple factors of $G_{\mathbb{C}}$. If $\mathfrak{L}$ is either A_n, B_n, C_n, E_6, or E_7, then one say that $(G, \mathcal{X})$ is of $\mathfrak{L}$ Shimura type. If $\mathfrak{L}$ is D_n with $n \geq 4$, then there exist three disjoint possibilities for the type of $(G, \mathcal{X})$: they are $D_n^{\mathbb{H}}$, $D_n^{\mathbb{R}}$, and D_n^{mixed}. If $n \geq 5$, then $(G, \mathcal{X})$ is of $D_n^{\mathbb{H}}$ (resp. of $D_n^{\mathbb{R}}$) Shimura type if and only if each simple, non-compact factor of $G_{\mathbb{R}}$ is isomorphic to $\mathbf{SO}^*(2n)_{\mathbb{R}}^{\mathrm{ad}}$ (resp. to $\mathbf{SO}(2, 2n-2)_{\mathbb{R}}^{\mathrm{ad}}$). The only if part of the previous sentence holds even if $n = 4$.

We will not detail here the precise difference between the Shimura types $D_4^{\mathbb{H}}$, $D_4^{\mathbb{R}}$, and D_4^{mixed} (see [De2, p. 272]).

3.4. Shimura varieties of Hodge type

Let $(G, \mathcal{X})$ be a Shimura pair. We say that $\mathrm{Sh}(G, \mathcal{X})$ (or $(G, \mathcal{X})$) is of *Hodge type*, if there exists an injective map $f : (G, \mathcal{X}) \hookrightarrow (\mathbf{GSp}(W, \psi), \mathcal{S})$ into a Shimura pair that defines a Siegel modular variety. The Hodge $\mathbb{Q}$–structure on W defined by any $x \in \mathcal{X}$ is of type $\{(-1, 0), (0, -1)\}$, cf. (13b). This implies that $x(\mathbb{G}_m)$ is the group of scalar automorphisms of $\mathbf{GL}_{W \otimes_{\mathbb{Q}} \mathbb{R}}$. Therefore $Z(G)$ contains the group $\mathbb{G}_m = Z(\mathbf{GL}_W)$ of scalar automorphisms of W. The image of $Z(G)_{\mathbb{R}}$ in $\mathbf{GSp}(W, \psi)_{\mathbb{R}}^{\mathrm{ad}}$ is contained in the centralizer of the image of x in $\mathbf{GSp}(W, \psi)_{\mathbb{R}}^{\mathrm{ad}}$ and thus it is contained in a compact group. From the last two sentences we get that we have a short exact sequence

$$(20) \qquad\qquad 0 \to \mathbb{G}_m \to Z(G) \to Z(G)^{\mathrm{c}} \to 0,$$

where $Z(G)_{\mathbb{R}}^{\mathrm{c}}$ is a compact group of multiplicative type. In this way we get the only if part of the following Proposition (see [De2, Prop. 2.3.2 or Cor. 2.3.4]).

Proposition 2. *A Shimura pair $(G, \mathcal{X})$ is of Hodge type if and only if the following two properties hold:*

(i) *there exists a faithful representation $G \hookrightarrow \mathbf{GL}_W$ with the property that the Hodge $\mathbb{Q}$–structure on W defined by a (any) $x \in \mathcal{X}$ is of type $\{(-1, 0), (0, -1)\}$;*

(ii) *we have a short exact sequence as in (20).*

If $(G, \mathcal{X})$ is of Hodge type, then (16) becomes (cf. [De2, Cor. 2.1.11])

$$\mathrm{Sh}(G, \mathcal{X})(\mathbb{C}) = G(\mathbb{Q})\backslash(\mathcal{X} \times G(\mathbb{A}_f)).$$

3.4.1. Moduli interpretation

Let $f : (G, \mathcal{X}) \hookrightarrow (\mathbf{GSp}(W, \psi), \mathcal{S})$ be an injective map. We fix a family $(s_\alpha)_{\alpha \in \mathcal{J}}$ of tensors of $\mathcal{T}(W^*)$ such that G is the subgroup of $\mathbf{GSp}(W, \psi)$ that fixes s_α for all $\alpha \in \mathcal{J}$, cf. [De3, Prop. 3.1 (c)]. Let L be a $\mathbb{Z}$-lattice of W such that we have a perfect form $\psi : L \otimes_{\mathbb{Z}} L \to \mathbb{Z}$. We follow [Va1, Subsect. 4.1] to present the standard moduli interpretation of the complex Shimura variety $\mathrm{Sh}(G, \mathcal{X})_{\mathbb{C}}$ with respect to the $\mathbb{Z}$-lattice L of W and the family of tensors $(s_\alpha)_{\alpha \in \mathcal{J}}$.

We consider quadruples of the form $[A, \lambda_A, (v_\alpha)_{\alpha \in \mathcal{J}}, k]$ where:

(a) (A, λ_A) is a principally polarized abelian variety over $\mathbb{C}$;

(b) $(v_\alpha)_{\alpha \in \mathcal{J}}$ is a family of Hodge cycles on A;

(c) k is an isomorphism $H_1(A^{\mathrm{an}}, \mathbb{Z}) \otimes_{\mathbb{Z}} \widehat{\mathbb{Z}} \overset{\sim}{\to} L \otimes_{\mathbb{Z}} \widehat{\mathbb{Z}}$ whose tensorization with $\mathbb{Q}$ (denoted also by k) takes the Betti realization of v_α into s_α for all $\alpha \in \mathcal{J}$ and which induces a symplectic similitude isomorphism between $(H_1(A^{\mathrm{an}}, \mathbb{Z}) \otimes_{\mathbb{Z}} \widehat{\mathbb{Z}}, \lambda_A)$ and $(L \otimes_{\mathbb{Z}} \widehat{\mathbb{Z}}, \psi)$.

We define $\mathcal{A}(G, \mathcal{X}, W, \psi)$ to be the set of isomorphism classes of quadruples of the above form that satisfy the following two conditions:

(i) there exists a similitude isomorphism $(H_1(A^{\mathrm{an}}, \mathbb{Q}), \lambda_A) \overset{\sim}{\to} (W, \psi)$ that takes the Betti realization of v_α into s_α for all $\alpha \in \mathcal{J}$;

(ii) by composing the homomorphism $x_A : \mathbb{S} \to \mathbf{GSp}(H_1(A^{\mathrm{an}}, \mathbb{R}), \lambda_A)$ that defines the Hodge $\mathbb{R}$-structure on $H_1(A^{\mathrm{an}}, \mathbb{R})$ with an isomorphism of real groups $\mathbf{GSp}(H_1(A^{\mathrm{an}}, \mathbb{R}), \lambda_A) \overset{\sim}{\to} \mathbf{GSp}(W \otimes_{\mathbb{Q}} \mathbb{R}, \psi)$ induced naturally by an isomorphism as in (i), we get an element of $\mathcal{X}$.

We have a right action of $G(\mathbb{A}_f)$ on $\mathcal{A}(G, \mathcal{X}, W, \psi)$ defined by the rule:

$$[A, \lambda_A, (v_\alpha)_{\alpha \in \mathcal{J}}, k] \cdot g := [A', \lambda_{A'}, (v_\alpha)_{\alpha \in \mathcal{J}}, g^{-1}k].$$

Here A' is the abelian variety which is isogeneous to A and which is defined by the $\mathbb{Z}$-lattice $H_1(A'^{,\mathrm{an}}, \mathbb{Z})$ of $H_1(A'^{,\mathrm{an}}, \mathbb{Q}) = H_1(A^{\mathrm{an}}, \mathbb{Q})$ whose tensorization with $\widehat{\mathbb{Z}}$ is $(k^{-1} \circ g)(L \otimes_{\mathbb{Z}} \widehat{\mathbb{Z}})$, while $\lambda_{A'}$ is the only rational multiple of λ_A which produces a principal polarization of A' (see [De1, Thm. 4.7] for the theorem of Riemann used here). Here as well as in (e) below, we will identify a polarization with its Betti realization.

There exists a $G(\mathbb{A}_f)$-equivariant bijection

$$f_{(G,\mathcal{X},W,\psi)} : \mathrm{Sh}(G, \mathcal{X})(\mathbb{C}) \overset{\sim}{\to} \mathcal{A}(G, \mathcal{X}, W, \psi)$$

defined as follows. To $[x, g] \in \mathrm{Sh}(G, \mathcal{X})(\mathbb{C}) = G(\mathbb{Q}) \backslash (\mathcal{X} \times G(\mathbb{A}_f))$ we associate the quadruple $f_{(G,\mathcal{X},W,\psi)}([x, g]) := [A, \lambda_A, (v_\alpha)_{\alpha \in \mathcal{J}}, k]$ where:

(d) A is associated to the Hodge $\mathbb{Q}$–structure on W defined by x and to the unique $\mathbb{Z}$-lattice $H_1(A^{\mathrm{an}}, \mathbb{Z})$ of $H_1(A^{\mathrm{an}}, \mathbb{Q}) = W$ for which we have an isomorphism $k - g^{-1} : H_1(\Lambda^{\mathrm{an}}, \mathbb{Z}) \otimes_{\mathbb{Z}} \widehat{\mathbb{Z}} \xrightarrow{\sim} L \otimes_{\mathbb{Z}} \widehat{\mathbb{Z}}$ induced naturally by the automorphism g^{-1} of $W \otimes_{\mathbb{Q}} \mathbb{A}_f$; thus we have $A^{\mathrm{an}} = H_1(A^{\mathrm{an}}, \mathbb{Z}) \backslash (W \otimes_{\mathbb{Q}} \mathbb{C}) / W_x^{0,-1}$, where $W \otimes_{\mathbb{Q}} \mathbb{C} = W_x^{-1,0} \oplus W_x^{0,-1}$ is the Hodge decomposition defined by x;

(e) λ_A is the only rational multiple of ψ which gives birth to a principal polarization of A;

(f) for each $\alpha \in \mathcal{J}$, the Betti realization of v_α is s_α.

The inverse $g_{(G,\mathcal{X},W,\psi)}$ of $f_{(G,\mathcal{X},W,\psi)}$ is defined as follows. We consider a quadruple $[A, \lambda_A, (v_\alpha)_{\alpha \in \mathcal{J}}, k] \in \mathcal{A}(G, \mathcal{X}, W, \psi)$. We choose a symplectic similitude isomorphism $i_A : (H_1(A^{\mathrm{an}}, \mathbb{Q}), \lambda_A) \xrightarrow{\sim} (W, \psi)$ as in (i). It gives birth naturally to an isomorphism $\tilde{i}_A : \mathbf{GSp}(H_1(A^{\mathrm{an}}, \mathbb{Q}), \lambda_A) \to \mathbf{GSp}(W, \psi)$ of groups over $\mathbb{Q}$. We define $x \in \mathcal{X}$ to be $\tilde{i}_{A,\mathbb{R}} \circ x_A$ (with x_A as in Definition 2) and $g \in G(\mathbb{A}_f)$ to be the composite isomorphism $W \otimes_{\mathbb{Q}} \mathbb{A}_f \xrightarrow{k^{-1}} H_1(A^{\mathrm{an}}, \mathbb{Q}) \otimes_{\mathbb{Q}} \mathbb{A}_f \xrightarrow{i_A \otimes 1_{\mathbb{A}_f}} W \otimes_{\mathbb{Q}} \mathbb{A}_f$. Then

$$g_{(G,\mathcal{X},W,\psi)}([A, \lambda_A, (v_\alpha)_{\alpha \in \mathcal{J}}, k]) := [x, g].$$

Taking $(G, \mathcal{X}) = (\mathbf{GSp}(W, \psi), \mathcal{S})$ and $\mathcal{J} - \emptyset$, we get a bijection between the set $\mathrm{Sh}(\mathbf{GSp}(W, \psi), \mathcal{S})(\mathbb{C})$ and the set of isomorphism classes of principally polarized abelian varieties over $\mathbb{C}$ of dimension $\frac{1}{2} \dim_{\mathbb{Q}}(W)$ that have (compatibly) level-N symplectic similitude structures for all positive integers N. Thus to give a $\mathbb{C}$-valued point of $\mathrm{Sh}(\mathbf{GSp}(W, \psi), \mathcal{S})$ is the same thing as to give a triple $[A, \lambda_A, (l_N)_{N \in \mathbb{N}}]$, where (A, λ_A) is a principally polarized abelian variety over $\mathbb{C}$ of dimension $\frac{1}{2} \dim_{\mathbb{Q}}(W)$ and where $l_N : (L/NL, \psi) \xrightarrow{\sim} (H_1(A^{\mathrm{an}}, \mathbb{Z}/N\mathbb{Z}), \lambda_A)$'s are forming a compatible system of symplectic similitude isomorphisms. The compatibility means here that if N_1 and N_2 are positive integers such that $N_1 | N_2$, then l_{N_1} is obtained from l_{N_2} by tensoring with $\mathbb{Z}/N_1\mathbb{Z}$.

3.4.2. Canonical models

Let $r := \frac{1}{2} \dim_{\mathbb{Q}}(W) \in \mathbb{N}$. Let $N \geq 3$ be a positive integer. Let $\mathcal{A}_{r,1,N}$ be the Mumford–moduli scheme over $\mathbb{Z}[\frac{1}{N}]$ that parametrizes isomorphism classes of principally polarized abelian schemes over $\mathbb{Z}[\frac{1}{N}]$-schemes that have level-N symplectic similitude structure and that have relative dimension r, cf. [MFK, Thms. 7.9 and 7.10]. We consider the $\mathbb{Q}$–scheme

$$\mathcal{A}_{r,1,\mathrm{all}} := \mathrm{proj.lim.}_{N \in \mathbb{N}} \mathcal{A}_{r,1,N}.$$

We have a natural identification $\mathrm{Sh}(\mathbf{GSp}(W, \psi), \mathcal{S})(\mathbb{C}) = \mathcal{A}_{r,1,\mathrm{all}}(\mathbb{C})$ of sets, cf. end of Subsubsection 3.4.1. One can easily check that this identification is in fact an isomorphism of complex manifolds. From the very definition of the algebraic structure on $\mathrm{Sh}(\mathbf{GSp}(W, \psi), \mathcal{S})_{\mathbb{C}}$ (obtained based on [BB, Thm. 10.11]), one gets that there exists a natural identification $\mathrm{Sh}(\mathbf{GSp}(W, \psi), \mathcal{S})_{\mathbb{C}} = \mathcal{A}_{r,1,\mathrm{all},\mathbb{C}}$ of $\mathbb{C}$-schemes. Classical works of Shimura, Taniyama, etc., show that the last identification is the extension to $\mathbb{C}$ of an identification $\mathrm{Sh}(\mathbf{GSp}(W, \psi), \mathcal{S}) = \mathcal{A}_{r,1,\mathrm{all}}$ of $\mathbb{Q}$–schemes.

The reflex field $E(G, \mathcal{X})$ is the *smallest* number field such that the closed subscheme $\mathrm{Sh}(G, \mathcal{X})_{\mathbb{C}}$ of $\mathrm{Sh}(\mathbf{GSp}(W, \psi), \mathcal{S})_{\mathbb{C}}$ is defined over $E(G, \mathcal{X})$. In other words, we have a natural closed embedding (cf. end of Subsection 3.2)

$$(21) \qquad f : \mathrm{Sh}(G, \mathcal{X}) \hookrightarrow \mathcal{A}_{r,1,\mathrm{all},E(G,\mathcal{X})} = \mathrm{Sh}(\mathbf{GSp}(W, \psi), \mathcal{S})_{E(G,\mathcal{X})}.$$

The pull back $(\mathcal{V}, \Lambda_{\mathcal{V}})$ to $\mathrm{Sh}(G, \mathcal{X})$ of the universal principally polarized abelian scheme over $\mathcal{A}_{r,1,\mathrm{all},E(G,\mathcal{X})}$ is such that there exists naturally a family of Hodge cycles $(v_{\alpha}^{\mathcal{V}})_{\alpha \in \mathcal{J}}$ on the abelian scheme $\mathcal{V}$.

If $y := [x, g] \in \mathrm{Sh}(G, \mathcal{X})(\mathbb{C})$ and if $f_{(G,\mathcal{X},W,\psi)}([x, g]) = [A, \lambda_A, (v_{\alpha})_{\alpha \in \mathcal{J}}, k]$, then each $y^*(v_{\alpha}^{\mathcal{V}})$ is the Hodge cycle v_{α} on $A = y^*(\mathcal{V})$.

Definition 4. *Let $(G_1, \mathcal{X}_1)$ be a Shimura pair. We say that $(G_1, \mathcal{X}_1)$ is of preabelian type, if there exists a Shimura pair $(G, \mathcal{X})$ of Hodge type such that we have an isomorphism $(G^{\mathrm{ad}}, \mathcal{X}^{\mathrm{ad}}) \xrightarrow{\sim} (G_1^{\mathrm{ad}}, \mathcal{X}_1^{\mathrm{ad}})$ of adjoint Shimura pairs. If moreover this isomorphism $(G^{\mathrm{ad}}, \mathcal{X}^{\mathrm{ad}}) \xrightarrow{\sim} (G_1^{\mathrm{ad}}, \mathcal{X}_1^{\mathrm{ad}})$ is induced naturally by an isogeny $G^{\mathrm{der}} \to G_1^{\mathrm{der}}$, then we say that $(G_1, \mathcal{X}_1)$ is of abelian type.*

Remark 1. *Let $(G_1, \mathcal{X}_1)$ be an arbitrary Shimura variety. Let $\rho_1 : G_1 \hookrightarrow \mathbf{GL}_{W_1}$ be a faithful representation. As in Subsubsection 3.4.1, one checks that $\mathrm{Sh}(G_1, \mathcal{X}_1)_{\mathbb{C}}$ is a moduli space of Hodge $\mathbb{Q}$–structures on W_1 equipped with extra structures. If moreover $(G_1, \mathcal{X}_1)$ is of abelian type, then $\mathrm{Sh}(G_1, \mathcal{X}_1)_{\mathbb{C}}$ is in fact a moduli scheme of polarized abelian motives endowed with Hodge cycles and certain compatible systems of level structures (cf. [Mi3]).*

3.4.3. Classification

Let $(G_1, \mathcal{X}_1)$ be a simple, adjoint Shimura pair. Then $(G_1, \mathcal{X}_1)$ is of abelian type if and only if $(G_1, \mathcal{X}_1)$ is of A_n, B_n, C_n, $D_n^{\mathbb{H}}$, or $D_n^{\mathbb{R}}$ Shimura type. For this classical result due to Satake and Deligne we refer to [Sa1], [Sa2, Part III], and [De2, Table 2.3.8]. There exists a Shimura pair $(G, \mathcal{X})$ of Hodge type whose adjoint is isomorphic to $(G_1, \mathcal{X}_1)$ and whose derived group G^{der} is simply connected if and only if $(G_1, \mathcal{X}_1)$ is of A_n, B_n, C_n, or $D_n^{\mathbb{R}}$ Shimura type (cf. [De2, Table 2.3.8]).

4. Integral models

In this Section we follow [Mi2] and [Va1] to define different integral models of Shimura varieties. Let $p \in \mathbb{N}$ be a prime. Let $\mathbb{Z}_{(p)}$ be the location of $\mathbb{Z}$ at its prime ideal (p). Let $\mathbb{A}_f^{(p)}$ be the ring of finite adèles with the p-component omitted; we have $\mathbb{A}_f = \mathbb{Q}_p \times \mathbb{A}_f^{(p)}$. Let $(G, \mathcal{X})$ be a Shimura pair. Let v be a prime of $E(G, \mathcal{X})$ that divides p. Let $O_{(v)}$ be the local ring of v.

4.1. Basic definitions

(a) Let H be a compact, open subgroup of $G(\mathbb{Q}_p)$. By an *integral model* of $\mathrm{Sh}_H(G, \mathcal{X})$ over $O_{(v)}$ we mean a faithfully flat scheme $\mathcal{N}$ over $O_{(v)}$ together with a $G(\mathbb{A}_f^{(p)})$-continuous action on it and a $G(\mathbb{A}_f^{(p)})$-equivariant isomorphism

$$\mathcal{N}_{E(G,\mathcal{X})} \xrightarrow{\sim} \mathrm{Sh}_H(G,\mathcal{X}).$$

When the $G(\mathbb{A}_f^{(p)})$-action on $\mathcal{N}$ is obvious, by abuse of language, we say that the $O_{(v)}$-scheme $\mathcal{N}$ is an integral model. The integral model $\mathcal{N}$ is said to be *smooth* (resp. *normal*) if there exists a compact, open subgroup H_0 of $G(\mathbb{A}_f^{(p)})$ such that for every inclusion $H_2 \subseteq H_1$ of compact, open subgroups of H_0, the natural morphism $\mathcal{N}/H_2 \to \mathcal{N}/H_1$ is a finite étale morphism between smooth schemes (resp. between normal schemes) of finite type over $O_{(v)}$. In other words, there exists a compact open subgroup H_0 of $G(\mathbb{A}_f^{(p)})$ such that $\mathcal{N}$ is a pro-étale cover of the smooth (resp. the normal) scheme $\mathcal{N}/H_0$ of finite type over $O_{(v)}$.

(b) A regular, faithfully flat $O_{(v)}$-scheme Y is called *p-healthy* (resp. *healthy*) regular, if for each open subscheme U of Y which contains $Y_{\mathbb{Q}}$ and all points of Y of codimension 1, every p-divisible group (resp. every abelian scheme) over U extends uniquely to a p-divisible group (resp. extends to an abelian scheme) over Y.

(c) A scheme Z over $O_{(v)}$ is said to have the *extension property* if for each healthy regular scheme Y over $O_{(v)}$, every $E(G,\mathcal{X})$-morphism $Y_{E(G,\mathcal{X})} \to Z_{E(G,\mathcal{X})}$ extends uniquely to an $O_{(v)}$-morphism $Y \to Z$.

(d) A smooth integral model of $\mathrm{Sh}_H(G,\mathcal{X})$ over $O_{(v)}$ that has the extension property is called an *integral canonical model* of $\mathrm{Sh}(G,\mathcal{X})/H$ over $O_{(v)}$.

(e) Let D be a Dedekind domain. Let K be the field of fractions of D. Let Z_K be a smooth scheme of finite type over K. By a *Néron model* of Z_K over D we mean a smooth scheme of finite type Z over D whose generic fibre is Z_K and which is uniquely determined by the following universal property: for each smooth scheme Y over D, every K-morphism $Y_K \to Z_K$ extends uniquely to a D-morphism $Y \to Z$.

(f) The group $G_{\mathbb{Q}_p}$ is called *unramified* if and only if extends to a reductive group scheme $G_{\mathbb{Z}_p}$ over $\mathbb{Z}_p$. In such a case, each compact, open subgroup of $G_{\mathbb{Q}_p}(\mathbb{Q}_p)$ of the form $G_{\mathbb{Z}_p}(\mathbb{Z}_p)$ is called a *hyperspecial* subgroup of $G_{\mathbb{Q}_p}(\mathbb{Q}_p)$.

(g) Let Z be a flat $O_{(v)}$-scheme and let Y be a closed subscheme of $Z_{k(v)}$. The *dilatation* W of Z centered on Y is an affine Z-scheme defined as follows. To define W, we can work locally in the Zariski topology of Z and therefore we van assume that $Z = \mathrm{Spec}(C)$ is an affine scheme. Let I be the ideal of C that defines Y and let π_v be a uniformizer of $O_{(v)}$. Then W is the spectrum of the C-subalgebra of $C[\frac{1}{\pi_v}]$ generated by $\frac{i}{\pi_v}$ with $i \in I$. The affine morphism $W \to Z$ of $O_{(v)}$-schemes enjoys the following universal property. Let $q : \tilde{Z} \to Z$ be a morphism of flat $O_{(v)}$-schemes. Then q factors uniquely through a morphism $\tilde{Z} \to W$ of Z-schemes if and only if $q_{k(v)} : \tilde{Z}_{k(v)} \to Z_{k(v)}$ factors through Y (i.e., $q_{k(v)}$ is a composite morphism $\tilde{Z}_{k(v)} \to Y \hookrightarrow Z_{k(v)}$).

4.2. Classical example

Let $f : (G,\mathcal{X}) \hookrightarrow (\mathbf{GSp}(W,\psi),\mathcal{S})$ be an injective map. Let L be a $\mathbb{Z}$-lattice of W such that ψ induces a perfect form $\psi : L \otimes_{\mathbb{Z}} L \to \mathbb{Z}$. Let $N \geq 3$ be a natural number which is prime to p. Let

$$K(N) := \{g \in \mathbf{GSp}(L,\psi)(\widehat{\mathbb{Z}})|g \bmod N \text{ is identity}\} \text{ and } K_p := \mathbf{GSp}(L,\psi)(\mathbb{Z}_p).$$

We have an identity $K_p = K(N) \cap \mathbf{GSp}(W,\psi)(\mathbb{Q}_p)$. Let

$$\mathcal{M} := \text{proj.lim.}_{N \in \mathbb{N}, g.c.d.(N,p)=1} \mathcal{A}_{r,1,N};$$

it is a $\mathbb{Z}_{(p)}$-scheme that parametrizes isomorphism classes of principally polarized abelian schemes over $\mathbb{Z}_{(p)}$-schemes that have compatible level-N symplectic similitude structures for all $N \in \mathbb{N}$ prime to p and that have relative dimension r.

The totally discontinuous, locally compact group $\mathbf{GSp}(W,\psi)(\mathbb{A}_f^{(p)})$ acts continuously on $\mathcal{M}$ and moreover $\mathcal{M}$ is a pro-étale cover of $\mathcal{A}_{r,1,N,\mathbb{Z}_{(p)}}$ for all $N \in \mathbb{N}$ prime to p. From (21) we get that we can identify $\text{Sh}_{K(N)}(\mathbf{GSp}(W,\psi),\mathcal{S}) = \mathcal{A}_{r,1,N,\mathbb{Q}}$ and $\text{Sh}_{K_p}(\mathbf{GSp}(W,\psi),\mathcal{S}) = \mathcal{M}_{\mathbb{Q}}$. From the last two sentences we get that $\mathcal{M}$ is a smooth integral model of $\text{Sh}_{K_p}(\mathbf{GSp}(W,\psi),\mathcal{S})$ over $\mathbb{Z}_{(p)}$.

Let $G_{\mathbb{Z}_{(p)}}$ be the Zariski closure of G in $\mathbf{GL}_{L \otimes_{\mathbb{Z}} \mathbb{Z}_{(p)}}$; it is an affine, flat group scheme over $\mathbb{Z}_{(p)}$ whose generic fibre is G. Let $H(N) := K(N) \cap G(\mathbb{A}_f)$ and $H_p := H(N) \cap G(\mathbb{Q}_p)$. From (21) we easily get that we have finite morphisms

$$(22a) \qquad\qquad f(N) : \text{Sh}_{H(N)}(G,\mathcal{X}) \to \text{Sh}_{K(N)}(\mathbf{GSp}(W,\psi),\mathcal{S})$$

and

$$(22b) \qquad\qquad f_p : \text{Sh}_{H_p}(G,\mathcal{X}) \to \text{Sh}_{K_p}(\mathbf{GSp}(W,\psi),\mathcal{S}).$$

As $N \geq 3$, a principally polarized abelian scheme with level-N structure has no automorphism (see [Mu, Ch. IV, 21, Thm. 5] for this result of Serre). This implies that $K(N)$ acts freely on $\mathcal{A}_{r,1,\text{all},E(G,\mathcal{X})} = \text{Sh}(\mathbf{GSp}(W,\psi),\mathcal{S})_{E(G,\mathcal{X})}$. From this and (21) we get that $H(N)$ acts freely on $\text{Sh}(G,\mathcal{X})$. Therefore the $E(G,\mathcal{X})$-scheme $\text{Sh}_{H(N)}(G,\mathcal{X})$ is smooth and thus $\text{Sh}_{H_p}(G,\mathcal{X})$ is a regular scheme which is formally smooth over $E(G,\mathcal{X})$.

Let $\mathcal{N}(N)$ be the normalization of $\mathcal{A}_{r,1,N}$ in the ring of fractions of $\text{Sh}_{H(N)}(G,\mathcal{X})$ and let $\mathcal{N}_p$ be the normalization of $\mathcal{M}$ in the ring of fractions of $\text{Sh}_{H_p}(G,\mathcal{X})$. [Comment: the role of the integral model $\mathcal{N}$ used in Section 1, will be played in what follows by $\mathcal{N}(N)$.] Let $O(G,\mathcal{X})$ be the ring of integers of $E(G,\mathcal{X})$. Let $O(G,\mathcal{X})_{(p)} := O(G,\mathcal{X}) \otimes_{\mathbb{Z}} \mathbb{Z}_{(p)}$; it is the normalization of $\mathbb{Z}_{(p)}$ in $E(G,\mathcal{X})$. The scheme $\mathcal{N}(N)$ is a faithfully flat $O(G,\mathcal{X})[\frac{1}{N}]$-scheme which is normal and of finite type and whose generic fibre is $\text{Sh}_{H(N)}(G,\mathcal{X})$ (the finite type part is implied by the fact that $O_{(v)}$ is an excellent ring). The scheme $\mathcal{N}_p$ is a faithfully flat $O(G,\mathcal{X})_{(p)}$-scheme which is normal and whose generic fibre is $\text{Sh}_{H_p}(G,\mathcal{X})$.

One gets the existence of a finite map

$$f(N) : \mathcal{N}(N) \to \mathcal{A}_{r,1,N}$$

and of a pro-finite map

$$f_p : \mathcal{N}_p \to \mathcal{M}$$

that extends naturally (22a) and (22b) (respectively). Moreover, the totally discontinuous, locally compact group $G(\mathbb{A}_f^{(p)})$ acts continuously on $\mathcal{N}_p$. Let

$$\mathcal{N}_v := \mathcal{N}_p \otimes_{O(G,\mathcal{X})_{(p)}} O_{(v)}.$$

Proposition 3. **(a)** *The $O_{(v)}$-scheme $\mathcal{N}_v$ is a normal integral model of $\mathrm{Sh}(G, \mathcal{X})$ over $O_{(v)}$. Moreover, $\mathcal{N}_v$ is a pro-étale cover of $\mathcal{N}(N)_{O_{(v)}}$.*

(b) *The morphism $f_p : \mathcal{N}_p \to \mathcal{M}$ is finite.*

Proof: Let H_0 be a compact, open subgroup of $G(\mathbb{A}_f^{(p)})$ such that $H_p \times H_0$ is a compact, open subgroup of $H(N)$. As $H(N)$ acts freely on $\mathcal{M}$, it also acts freely on $\mathcal{N}_p$. This implies that $\mathcal{N}_v$ is a pro-étale cover of both $\mathcal{N}(N)_{O_{(v)}}$ and $\mathcal{N}_v/H_0$. Therefore for all open subgroups H_1 and H_2 of H_0 with $H_1 \leqslant H_2$, the morphism $\mathcal{N}_v/H_1 \to \mathcal{N}_v/H_2$ is a finite morphism between étale covers of $\mathcal{N}(N)_{O_{(v)}}$ and therefore it is an étale cover. Based on this, one easily checks that the right action of $G(\mathbb{A}_f^{(p)})$ on $\mathcal{N}_v$ is continuous. Thus (a) holds.

Part (b) is an easy consequence of the fact that $\mathcal{N}_p/H_0$ is a finite scheme over $\mathcal{M}_{O(G,\mathcal{X})_{(p)}}/H_0$. $\square$

4.2.1. PEL type Shimura varieties

Let $\mathcal{B}$ be the $\mathbb{Q}$–subalgebra of $\mathrm{End}(W)$ formed by elements fixed by G. We consider two axioms:

(*) the group G is the identity component of the centralizer of $\mathcal{B}$ in $\mathbf{GSp}(W, \psi)$;

(**) the group G is the centralizer of $\mathcal{B}$ in $\mathbf{GSp}(W, \psi)$.

If the axiom (*) holds, then one calls $\mathrm{Sh}(G, \mathcal{X})$ a Shimura variety of *PEL type*. If the axiom (**) holds, then one calls $\mathrm{Sh}(G, \mathcal{X})$ a Shimura variety of PEL type of either *A or C type*. Here PEL stands for polarizations, endomorphisms, and level structures while the A and C types refer to the fact that all simple factors of $G_{\mathbb{C}}^{\mathrm{ad}}$ are (under the axiom (**)) of some A_n or C_n Lie type (and not of D_n Lie type with $n \geq 4$).

If the axiom (**) holds, then we can choose the family $(v_\alpha)_{\alpha \in \mathcal{J}}$ to be exactly the family of all elements of $\mathcal{B}$. In such a case, all Hodge cycles mentioned in Subsection 3.4.2 are defined by endomorphisms. Let $\mathcal{B}_{(p)} := \mathcal{B} \cap \mathrm{End}(L_{\otimes_{\mathbb{Z}}\mathbb{Z}_{(p)}})$; it is a $\mathbb{Z}_{(p)}$-order of $\mathcal{B}$.

4.2.2. Example

Suppose that $\mathcal{B}_{(p)}$ is a semisimple $\mathbb{Z}_{(p)}$-algebra and that $G_{\mathbb{Z}_{(p)}}$ is the centralizer of $\mathcal{B}_{(p)}$ in the group scheme $\mathbf{GSp}(L \otimes_{\mathbb{Z}} \mathbb{Z}_{(p)}, \psi)$. Then $G_{\mathbb{Z}_{(p)}}$ is a reductive group scheme and moreover $\mathcal{N}_p$ (resp. $\mathcal{N}_v$) is a moduli scheme of principally polarized abelian schemes which are over $O(G, \mathcal{X})_{(p)}$-schemes (resp. over $O_{(v)}$-schemes), which have relative dimension r, which have compatible level-N symplectic similitude structures for all $N \in \mathbb{N}$ prime to p, which are endowed with a $\mathbb{Z}_{(p)}$-algebra $\mathcal{B}_{(p)}$ of $\mathbb{Z}_{(p)}$-endomorphisms, and which satisfy certain axioms that are related to the properties (d) to (f) of Subsubsection 3.4.2. Unfortunately, presently this is the *only case* when $\mathcal{N}_p$ (resp. when $\mathcal{N}_v$) has a good moduli interpretation. This explains the difficulties one encounters in getting as well as of stating results pertaining to either $\mathcal{N}_p$ or $\mathcal{N}_v$.

4.3. Main problems

Here is a list of six main problems in the study of $\mathcal{N}(N)$, $\mathcal{N}_p$, and $\mathcal{N}_v$. For simplicity, these problems will be stated here only in terms of $\mathcal{N}(N)$.

(a) Determine when $\mathcal{N}(N)$ is uniquely determined up to isomorphism by its generic fibre $\mathrm{Sh}_{H(N)}(G, \mathcal{X})$ and by a suitable universal property.

(b) Determine when $\mathcal{N}(N)$ is a smooth $O(G, \mathcal{X})[\frac{1}{N}]$-scheme.

(c) Determine when $\mathcal{N}(N)$ is a projective $O(G, \mathcal{X})[\frac{1}{N}]$-scheme.

(d) Identify and study different stratifications of the special fibres of $\mathcal{N}(N)$.

(e) Describe the points of $\mathcal{N}(N)$ with values in finite fields.

(f) Describe the points of $\mathcal{N}(N)$ with values in $O[\frac{1}{N}]$, where O is the ring of integers of some finite field extension of $E(G, \mathcal{X})$.

In the next four Sections we will study the first four problems one by one, in a way that could be useful towards the partial solutions of the problem (e). Any approach to the problem (f) would require a very good understanding of the first five problems and this is the reason (as well as the main motivation) for why the six problems are listed together.

5. Uniqueness of integral models

Until the end we will use the following notations introduced in Section 4:

$$f : (G, \mathcal{X}) \hookrightarrow (\mathbf{GSp}(W, \psi), \mathcal{S}), \quad L, \quad N, \quad K(N), \quad K_p, \quad H(N), \quad H_p, \quad O(G, \mathcal{X}),$$

$$O(G, \mathcal{X})_{(p)}, \quad v, \quad O_{(v)}, \quad f(N) : \mathcal{N}(N) \to \mathcal{A}_{r,1,N}, \quad f_p : \mathcal{N}_p \to \mathcal{M}, \quad \mathcal{N}_v, \quad G_{\mathbb{Z}_{(p)}}.$$

Let e_v be the index of ramification of v. Let $k(v)$ be the residue field of v. Let

$$\mathcal{L}(N)_v := \mathcal{N}(N) \otimes_{O(G, \mathcal{X})[\frac{1}{N}]} k(v) \quad \text{and} \quad \mathcal{L}_v := \mathcal{N}_v \otimes_{O_{(v)}} k(v).$$

In this Section we study when the $k(v)$-scheme $\mathcal{L}(N)_v$ (resp. $\mathcal{L}_v$) is uniquely determined in some sensible way by $\mathrm{Sh}_{H(N)}(G, \mathcal{X})$ (resp. by $\mathrm{Sh}_{H_p}(G, \mathcal{X})$) and by the prime v of $E(G, \mathcal{X})$. Whenever one gets such a uniqueness property, one can call $\mathcal{L}(N)_v$ (resp. $\mathcal{L}_v$) as the *canonical fibre model* of $\mathrm{Sh}_{H(N)}(G, \mathcal{X})$ (resp. $\mathrm{Sh}_{H_p}(G, \mathcal{X})$) at v (or over $k(v)$).

Milne's original insight (see [Mi2] and [Va1]) was to prove in many cases the uniqueness of $\mathcal{N}_v$ and $\mathcal{L}_v$ by showing first that:

(i) the $O_{(v)}$-scheme $\mathcal{N}_v$ has the extension property, and

(ii) $\mathcal{N}_v$ is a healthy regular scheme in the sense of Definition 4.1 (b).

While (i) always holds (see Proposition 4 below), it is very hard in general to decide if (ii) holds. However, results of [Va1], [Va2], and [Va11] allow us to get that (ii) holds in many cases of interest (see Subsection 5.1). Subsection 5.2 shows how one gets the uniqueness of $\mathcal{N}(N)$ (and therefore also of $\mathcal{L}(N)_v$) via (the uniqueness of) Néron models.

Proposition 4. *The $O_{(v)}$-scheme $\mathcal{N}_v$ has the extension property.*

Proof: Let Y be a healthy regular scheme over $O_{(v)}$. Let $q : Y_{E(G,\mathcal{X})} \to \mathrm{Sh}_{H_p}(G, \mathcal{X})$ be a morphism of $E(G, \mathcal{X})$-schemes. Let $(\mathcal{U}, \lambda_{\mathcal{U}})$ be the pull back to $Y_{E(G,\mathcal{X})}$ of the universal principally polarized abelian scheme over $\mathcal{M}$ (via the composite morphism $Y_{E(G,\mathcal{X})} \to \mathrm{Sh}_{H_p}(G, \mathcal{X}) \to \mathrm{Sh}_{K_p}(\mathbf{GSp}(W, \psi), S) = \mathcal{M}_{\mathbb{Q}}$). As the universal principally polarized abelian scheme over $\mathcal{M}$ has a level-N symplectic similitude structure for all $N \in \mathbb{N}$ prime to p, the same holds for $(\mathcal{U}, \lambda_{\mathcal{U}})$. From this and the Néron–Ogg–Shafarevich criterion of good reduction (see [BLR, Ch. 7, 7.4, Thm. 5]) we get that $\mathcal{U}$ extends to an abelian scheme $\mathcal{U}_U$ over an open subscheme U of Y which contains $Y_{\mathbb{Q}} = Y_{E(G,\mathcal{X})}$ and for which we have $\mathrm{codim}_Y(Y \setminus U) \geq 2$. Thus $\mathcal{U}_U$ extends to an abelian scheme $\mathcal{U}_Y$ over Y, cf. the very definition of a healthy regular scheme. The polarization $\lambda_{\mathcal{U}}$ extends as well to a polarization $\lambda_{\mathcal{U}_Y}$ of $\mathcal{U}_Y$, cf. [Mi2, Prop. 2.14]. Moreover, each level-N symplectic similitude structure of $(\mathcal{U}, \lambda_{\mathcal{U}})$ extends to a level-N symplectic similitude structure of $(\mathcal{U}_Y, \lambda_{\mathcal{U}_Y})$. This implies that the composite of q with the finite morphism $\mathrm{Sh}_{H_p}(G, \mathcal{X}) \to \mathcal{M}_{E(G,\mathcal{X})}$ extends uniquely to a morphism $Y \to \mathcal{M}_{O_{(v)}}$. As Y is a regular scheme and thus also normal and as $\mathcal{N}_v \to \mathcal{M}_{O_{(v)}}$ is a finite morphism, we get that the morphism $Y \to \mathcal{M}_{O_{(v)}}$ factors uniquely through $\mathcal{N}_v$ (as it does so generically). This implies that $q : Y_{E(G,\mathcal{X})} \to \mathrm{Sh}_{H_p}(G, \mathcal{X})$ extends uniquely to a morphism $q_Y : Y \to \mathcal{N}_v$ of $O_{(v)}$-schemes. From this the Proposition follows. $\square$

Proposition 5. *Let Y be a regular scheme which is faithfully flat over $\mathbb{Z}_{(p)}$. Then the following two properties hold:*

(a) *Let U be an open subscheme of Y which contains $Y_{\mathbb{Q}}$ and the generic points of $Y_{\mathbb{F}_p}$. Let A_U be an abelian scheme over U with the property that its p-divisible group D_U extends to a p-divisible group D over Y. Then A_U extends to an abelian scheme A over U.*

(b) *If Y is a p-healthy regular scheme, then it is also a healthy regular scheme.*

Proof: Part (b) follows from (a) and the very definitions. To prove (a) we follow [Va2, Prop. 4.1]. Let $N \geq 3$ be a positive integer relatively prime to p.

To show that A exists, we can assume that Y is local, complete, and strictly henselian, that U is the complement of the maximal point y of Y, that A_U has a principal polarization λ_{A_U}, and that (A_U, λ_{A_U}) has a level-N symplectic similitude structure $l_{U,N}$ (see [FC, (i)-(iii) of pp. 185, 186]). We write $Y = \mathrm{Spec}(R)$. Let λ_{D_U} be the principal quasi-polarization of D_U defined naturally by λ_{A_U}; it extends to a principal quasi-polarization λ_D of D (cf. Tate's theorem [Ta, Thm. 4]). Let r be the relative dimension of A_U. Let $(\mathcal{A}, \Lambda_{\mathcal{A}})$ be the universal principally polarized abelian scheme over $\mathcal{A}_{r,1,N}$.

Let $m_U : U \to \mathcal{A}_{r,1,N}$ be the morphism defined by $(A_U, \lambda_{A_U}, l_{U,N})$. We show that m_U extends to a morphism $m : Y \to \mathcal{A}_{r,1,N}$.

Let $N_0 \in \mathbb{N}$ be prime to p. From the classical purity theorem we get that the étale cover $A_U[N_0] \to U$ extends to an étale cover $Y_{N_0} \to Y$. But as Y is strictly henselian, Y has no connected étale cover different from Y. Thus each Y_{N_0} is a disjoint union of N_0^{2r}-copies of Y. From this we get that (A_U, λ_{A_U}) has a level-N_0 symplectic similitude structure l_{U,N_0} for every $N_0 \in \mathbb{N}$ prime to p.

Let $\overline{\mathcal{A}}_{r,1,N}$ be a projective, toroidal compactification of $\mathcal{A}_{r,1,N}$ such that (cf. [FC, Chap. IV, Thm. 6.7]):

(a) the complement of $\mathcal{A}_{r,1,N}$ in $\overline{\mathcal{A}}_{r,1,N}$ has pure codimension 1 in $\overline{\mathcal{A}}_{r,1,N}$ and

(b) there exists a semi-abelian scheme over $\overline{\mathcal{A}}_{r,1,N}$ that extends $\mathcal{A}$.

Let $\tilde{Y}$ be the normalization of the Zariski closure of U in $Y \times_{\mathbb{Z}} \overline{\mathcal{A}}_{r,1,N}$. It is a projective, normal, integral Y-scheme which has U as an open subscheme. Let C be the complement of U in $\tilde{Y}$ endowed with the reduced structure; it is a reduced, projective scheme over the residue field k of y. The $\mathbb{Z}$-algebras of global functions of Y, U, and $\tilde{Y}$ are all equal to R (cf. [Ma, Thm. 38] for U). Thus C is a connected k-scheme, cf. [Ha, Ch. III, Cor. 11.3] applied to $\tilde{Y} \to Y$.

Let $\overline{A}_{\tilde{Y}}$ be the semi-abelian scheme over $\tilde{Y}$ that extends A_U (it is unique, cf. [FC, Chap. I, Prop. 2.7]). Due to the existence of the l_{U,N_0}'s, the Néron–Ogg–Shafarevich criterion implies that $\overline{A}_{\tilde{Y}}$ is an abelian scheme in codimension at most 1. Therefore, since the complement of $\mathcal{A}_{r,1,N}$ in $\overline{\mathcal{A}}_{r,1,N}$ has pure codimension 1 in $\overline{\mathcal{A}}_{r,1,N}$, it follows that $\overline{A}_{\tilde{Y}}$ is an abelian scheme. Thus m_U extends to a morphism $m_{\tilde{Y}} : \tilde{Y} \to \mathcal{A}_{r,1,N}$. Let $\lambda_{\overline{A}_{\tilde{Y}}} := m_{\tilde{Y}}^*(\Lambda_{\mathcal{A}})$. Tate's theorem implies that the principally quasi-polarized p-divisible group of $(\overline{A}_{\tilde{Y}}, \lambda_{\overline{A}_{\tilde{Y}}})$ is the pull-back $(D_{\tilde{Y}}, \lambda_{D_{\tilde{Y}}})$ of (D, λ_D) to $\tilde{Y}$. Hence the pull back (D_C, λ_{D_C}) of $(D_{\tilde{Y}}, \lambda_{D_{\tilde{Y}}})$ to C is constant i.e., it is the pull back to C of a principally quasi-polarized p-divisible group over k.

We check that the image $m_{\tilde{Y}}(C)$ of C through $m_{\tilde{Y}}$ is a point $\{y_0\}$ of $\mathcal{A}_{r,1,N}$. Since C is connected, to check this it suffices to show that, if $\widehat{O_c}$ is the completion of the local ring O_c of C at an arbitrary point c of C, then the morphism $\mathrm{Spec}(\widehat{O_c}) \to \mathcal{A}_{r,1,N}$ defined naturally by $m_{\tilde{Y}}$ is constant. But as (D_C, λ_{D_C}) is constant, this follows from Serre–Tate deformation theory (see [Me, Chaps. 4, 5]). Thus $m_{\tilde{Y}}(C)$ is a point $\{y_0\}$ of $\mathcal{A}_{r,1,N}$.

Let R_0 be the local ring of $\mathcal{A}_{r,1,N}$ at y_0. Because Y is local and $\tilde{Y}$ is a projective Y-scheme, each point of $\tilde{Y}$ specializes to a point of C. Hence each point of the image of $m_{\tilde{Y}}$ specializes to y_0 and thus $m_{\tilde{Y}}$ factors through the natural morphism $\mathrm{Spec}(R_0) \to \mathcal{A}_{r,1,N}$. Since R is the ring of global functions of $\tilde{Y}$, the resulting morphism $\tilde{Y} \to \mathrm{Spec}(R_0)$ factors through a morphism $\mathrm{Spec}(R) \to \mathrm{Spec}(R_0)$. Therefore $m_{\tilde{Y}}$ factors through a morphism $m : Y \to \mathcal{A}_{r,1,N}$ that extends m_U. This ends the argument for the existence of m. We conclude that $A := m^*(\mathcal{A})$ is an abelian scheme over Y which extends A_U. Thus (a) holds. $\qquad\square$

5.1. Examples of healthy regular schemes

In (the proofs of) [FC, Ch. IV, Thms. 6.4, 6.4', and 6.8] was claimed that every regular scheme which is faithfully flat over $\mathbb{Z}_{(p)}$ is p-healthy regular as well as healthy regular. In turns out that this claim is far from being true. For instance, an example of Raynaud–Gabber–Ogus (see [dJO1, Sect. 6]) shows that the regular scheme $\mathrm{Spec}(W(k)[[x,y]]/((xy)^{p-1} - p))$ is neither p-healthy nor healthy regular. Here $W(k)$ is the ring of Witt vectors with coefficients in a perfect field k of characteristic p.

Based on Proposition 5 (b) and a theorem of Raynaud (see [Ra, Thm. 3.3.3]), one easily checks that if $e_v < p - 2$, then each regular scheme which is formally smooth over $O_{(v)}$ is a healthy regular scheme (see [Va1, Subsubsect. 3.2.17]). In [Va2, Thm. 1.3] it is proved that the same holds provided $e_v = 1$. In [Va11] it is proved that the same holds provided $e_v = p - 1$. Even more, in [Va11, Thm. 1.3 and Cor. 1.5] it is proved that:

Theorem 1. (a) *Suppose that $p > 2$. Each regular scheme which is formally smooth over $O_{(v)}$ is healthy regular if and only if the following inequality holds $e_v \leq p-1$.*

(b) *Suppose that $p = 2$ and $e_v = 1$. Then each regular scheme which is formally smooth over $O_{(v)}$ is healthy regular.*

Part (a) also holds for $p = 2$ but this is not checked loc. cit. and this is why above for $p = 2$ we stated only one implication in the form of (b). From Theorem 1 and Propositions 3 (a) and 4 we get the following answer to the problem 4.3 (a):

Corollary 1. *Suppose that $e_v \leq p - 1$ and that $\mathcal{N}_v$ is a regular scheme which is formally smooth over $O_{(v)}$ (i.e., and that $\mathcal{N}(N)_{O_{(v)}}$ is a smooth $O_{(v)}$-scheme). Then $\mathcal{N}_v$ is the integral canonical model of $\mathrm{Sh}_{H_p}(G, \mathcal{X})$ over $O_{(v)}$ and it is uniquely determined up to unique isomorphism. Thus also $\mathcal{L}(N)_v$ and $\mathcal{L}_v$ are uniquely determined by $\mathrm{Sh}_{H(N)}(G, \mathcal{X})$ and $\mathrm{Sh}_{H_p}(G, \mathcal{X})$ (respectively) and v.*

5.1.1. Example

The integral canonical model of $\mathrm{Sh}_{K_p}(\mathbf{GSp}(W, \psi), \mathcal{S})$ over $\mathbb{Z}_{(p)}$ is $\mathcal{M}$.

5.2. Integral models as Nèron models

In [Ne] it is showed that each abelian variety over the field of fractions K of a Dedekind domain D has a Néron model over D. In [BLR] it is checked that many other closed subschemes of torsors of certain commutative group varieties over K, have Néron models over D. But most often, for $N >> 0$ the $E(G, \mathcal{X})$-scheme $\mathrm{Sh}_{H(N)}(G, \mathcal{X})$ can not be embedded in such torsors; we include one basic example.

5.2.1. Example

Suppose that $G_{\mathbb{R}}^{\mathrm{ad}}$ is isomorphic to $\mathbf{SU}(a,b)_{\mathbb{R}}^{\mathrm{ad}} \times_{\mathbb{R}} \mathbf{SU}(a+b,0)_{\mathbb{R}}^{\mathrm{ad}}$ for some positive integers $a \geq 3$ and $b \geq 3$. One has $H^{1,0}(\mathcal{C}(\mathbb{C}),\mathbb{C}) = 0$ for each connected component $\mathcal{C}$ of $\mathrm{Sh}_{H(N)}(G,\mathcal{X})_{\mathbb{C}}$, cf. [Pa, Thm. 2, 2.8 (i)]. The analytic Lie group $\mathrm{Alb}(\mathcal{C})^{\mathrm{an}}$ associated to the albanese variety $\mathrm{Alb}(\mathcal{C})$ is isomorphic to $[\mathrm{Hom}(H^{1,0}(\mathcal{C}(\mathbb{C}),\mathbb{C}),\mathbb{C})]/H_1(\mathcal{C},\mathbb{Z})$ and therefore it is 0. The $\mathbb{Q}$–rank of G^{ad} is 0 and this implies that $\mathrm{Sh}_{H(N)}(G,\mathcal{X})$ is a projective $E(G,\mathcal{X})$-scheme, cf. [BHC, Thm. 12.3 and Cor. 12.4]. From the last two sentences one gets that $\mathcal{C}$ is a connected, projective variety over $\mathbb{C}$ whose albanese variety $\mathrm{Alb}(\mathcal{C})$ is trivial. Thus $\mathcal{C}$ can not be embedded into commutative group varieties over $\mathbb{C}$. Therefore the connected components of the $E(G,\mathcal{X})$-scheme $\mathrm{Sh}_{H(N)}(G,\mathcal{X})$ can not be embedded into torsors of commutative group varieties over $E(G,\mathcal{X})$.

Based on the previous example we get that the class of Néron models introduced below is new (cf. [Va4, Prop. 4.4.1] and [Va11, Thm. 4.3.1]).

Theorem 2. *Suppose that for each prime p that does not divide N and for every prime v of $E(G,\mathcal{X})$ that divides p, we have $e_v \leq p-1$. Suppose that $\mathcal{N}(N)$ is a smooth, projective $O(G,\mathcal{X})[\frac{1}{N}]$-scheme. Then $\mathcal{N}(N)$ is the Néron model of its generic fibre $\mathrm{Sh}_{H(N)}(G,\mathcal{X})$ over $O(G,\mathcal{X})[\frac{1}{N}]$ (and thus it is uniquely determined by $\mathrm{Sh}_{H(N)}(G,\mathcal{X})$ and N).*

Theorem 2 provides a better answer to problem 4.3 (a) than Corollary 1, *provided* in addition we know that $\mathcal{N}(N)$ is a projective $O(G,\mathcal{X})[\frac{1}{N}]$-scheme.

6. Smoothness of integral models

We will use the notations listed at the beginning of Section 5. In this Section we study the smoothness of $\mathcal{N}_v$ and $\mathcal{N}(N)_v$. Let $(G_1,\mathcal{X}_1)$ be a Shimura pair such that the group $G_{1,\mathbb{Q}_p}$ is unramified. Let H_1 be a hyperspecial subgroup of $G_1(\mathbb{Q}_p) = G_{1,\mathbb{Q}_p}(\mathbb{Q}_p)$, cf. Definition 4.1 (f). In 1976 Langlands conjectured the existence of a good integral model of $\mathrm{Sh}_{H_1}(G_1,\mathcal{X}_1)$ over each local ring $O_{(v_1)}$ of $E(G_1,\mathcal{X}_1)$ at a prime v_1 of $E(G_1,\mathcal{X}_1)$ that divides p (see [La, p. 411]); unfortunately, Langlands did not explain what good is supposed to stand for. We emphasize that the assumption that $G_{1,\mathbb{Q}_p}$ is unramified implies that $E(G_1,\mathcal{X}_1)$ is unramified above p (see [Mi3, Cor. 4.7 (a)]); thus the index of ramification e_{v_1} of v_1 is 1.

In 1992 Milne made the following conjecture (slight reformulation made by us, as in [Va1, Conj. 3.2.5]; strictly speaking, both Langlands and Milne stated their conjectures over the completion of $O_{(v_1)}$).

Conjecture 1. *There exists an integral canonical model of $\mathrm{Sh}_{H_1}(G_1,\mathcal{X}_1)$ over $O_{(v_1)}$.*

From the classical works of Zink, Rapoport–Langlands, and Kottwitz one gets (see [Zi1], [LR], and [Ko2]):

Theorem 3. (a) *The Milne conjecture holds if $p \geq 3$ and $\mathrm{Sh}(G_1, \mathcal{X}_1)$ is a Shimura variety of PEL type.*

(b) *The Milne conjecture holds if $\mathrm{Sh}(G_1, \mathcal{X}_1)$ is a Shimura variety of PEL type of either A or C type.*

The main results of [Va1] and [Va6] say (see [Va1, 1.4, Thm. 2, and Thm. 6.4.1] and [Va6, Thm. 1.3]):

Theorem 4. *Suppose one of the following two conditions holds:*

(a) $p \geq 5$ *and* $\mathrm{Sh}_{H_1}(G_1, \mathcal{X}_1)$ *is of abelian type;*

(b) p *is arbitrary and* $\mathrm{Sh}_{H_1}(G_1, \mathcal{X}_1)$ *is a unitary Shimura variety.*

Then the Milne conjecture holds. Moreover, for each prime v_1 of $E(G_1, \mathcal{X}_1)$ that divides p, the integral canonical model of $\mathrm{Sh}_{H_1}(G_1, \mathcal{X}_1)$ over $O_{(v_1)}$ is a proétale cover of a smooth, quasi-projective $O_{(v_1)}$-scheme.

Remark 2. *See [Va2, Thm. 1.3] and [Va6, Thm. 1.3] for two corrections to the proof of Theorem 4 under the assumption that condition (a) holds. More precisely:*

- *the original argument of Faltings for the proof of Proposition 5 was incorrect and it has been corrected in [Va2] (cf. proof of Proposition 5);*

- *the proof of Theorem 4 for the cases when $G_{1,\mathbb{C}}^{\mathrm{ad}}$ has simple factors isomorphic to $\mathbf{PGL}_{p^m}$ for some $m \in \mathbb{N}$ was partially incorrect and it has been corrected by [Va6, Thm. 1.3] (cf. [Va6, Appendix, E.3]).*

6.1. Strategy of the proof of Theorem 4, part a

To explain the four main steps of the proof of the (a) part of Theorem 4, we will sketch the argument why the assumptions that $p > 3$ and that $G_{\mathbb{Z}_{(p)}}$ is a reductive group scheme over $\mathbb{Z}_{(p)}$ imply that $\mathcal{N}_p$ is a formally smooth scheme over $\mathbb{Z}_{(p)}$. Let $W(\mathbb{F})$ be the ring of Witt vectors with coefficients in an algebraic closure $\mathbb{F}$ of $\mathbb{F}_p$. Let $B(\mathbb{F}) := W(\mathbb{F})[\frac{1}{p}]$.

Let $y : \mathrm{Spec}(\mathbb{F}) \to \mathcal{N}_p$ and let $z : \mathrm{Spec}(V) \to \mathcal{N}_p$ be a lift of y, where V is a finite, discrete valuation ring extension of $W(\mathbb{F})$. Let e be the index of ramification of V. Let R_e be the p-adic completion of the $W(\mathbb{F})[[x]]$-subalgebra of $B(\mathbb{F})[[x]]$ generated by $\frac{x^{en}}{n!}$ with $n \in \mathbb{N}$. Let Φ_e be the Frobenius endomorphism of R_e which is compatible with the Frobenius automorphism of $W(\mathbb{F})$ and which takes x to x^p. We have a natural $W(\mathbb{F})$-epimorphism $m_\pi : R_e \twoheadrightarrow V$ which maps x to a fixed uniformizer π of V. The kernel J_π of m_π has divided power structures and thus we can speak about the evaluation of F-crystals at the thickening $\mathrm{Spec}(V) \hookrightarrow \mathrm{Spec}(R_e)$ defined naturally by m_π. We now consider the principally quasi-polarized, filtered F-crystal of the pull back (A_V, λ_{A_V}) to $\mathrm{Spec}(V)$ of the universal principally abelian scheme over $\mathcal{M}$ (via $f_p \circ z$). Its evaluation at the thickening $\mathrm{Spec}(V) \hookrightarrow \mathrm{Spec}(R_e)$ is of the form

$$(M_{R_e}, F_V^1, \phi_{M_{R_e}}, \nabla_{M_{R_e}}, \psi_{M_{R_e}}),$$

where M_{R_e} is a free R_e-module of rank $2r$, F_V^1 is a direct summand of $H^1_{\mathrm{dR}}(A_V/V) = M_{R_e}/J_\pi M_{R_e}$ of rank r, $\phi_{M_{R_e}}$ is a Φ_e-linear endomorphism of M_{R_e}, $\nabla_{M_{R_e}}$ is an integrable, nilpotent modulo p connection on M_{R_e}, and $\psi_{M_{R_e}}$ is a perfect alternating form on M_{R_e}. The generic fibre of A_V is equipped with a family of Hodge cycles whose de Rham realizations belong to $\mathcal{T}(M_{R_e}[\frac{1}{p}]/J_\pi M_{R_e}[\frac{1}{p}])$ and lift naturally to define a family of tensors $(t_{z,\alpha})_{\alpha\in\mathcal{J}}$ of $\mathcal{T}(M_{R_e}[\frac{1}{p}])$.

The **first main step** is to show that, under some conditions on the closed embedding homomorphism $G_{\mathbb{Z}_{(p)}} \hookrightarrow \mathbf{GSp}(L \otimes_{\mathbb{Z}} \mathbb{Z}_{(p)}, \psi)$ and under the assumption that $p > 3$, the Zariski closure in $\mathbf{GSp}(M_{R_e}, \psi_{M_{R_e}})$ of the subgroup of $\mathbf{GSp}(M_{R_e}[\frac{1}{p}], \psi_{M_{R_e}})$ that fixes $t_{z,\alpha}$ for all $\alpha \in \mathcal{J}$, is a reductive group scheme $\tilde{G}_{R_e}$ over R_e. See [Va1, Subsect. 5.2] for more details and see [Va1, (5.2.12)] for the fact that the reductive group scheme $\tilde{G}_{R_e}$ is isomorphic to $G_{\mathbb{Z}_{(P)}} \times_{\mathbb{Z}_{(p)}} R_e$.

The **second main step** is to show that we can lift F_V^1 to a direct summand $F_{R_e}^1$ of M_{R_e} in such a way that $\psi_{M_{R_e}}(F_{R_e}^1 \otimes F_{R_e}^1) = 0$ and that for each element $\alpha \in \mathcal{J}$ the tensor $t_{z,\alpha}$ belongs to the F^0-filtration of $\mathcal{T}(M_{R_e}[\frac{1}{p}])$ defined by $F_{R_e}^1[\frac{1}{p}]$. The essence of this second main step is the classical theory of infinitesimal liftings of cocharacters of smooth group schemes (see [DG, Exp. IX]). Due to the existence of $F_{R_e}^1$, the morphism $z : \mathrm{Spec}(V) \to \mathcal{N}_p$ lifts to a morphism $w : \mathrm{Spec}(R_e) \to \mathcal{N}_p$. The reduction of w modulo the ideal $R_e \cap xB(\mathbb{F})[[x]]$ of R_e is a lift $z_0 : \mathrm{Spec}(W(\mathbb{F})) \to \mathcal{N}_p$ of y. Thus, by replacing z with z_0 we can assume that $V = W(\mathbb{F})$. See [Va1, Subsect. 5.3] for more details.

The **third main step** uses the lift $z_0 : \mathrm{Spec}(W(\mathbb{F})) \to \mathcal{N}_p$ of y and Faltings deformation theory (see [Fa, Sect. 7]) to show that $\mathcal{N}_p$ is formally smooth over $\mathbb{Z}_{(p)}$ at its $\mathbb{F}$-valued point defined by y. See [Va1, Subsect. 5.4] for more details.

The **fourth main step** shows that for $p > 3$ the mentioned conditions on the closed embedding homomorphism $G_{\mathbb{Z}_{(p)}} \hookrightarrow \mathbf{GSp}(L \otimes_{\mathbb{Z}} \mathbb{Z}_{(p)}, \psi)$ always hold, *provided* we replace $f : (G, \mathcal{X}) \hookrightarrow (\mathbf{GSp}(W, \psi), \mathcal{S})$ by a suitable other injective map $f_1 : (G_1, \mathcal{X}_1) \hookrightarrow (\mathbf{GSp}(W_1, \psi_1), \mathcal{S}_1)$ with the property that $(G^{\mathrm{ad}}, \mathcal{X}^{\mathrm{ad}}) = (G_1^{\mathrm{ad}}, \mathcal{X}_1^{\mathrm{ad}})$. See [Va1, Subsects. 6.5 and 6.6] for more details.

Remark 3. *In [Va7] it is shown that Theorem 4 holds even if $p \in \{2,3\}$ and $\mathrm{Sh}(G_1, \mathcal{X}_1)$ is of abelian type. In [Ki] it is claimed that Theorem 4 holds for $p \geq 3$. The work [Ki] does not bring any new conceptual ideas to [Va1], [Va6], and [Va7]. In fact, the note [Ki] is only a variation of [Va1], [Va6], and [Va7]. This variation is made possible due to recent advances in the theory of crystalline representations achieved by Fontaine, Breuil, and Kisin. We emphasize that [Ki] does not work for $p = 2$ while [Va7] works as well for $p = 2$.*

6.2. *Strategy of the proof of Theorem 4, part b*

To explain the three main steps of the proof of the (b) part of Theorem 4, in this Subsection we will assume that $(G_1, \mathcal{X}_1)$ is a simple, adjoint, unitary Shimura pair of isotypic A_n Dynkin type. In [De2, Prop. 2.3.10] it is proved the existence of an injective map $f : (G, \mathcal{X}) \hookrightarrow (\mathbf{GSp}(W, \psi), \mathcal{S})$ of Shimura pairs such that we have $(G^{\mathrm{ad}}, \mathcal{X}^{\mathrm{ad}}) = (G_1, \mathcal{X}_1)$.

The **first step** uses a modification of the proof of [De2, Prop. 2.3.10] to show that we can choose f such that $G_{\mathbb{Z}_{(p)}}$ is the subgroup of $\mathbf{GSp}(L \otimes_{\mathbb{Z}} \mathbb{Z}_{(p)}, \psi)$ that fixes a semisimple $\mathbb{Z}_{(p)}$–subalgebra $\mathcal{B}_{(p)}$ of $\mathrm{End}(W)$ (see [Va6, Prop. 3.2]). Let $H_{1,p} := G^{\mathrm{ad}}_{\mathbb{Z}_{(p)}}(\mathbb{Z}_p)$; it is a hypersecial subgroup of $G_{1,\mathbb{Q}_p}(\mathbb{Q}_p) = G^{\mathrm{ad}}_{\mathbb{Q}_p}(\mathbb{Q}_p)$.

The **second step** only applies Theorem 3 to conclude that $\mathcal{N}_p$ is a formally smooth $O(G, \mathcal{X})_{(p)}$-scheme.

The **third step** uses the standard moduli interpretation of $\mathcal{N}_p$ to show that the analogue $\mathcal{N}_{1,p}$ of $\mathcal{N}_p$ but for $(G_1, \mathcal{X}_1, H_{1,p})$ instead of for $(G, \mathcal{X}, H_p)$ exists as well (see [Va6, Thm. 4.3 and Cor. 4.4]). If we fix a $\mathbb{Z}_{(p)}$-monomorphism $O(G, \mathcal{X})_{(p)} \hookrightarrow W(\mathbb{F})$, then every connected component $\mathcal{C}_1$ of $\mathcal{N}_{1,W(\mathbb{F})}$ will be isomorphic to the quotient of a connected component $\mathcal{C}$ of $\mathcal{N}_{W(\mathbb{F})}$ by a suitable group action $\mathfrak{T}$ whose generic fibre is free and which involves a torsion group. The key point is to show that the action $\mathfrak{T}$ itself is free (i.e., $\mathcal{C}_1$ is a formally smooth $W(\mathbb{F})$-scheme). If $p > 2$ and p does not divide $n + 1$, then the torsion group of the action $\mathfrak{T}$ has no elements of order p and thus the action $\mathfrak{T}$ is free (cf. proof of [Va1, Thm. 6.2.2 b)]). In [Va6] it is checked that the action $\mathfrak{T}$ is always free i.e., it is free even for the harder cases when either $p = 2$ or p divides $n + 1$. The proof relies on the moduli interpretation of $\mathcal{N}_p$ which makes this group action quite explicit. The cases $p = 2$ and p divides $n + 1$ are the hardest due to the following two reasons.

(i) If $p = 2$ and if A is an abelian variety over $\mathbb{F}$ whose 2-rank a is positive, then the group $(\mathbb{Z}/2\mathbb{Z})^{a^2}$ is naturally a subgroup of the group of automorphisms of the formal deformation space $\mathrm{Def}(A)$ of A in such a way that the filtered Dieudonné module of a lift $\star$ of A to $\mathrm{Spf}(W(\mathbb{F}))$ depends only on the orbit under this action of the $\mathrm{Spf}(W(\mathbb{F}))$-valued point of $\mathrm{Def}(A)$ defined by $\star$.

(ii) For a positive integer m divisible by $p - 1$ there exist actions of $Z/p\mathbb{Z}$ on $\mathbb{Z}_p[[x_1, \ldots, x_m]]$ such that the induced actions on $\mathbb{Z}_p[[x_1, \ldots, x_m]][\frac{1}{p}]$ are free.

Theorem 5. *We assume that either 6 divides N or $\mathrm{Sh}(G, \mathcal{X})$ is a unitary Shimura variety. We also assume that the Zariski closure of G in $\mathbf{GL}_{L \otimes_{\mathbb{Z}} \mathbb{Z}[\frac{1}{N}]}$ is a reductive group scheme over $\mathbb{Z}[\frac{1}{N}]$. Then $\mathcal{N}(N)$ is a smooth scheme over either $O(G, \mathcal{X})[\frac{1}{N}]$ or $\mathbb{Z}[\frac{1}{N}]$.*

Proof: Let p be an arbitrary prime that does not divide N and let v be a prime of $E(G, \mathcal{X})$ that divides p. The group scheme $G_{\mathbb{Z}_{(p)}}$ is reductive. Thus the group $G_{\mathbb{Q}_p}$ is unramified. This implies that $E(G, \mathcal{X})$ is unramified over p and that H_p is a hyperspecial subgroup of $G_{\mathbb{Q}_p}(\mathbb{Q}_p)$. Therefore $O(G, \mathcal{X})[\frac{1}{N}]$ is an étale $\mathbb{Z}[\frac{1}{N}]$-algebra. From this and Proposition 3 (a) we get that to prove the Theorem it suffices to show that each scheme $\mathcal{N}_v$ is regular and formally smooth over $O_{(v)}$.

Let $\mathcal{I}_v$ be the integral canonical model of $\mathrm{Sh}_{H_p}(G, \mathcal{X})$ over $O_{(v)}$, cf. Theorem 4. As $\mathcal{I}_v$ is a healthy regular scheme (cf. Theorem 1), from Proposition 4 we get that we have an $O_{(v)}$-morphism $a : \mathcal{I}_v \to \mathcal{N}_v$ whose generic fibre is the identity automorphism of $\mathrm{Sh}_{H_p}(G, \mathcal{X})$. The morphism a is a pro-étale cover of a morphism $a_{H_0} : \mathcal{I}_v/H_0 \to \mathcal{N}_v/H_0$ of normal $O_{(v)}$-schemes of finite type, where H_0 is a small enough compact, open subgroup of $G(\mathbb{A}_f^{(p)})$. From Theorem 4 we get that $\mathcal{I}_v/H_0$ is a quasi-projective $O_{(v)}$-scheme. Thus a_{H_0} is a quasi-projective morphism between flat $O_{(v)}$-schemes. As each discrete valuation ring of mixed characteristic $(0, p)$ is a healthy regular scheme, the morphism a satisfies the valuative criterion of properness with respect to such discrete valuation rings. From the last two sentences we get that a_{H_0} is in fact a projective morphism.

We consider an open subscheme $\mathcal{V}_v$ of $\mathcal{N}_v$ which contains $\mathrm{Sh}_{H_p}(G, \mathcal{X})$ and for which the morphism $a^{-1}(\mathcal{V}_v) \to \mathcal{V}_v$ is an isomorphism. As $\mathcal{I}_v$ has the extension property (cf. Definition 4.1 (d)), from Theorem 4 we easily get that we can assume that $\mathcal{V}_v$ contains the formally smooth locus of $\mathcal{N}_v$ over $O_{(v)}$. As a_{H_0} is projective, from Proposition 3 (a) we get that we can also assume that we have an inequality $\mathrm{codim}_{\mathcal{N}_v}(\mathcal{N}_v \setminus \mathcal{V}_v) \geq 2$. Obviously we can assume that $\mathcal{V}_v$ is H_0-invariant. Thus the projective morphism $a_{H_0} : \mathcal{I}_v/H_0 \to \mathcal{N}_v/H_0$ is an isomorphism above $\mathcal{V}_v/H_0$.

To check that $\mathcal{N}_v$ is a regular scheme which is formally smooth over $O_{(v)}$ it suffices to show that a_{H_0} is an isomorphism. To check that a_{H_0} is an isomorphism, it suffices to show that $a_{H_0}^{-1}(\mathcal{V}_v/H_0)$ contains all points of $\mathcal{I}_v/H_0$ of codimension 1 (this is so as the projective morphism a_{H_0} is a blowing up of a closed subscheme of $\mathcal{N}_v/H_0$; the proof of this is similar to [Ha, Ch. II, Thm. 7.17]). We show that the assumption that there exists a point y of $\mathcal{I}_v/H_0$ of codimension 1 which does not belong to $a_{H_0}^{-1}(\mathcal{V}_v/H_0) \tilde{\to} \mathcal{V}_v/H_0$ leads to a contradiction.

Let $\mathcal{C}$ be the open subscheme of $\mathcal{I}_v/H_0$ that contains: (i) the generic fibre of $\mathcal{I}_v/H_0$ and (ii) the connected component $\mathcal{E}$ of the special fibre of $\mathcal{I}_v/H_0$ whose generic point is y. The image $\mathcal{E}_0 := a_{H_0}(\mathcal{E})$ has dimension less than $\mathcal{E}$ and is contained in the non-smooth locus of $\mathcal{N}_v/H_0$. The morphism $\mathcal{C} \to \mathcal{N}_v/H_0$ factors through the dilatation $\mathcal{D}$ of $\mathcal{N}_v/H_0$ centered on the reduced scheme of the non-smooth locus of $\mathcal{N}_v/H_0$, cf. the universal property of dilatations (see Definition 4.1 (g) or [BLR, Ch. 3, 3.2, Prop. 3.1 (b)]). But $\mathcal{D}$ is an affine $\mathcal{N}_v/H_0$-scheme and thus the image of the projective $\mathcal{N}_v/H_0$-scheme $\mathcal{E}$ in $\mathcal{D}$ has the same dimension as $\mathcal{E}_0$. By repeating the process we get that the image of $\mathcal{E}$ in a smoothening $\mathcal{D}_\infty$ of $\mathcal{N}_v/H_0$ obtaining via a sequence of blows up centered on non-smooth loci (see [BLR, Ch. 3, Thm. 3 of 3.1 and Thm. 2 of 3.4]), has dimension $\dim(\mathcal{E}_0)$ and thus it has dimension less than $\mathcal{E}$. But each discrete valuation ring of $\mathcal{D}_\infty$ dominates a local ring of $\mathcal{I}_v/H_0$ (as a_{H_0} is a projective morphism) and therefore it is also a local ring of $\mathcal{I}_v/H_0$. As $\mathcal{D}_\infty$ has at least one discrete valuation ring which is not a local ring of $\mathcal{V}_v/H_0$, we get that this discrete valuation ring is the local ring of y. Thus the image of $\mathcal{E}$ in $\mathcal{D}_\infty$ has the same dimension as $\mathcal{E}$. Contradiction. $\square$

7. Projectiveness of integral models

The $\mathbb{C}$-scheme $\mathrm{Sh}_{H(N)}(G, \mathcal{X})_\mathbb{C}$ is projective if and only if the $\mathbb{Q}$–rank of G^{ad} is 0, cf. [BHC, Thm. 12.3 and Cor. 12.4]. Based on this Morita conjectured in 1975 that (see [Mo]):

Conjecture 2. *Suppose that the $\mathbb{Q}$–rank of G^{ad} is 0. Then for each $N \in \mathbb{N}$ with $N \geq 3$, the $O(G, \mathcal{X})[\frac{1}{N}]$-scheme $\mathcal{N}(N)$ is projective.*

Conjecture 3. *Let A_E be an abelian variety over a number field E. Let H_A be the Mumford–Tate group of some extension A of A_E to $\mathbb{C}$. If the $\mathbb{Q}$–rank of H_A^{ad} is 0, then A_E has potentially good reduction everywhere (i.e., there exists a finite field* *extension E_1 of E such that A_{E_1} extends to an abelian scheme over the ring of integers of E_1).*

7.1. On the equivalence of Conjectures 2 and 3

In [Mo] it is shown that Conjectures 2 and 3 are equivalent. We recall the argument for this. Suppose that Conjecture 2 holds. To check that Conjecture 3 holds, we can replace E by a finite field extension of it and we can replace A_E by an abelian variety over E which is isogeneous to it. Based on this and [Mu, Ch. IV, §23, Cor. 1], we can assume that A_E has a principal polarization λ_{A_E}. By enlarging E, we can also assume that all Hodge cycles on A are pull backs of Hodge cycles on A_E (cf. [De3, Prop. 2.9 and Thm. 2.11]) and that (A_E, λ_{A_E}) has a level-$l_1 l_2$ symplectic similitude structure. Here l_1 and l_2 are two distinct odd primes. By taking $G = H_A$ and x_A to belong to $\mathcal{X}_A$, we can assume that (A_E, λ_{A_E}) is the pulls back of the universal principally polarized abelian schemes over $\mathcal{N}(l_1)$ and $\mathcal{N}(l_2)$. As $\mathcal{N}(l_1)$ and $\mathcal{N}(l_2)$ are projective schemes over $O(G, \mathcal{X})[\frac{1}{l_1}]$ and $O(G, \mathcal{X})[\frac{1}{l_2}]$ (respectively), we get that (A_E, λ_{A_E}) extends to a principally polarized abelian scheme over the ring of integers of E. Thus Conjecture 2 implies Conjecture 3.

The arguments of the previous paragraph can be reversed to show that Conjecture 3 implies Conjecture 2.

Definition 5. *We say A_E (resp. $(G, \mathcal{X})$) has compact factors, if for each simple factor $\dagger$ of H_A^{ad} (resp. of G^{ad}) there exists a simple factor of $\dagger_{\mathbb{R}}$ which is compact.*

In [Va4, Thm. 1.2 and Cor. 4.3] it is proved that:

Theorem 6. *Suppose that $(G, \mathcal{X})$ (resp. A_E) has compact factors. Then Conjecture 2 (resp. 3) holds.*

7.2. Different approaches

Let $L_A := H_1(A^{\mathrm{an}}, \mathbb{Z})$ and $W_A := L_A \otimes_{\mathbb{Z}} \mathbb{Q}$. We present different approaches to prove Conjectures 2 and 3 developed by Grothendieck, Morita, Paugam, and us.

(a) Suppose that there exists a prime p such that the group $H_{\mathbb{Q}_p}^{\mathrm{ad}}$ is anisotropic (i.e., its $\mathbb{Q}_p$-rank is 0). Then Conjectures 2 and 3 are true (see [Mo] for the potentially good reduction outside of those primes dividing p; see [Pau] for the potentially good reduction even at the primes dividing p).

(b) Let $\mathcal{B}$ be as in Subsubsection 4.2.1 (resp. be the centralizer of H_A in $\mathrm{End}(W_A)$). We assume that the centralizer of $\mathcal{B}$ in $\mathrm{End}(W)$ (resp. in $\mathrm{End}(W_A)$) is a central division algebra over $\mathbb{Q}$. Then Conjecture 2 (resp. 3) holds (see [Mo]).

(c) By replacing E with a finite field extension of it, we can assume that A_E has everywhere semi-abelian reduction. Let $l \in \mathbb{N}$ be a prime different from p. Let $T_l(A_E)$ be the l adic Tate–module of A_E. As $\mathbb{Z}_l$-modules we can identify $T_l(A_E) = H_1(A^{\mathrm{an}}, \mathbb{Z}) \otimes_{\mathbb{Z}} \mathbb{Z}_l = L_A \otimes_{\mathbb{Z}} \mathbb{Z}_l$. By replacing E with a finite field extension of it, we can assume that for each prime $l \in \mathbb{N}$ the l-adic representation $\rho_l : \mathrm{Gal}(E) \to \mathbf{GL}_{T_l(A_E)}(\mathbb{Q}_l)$ factors through $H_A(\mathbb{Q}_l)$. Let w be a prime of E that divides p. If A_E does not have good reduction at w, then there exists a $\mathbb{Z}_l$-submodule T of $T_l(A_E)$ such that the inertia group of w acts trivially on T and $T_l(A_E)/T$ and non-trivially on $T_l(A_E)$ (see [SGA7, Vol. I, Exp. IX, Thm. 3.5]). This implies that $H_A(\mathbb{Q}_l)$ has unipotent elements of unipotent class 2.

In [Pau] is is shown that if $H_A(\mathbb{Q}_l)$ has no unipotent element of unipotent class 2, then Conjecture 3 holds for A. Using this, Conjectures 2 and 3 are proved in [Pau] in many cases. These cases are particular cases of either Theorem 6 or (a).

(d) We explain the approach used in [Va4] to prove Theorem 6. Let B_E be another abelian variety over E. We say that A_E and B_E are *adjoint-isogeneous*, if the adjoint groups of the Mumford–Tate groups H_A, H_B, and $H_{A\times_{\mathbb{C}}B}$ are isomorphic (more precisely, the standard monomorphism $H_{A\times_{\mathbb{C}}B} \hookrightarrow H_A \times_{\mathbb{Q}} H_B$ induces naturally isomorphisms $H^{\mathrm{ad}}_{A\times_{\mathbb{C}}B}\xrightarrow{\sim}H^{\mathrm{ad}}_A$ and $H^{\mathrm{ad}}_{A\times_{\mathbb{C}}B}\xrightarrow{\sim}H^{\mathrm{ad}}_B$).

To prove Conjecture 3 for A_E it is the same thing as to prove Conjecture 3 for B_E. Based on this, to prove Conjecture 3, one can replace the monomorphism $H_A \hookrightarrow \mathbf{GL}_{W_A}$ by another one $H_B \hookrightarrow \mathbf{GL}_{W_B}$ which is simpler. Based on this and Subsection 7.1, to prove Conjectures 2 and 3 it suffices to prove Conjecture 2 in the cases when:

(i) the adjoint group G^{ad} is a simple $\mathbb{Q}$–group;

(ii) if F is a totally real number field such that $G^{\mathrm{ad}} = \mathrm{Res}_{F/\mathbb{Q}}\, G^{\mathrm{ad},F}$, with $G^{\mathrm{ad},F}$ an absolutely simple adjoint group over F, then F is naturally a $\mathbb{Q}$–subalgebra of the semisimple $\mathbb{Q}$–algebra $\mathcal{B}$ we introduced in Subsubsection 4.2.1;

(iii) the monomorphism $G \hookrightarrow \mathbf{GL}_W$ is simple enough.

Suppose that $(G, \mathcal{X})$ has compact factors. By considering a large field that contains both $\mathbb{R}$ and $\mathbb{Q}_p$, one obtains naturally an identification $\mathrm{Hom}(F, \mathbb{R}) = \mathrm{Hom}(F, \overline{\mathbb{Q}_p})$. Thus we can speak about the p-adic field F_{j_0} which is the factor of

$$(23a) \qquad\qquad F \otimes_{\mathbb{Q}} \mathbb{Q}_p = \prod_{j \in J} F_j$$

that corresponds (via the mentioned identification) to a simple, compact factor of $G^{\mathrm{ad}}_{\mathbb{R}} = \prod_{i \in \mathrm{Hom}(F,\mathbb{R})} G^{\mathrm{ad},F} \times_{F,i} \mathbb{R}$. The existence of such a simple, compact factor is guaranteed by Definition 5.

To prove Theorem 6, it suffices to show that each morphism $c : \mathrm{Spec}(k((x))) \to \mathcal{N}(N)$, with k an algebraically closed field of prime characteristic p that does not divide N, extends to a morphism $\mathrm{Spec}(k[[x]]) \to \mathcal{N}(N)$.

We outline the argument for why the assumption that there exists such a morphism $c : \mathrm{Spec}(k((x))) \to \mathcal{N}(N)$ which does not extend, leads to a contradiction. The morphism c gives birth naturally to an abelian variety E of dimension r over $k((x))$. We can assume that E extends to a semi-abelian scheme $E_{k[[x]]}$ over $k[[x]]$ whose special fibre E_k is not an abelian variety. Let T_k be the maximal torus of C_k. The field F acts naturally on $X^*(T_k) \otimes_{\mathbb{Z}} \mathbb{Q}$, where $X^*(T_k)$ is the abelian group of characters of T_k. Let k_1 be an algebraic closure of $k((x))$. Let (M, ϕ)

be the contravariant Dieudonné module of E_{k_1}. Due to (ii), one has a natural decomposition of $F \otimes_{\mathbb{Q}} \mathbb{Q}_p$-modules

$$(23b) \qquad (M[\frac{1}{p}], \phi) = \oplus_{j \in J}(M_j, \phi).$$

For each $m \in \mathbb{N}$, the composite monomorphism $T_k[p^m] \hookrightarrow E_k[p^m] \hookrightarrow E_k$ lifts uniquely to a homomorphism $(T_k[p^m])_{k[[x]]} \to E_{k[[x]]}$ (see [DG, Exp. IX, Thms. 3.6 and 7.1]) which due to Nakayama's lemma is a closed embedding. This implies that we have a monomorphism $(T_k[p^m])_{k((x))} \hookrightarrow E[p^m]$. Taking $m \to \infty$, at the level of Dieudonné modules over k_1 we get an epimorphism

$$(23c) \qquad \theta : (M[\frac{1}{p}], \phi) \twoheadrightarrow (X^*(T_k) \otimes_{\mathbb{Z}} B(k_1), 1_{X^*(T_k)} \otimes p\sigma_{k_1})$$

which (due to the uniqueness part of this paragraph) is compatible with the natural F-actions. Here σ_{k_1} is the Frobenius automorphism of the field of fractions $B(k_1)$ of the ring $W(k_1)$ of Witt vectors with coefficients in k_1.

From (23b) and (23c) we get that each (M_j, ϕ) has Newton polygon slope 1. But based on (iii) one can assume that the F-isocrystal (M_{j_0}, ϕ) has no integral Newton polygon slope. Contradiction.

8. Stratifications

We will use the notations listed in the beginning of Section 5. Let $\mathcal{N}(N)_v^s$ be the smooth locus of $\mathcal{N}(N)_v$ over $O_{(v)}$; its generic fibre is $\mathrm{Sh}_{H(N)}(G, \mathcal{X})$ (cf. Subsection 4.2). In this Section we will study different stratifications of the special fibre $\mathcal{L}(N)_v^s$ of $\mathcal{N}(N)_v^s$. We begin with few extra notations.

Let ψ^* be the perfect alternating form on L^* induced naturally by ψ. Let $\mathcal{H}_{\mathbb{Z}_{(p)}}$ be the flat, closed subgroup scheme of $G_{\mathbb{Z}_{(p)}}$ which fixes ψ^*; its generic fibre is a connected group $\mathcal{H}_{\mathbb{Q}}$. Let $(s_\alpha)_{\alpha \in \mathcal{J}} \subseteq \mathcal{T}(W^*)$ be a family of tensors as in Subsection 3.4.2. We denote also by $(\mathcal{V}, \Lambda_\mathcal{V})$ the pull back to $\mathcal{N}(N)$ of the universal principally polarized abelian scheme over $\mathcal{A}_{r,1,N}$ (to be compared with Subsubsection 3.4.2). By replacing N with an integral power of itself, we can speak about a family $(v_\alpha^\mathcal{V})_{\alpha \in \mathcal{J}}$ of Hodge cycles on $\mathcal{V}_{\mathbb{Q}}$ obtained as in Subsubsection 3.4.2. Such a replacement is irrelevant for this Section as we are interested in points of $\mathcal{N}(N)$ with values in k, $W(k)$, or $B(k)$. Here k is an algebraically closed field of characteristic p, $W(k)$ is the ring of Witt vectors with coefficients in k, and $B(k) = W(k)[\frac{1}{p}]$ is the field of fractions of $W(k)$. Let σ_k be the Frobenius automorphism of k, $W(k)$, or $B(k)$.

All the results of Section 5 involve finite primes unramified over p. Due to this in this Section we will assume that:

(*) the prime v of $E(G, \mathcal{X})$ is unramified over p and the $k(v)$-scheme $\mathcal{L}(N)_v^s$ is non-empty.

See [Va7, Lem. 4.1] for a general criterion on when (*) holds.

8.1. *F-crystals with tensors*

Let $y : \mathrm{Spec}(k) \to \mathcal{L}(N)_v^{\mathrm{s}}$. Let $z : \mathrm{Spec}(W(k)) \to \mathcal{N}(N)_v^{\mathrm{s}}$ be a lift of y, cf. (*). Let $(A, \lambda_A) := z^*((\mathcal{V}, \Lambda_{\mathcal{V}})_{\mathcal{N}(N)_v^{\mathrm{s}}})$. Let

$$(M, \phi, \psi_M)$$

be the principally quasi-polarized Dieudonné module of $(A, \lambda_A)_k$. Thus ψ_M is a perfect alternating form on M such that we have $\psi(\phi(a) \otimes \phi(b)) = p\sigma_k(\psi(a \otimes b))$ for all $a, b \in M$. The σ_k-linear automorphism $\phi : M[\frac{1}{p}] \xrightarrow{\sim} M[\frac{1}{p}]$ extends naturally to a σ_k-linear automorphism $\phi : \mathcal{T}(M[\frac{1}{p}]) \xrightarrow{\sim} \mathcal{T}(M[\frac{1}{p}])$.

The abelian variety $A_{B(k)}$ is endowed naturally with a family $(v_\alpha)_{\alpha \in \mathcal{J}}$ of Hodge cycles (it is obtained from the family $(v_\alpha^\vee)_{\alpha \in \mathcal{J}}$ of Hodge cycles on $\mathcal{V}_{\mathbb{Q}}$ via a natural pull back process). Let $t_\alpha \in \mathcal{T}(M[\frac{1}{p}])$ be the de Rham component of v_α.

Let F^1 be the Hodge filtration of M defined by the lift A of A_k. We have $\phi(\frac{1}{p}F^1 + M) = M$. Let $\mu_z : \mathbb{G}_m \to \mathbf{GL}_M$ be the inverse of the canonical split cocharacter of (M, F^1, ϕ) defined in [We, p. 512]. It gives birth to a direct sum decomposition $M = F^1 \oplus F^0$ such that $\mathbb{G}_m$ acts via μ_z trivially on F^0 and via the inverse of the identical character of $\mathbb{G}_m$ on F^1.

It is known that the element t_α of $\mathcal{T}(M[\frac{1}{p}])$ is a de Rham and thus also crystalline cycle. If the abelian variety $A_{B(k)}$ is definable over a number subfield of $B(k)$, then this result was known since long time (for instance, see [Bl, Thm. (0.3)]). The general case follows from loc. cit. and [Va1, Principle B of 5.2.16] (in the part of [Va1, Subsect. 5.2] preceding the Principle B an odd prime is used; however the proof of loc. cit. applies to all primes). The fact that t_α is a crystalline cycle means that:

(i) the tensor t_α belongs to the F^0-filtration of $\mathcal{T}(M[\frac{1}{p}])$ defined by $F^1[\frac{1}{p}]$ and it is fixed by ϕ.

Let $\mathcal{G}_{B(k)}$ be the subgroup of $\mathbf{GSp}(M[\frac{1}{p}], \psi_M)$ that fixes t_α for all $\alpha \in \mathcal{J}$. Let $\mathcal{G}$ be the Zariski closure of $\mathcal{G}_{B(k)}$ in $\mathbf{GSp}(M, \psi_M)$ (or $\mathbf{GL}_M$); it is an affine, flat group scheme over $W(k)$. We refer to the quadruple

$$\mathcal{C}_y := (M, \phi, (t_\alpha)_{\alpha \in \mathcal{J}}, \psi_M)$$

as the *principally quasi-polarized F-crystal with tensors* attached to $y \in \mathcal{N}(N)_v^{\mathrm{s}}$. It is easy to see that this terminology makes sense (i.e., t_α depends only on $y : \mathrm{Spec}(k) \to \mathcal{L}(N)_v^{\mathrm{s}}$ and not on the choice of the lift $z : \mathrm{Spec}(W(k)) \to \mathcal{N}(N)_v^{\mathrm{s}}$ of y). We note down that $\mathcal{G}$ is uniquely determined by $\mathcal{C}_y$. We refer to the quadruple

$$\mathcal{R}_y := (M[\frac{1}{p}], \phi, (t_\alpha)_{\alpha \in \mathcal{J}}, \psi_M)$$

as the *principally quasi-polarized F-isocrystal with tensors* attached to $y \in \mathcal{N}(N)_v^{\mathrm{s}}$. From (i) and the functorial aspects of [Wi, p. 513] we get that each tensor t_α is fixed by μ_z. This implies that:

(ii) the cocharacter $\mu_z : \mathbb{G}_m \to \mathbf{GL}_M$ factors through $\mathcal{G}$ and we denote also by $\mu_z : \mathbb{G}_m \to \mathcal{G}$ this factorization.

If $y_i : \mathrm{Spec}(k) \to \mathcal{L}(N)_v^{\mathrm{s}}$ is a point indexed by the elements i of some set, then we will use the index i to write down $\mathcal{C}_{y_i} = (M_i, \phi_i, (t_{i,\alpha})_{\alpha \in \mathcal{J}}, \psi_{M_i})$ as well as $\mathcal{R}_{y_i} = (M_i[\frac{1}{p}], \phi_i, (t_{i,\alpha})_{\alpha \in \mathcal{J}}, \psi_{M_i})$.

If $y_i : \mathrm{Spec}(k) \to \mathcal{L}(N)_v$ does not factor through $\mathcal{L}(N)_v^{\mathrm{s}}$, then we define $\mathcal{C}_{y_i} := (M_i, \phi_i, \psi_{M_i})$ to be the principally quasi-polarized Dieudonné module of $y_i^*((\mathcal{V}, \Lambda_{\mathcal{V}})_{\mathcal{L}(N)_v^{\mathrm{s}}})$. Similarly we define $\mathcal{R}_{y_i} := (M_i[\frac{1}{p}], \phi_i, \psi_{M_i})$.[1]

Before studying different stratifications of $\mathcal{L}(N)_v^{\mathrm{s}}$ defined naturally by basic properties of the $\mathcal{C}_y$'s, we will first present basic definitions on stratifications of reduced schemes over fields.

8.2. Types of stratifications

Let K be a field. By a *stratification* $\mathfrak{S}$ of a reduced $\mathrm{Spec}(K)$-scheme X *(in potentially an infinite number of strata)*, we mean that:

(i) for each field l that is either K or an algebraically closed field which contains K and that has countable transcendental degree over K, a set $\mathfrak{S}_l$ of disjoint reduced, locally closed subschemes of X_l is given such that each point of X_l with values in an algebraic closure of l factors through some element of $\mathfrak{S}_l$;

(ii) if $i_{12} : l_1 \hookrightarrow l_2$ is an embedding between two fields as in (a), then the reduced scheme of the pull back to l_2 of every member of $\mathfrak{S}_{l_1}$, is an element of $\mathfrak{S}_{l_2}$; thus we have a natural pull back injective map $\mathfrak{S}(i_{12}) : \mathfrak{S}_{l_1} \hookrightarrow \mathfrak{S}_{l_2}$.

Each element $\mathfrak{s}$ of some set $\mathfrak{S}_l$ is referred as a *stratum* of $\mathfrak{S}$; we denote by $\bar{\mathfrak{s}}$ the Zariski closure of $\mathfrak{s}$ in X_l. If all maps $\mathfrak{S}(i_{12})$'s are bijections, then we identify $\mathfrak{S}$ with $\mathfrak{S}_K$ and we say $\mathfrak{S}$ is *of finite type*.

Definition 6. *We say that the stratification $\mathfrak{S}$ has (or satisfies):*

(a) *the strong purity property if for each field l as in (i) above and for every stratum $\mathfrak{s}$ of $\mathfrak{S}_l$, locally in the Zariski topology of $\bar{\mathfrak{s}}$ we have $\mathfrak{s} = \bar{\mathfrak{s}}_a$, where a is some global function of $\bar{\mathfrak{s}}$ and where $\bar{\mathfrak{s}}_a$ is the largest open subscheme of $\bar{\mathfrak{s}}$ over which a is an invertible function;*

(b) *the purity property if for each field l as in (i) above, every element of $\mathfrak{S}_l$ is an affine X_l-scheme (thus $\mathfrak{S}$ has the purity property if and only if each stratum of it is an affine X-scheme);*

(c) *the weak purity property if for each field l as in (i) above and for every stratum $\mathfrak{s}$ of $\mathfrak{S}_l$, the scheme $\bar{\mathfrak{s}}$ is noetherian and the complement of $\mathfrak{s}$ in $\bar{\mathfrak{s}}$ is either empty or has pure codimension 1 in $\bar{\mathfrak{s}}$.*

---

[1] For each lift of y_i to a point of $\mathcal{N}(N)_v$ with values in a finite discrete valuation ring extension of $W(k)$, one defines naturally a family of tensors $(t_{i,\alpha})_{\alpha \in \mathcal{J}}$ of $\mathcal{T}(M_i[\frac{1}{p}])$. We do not know if this family of tensors: (i) does not depend on the choice of the lift and (ii) can be used in Subsections 8.4 and 8.5 in the same way as the families of tensors attached to k-valued points of $\mathcal{L}(N)_v^{\mathrm{s}}$.

As the terminology suggests, the strong purity property implies the purity property and the purity property implies the weak purity property. The converses of these two statements do not hold. For instance, there exist affine, integral, noetherian schemes Y which have open subschemes whose complements in Y have pure codimension 1 in Y but are not affine (see [Va3, Rm. 6.3 (a)]).

8.2.1. Example

Suppose that $K = k$, that X is an integral k-scheme, and that there exists a Barsotti–Tate group D of level 1 over X which generically is ordinary. Let $\mathfrak{D}$ be the stratification of X of finite type which has two strata: the ordinary locus $\mathfrak{s}_0$ of D and the non-ordinary locus $\mathfrak{s}_n$ of D. We have $\bar{\mathfrak{s}}_0 = X$ and $\bar{\mathfrak{s}}_n = \mathfrak{s}_n$. Moreover locally in the Zariski topology of X we have an identity $\mathfrak{s}_0 = X_a$, where a is the global function on X which is the determinant of the Hasse–Witt map of D. Thus the stratification $\mathfrak{D}$ has the strong purity property.

8.3. Newton polygon stratification

Let $\mathfrak{N}$ be the stratification of $\mathcal{L}(N)_v$ of finite type with the property that two geometric points $y_1, y_2 : \operatorname{Spec}(k) \to \mathcal{L}(N)_v$ factor through the same stratum if and only if the Newton polygons of (M_1, ϕ_1) and (M_2, ϕ_2) coincide. In [dJO2] it is shown that $\mathfrak{N}$ has the weak purity property (see [Zi2] for a more recent and nice proof of this).

Theorem 7. *The stratification $\mathfrak{N}$ of $\mathcal{L}(N)_v$ has the purity property.*

Proof: The stratification $\mathfrak{N}$ is the Newton polygon stratification of $\mathcal{L}(N)_v$ defined by the F-crystal over $\mathcal{L}(N)_v$ associated to the p-divisible group of $\mathcal{V}_{\mathcal{L}(N)_v}$. Thus the Theorem is a particular case of [Va3, Main Thm. B]. $\square$

8.4. Rational stratification

Let $\mathfrak{R}$ be the stratification of $\mathcal{L}(N)_v^{\mathrm{s}}$ with the property that two geometric points $y_1, y_2 : \operatorname{Spec}(k) \to \mathcal{L}(N)_v^{\mathrm{s}}$ factor through the same stratum if and only if there exists an isomorphism $\mathcal{R}_{y_1} \tilde{\to} \mathcal{R}_{y_2}$ to be called a *rational isomorphism*.

Theorem 8. *The following three properties hold:*

    (a) *Each stratum of $\mathfrak{R}$ is an open closed subscheme of a stratum of the restriction $\mathfrak{N}^{\mathrm{s}}$ of $\mathfrak{N}$ to $\mathcal{L}(N)_v^{\mathrm{s}}$.*

    (b) *The stratification $\mathfrak{R}$ of $\mathcal{L}(N)_v^{\mathrm{s}}$ is of finite type.*

    (c) *The stratification $\mathfrak{R}$ of $\mathcal{L}(N)_v^{\mathrm{s}}$ has the purity property.*

Proof: We use left lower indices to denote pulls back of F-crystals. Let l be either $k(v)$ or an algebraically closed field that contains $k(v)$ and that has countable transcendental degree over $k(v)$. Let S_0 be a stratum of $\mathfrak{N}^{\mathrm{s}}$ contained in $\mathcal{L}(N)_{v,l}^{\mathrm{s}}$. Let S_1 be an irreducible component of S_0. To prove the part (a) it suffices to show that for each two geometric points y_1 and y_2 of S_1 with values in the same algebraically closed field k, there exists a rational isomorphism $\mathcal{R}_{y_1} \tilde{\to} \mathcal{R}_{y_2}$. We can assume that k is an algebraic closure of $\bar{l}((x))$ and that y_1 and y_2 factor through the generic point and the special point (respectively) of a morphism $m : \mathrm{Spec}(\bar{l}[[x]]) \to \mathcal{L}(N)^{\mathrm{s}}_{v,l}$ of l-schemes. Here x is an independent variable. We denote also by y_1 and y_2, the k-valued points of $\mathrm{Spec}(\bar{l}[[x]])$ or of its perfection $\mathrm{Spec}(\bar{l}[[x]]^{\mathrm{perf}})$ defined naturally by the factorizations of y_1 and y_2 through m.

Let Φ be the Frobenius lift of $W(\bar{l})[[x]]$ that is compatible with $\sigma_{\bar{l}}$ and that takes x to x^p. Let $\mathfrak{V} = (V, \phi_V, \psi_V, \nabla_V)$ be the principally quasi-polarized F-crystal over $\bar{l}[[x]]$ of $m^*((\mathcal{V}, \Lambda_{\mathcal{V}})_{\mathcal{L}(N)^{\mathrm{s}}_{v,l}})$. Thus V is a free $W(\bar{l})[[x]]$-module of rank $2r$ equipped with a perfect alternating form ψ_V, $\phi_V : V \to V$ is a Φ-linear endomorphism, and $\nabla_V : V \to Vdx$ is a connection. Let $t^V_\alpha \in \mathcal{T}(V[\frac{1}{p}])$ be the de Rham realization of the Hodge cycle $n^*_{B(\bar{l})}(v^{\mathcal{V}}_\alpha)$ on $n^*_{B(k)}((\mathcal{V})_{\mathcal{N}(N)^{\mathrm{s}}_{W(l)}})$, where $n : \mathrm{Spec}(W(\bar{l})[[x]]) \to \mathcal{N}(N)^{\mathrm{s}}_{W(l)}$ is a lift of m.

Fontaine's comparison theory (see [Fo]) assures us that there exists an isomorphism $(M_1[\frac{1}{p}], (t_{1,\alpha})_{\alpha \in \mathcal{J}}, \psi_{M_1}) \tilde{\to} (W^* \otimes_{\mathbb{Q}} B(k), (s_\alpha)_{\alpha \in \mathcal{J}}, \psi^*)$.

Based on this and [Ko1] we get that $\mathcal{R}_{y_1}$ is isomorphic to the pull back to k of the principally quasi-polarized F-isocrystal $\mathcal{R}_1$ with tensors defined naturally by a principally quasi-polarized F-crystal $\mathcal{C}_1$ with tensors over an algebraic closure $\overline{k(v)}$ of $k(v)$. Strictly speaking [Ko1] uses a language of σ_k-conjugacy classes of sets of the form $G(B(k))$ or $\mathcal{H}_{\mathbb{Q}}(B(k))$ and not a language which involves polarizations and tensors (and thus which involves σ_k-conjugacy classes of sets of the form $\mathcal{H}_{\mathbb{Q}}(B(k))s_0$, where $s_0 \in G(B(k))$ is an element whose image in $(G/\mathcal{H}_{\mathbb{Q}})(B(k)) = \mathbb{G}_m(B(k))$ belongs to $(G/\mathcal{H}_{\mathbb{Q}})(\mathbb{Q}_p) = \mathbb{G}_m(\mathbb{Q}_p)$); but the arguments of [Ko1] apply entirely in the present principally quasi-polarized context which involves sets of the form $\mathcal{H}_{\mathbb{Q}}(B(k))s_0 \subseteq G(B(k))$. Here s_0 is $\mu_0[\frac{1}{p}]$, where $\mu_0 : \mathbb{G}_m \to G_{B(k)}$ is an arbitrary cocharacter whose extension to $\mathbb{C}$ via an $O_{(v)}$-monomorphism $B(k) \hookrightarrow \mathbb{C}$ is $G(\mathbb{C})$-conjugate to the cocharacters $\mu_x : \mathbb{G}_m \to G_{\mathbb{C}}$ with $x \in \mathcal{X}$.

Let $\mathcal{C}_1^-$ be $\mathcal{C}_1$ but viewed only as an F-crystal over $\overline{k(v)}$. Let $M_{1,1}$ be the $W(k)$-lattice of $M_1[\frac{1}{p}]$ that corresponds naturally to $\mathcal{C}_{1,k}^-$ via an isomorphism $i_{1,1} : \mathcal{R}_{1,k} \tilde{\to} \mathcal{R}_{y_1}$.

From [Ka, Thm. 2.7.4] we get the existence of an isogeny $i_0 : \mathfrak{V}_0 \to \mathfrak{V}$, where $\mathfrak{V}_0$ is an F-crystal over $\bar{l}[[x]]$ whose extension to the $\bar{l}[[x]]$-subalgebra $\bar{l}[[x]]^{\mathrm{perf}}$ of k is constant (i.e., it is the pull back of an F-crystal over $\bar{l}$). Let $i_{0,k} : M_0 \to M_1$ be the $W(k)$-linear monomorphism that defines $y_1^*(i_0)$. We can assume that $i_{0,k}(M_0)$ is contained in $M_{1,1}$. The inclusion $i_{0,k}(M_0) \subseteq M_{1,1}$ gives birth to a morphism $j_0 : \mathfrak{V}_{0,k} \to \mathcal{C}_{1,k}^-$ of F-crystals over k. It is the extension to k of a morphism $j_0^{\mathrm{perf}} : \mathfrak{V}_{0,\bar{l}[[x]]^{\mathrm{perf}}} \to \mathcal{C}_{1,\bar{l}[[x]]^{\mathrm{perf}}}^-$, cf. [RR, Lem. 3.9] and the fact that $\mathfrak{V}_{0,\bar{l}[[x]]^{\mathrm{perf}}}$ and $\mathcal{C}_{1,\bar{l}[[x]]^{\mathrm{perf}}}^-$ are constant F-crystals over $k[[x]]^{\mathrm{perf}}$. Let $j_1^{\mathrm{perf}} : \mathcal{C}_{1,\bar{l}[[x]]^{\mathrm{perf}}}^- \to \mathfrak{V}_{0,\bar{l}[[x]]^{\mathrm{perf}}}$ be a morphism of F-crystal such that $j_0^{\mathrm{perf}} \circ j_1^{\mathrm{perf}} = p^q 1_{\mathcal{C}_{1,\bar{l}[[x]]^{\mathrm{perf}}}^-}$ for some $q \in \mathbb{N}$.

By composing j_1^{perf} with $i_{0,k[[x]]^{\mathrm{perf}}}$ we get an isogeny $i_1 : \mathcal{C}_{1,\bar{l}[[x]]^{\mathrm{perf}}}^- \to \mathfrak{V}_{\bar{l}[[x]]^{\mathrm{perf}}}$ whose extension to k is defined by the inclusion $p^q M_{1,1} \subseteq M_1$. The isomorphism of F-isocrystals over $\mathrm{Spec}(k[[x]]^{\mathrm{perf}})$ defined by p^{-q} times i_1 takes $t_{1,\alpha}$ to t^V_α for all $\alpha \in \mathcal{J}$, as this is so generically. Thus $y_2^*(i_1)$ is an isogeny which when viewed as an isomorphism of F-isocrystals is p^q times an isomorphism $i_{1,2} : \mathcal{R}_{1,k} \tilde{\to} \mathcal{R}_{y_2}$. Thus there exists a rational isomorphism $i_{1,2} \circ i_{1,1}^{-1} : \mathcal{R}_{y_1} \tilde{\to} \mathcal{R}_{y_2}$. Thus (a) holds.

Part (b) follows from (a) and the fact that $\mathfrak{N}^{\mathrm{s}}$ is a stratification of finite type. Part (c) follows from (a) and Theorem 7. $\square$

Remark 4. *The proof of Theorem 8 (a) and (b) is in essence only a concrete variant of a slight refinement of [RR, Thm. 3.8]. The only new thing it brings to loc. cit., is that it weakens the hypotheses of loc. cit. (i.e., it considers the "Newton point" of only one faithful representation which is $G_{\mathbb{Q}_p} \hookrightarrow \boldsymbol{GL}_{W^* \otimes_{\mathbb{Q}} \mathbb{Q}_p}$).*

8.5. A quasi Shimura p-variety of Hodge type

Let $\mathcal{H} := \mathcal{H}_{\mathbb{Z}_{(p)}} \times_{\mathbb{Z}_{(p)}} W(k)$, where $\mathcal{H}_{\mathbb{Z}_{(p)}}$ is as in the beginning of this Section. The group $\mathcal{H}_{B(k)}$ is a connected group and we have a short exact sequence

$$(24) \qquad\qquad 1 \to \mathcal{H} \to G_{W(k)} \to \mathbb{G}_m \to 1.$$

We fix an $O_{(v)}$-embedding $W(k) \hookrightarrow \mathbb{C}$. Let ν be the set of cocharacters of $G_{W(k)}$ whose extension to $\mathbb{C}$ are $G(\mathbb{C})$-conjugate to any one of the cocharacters $\mu_x : \mathbb{G}_m \to G_{\mathbb{C}}$ with $x \in \mathcal{X}$. Let $\mu_z : \mathbb{G}_m \to \mathcal{G}$ be the cocharacter introduced in the property (ii) of Subsection 8.1.

Until the end we will also assume that the following three properties hold:

(**) the group scheme $G_{\mathbb{Z}_{(p)}}$ is smooth over $\mathbb{Z}_{(p)}$;

(***) for each algebraically closed field k of countable transcendental degree over $k(v)$ and for every point $y : \mathrm{Spec}(k) \to \mathcal{L}(N)^{\mathrm{s}}_v$, there exists an isomorphism

$$\rho_y : (M_0, (s_\alpha)_{\alpha \in \mathcal{J}}, \psi^*) \tilde{\to} (M, (t_\alpha)_{\alpha \in \mathcal{J}}, \psi_M);$$

(****) the set ν_0 of cocharacters of $G_{W(k)}$ formed by all cocharacters of the form $\rho_y^{-1} \mu_z \rho_y : \mathbb{G}_m \to G_{W(k)}$, with z running through all $W(k)$-valued points of $\mathcal{N}(N)^{\mathrm{s}}$ and with ρ_y running through all isomorphisms as in (**), is a $\mathcal{H}(W(k))$-conjugacy class of cocharacters of $G_{W(k)}$.

We fix an element $\mu_0 : \mathbb{G}_m \to G_{W(k)}$ of ν_0. Let

$$\mathcal{E}_0 := (M_0, \phi_0, (s_\alpha)_{\alpha \in \mathcal{J}}, \psi^*) := (L^* \otimes_{\mathbb{Z}} W(k), \mu_0(\tfrac{1}{p})(1 \otimes \sigma), \mathcal{H}, (s_\alpha)_{\alpha \in \mathcal{J}}, \psi^*)$$

Let $\vartheta_0 : M_0 \to M_0$ be the Verschiebung map of ϕ_0. We have $\vartheta_0 \phi_0 = \phi_0 \vartheta_0 = p 1_{M_0}$.

Let $[x_z, g_z] \in \mathrm{Sh}_{H(N)}(G, \mathcal{X})(\mathbb{C})$ be the complex point defined by the composite of the morphism $\mathrm{Spec}(\mathbb{C}) \to \mathrm{Spec}(W(k))$ with $z : \mathrm{Spec}(W(k)) \to \mathcal{N}(N)^{\mathrm{s}}_v$. Under the fixed $O_{(v)}$-embedding $W(k) \hookrightarrow \mathbb{C}$, we can identify:

— $M \otimes_{W(k)} \mathbb{C} = H^1_{\mathrm{dR}}(A^{\mathrm{an}}/\mathbb{C}) = H^1(A^{\mathrm{an}}, \mathbb{Q}) \otimes_{\mathbb{Q}} \mathbb{C} = W^* \otimes_{\mathbb{Q}} \mathbb{C}$ (cf. property (d) of Subsubsection 3.4.1 for the last identification);

— $F^1 \otimes_{W(k)} \mathbb{C}$ with the Hodge filtration of $H^1_{\mathrm{dR}}(A^{\mathrm{an}}/\mathbb{C}) = W^* \otimes_{\mathbb{Q}} \mathbb{C}$ defined by the point $x_z \in \mathcal{X}$;

— $t_\alpha = s_\alpha$ for all $\alpha \in \mathcal{J}$ and thus $\mathcal{G}_{\mathbb{C}} = G_{\mathbb{C}}$ (see [De3]).

Based on this we easily get that $\mu_{z,\mathbb{C}} : \mathbb{G}_m \to \mathcal{G}_{\mathbb{C}} = G_{\mathbb{C}}$ is $\mathbb{G}(\mathbb{C})$-conjugate to $\mu_{x_z} : \mathbb{G}_m \to G_{\mathbb{C}}$. From this we get that the cocharacter $\rho_y^{-1}\mu_z\rho_y : \mathbb{G}_m \to G_{W(k)}$ belongs to ν. Thus we have:

$$\nu_0 \subseteq \nu.$$

By composing ρ_y with an automorphism of $(L^* \otimes_{\mathbb{Z}} W(k), (s_\alpha)_{\alpha \in \mathcal{J}})$ defined by an element of $\mathcal{H}(W(k))$, we can assume that in fact we have $\rho_y^{-1}\mu_z\rho_y = \mu_0$ (cf. (****)). This implies that ρ_y gives birth to an isomorphism of the form

$$\rho_y : (M_0, g_y\phi_0, \mathcal{H}, (s_\alpha)_{\alpha \in \mathcal{J}}, \psi^*) \tilde{\to} \mathcal{C}_y$$

for some element $g_y \in G_{W(k)}(W(k))$. For $g \in \mathcal{H}(W(k))$, let

$$\mathcal{E}_g := (M_0, g\phi_0, (s_\alpha)_{\alpha \in \mathcal{J}}, \psi^*).$$

Therefore $\mathcal{E}_0 = \mathcal{E}_{1_{M_0}}$ and moreover

$$\mathcal{C}_y \text{ is isomorphic with } \mathcal{E}_{g_y}.$$

Let

$$\mathcal{F}_0 := \{\mathcal{E}_g | q \in \mathcal{H}(W(k))\};$$

it is a *family* of principally quasi-polarized F-crystals with tensors. We emphasize that, due to (***) and (****), the isomorphism class of the family $\mathcal{F}_0$ depends only on $\mathcal{L}(N)_v^{\mathrm{s}}$ and not on the choice of the element $\mu_0 : \mathbb{G}_m \to G_{W(k)}$ of ν_0.

Definition 7. *Let $m \in \mathbb{N}$. By the D-truncation of level m (or mod p^m) of $\mathcal{E}_g$ we mean the reduction $\mathcal{E}_g[p^m]$ of $(M_0, g\phi_0, \vartheta_0 g^{-1}, \mathcal{H}, \psi^*)$ modulo p^m (here it is more convenient to use $\mathcal{H}$ instead of $(s_\alpha)_{\alpha \in \mathcal{J}}$). For $g_1, g_2 \in \mathcal{H}(W(k))$, by an inner isomorphism between $\mathcal{E}_{g_1}[p^m]$ and $\mathcal{E}_{g_1}[p^m]$ we mean an isomorphism $\mathcal{E}_{g_1}[p^m] \tilde{\to} \mathcal{E}_{g_1}[p^m]$ defined by an element of $\mathcal{H}(W_m(k))$.*

Remark 5. (a) *Statement (***) is a more general form of a conjecture of Milne (made in 1995). In [Va8] it is shown that (***) holds if either $p > 2$ or $p = 2$ and $G_{\mathbb{Z}_{(p)}}$ is a torus. The particular case of this result when moreover $G_{\mathbb{Z}_{(p)}}$ is a reductive group scheme over $\mathbb{Z}_{(p)}$, is also claimed in [Ki].*

(b) If the statement (****) does not hold, then one has to work out what follows with a fixed connected component of $\mathcal{L}(N)_v^{\mathrm{s}}$ instead of with $\mathcal{L}(N)_v^{\mathrm{s}}$.

(c) In many cases one can choose the cocharacter $\mu_0 : \mathbb{G}_m \to G_{W(k)}$ in such a way that the quadruple $\mathcal{E}_0$ is a canonical object of the family $\mathcal{F}_0$. For instance, if $G_{\mathbb{Z}_{(p)}}$ is a reductive group scheme over $\mathbb{Z}_{(p)}$, then we have $\nu_0 = \nu$ and one can choose μ_0 as follows. Let $B_{\mathbb{Z}_p}$ be a Borel subgroup scheme of $G_{\mathbb{Z}_p} := G_{\mathbb{Z}_{(p)}} \times_{\mathbb{Z}_{(p)}} \mathbb{Z}_p$. Let $T_{\mathbb{Z}_p}$ be a maximal torus of $B_{\mathbb{Z}_p}$. Let $G_{W(k(v))} := G_{\mathbb{Z}_p} \times_{\mathbb{Z}_p} W(k(v))$. Let $\mu_{0,W(k(v))} : \mathbb{G}_m \to G_{W(k(v))}$ be the unique cocharacter whose extension $\mu_0 : \mathbb{G}_m \to G_{W(k)}$ to $W(k)$ belongs to the set ν, which factors through $T_{\mathbb{Z}_p} \times_{\mathbb{Z}_p} W(k(v))$, and

through which $\mathbb{G}_m$ *acts on* $\mathrm{Lie}(B_{\mathbb{Z}_p}) \otimes_{\mathbb{Z}_p} W(k(v))$ *via the trivial and the identical characters of* $\mathbb{G}_m$ *(cf. [Mi3, Cor. 4.7 (b)]). As pairs of the form* $(B_{\mathbb{Z}_p}, T_{\mathbb{Z}_p})$ *are* $G_{\mathbb{Z}_p}(\mathbb{Z}_p)$*-conjugate, the isomorphism class of* $\mathcal{E}_0$ *constructed via such a cocharacter* μ_0 *does not depend on the choice of* $(B_{\mathbb{Z}_p}, T_{\mathbb{Z}_p})$*. Thus* $\mathcal{E}_0$ *is a canonical object of the family* $\mathcal{F}_0$*.*

Theorem 9. *Under the assumptions (*) to (****) of this Section, the* $\mathcal{A}_{r,1,N,k(v)}$*-scheme* $\mathcal{L}(N)^{\mathrm{s}}_v$ *is a quasi Shimura p-variety of Hodge type relative to* $\mathcal{F}_0$ *in the sense of [Va5, Def. 4.2.1].*

Proof: As [Va5, Def. 4.2.1] is a very long definition, the essence of its parts will be pointed out at the right time in this proof. We emphasize that due to (**) and (24), the group scheme $\mathcal{H}$ is smooth over $W(k)$ and therefore the statement of the Theorem makes sense.

Let

$$(25a) \qquad\qquad M_0 = F_0^1 \oplus F_0^0$$

be the direct sum decomposition such that $\mathbb{G}_m$ acts through $\mu_0 : \mathbb{G}_m \to G_{W(k)}$ trivially on F_0^0 and via the inverse of the identical character of $\mathbb{G}_m$ on F_0^1. To (25a) corresponds a direct sum decomposition

$$(25b) \qquad \mathrm{End}(M_0) = \mathrm{Hom}(F_0^0, F_0^1) \oplus \mathrm{End}(F_0^0) \oplus \mathrm{End}(F_0^1) \oplus \mathrm{Hom}(F_0^1, F_0^0)$$

of $W(k)$-modules. Let $\mathrm{Lie}(\mathcal{H}) = \oplus_{i=-1}^{i} \tilde{F}_0^i(\mathrm{Lie}(\mathcal{H}))$ be the direct sum decomposition such that $\mathbb{G}_m$ acts trough μ_0 on $\tilde{F}_0^i(\mathrm{Lie}(\mathcal{H}))$ via the $-i$-th power of the identity character of $\mathbb{G}_m$. Thus we have an identity

$$(25c) \qquad\qquad \tilde{F}_0^{-1}(\mathrm{Lie}(\mathcal{H})) = \mathrm{Hom}(F_0^1, F_0^0) \cap \mathrm{Lie}(\mathcal{H}),$$

the intersection being taken inside $\mathrm{End}(M_0)$. Let $\mathcal{U}$ be the connected, smooth, unipotent subgroup scheme of $\mathcal{H}$ defined by the following rule: if C is a commutative $W(k)$-algebra, then $\mathcal{U}(C) = 1_{M_0 \otimes_{W(k)} C} + \tilde{F}_0^{-1}(\mathrm{Lie}(\mathcal{H})) \otimes_{W(k)} C$.

The smooth $k(v)$-scheme $\mathcal{L}(N)^{\mathrm{s}}_v$ is equidimensional of dimension d. As μ_0 belongs to ν_0 and thus to ν, from Formula (17) we get that the rank e_- of $\tilde{F}_0^{-1}(\mathrm{Lie}(\mathcal{H}))$ is precisely d. Thus the smooth $k(v)$-scheme $\mathcal{L}(N)^{\mathrm{s}}_v$ is equidimensional of dimension e_-. In other words, the axiom (i) of [Va5, Def. 4.2.1] holds.

Let R_y be the completion of the local ring of $\mathcal{N}(N)^{\mathrm{s}}_{W(k)}$ at its k-valued point defined by y. We fix an identification $R_y = W(k)[[x_1, \ldots, x_d]]$. Let Φ be the Frobenius lift of R_y which is compatible with σ_k and which takes x_i to x_i^p for all $i \in \{1, \ldots, d\}$. We have a natural morphism $\mathrm{Spec}(R_y) \to \mathcal{N}(N)^{\mathrm{s}}$ which is formally étale. The principally quasi-polarized filtered F-crystal over R_y/pR_y of the pull back to $\mathrm{Spec}(R_y)$ of $(\mathcal{V}, \Lambda_{\mathcal{V}})$ is isomorphic to

$$(26a) \qquad\qquad (M_0 \otimes_{W(k)} R_y, F_0^1 \otimes_{W(k)} R_y, h_y(g_y \phi_0 \otimes \Phi), \psi^*, \nabla_y),$$

where $h_y \in \mathcal{H}(R_y)$ is such that modulo the ideal $(x_1, \ldots, x_d)$ of R_y is the identity element of $\mathcal{H}(W(k))$ and where ∇_y is an integrable, nilpotent modulo p connection on $M_0 \otimes_{W(k)} R_y$. We have:

(i) for each element $\alpha \in \mathcal{J}$, the tensor $t_\alpha \in \mathcal{T}(M_0[\frac{1}{p}]) \otimes_{B(k)} R_y[\frac{1}{p}] = \mathcal{T}(M_0 \otimes_{W(k)} R_y[\frac{1}{p}])$ is the de Rham realization of the pull back to $\mathrm{Spec}(R_y[\frac{1}{p}])$ of the Hodge cycle $v_\alpha^\vee$ on $\mathcal{V}_{\mathbb{Q}}$ and therefore it is annihilated by ∇_y;

(ii) the connection ∇_y is versal.

The two properties (i) and (ii) hold as, up to $W(k)$-automorphisms of R_y that leave invariant its ideal $(x_1, \ldots, x_d)$, we can choose the morphism $h_y : \mathrm{Spec}(R_y) \to \mathcal{H}$ to factor through a formally étale morphism $h_y \to \mathcal{U}$ (i.e., we can choose h_y to be the universal element of the completion of $\mathcal{U}$). If $G_{\mathbb{Z}_{(p)}}$ is a reductive group scheme, then the fact that such a choice of h_y is possible follows from [Va1, Subsect. 5.4]. The general case is entirely the same (for instance, cf. [Va7, Subsects. 3.3 and 3.4]).

We recall the standard argument that ∇_y annihilates each t_α with $\alpha \in \mathcal{J}$. We view $\mathcal{T}(M_0)$ as a module over the Lie algebra (associated to) $\mathrm{End}(M_0)$ and accordingly we denote also by ∇_y the connection on $\mathcal{T}(M_0 \otimes_{W(k)} R_y[\frac{1}{p}])$ which extends naturally the connection ∇_y on $M_0 \otimes_{W(k)} R_y$. The Φ-linear action of $h_y(g_y\phi_0 \otimes \Phi)$ on $M_0 \otimes_{W(k)} R_y$ extends to a Φ-linear action of $h_y(g_y\phi_0 \otimes \Phi)$ on $\mathcal{T}(M_0 \otimes_{W(k)} R_y[\frac{1}{p}])$. For instance, if $a \in M_0^* \otimes_{W(k)} R_y = (M_0 \otimes_{W(k)} R_y)^*$ and if $b \in M_0 \otimes_{W(k)} R_y$, then $[h_y(g_y\phi_0\otimes\Phi)](a) \in M_0^* \otimes_{W(k)} R_y[\frac{1}{p}]$ maps $[h_y(g_y\phi_0\otimes\Phi)](b)$ to $\Phi(a(b))$. As ϕ_0, g_y, and h_y fix t_α, the tensor $t_\alpha \in \mathcal{T}(M_0\otimes_{W(k)}R_y[\frac{1}{p}])$ is also fixed by $h_y(g_y\phi_0 \otimes \Phi)$. The connection ∇_y is the unique connection on $M_0 \otimes_{W(k)} R_y$ such that we have an identity

$$\nabla_y \circ [h_y(g_y\phi_0 \otimes \Phi)] = [h_y(g_y\phi_0 \otimes \Phi)] \otimes d\Phi) \circ \nabla_y,$$

cf. [Fa2, Thm. 10]. From the last two sentences we get that

$$\nabla_y(t_\alpha) = [h_y(g_y\phi_0 \otimes \Phi) \otimes d\Phi](\nabla_y(t_\alpha)).$$

As we have $d\Phi(x_i) = px_i^{p-1}dx_i$ for all $i \in \{1, \ldots, d\}$, by induction on $q \in \mathbb{N}$ we get that $\nabla_y(t_\alpha) \in \mathcal{T}(M_0) \otimes_{W(k)} (x_1, \ldots, x_d)^q \Omega^\wedge_{R_y/W(k)}[\frac{1}{p}]$. Here $\Omega^\wedge_{R_y/W(k)}$ is the p-adic completion of the sheaf of relative 1-differential forms. As R_y is complete with respect to the $(x_1, \ldots, x_d)$-topology, we have $\nabla_y(t_\alpha) = 0$.

Due to the property (ii), the morphism $\mathcal{L}(N)_v^{\mathrm{s}} \to \mathcal{A}_{d,1,N,k}$ induces k-epimorphisms at the level of complete, local rings of residue field k i.e., it is a formal closed embedding at all k-valued points (this is precisely the statement of [Va7, Part I, Thm. 1.5 (b)]). Thus the axiom (ii) of [Va5, Def. 4.2.1] holds.

Based on the property (i) and a standard application of Artin's approximation theorem, we get that there exists an étale map $\eta_y : \mathrm{Spec}(E_y) \to \mathcal{N}(N)_{W(k)}^{\mathrm{s}}$ whose image contains the k-valued point of $\mathcal{N}(N)_{W(k)}^{\mathrm{s}}$ defined naturally by y and for which the following three properties hold:

(iii) the p-adic completion $E_y^\wedge$ of E_y has a Frobenius lift Φ_{E_y};

(iv) the principally quasi-polarized filtered F-crystal over E_y/pE_y of the pull back to $\mathrm{Spec}(E_y/pE_y)$ of $(\mathcal{V}, \Lambda_\mathcal{V})$ is isomorphic to

$$(26b) \qquad (M_0 \otimes_{W(k)} E_y^\wedge, F_0^1 \otimes_{W(k)} E_y^\wedge, j_y(g_y \phi_0 \otimes \Phi_{E_y}), \psi^*, \nabla_y^{\mathrm{alg}}),$$

where $j_y \in \mathcal{H}(E_y^\wedge)$ and where ∇_y^{alg} is an integrable, nilpotent modulo p connection on $M_0 \otimes_{W(k)} R_y$ which is versal at each k-valued point of $\mathrm{Spec}(E_y^\wedge)$;

(v) for each element $\alpha \in \mathcal{J}$, the tensor $t_\alpha \in \mathcal{T}(M_0[\frac{1}{p}] \otimes_{B(k)} E_y^\wedge[\frac{1}{p}]) = \mathcal{T}(M_0 \otimes_{W(k)} E_y^\wedge[\frac{1}{p}])$ is the de Rham realization of the pull back to $\mathrm{Spec}(E_y^\wedge[\frac{1}{p}])$ of the Hodge cycle $v_\alpha^\vee$ on $\mathcal{V}_\mathbb{Q}$ and therefore it is annihilated by ∇_y^{alg}.

Let $\bar{\eta}_y : \mathrm{Spec}(E_y/pE_y) \to \mathcal{L}(N)_{v,k}^{\mathrm{s}}$ be the étale map defined naturally by η_y. Let I_k be a finite set of k-valued points of $\mathcal{L}(N)_v^{\mathrm{s}}$ such that we have an identity

$$\cup_{\tilde{y} \in I_k} \mathrm{Im}(\bar{\eta}_{\tilde{y}}) = \mathcal{L}(N)_v^{\mathrm{s}}.$$

This means that the axiom (iii.a) of [Va5, Def. 4.2.1] holds for the family of étale maps $(\bar{\eta}_{\tilde{y}})_{\tilde{y} \in I_k}$.

Let $\mathcal{W}_{+0}$ be the maximal parabolic subgroup scheme of $\mathbf{GL}_{M_0}$ that normalizes F_0^1. Let $\mathcal{W}_{+0}^{\mathcal{H}} := \mathcal{H} \cap \mathcal{W}_{+0}$; it is a smooth subgroup scheme of $\mathcal{H}$ (cf. [Va5, Lem. 4.1.2]). As ∇_y^{alg} is versal at each k-valued point of $\mathrm{Spec}(E_y^\wedge)$, we have:

(vi) the reduction modulo p of j_y is a morphism $\mathrm{Spec}(E_y/pE_y) \to \mathcal{H}_k$ whose composite with the quotient morphism $\mathcal{H}_k \twoheadrightarrow \mathcal{H}_k/\mathcal{W}_{+0,k}^{\mathcal{H}}$ is étale.

Property (vi) implies that the axiom (iii.b) of [Va5, Def. 4.2.1] holds for $(\bar{\eta}_{\tilde{y}})_{\tilde{y} \in I_k}$.

Based on properties (iv) and (v), it is easy to see that the axiom (iii.c) of [Va5, Def. 4.2.1] holds for $(\bar{\eta}_{\tilde{y}})_{\tilde{y} \in I_k}$.

The fact that the axiom (iii.d) of [Va5, Def. 4.2.1] holds as well for $(\bar{\eta}_{\tilde{y}})_{\tilde{y} \in I_k}$ is only a particular case of Faltings' deformation theory [Fa, §7, Thm. 10 and Rm. i) to iii) after it], cf. the versality part of the property (iv). More precisely, if $\omega \in \mathrm{Ker}(\mathcal{H}(R_y) \twoheadrightarrow \mathcal{H}(R_y/(x_1, \ldots, x_d)))$ is such that the composite of ω modulo p with the quotient morphism $\mathcal{H}_k \twoheadrightarrow \mathcal{H}_k/\mathcal{W}_{+0,k}^{\mathcal{H}}$ is formally étale, then there exists an $W(k)$-automorphism $a_y : R_y \tilde{\to} R_y$ that leaves invariant the ideal $(x_1, \ldots, x_d)$ and for which the extension of (26a) via a_y is isomorphic to

$$(M_0 \otimes_{W(k)} R_y, F_0^1 \otimes_{W(k)} R_y, \omega(g_y \phi_0 \otimes \Phi), \psi^*, \nabla_y)$$

under an isomorphism defined by an element of $\mathrm{Ker}(\mathcal{H}(R_y) \twoheadrightarrow \mathcal{H}(R_y/(x_1, \ldots, x_d)))$. Thus axioms (i) to (iii) of [Va5, Def. 4.2.1] hold i.e., $\mathcal{L}(N)_v^{\mathrm{s}}$ is a quasi Shimura p-variety of Hodge type relative to $\mathcal{F}_0$ in the sense of [Va5, Def. 4.2.1]. $\qquad \square$

8.6. Level m stratification

We assume that properties (*) to (****) of this Section hold. Let m be a positive integer. From Theorem 9 and [Va5, Cor. 4.3] we get that there exists a stratification $\mathfrak{L}_m$ of $\mathcal{L}(N)_v^{\mathrm{s}}$ with the property that two geometric points $y_1, y_2 : \mathrm{Spec}(k) \to \mathcal{L}(N)_v^{\mathrm{s}}$ factor through the same stratum if and only $\mathcal{E}_{g_{y_1}}[p^m]$ is inner isomorphic to $\mathcal{E}_{g_{y_2}}[p^m]$. We call $\mathfrak{L}_m$ as the level m stratification of $\mathcal{L}(N)_v^{\mathrm{s}}$. Among its many properties we list here only three:

Proposition 6. *Let l be either $k(v)$ or an algebraically closed field of countable transcendental degree over $k(v)$. Let $\mathfrak{n}$ be a stratum of $\mathfrak{N}^s$ which is a locally closed subscheme of $\mathcal{L}(N)^s_{v,l}$. Then we have:*

(a) there exists a family $(\mathfrak{l}_i)_{i \in L(\mathfrak{n})}$ of strata of $\mathfrak{L}_{\lceil \frac{r}{2} \rceil}$ which are locally closed subschemes of $\mathcal{L}(N)^s_{v,l}$ and such that we have an identity

$$(45) \qquad\qquad \mathfrak{n}(\bar{l}) = \cup_{i \in L(\mathfrak{n})} \mathfrak{l}_i(\bar{l});$$

(b) the scheme $\mathfrak{n}$ is regular and equidimensional;

(c) the $\mathcal{L}(N)^s_{v,l}$-scheme $\mathfrak{n}$ is quasi-affine.

Proof: The Newton polygon of a p-divisible group D over k of codimension c and dimension d is uniquely determined by $D[p^{\lceil \frac{cd}{c+d} \rceil}]$, cf. [NV2, Thm. 1.2]. Thus the Newton polygon of (M, ϕ) is uniquely determined by the inner isomorphism class of $\mathcal{E}_{g_y}[p^{\lceil \frac{r}{2} \rceil}]$. From this the part (a) follows.

Parts (b) and (c) are particular cases of [Va5, Cor. 4.3]. $\qquad\qquad\square$

Remark 6. *(a) For PEL type Shimura varieties, the idea of level m stratifications shows up first in [We]. The level 1 stratifications generalize the Ekedahl–Oort stratifications studied extensively by Kraft, Ekedahl, Oort, Wedhorn, Moonen, and van der Geer.*

(b) Suppose that $G_{\mathbb{Z}_{(p)}}$ is a reductive group scheme and that the properties (*) and (***) of this Section hold. As $G_{\mathbb{Z}_{(p)}}$ is a reductive group scheme, it is easy to see that the properties (**) and (****) hold as well. Thus the level m stratification $\mathfrak{L}_m$ exists. It is known that $\mathfrak{L}_1$ has a finite number of strata (see [Va10, Sect. 12]).

8.6.1. Problem

Study when $\mathfrak{L}_m$ has the purity property.

8.7. Traverso stratifications

We continue to assume that properties (*) to (****) of this Section hold. Let

$$n_v \in \mathbb{N}$$

be the smallest positive integer such that for all elements $g \in \mathcal{H}(W(k))$ and $g_1 \in \operatorname{Ker}(\mathcal{H}(W(k)) \to \mathcal{H}(W_{n_v}(k)))$, the quadruples $\mathcal{E}_g$ and $\mathcal{E}_{gg_1}$ are isomorphic. The existence of n_v is implied by [Va3, Main Thm. A].

Lemma 1. *We assume that the assumptions (*) to (****) of this Section hold. Let $g_1, g_2 \in \mathcal{H}(W(k))$. The D-truncations of level m of $\mathcal{E}_{g_1}$ and $\mathcal{E}_{g_2}$ are inner isomorphic if and only if there exists $g_3 \in \mathcal{H}(W(k))$ such that we have $g_3 g_2 \phi_0 g_3^{-1} = g_0 g_1 \phi_0$ for some element $g_0 \in \operatorname{Ker}(\mathcal{H}(W(k)) \to \mathcal{H}(W_m(k)))$.*

Proof: This is only a principal quasi-polarized variant of [Va3, Lem. 3.2.2]. Its proof is entirely the same as of loc. cit. $\qquad\square$

Due to Lemma 1, from the very definition of n_v we get that for every two elements $g_1, g_2 \in \mathcal{H}(W(k))$ we have the following equivalence:

(i) $\mathcal{E}_{g_1}$ is isomorphic to $\mathcal{E}_{g_2}$ if and only if $\mathcal{E}_{g_1}[p^{n_y}]$ is isomorphic to $\mathcal{E}_{g_2}[p^{n_y}]$.

Due to the property (i), for $m \geq n_v$ we have an identity

$$\mathfrak{L}_m = \mathfrak{L}_{n_v}.$$

We refer to

$$\mathfrak{T} := \mathfrak{L}_{n_v}$$

as the Traverso stratification of $\mathcal{L}(N)^{\mathrm{s}}_v$. Such stratifications were studied in [Tr1] to [Tr2] (using the language of group actions), in [Oo] (using the language of foliations), and in [Va3] and [Va5] (using the language of ultimate or Traverso stratifications). Based on Theorem 9, the next Theorem is only a particular case of [Va5, Cor. 4.3.1 (b)].

Theorem 10. *Under the assumptions (*) to (****) of this Section, the Traverso stratification $\mathfrak{T}$ of $\mathcal{L}(N)^{\mathrm{s}}_v$ has the purity property.*

8.7.1. Problems

1. Find upper bounds for n_v which are sharp.

2. Study the dependence of n_v on v.

8.7.2. Example

We assume that f is an isomorphism i.e., we have an identification $(G, \mathcal{X}) = (\mathbf{GSp}(W, \psi), \mathcal{S})$. We have $v = p$ and thus we will denote n_v by n_p. We also assume that y is a supersingular point i.e., all Newton polygon slopes of (M, ϕ) are $\frac{1}{2}$. The isomorphism class of (M, ϕ, ψ_M) is uniquely determined by $\mathcal{E}_{g_y}[p^r]$, cf. [NV1, Thm. 1.3]. Moreover, in general we can not replace in the previous sentence $\mathcal{E}_{g_y}[p^r]$ by $\mathcal{E}_{g_y}[p^{r-1}]$ (cf. [NV1, Example 3.3] and the result [Va3, Prop. 5.3.3] which says that each principally quasi-polarized Dieudonné module over k is the one attached to a principally polarized abelian variety over k).

Therefore, the restrictions of $\mathfrak{T}$ and $\mathfrak{L}_r$ to the (reduced) supersingular locus of $\mathcal{A}_{r,1,N,\mathbb{F}_p} = \mathcal{L}(N)_p = \mathcal{L}(N)^{\mathrm{s}}_p$ coincide and we have an inequality

$$n_p \geq r.$$

Based on Traverso's isomorphism conjecture (cf. [Tr3, §40, Conj. 4] or [NV1, Conj. 1.1]), one would be inclined to expect that n_p is in fact exactly r. However, we are not at all at the point where we could state this as a solid expectation.

References

[Bl]   D. Blasius, *A p-adic property of Hodge cycles on abelian varieties*, Motives (Seattle, WA, 1991), pp. 293–308, Proc. Sympos. Pure Math., **55**, Part 2, Amer. Math. Soc., Providence, RI, 1994.

[Bo]   A. Borel, *Linear algebraic groups. Second edition*, Grad. Texts in Math., Vol. **126**, Springer-Verlag, New York, 1991.

[BB]   W. Baily and A. Borel, *Compactification of arithmetic quotients of bounded symmetric domains*, Ann. of Math. (2) **84** (1966), no. 3, pp. 442–528.

[BHC]   A. Borel and Harish-Chandra, *Arithmetic subgroups of algebraic groups*, Ann. of Math. (2) **75** (1962), no. 3, pp. 485–535.

[BLR]   S. Bosch, W. Lütkebohmert, and M. Raynaud, *Néron models*, Ergebnisse der Mathematik und ihrer Grenzgebiete (3), Vol. **21**, Springer-Verlag, Berlin, 1990.

[dJO1]   J. de Jong and F. Oort, *On extending families of curves*, J. Algebraic Geom. **6** (1997), pp. 545–562.

[dJO2]   J. de Jong and F. Oort, *Purity of the stratification by Newton polygons*, J. of Amer. Math. Soc. **13** (2000), no. 1, pp. 209–241.

[De1]   P. Deligne, *Travaux de Shimura*, Séminaire Bourbaki, 23ème année (1970/71), Exp. No. 389, pp. 123–165, Lecture Notes in Math., Vol. **244**, Springer-Verlag, 1971.

[De2]   P. Deligne, *Variétés de Shimura: interprétation modulaire, et techniques de construction de modèles canoniques*, Automorphic forms, representations and L-functions (Oregon State Univ., Corvallis, OR, 1977), Part 2, pp. 247–289, Proc. Sympos. Pure Math., Vol. **33**, Amer. Math. Soc., Providence, RI, 1979.

[De3]   P. Deligne, *Hodge cycles on abelian varieties*, Hodge cycles, motives, and Shimura varieties, Lecture Notes in Math., Vol. **900**, pp. 9–100, Springer-Verlag, 1982.

[DG]   M. Demazure, A. Grothendieck, et al., *Schémas en groupes, Vol. II*, Lecture Notes in Math., Vol. **152**, Springer-Verlag, 1970.

[Fa]   G. Faltings, *Integral crystalline cohomology over very ramified valuation rings*, J. of Amer. Math. Soc. **12** (1999), no. 1, pp. 117–144.

[Fo]   J.-M. Fontaine, *Représentations p-adiques semi-stables*, Périodes p-adiques (Bures-sur-Yvette, 1988), J. Astérisque **223**, pp. 113–184, Soc. Math. de France, Paris, 1994.

[FC]   G. Faltings and C.-L. Chai, *Degeneration of abelian varieties*, Ergebnisse der Mathematik und ihrer Grenzgebiete (3), Vol. **22**, Springer-Verlag, Berlin, 1990.

[Ha]   R. Hartshorne, *Algebraic geometry*, Grad. Texts in Math., **52**, Springer-Verlag, Berlin, 1977.

[Ka]   N. Katz, *Slope filtration of F-crystals*, Journées de Géométrie algébrique de Rennes 1978, J. Astérisque **63**, Part 1, pp. 113–163, Soc. Math. de France, Paris, 1979.

[Ki]   M. Kisin, *Integral canonical models of Shimura varieties*, 10 pages note available at http://www.math.uchicago.edu/~kisin/preprints.html.

[Ko1]   R. E. Kottwitz, *Isocrystals with additional structure*, Comp. Math. **56** (1985), no. 2, pp. 201–220.

[Ko2]   R. E. Kottwitz, *Points on some Shimura Varieties over finite fields*, J. of Amer. Math. Soc. **5** (1992), no. 2, pp. 373–444.

[La]   R. Langlands, *Some contemporary problems with origin in the Jugendtraum*, Mathematical developments arising from Hilbert problems (Northern Illinois Univ., De Kalb, IL, 1974), pp. 401–418, Proc. Sympos. Pure Math., Vol. **28**, Amer. Math. Soc., Providence, RI, 1976.

[LR]   R. Langlands and M. Rapoport, *Shimuravarietäten und Gerben*, J. reine angew. Math. 378 (1987), pp. 113–220.

[Ma]   H. Matsumura, *Commutative algebra. Second edition,*, The Benjamin/Cummings Publishing Co., Inc., 1980.

[Me]   W. Messing, *The crystals associated to Barsotti–Tate groups, with applications to abelian schemes*, Lecture Notes in Math., Vol. **264**, Springer-Verlag, 1972.

[Mi1]   J. S. Milne, *Canonical models of (mixed) Shimura varieties and automorphic vector bundles*, Automorphic Forms, Shimura varieties and L-functions, Vol. I (Ann Arbor, MI, 1988), pp. 283–414, Perspectives in Math., Vol. **10**, Acad. Press, Boston, MA, 1990.

[Mi2]   J. S. Milne, *The points on a Shimura variety modulo a prime of good reduction*, The Zeta function of Picard modular surfaces, pp. 153–255, Univ. Montréal, Montreal, Quebec, 1992.

[Mi3]   J. S. Milne, *Shimura varieties and motives*, Motives (Seattle, WA, 1991), Part 2, pp. 447–523, Proc. Symp. Pure Math., Vol. **55**, Amer. Math. Soc., Providence, RI, 1994.

[Mi4]   J. S. Milne, *Descent for Shimura varieties*, Mich. Math. J. **46** (1999), no. 1, pp. 203–208.

[Mo]   Y. Morita, *On potential good reduction of abelian varieties*, J. Fac. Sci. Univ. Tokyo Sect. I A Math. **22** (1975), no. 3, pp. 437–447.

[Mu]   D. Mumford, *Abelian varieties*, Tata Inst. of Fund. Research Studies in Math., No. **5**, Published for the Tata Institute of Fundamental Research, Bombay; Oxford Univ. Press, London, 1970 (reprinted 1988).

[MFK]   D. Mumford, J. Fogarty, and F. Kirwan, *Geometric invariant theory. Third enlarged edition*, Ergebnisse der Mathematik und ihrer Grenzgebiete (2), Vol. **34**, Springer-Verlag, Berlin, 1994.

[Ne]   A. Néron, *Modèles minimaux des variétés abéliennes*, Inst. Hautes Études Sci. Publ. Math., Vol. **21**, 1964.

[NV1]   M.-H. Nicole and A. Vasiu, *Minimal truncations of supersingular p-divisible groups*, 9 pages, to appear in Indiana Univ. Math. J., `xxx.arxiv.org/abs/math/0606777`.

[NV2]   M.-H. Nicole and A. Vasiu, *Traverso's isogeny conjecture for p-divisible groups*, 8 pages, to appear in Rend. Semin. Mat. U. Padova, `xxx.arxiv.org/abs/math/0606780`.

[Oo]   F. Oort, *Foliations in moduli spaces of abelian varieties*, J. of Amer. Math. Soc. **17** (2004), no. 2, pp. 267–296.

[Pa]   R. Parthasarathy, *Holomorphic forms in $\Gamma\backslash G/K$ and Chern classes*, Topology **21** (1982), no. 2, pp. 157–178.

[Pau]   F. Paugam, *Galois representations, Mumford–Tate groups and good reduction of abelian varieties*, Math. Ann. **329** (2004), no. 1, pp. 119–160. Erratum: Math. Ann. **332** (2004), no. 4, p. 937.

[Ra]   M. Raynaud, *Schémas en groupes de type (p,...,p)*, Bull. Soc. Math. France **102** (1974), pp. 241–280.

[RR]   M. Rapoport and M. Richartz, *On the classification and specialization of F-isocrystals with additional structure*, Comp. Math. **103** (1996), no. 2, pp. 153–181.

[Sa1]   I. Satake, *Holomorphic imbeddings of symmetric domains into a Siegel space*, Amer. J. Math. **87** (1965), pp. 425–461.

[Sa2]   I. Satake, *Symplectic representations of algebraic groups satisfying a certain analyticity condition*, Acta Math. **117** (1967), pp. 215–279.

[SGA7]   A. Grothendieck, *Groupes de monodromie en géométrie algébrique. I*, Séminaire de Géométrie Algébrique du Bois-Marie 1967–1969 (SGA 7 I), Lecture Notes in Math., Vol. **288**, Springer-Verlag, Berlin-New York, 1972.

[Ta]   J. Tate, *p-divisible groups*, Proceedings of a conference on local fields (Driesbergen, 1966), pp. 158–183, Springer-Verlag, Berlin, 1967.

[Ti]   J. Tits, *Classification of algebraic semisimple groups*, Algebraic Groups and Discontinuous Subgroups (Boulder, CO, 1965), pp. 33–62, Proc. Sympos. Pure Math., Vol. **9**, Amer. Math. Soc., Providence, RI, 1966.

[Tr1]   C. Traverso, *Sulla classificazione dei gruppi analitici di caratteristica positiva*, Ann. Scuola Norm. Sup. Pisa **23** (1969), no. 3, pp. 481–507.

[Tr2]   C. Traverso, *p-divisible groups over fields*, Symposia Mathematica, Vol. **XI** (Convegno di Algebra Commutativa, INDAM, Rome, 1971), pp. 45–65, Academic Press, London, 1973.

[Tr3]   C. Traverso, *Specializations of Barsotti–Tate groups*, Symposia Mathematica, Vol. **XXIV** (Sympos., INDAM, Rome, 1979), pp. 1–21, Acad. Press, London-New York, 1981.

[Va1]   A. Vasiu, *Integral canonical models of Shimura varieties of preabelian type*, Asian J. Math. **3** (1999), no. 2, pp. 401–518.

[Va2]   A. Vasiu, *A purity theorem for abelian schemes*, Mich. Math. J. **54** (2004), no. 1, pp. 71–81.

[Va3]   A. Vasiu, *Crystalline boundedness principle*, Ann. Sci. l'École Norm. Sup. **39** (2006), no. 2, pp. 245–300.

[Va4]   A. Vasiu, *Projective integral models of Shimura varieties of Hodge type with compact factors*, 24 pages, to appear in Crelle, `xxx.arxiv.org/abs/math/0408421`.

[Va5]   A. Vasiu, *Level m stratifications of versal deformations of p-divisible groups*, 35 pages, to appear in J. Alg. Geom., `xxx.arxiv.org/abs/math/0505507`.

[Va6]   A. Vasiu, *Integral canonical models of unitary Shimura varieties*, 27 pages, to appear in Asian J. Math., `xxx.arxiv.org/abs/math/0608032`.

[Va7]   A. Vasiu, *Good Reductions of Shimura varieties of Hodge type in arbitrary unramified mixed characteristic, Parts I and II*, Part I is available at `xxx.arxiv.org/abs/0707.1668`.

[Va8]   A. Vasiu, *A motivic conjecture of Milne*, `xxx.arxiv.org/abs/math/0308202`.

[Va9]   *A. Vasiu, Manin problems for Shimura varieties of Hodge type*, `xxx.arxiv.org/abs/math/0209410`.

[Va10]   *A. Vasiu, Mod p classification of Shimura F-crystals*, `xxx.arxiv.org/abs/math/0304030`.

[Va11]   *A. Vasiu, Purity results for finite flat group schemes over ramified bases*, manuscript available in the archive of preprints.

[We]   T. Wedhorn, *The dimension of Oort strata of Shimura varieties of PEL-type*, Moduli of abelian varieties (Texel Island, 1999), pp. 441–471, Progr. of Math., Vol. **195**, Birkhäuser, Basel, 2001.

[Wi]   J.-P. Wintenberger, *Un scindage de la filtration de Hodge pour certaines variétés algébriques sur les corps locaux*, Ann. of Math. (2) **119** (1984), no. 3, pp. 511–548.

[Zi1]   T. Zink, *Isogenieklassen von Punkten von Shimuramannigfaltigkeiten mit Werten in einem endlichen Körper*, Math. Nachr. **112** (1983), pp. 103–124.

[Zi2]   T. Zink, *de Jong-Oort purity for p-divisible groups*, available at `www.math.uni-bielefeld.de/~zink`.

Higher-Dimensional Geometry over Finite Fields
D. Kaledin and Y. Tschinkel (Eds.)
IOS Press, 2008

Lectures on zeta functions over finite fields

Daqing Wan

Department of Mathematics, University of California, Irvine, CA92697-3875
e-mail: dwan@math.uci.edu

Abstract. We give an introduction to zeta functions over finite fields, focusing on moment zeta functions and zeta functions of affine toric hypersurfaces.

1. Introduction

These are the notes from the summer school in Göttingen sponsored by NATO Advanced Study Institute on Higher-Dimensional Geometry over Finite Fields that took place in 2007. The aim was to give a short introduction to zeta functions over finite fields, focusing on moment zeta functions and zeta functions of affine toric hypersurfaces. Along the way, both concrete examples and open problems are presented to illustrate the general theory. For simplicity, we have kept the original lecture style of the notes. It is a pleasure to thank Phong Le for taking the notes and for his help in typing up the notes.

2. Zeta Functions over Finite Fields

Definitions and Examples

Let p be a prime, $q = p^a$ and $\mathbb{F}_q$ be the finite field of q elements. For the affine line $\mathbb{A}^1$, we have $\mathbb{A}^1(\mathbb{F}_q) = \mathbb{F}_q$ and $\#\mathbb{A}^1(\mathbb{F}_q) = q$.

Fix an algebraic closure $\overline{\mathbb{F}_q}$. $\mathrm{Frob}_q : \overline{\mathbb{F}_q} \mapsto \overline{\mathbb{F}_q}$, defined by $\mathrm{Frob}_q(x) = x^q$. For $k \in \mathbb{Z}_{>0}$,

$$\mathbb{F}_{q^k} = \mathrm{Fix}\left(\mathrm{Frob}_q^k | \overline{\mathbb{F}_q}\right), \quad \mathbb{A}^1(\overline{\mathbb{F}_q}) = \overline{\mathbb{F}_q} = \bigcup_{k=1}^{\infty} \mathbb{F}_{q^k}.$$

Given a geometric point $x \in \overline{\mathbb{F}_q}$, the orbit $\{x, x^q, \ldots, x^{q^{\deg(x)-1}}\}$ of x under Frob_q is called the closed point of $\mathbb{A}^1$ containing x. The length of the orbit is called the degree of the closed point. We may correspond this uniquely to the monic irreducible polynomial $(t - x)(t - x^q) \ldots (t - x^{q^{\deg(x)-1}})$. Let $|\mathbb{A}^1|$ denote the set of closed points of $\mathbb{A}^1$ over $\mathbb{F}_q$. Similarly, let $|\mathbb{A}^1|_k$ denote the set of closed points of $\mathbb{A}^1$ of degree k. Hence

$$|\mathbb{A}^1| = \bigsqcup_{k=1}^{\infty} |\mathbb{A}^1|_k.$$

Example 2.1. The zeta function of $\mathbb{A}^1$ over $\mathbb{F}_q$ is

$$\begin{aligned}
Z(\mathbb{A}^1, T) &= \exp\left(\sum_{k=1}^{\infty} \frac{T^k}{k} \# \mathbb{A}^1(\mathbb{F}_{q^k})\right) \\
&= \exp\left(\sum_{k=1}^{\infty} \frac{T^k}{k} q^k\right) \\
&= \frac{1}{1-qT} \in \mathbb{Q}(T).
\end{aligned}$$

The reciprocal pole is a Weil q-number. There is also a product decomposition

$$Z(\mathbb{A}^1, T) = \prod_{k=1}^{\infty} \frac{1}{(1 - T^k)^{\# |\mathbb{A}^1|_k}}.$$

More generally, let X be quasi-projective over $\mathbb{F}_q$, or a scheme of finite type over $\mathbb{F}_q$. By birational equivalence and induction, one can often (but not always) assume that X is a hypersurface $\{f(x_1, \ldots, x_n) = 0 | x_i \in \overline{\mathbb{F}_q}\}$. Consider the Frobenius action on $X(\overline{\mathbb{F}_q})$. Let $|X|$ be the set of all closed points of X and $|X|_k$ be the set of closed points on X of degree k. As in the previous case, we have

$$X(\overline{\mathbb{F}_q}) = \bigsqcup_{k=1}^{\infty} X(\mathbb{F}_{q^k}), \quad |X| = \bigsqcup_{k=1}^{\infty} |X|_k.$$

Definition 2.2. The zeta functions of X is

$$Z(X, T) = \exp\left(\sum_{k=1}^{\infty} \frac{T^k}{k} \# X(\mathbb{F}_{q^k})\right)$$

$$= \prod_{k=1}^{\infty} \frac{1}{(1 - T^k)^{\# |X|_k}} \in 1 + T\mathbb{Z}[[T]].$$

Question 2.3. What does $Z(X, T)$ look like?

The answer was proposed by André Weil in his celebrated Weil conjectures. More precisely, Dwork [7] proved that $Z(X, T)$ is a rational function. Deligne [6] proved that the reciprocal zeros and poles of $Z(X, T)$ are Weil q−numbers.

Moment Zeta Functions

Let $f : X \mapsto Y/\mathbb{F}_q$. One has

$$X(\overline{\mathbb{F}_q}) = \bigsqcup_{y \in Y(\overline{\mathbb{F}_q})} f^{-1}(y)(\overline{\mathbb{F}_q}).$$

Similarly

$$X(\mathbb{F}_q) = \bigsqcup_{y \in Y(\mathbb{F}_q)} f^{-1}(y)(\mathbb{F}_q).$$

From this we get

$$\#X(\mathbb{F}_{q^k}) = \sum_{y \in Y(\mathbb{F}_{q^k})} \#f^{-1}(y)(\mathbb{F}_{q^k})$$

for $k = 1, 2, 3, \ldots$. This number is known as the first moment of f over $\mathbb{F}_{q^k}$.

Definition 2.4. For $d \in \mathbb{Z}_{>0}$, the d-th moment of f over $\mathbb{F}_{q^k}$ is

$$M_d(f \otimes \mathbb{F}_{q^k}) = \sum_{y \in Y(\mathbb{F}_{q^k})} \#f^{-1}(y)(\mathbb{F}_{q^{dk}})$$

$k = 1, 2, 3, \ldots$

Definition 2.5. The d-th moment zeta function of f over $\mathbb{F}_q$ is

$$\begin{aligned}
Z_d(f, T) &= \exp\left(\sum_{k=1}^{\infty} \frac{T^k}{k} M_d(f \otimes \mathbb{F}_{q^k})\right) \\
&= \prod_{y \in |Y|} Z\left(f^{-1}(y) \otimes_{\mathbb{F}_{q^{\deg(y)}}} \mathbb{F}_{q^{d \times \deg(y)}}, T^{\deg(y)}\right) \in 1 + T\mathbb{Z}[[T]].
\end{aligned}$$

Geometrically $M_d(f \otimes \mathbb{F}_{q^k})$ can be thought of as certain point counting along the fibres of f. Note that $M_d(f, k)$ will increase as d increases. Figure 2 illustrates this. The sequence of moment zeta functions $Z_d(f, T)$ measures the arithmetic variation of rational points along the fibres of f. It naturally arises from the study of Dwork's unit root conjecture [28].

Question 2.6.

1. For a given f, what is $Z_d(f, T)$?
2. How does $Z_d(f, T)$ vary with d?

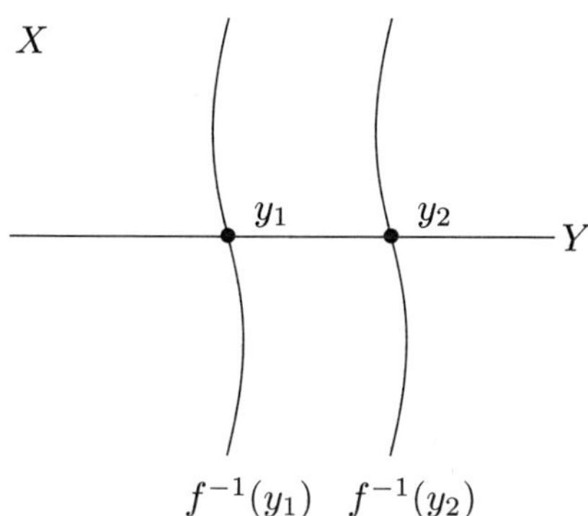

Figure 1. $f^{-1}(y)$

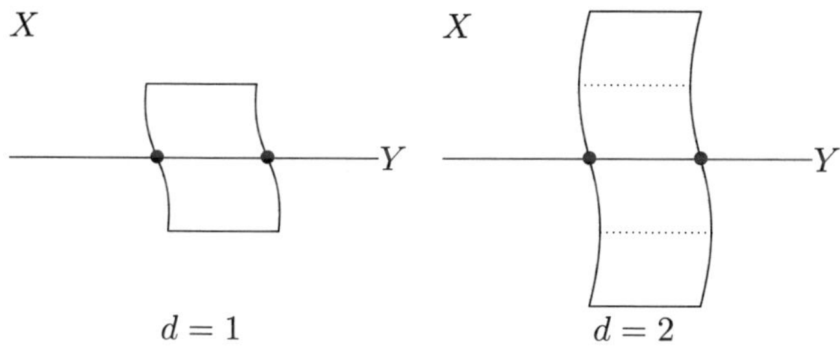

Figure 2. $f^{-1}(y)(\mathbb{F}_{q^d})$
As d increases the area where we count points will also increase.

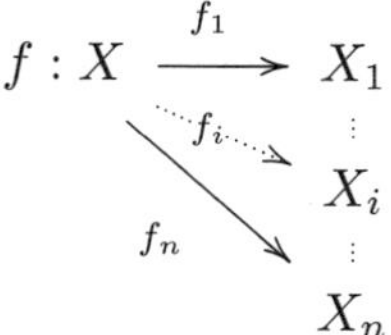

Figure 3. $f : X \mapsto X_1 \times \ldots \times X_n$

Partial Zeta Functions

Assume $f : X \mapsto X_1 \times \ldots \times X_n$ defined by $x \mapsto (f_1(x), \ldots, f_n(x))$ is an embedding. There are many ways to satisfy this property. For example the addition of the identity function $f_n : X \mapsto X$ will assure f is an embedding.

Let $d_1, \ldots, d_n \in \mathbb{Z}_{>0}$. For $k = 1, 2, 3, \ldots$, let

$$M_{d_1,\ldots,d_n}(f \otimes \mathbb{F}_{q^k}) :=$$
$$\#\{x \in X(\overline{\mathbb{F}_q}) \mid f_1(x) \in X_1(\mathbb{F}_{q^{d_1 k}}), \ldots, f_n(x) \in X_n(\mathbb{F}_{q^{d_n k}})\} < \infty$$

Definition 2.7. Define the partial zeta function of f over $\mathbb{F}_q$ to be

$$Z_{d_1,\ldots,d_n}(f,T) = \exp\left(\sum_{k=1}^{\infty} \frac{T^k}{k} M_{d_1,\ldots,d_n}(f \otimes \mathbb{F}_{q^k})\right).$$

The partial zeta function measures the distribution of rational points of X independently along the fibres of the n-tuple of morphisms $(f_1, \cdots, f_n)$.

Example 2.8. If $f_1 : X \mapsto X_1$ and $f_2 = \mathrm{Id} : X \mapsto X$, then $Z_{1,d}(f,T) = Z_d(f_1,T)$.

Thus, partial zeta functions are generalizations of moment zeta functions.

Question 2.9.

1. What is $Z_{d_1,\ldots,d_n}(f,T)$?
2. How does $Z_{d_1,\ldots,d_n}(f,T)$ vary as $\{d_1,\ldots,d_n\}$ varies?

We have

Theorem 2.10 ([26]). *The partial zeta function* $Z_{d_1,\ldots,d_n}(f,T)$ *is a rational function. Furthermore, its reciprocal zeros and poles are Weil q-numbers.*

3. General Properties of $Z(f,T)$.

Trace Formula

By Grothendieck [14], $Z(X,T)$ can be expressed in terms of l-adic cohomology. More precisely, let $\overline{X} = X \otimes_{\mathbb{F}_q} \overline{\mathbb{F}_q}$. Then,

Theorem 3.1. *There are finite dimensional vector spaces $H_c^i(X)$ with invertible linear action by* Frob_q *such that*

$$Z(X,T) = \prod_{i=0}^{2\dim(X)} \det(I - \mathrm{Frob}_q^{-1}T \mid H_c^i(X))^{(-1)^{i-1}},$$

where

$$H_c^i(X) = \begin{cases} H_c^i(\overline{X}, \mathbb{Q}_l), & l \neq p, prime \\ H_{rig,c}(X, \mathbb{Q}_p), & l = p. \end{cases}$$

This is used to show that $Z(X,T) \in \mathbb{Q}(T)$. One should note:

1. $Z(X,T)$ is independent of the choice of l.
2. $\det(I - \mathrm{Frob}_q^{-1}T \mid H_c^i(X))$ may depend on the choice of l due to possible cancellation. The conjectural independence on l is still open in general.

Riemann Hypothesis

Fix an embedding of $\overline{\mathbb{Q}_l} \hookrightarrow \mathbb{C}$. Let $b_i = \dim H_c^i(X)$. Consider the factorization

$$\det(I - \mathrm{Frob}_q^{-1} T | H_c^i(X)) = \prod_{j=1}^{b_i}(1 - \alpha_{ij}T), \alpha_{ij} \in \mathbb{C}.$$

The α_{ij}'s are Weil q-numbers, that is,

1. The α_{ij}'s are algebraic integers over $\mathbb{Q}$.
2. For $\sigma \in \mathrm{Gal}(\overline{\mathbb{Q}}/\mathbb{Q}), |\alpha_{ij}| = |\sigma(\alpha_{ij})| = \sqrt{q}^{\omega_{ij}}$ for some integer ω_{ij}, called the weight of α_{ij} with $0 \leq \omega_{ij} \leq i, \forall j = 1, \dots b_i$.

The $l \neq p$ case was proved by Deligne [6] and the $l = p$ case by Kedlaya [19].

Slopes (p-adic Riemann Hypothesis)

Consider an embedding $\overline{\mathbb{Q}_l} \hookrightarrow \mathbb{C}_p$. Then what is the $\mathrm{ord}_q(\alpha_{ij}) \in \mathbb{Q}_{\geq 0}$? This is referred to as the slope of α_{ij}.

By Riemann Hypothesis,

$$\alpha_{ij}\overline{\alpha_{ij}} = q^{\omega_{ij}},$$

$$0 \leq \mathrm{ord}_q(\alpha_{ij}) \leq \mathrm{ord}_q(\alpha_{ij}\overline{\alpha_{ij}}) = \omega_{ij} \leq i,$$

Further, Deligne's integrality theorem implies that

$$i - \dim(X) \leq \mathrm{ord}_q(\alpha_{ij}).$$

Question 3.2. Given $X/\mathbb{F}_q$, the following questions arise:

1. What is $b_{i,l} := b_i$?
2. What is ω_{ij}?
3. What is the slope $\mathrm{ord}_q(\alpha_{ij})$?

Example 3.3. If X is a smooth projective variety over $\mathbb{F}_q$, then:

1. $H_c^i(X)$ is pure of weight i, i.e. $\omega_{ij} = i$ for $1 \leq j \leq b_i$. Thus $b_{i,l}$ is independent of l.
2. The q-adic Newton polygon (NP) of $\det(I - \mathrm{Frob}_q^{-1} T | H_c^i(X)) \in \mathbb{Z}[[T]]$ lies above the Hodge polygon of $H_c^i(X)$. This was conjectured by Katz [17] and proven by Mazur [20] and Ogus [2]. We will discuss this more later.

4. Moment Zeta Functions

Let $f : X \to Y/\mathbb{F}_q$. For $d \in \mathbb{Z}_{>0}$, recall the d-th moment of $f \otimes \mathbb{F}_{q^k}$ is

$$M_d(f \otimes \mathbb{F}_{q^k}) = \sum_{y \in Y(\mathbb{F}_{q^k})} \#f^{-1}(y)(\mathbb{F}_{q^{dk}}).$$

Question 4.1.

1. How does $M_d(f \otimes \mathbb{F}_{q^k})$ vary as k varies?
2. How does $M_d(f \otimes \mathbb{F}_{q^k})$ vary with d?
3. How does $M_d(f \otimes \mathbb{F}_{q^k})$ vary with both d and k?

Definition 4.2. Define the d-th moment zeta function of f to be

$$Z_d(f, T) = \exp\left(\sum_{k=1}^{\infty} \frac{T^k}{k} M_d(f \otimes \mathbb{F}_{q^k})\right).$$

Observe for $d = 1$ we have $Z_1(f, T) = Z(X, T)$. Recall that $Z_d(f, T) \in \mathbb{Q}(T)$ and its reciprocal zeros and poles are Weil q-numbers. This follows from the following more precise cohomological formula.

Theorem 4.3. *Let $l \neq p$. Let $\mathfrak{F}^i = R^i f_! \mathbb{Q}_l$ be the i-th relative l-adic cohomology with compact support. Let $\sigma_{d,j,i} = Sym^{d-j}\mathfrak{F}^i \otimes \bigwedge^j \mathfrak{F}^i$. Then $Z_d(f, T) =$*

$$\prod_{i=0}^{2\dim(X/Y)} \prod_{j=0}^{d} \prod_{k=0}^{2\dim(Y)} \det\left(I - \mathrm{Frob}_q^{-1} T | H_c^k(\overline{Y}, \sigma_{d,j,i})\right)^{(-1)^{i+j+k-1}(j-1)}$$

Proof. For an l-adic sheaf $\mathfrak{F}$ on Y, let $L(\mathfrak{F}, T)$ denote the L-function of $\mathfrak{F}$. The trace formula in [14] applies to the L-function $L(\mathfrak{F}, T)$:

$$L(\mathfrak{F}, T) = \prod_{i=0}^{2\dim(Y)} \det(I - \mathrm{Frob}_q^{-1} T | H_c^i(\overline{Y}, \mathfrak{F}))^{(-1)^{i-1}}.$$

The d-th Adams operation of a sheaf $\mathfrak{F}$ can be written as the virtual sheaf [23]

$$[\mathfrak{F}]^d = \sum_{j \geq 0} (-1)^j (j - 1) \left[Sym^{d-j}\mathfrak{F} \otimes \bigwedge^j \mathfrak{F} \right].$$

It follows that

$$Z_d(f,T) = \prod_{y \in |Y|} Z\left(f^{-1}(y) \otimes_{\mathbb{F}_{q^{\deg(y)}}} \mathbb{F}_{q^{\deg(y)d}}, T^{\deg(y)}\right)$$

$$= \prod_{y \in |Y|} \prod_{i \geq 0} \det\left(I - (\mathrm{Frob}_{q^{\deg(y)}}^{-1})^d T^{\deg(y)} | \mathfrak{F}_y^i\right)^{(-1)^{i-1}}$$

$$= \prod_{i \geq 0} \prod_{y \in |Y|} \det\left(I - T^{\deg(y)}(\mathrm{Frob}_{q^{\deg(y)}}^{-1}) | [\mathfrak{F}_y^i]^d\right)^{(-1)^{i-1}}$$

$$= \prod_{i \geq 0} L([\mathfrak{F}^i]^d / Y, T)^{(-1)^i}$$

$$= \prod_{i \geq 0} \prod_{j \geq 0} L(\sigma_{d,j,i}, T)^{(-1)^{i+j}(j-1)}$$

$$= \prod_k \prod_{i \geq 0} \prod_{j \geq 0} \det\left(I - T\mathrm{Frob}_q^{-1} | H_c^k(\overline{Y}, \sigma_{d,j,i}, T)\right)^{(-1)^{i+j+k-1}(j-1)}.$$

$$\square$$

To use this formula, one needs to know:

1. The total degree of $Z_d(f,T)$: number of zeros + number of poles.
2. The high weight trivial factor which gives the main term.
3. The vanishing of nontrivial high weight term which gives a good error bound.

Note:

1. There is an explicit upper bound for the total degree of $Z_d(f,T)$, which grows exponentially in d, see [9].
2. There exists a total degree bound of the form $c_1 d^{c_2}$ which is a polynomial in d. However, the constant c_1 is not yet known to be effective if $\dim Y \geq 2$, see [9].

Question 4.4. How do we make c_1 effective?

Example: Artin-Schreier hypersurfaces

Let

$$g(x_1, \ldots, x_n, y_1, \ldots, y_{n'}) \in \mathbb{F}_q[x_1, \ldots, x_n, y_1, \ldots, y_{n'}].$$

We may also rewrite this as $g = g_m + g_{m-1} + \ldots + g_0$, where g_i is the homogeneous part of degree i and $g_m \neq 0$.

Consider:

$$X : \{x_0^p - x_0 = g(x_1, \ldots, x_n, y_1, \ldots, y_{n'})\} \hookrightarrow \mathbb{A}^{n+n'+1}$$
$$Y : \mathbb{A}^{n'}$$
$$f : X \mapsto Y, (x_0, x_1, \ldots, x_n, y_1, \ldots, y_{n'}) \mapsto (y_1, \ldots, y_{n'})$$

One may then ask:

$$M_d(f) = \#\{x_i \in \mathbb{F}_{q^d}, y_i \in \mathbb{F}_q | x_0^p - x_0 = g(x,y)\} = ?$$

Ideally for nice g, one hopes:

$$M_d(f) = q^{dn+n'} + O(q^{(dn+n')/2})$$

Theorem 4.5 (Deligne, [5]). *Assume that g is a Deligne polynomial of degree m, i.e., the leading form g_m is a smooth projective hypersurface in $\mathbb{P}^{n+n'}$ and $p \nmid m$. Then*

$$|M_1(f) - q^{n+n'}| \leq (p-1)(m-1)^{n+n'} q^{\frac{n+n'}{2}}.$$

For $d > 1$, a similar estimate can be obtained in some cases.

Assume $f^{-1}(y)$ is a Deligne polynomial of degree m for all $y \in \mathbb{A}^{n'}(\mathbb{F}_q)$. Then, applying Deligne's estimate fibre by fibre, one deduces

$$\# f^{-1}(y)(\mathbb{F}_{q^d}) = q^{dn} + E_y(d),$$

$$|E_y(d)| \leq (p-1)(m-1)^n q^{\frac{dn}{2}},$$

where $E_y(d)$ is some error term. From this, we get

$$M_d(f) = \sum_{y \in \mathbb{A}^{n'}(\mathbb{F}_q)} \# f^{-1}(y)(\mathbb{F}_{q^d})$$
$$= q^{dn+n'} + \sum_{y \in \mathbb{A}^{n'}(\mathbb{F}_q)} E_y(d)$$

Thus, we get the "trivial" estimate:

$$|M_d(f) - q^{dn+n'}| \leq (p-1)(m-1)^n q^{\frac{dn}{2}+n'}$$

Ideally, one would hope to replace n' by $n'/2$ in the above error bound.

If one applies the Katz type estimate via monodromy calculation as in [18], one gets $\sqrt{q}$ savings in good cases, i.e., with error term $O(q^{\frac{dn}{2}+n'-\frac{1}{2}})$. This is still far from the expected error bound $O(q^{\frac{dn+n'}{2}})$ if $n' \geq 2$.

Definition 4.6. The d-th fibered sum of g is

$$\bigoplus_{Y}^{d} g = g(x_{11}, \ldots, x_{1n}, y_1, \ldots, y_{n'}) + \ldots + g(x_{d1}, \ldots, x_{dn}, y_1, \ldots, y_{n'}).$$

Observe the y_i values remain the same while the x_{ij} values vary.

Theorem 4.7 (Fu-Wan, [9]). *Assume $\bigoplus_{Y}^{d} g$ is a Deligne polynomial of degree m. Then*

1. $|M_d(f) - q^{dn+n'}| \leq (p-1)(m-1)^{dn+n'} q^{\frac{dn+n'}{2}}$
2. $|M_d(f) - q^{dn+n'}| \leq c(p,n,n') d^{3(m+1)^n - 1} q^{\frac{dn+n'}{2}}$

The constant c is not known to be effective if $n' \geq 2$.

If p does not divide d, then $\bigoplus_{Y}^{d} g$ is a Deligne polynomial for a generic g of degree m. Thus, the assumption is satisfied for many g if p does not divide d. However, if $p \mid d$, there are no such g.

Question 4.8. If $p|d$, what would be the best estimate $M_d(f)$?

Example: Toric Calabi-Yau hypersurfaces

This geometric example is studied in a joint work with A. Rojas-Leon [21]. Let $n \geq 2$. We consider

$$X : \{x_1 + \ldots + x_n + \frac{1}{x_1 \ldots x_n} - y = 0\} \hookrightarrow \mathbb{G}_m^n \times \mathbb{A}^1,$$

$$Y = \mathbb{A}^1,$$

$$f : (x_1, \cdots, x_n, y) \longrightarrow y.$$

For $y \neq (n+1)\zeta$, with $\zeta^{n+1} = 1$, we have

$$f^{-1}(y) : x_1 + \ldots + x_n + \frac{1}{x_1 \ldots x_n} - y = 0$$

is an affine Calabi-Yau hypersurface in $\mathbb{G}_m^n$.

For $n = 2$, we have an elliptic curve. For $n = 3$, we have a K3 surface. For $n = 4$, we have a Calabi-Yau 3-fold. Recall

$$M_d(f) = \sum_{y \in \mathbb{F}_q} \# f^{-1}(y)(\mathbb{F}_{q^d}).$$

For $d = 1$, we have $M_1(f) = \# X(\mathbb{F}_q) = (q-1)^n$. For every $y \in \mathbb{F}_q$, we have

$$\# f^{-1}(y)(\mathbb{F}_{q^d}) = \frac{(q^d - 1)^n - (-1)^n}{q^d} + E_y(d),$$

where $E_y(d)$ is some error term with $|E_y(d)| \leq n q^{d(n-1)/2}$. Thus,

$$M_d(f) = q \frac{(q^d - 1)^n - (-1)^n}{q^d} + \sum_{y \in \mathbb{F}_q} E_y(d).$$

From this, we obtain the "trivial" estimate

$$|M_d(f) - \frac{(q^d - 1)^n - (-1)^n}{q^{d-1}}| \leq n q^{d(n-1)/2+1}.$$

Theorem 4.9 (Rojas-Leon and Wan, [21]). *If $p \nmid (n+1)$, then*

1. $|M_d(f) - \left(\frac{(q^d-1)^n - (-1)^n}{q^{d-1}} + \frac{1}{2}(1 + (-1)^d) q^{d(n-1)/2+1} \right)| \leq D q^{d(n-1)/2+\frac{1}{2}}$
   where D is an explicit constant depending only on n and d.
2. *The purity decomposition of $Z_d(f, T)$ is determined.*

Question 4.10. How do $M_d(f)$ and $Z_d(f, T)$ vary with d?

5. Zeta Functions of Fibres

We continue with the previous example.

Example 5.1. For $y \in \mathbb{F}_q$, let

$$f^{-1}(y) = x_1 + \ldots + x_n + \frac{1}{x_1 \ldots x_n} - y = 0 \hookrightarrow \mathbb{G}_m^n.$$

This is singular when $y \in \{(n+1)\zeta | \zeta^{n+1} = 1\}$. This family forms the mirror family of

$$\{x_0^{n+1} + \ldots + x_n^{n+1} - y x_0 \ldots x_n = 0\}.$$

Let $p \nmid (n+1), y \in \mathbb{F}_q \setminus \{(n+1)\zeta | \zeta^{n+1} = 1\}$. Then

$$Z(f^{-1}(y)/\mathbb{F}_q, T) = Z\left(\left\{\frac{(q^k - 1)^n - (-1)^n}{q^k}\right\}_{k=1}^{\infty}, T\right) P_y(T)^{(-1)^n},$$

where $P_y(T) \in 1 + T\mathbb{Z}[T]$ of degree n, pure of weight $(n-1)$. Write

$$P_y(T) = (1 - \alpha_1(y)T) \ldots (1 - \alpha_n(y)T), \quad |\alpha_i(y)| = \sqrt{q^{n-1}}.$$

Then we get the following:

Corollary 5.2.

$$\left|\#f^{-1}(y)(\mathbb{F}_q) - \frac{(q-1)^n - (-1)^n}{q}\right| \le n\sqrt{q^{n-1}}.$$

The star decomposition in [22], [27] implies

Theorem 5.3. *There is a nonzero polynomial $H_p(y) \in \mathbb{F}_p[y]$ such that if $H_p(y) \ne 0$ for some $y \in \mathbb{F}_q$, then $\mathrm{ord}_q(\alpha_i(y)) = i - 1$ for $1 \le i \le n$.*

Equivalently, this family of polynomials $f^{-1}(y)$ is generically ordinary. An alternative proof can be found in Yu [31].

Moment Zeta Functions

For $d > 0$, recall

$$M_d(f) = \sum_{y \in \mathbb{F}_q} \#f^{-1}(y)(\mathbb{F}_{q^d}),$$

$$M_d(f \otimes \mathbb{F}_{q^k}) = \sum_{y \in \mathbb{F}_{q^k}} \#f^{-1}(y)(\mathbb{F}_{q^{dk}}), k = 1, 2, 3, \ldots,$$

$$Z_d(f,T) = \exp\left(\sum_{k=1}^{\infty} \frac{T^k}{k} M_d(f \otimes \mathbb{F}_{q^k})\right) \in \mathbb{Q}(T).$$

Let

$$S_d(T) = \prod_{k=0}^{[\frac{n-2}{2}]} \frac{1 - q^{dk}T}{1 - q^{dk+1}T} \prod_{i=0}^{n-1} (1 - q^{di+1}T)^{(-1)^{i+1}\binom{n}{i+1}}.$$

Theorem 5.4 (Rojas-Leon and Wan, [21]). *Assume that $(n+1)$ divides $(q-1)$. Then, the d-th moment zeta function for the above one parameter toric CY family f has the following factorization*

$$Z_d(f,T)^{(-1)^{n-1}} = P_d(T)\left(\frac{Q_d(T)}{P(d,T)}\right)^{n+1} A_d(T)S_d(T).$$

We now explain each of the above factors. First, $P_d(T)$ is the non-trivial factor which has the form

$$P_d(T) = \prod_{a+b=d,\,0\leq b\leq n} P_{a,b}(T)^{(-1)^{b-1}(b-1)},$$

and each $P_{a,b}(T)$ is a polynomial in $1+T\mathbb{Z}[T]$, pure of weight $d(n-1)+1$, whose degree r is given explicitly and which satisfies the functional equation

$$P_{a,b}(T) = \pm T^r q^{(d(n-1)+1)r/2} P_{a,b}(1/q^{d(n-1)+1}T).$$

Second, $P(d,T) \in 1+T\mathbb{Z}[T]$ is the d-th Adams operation of the "non-trivial" factor in the zeta function of a singular fibre X_t, where $t = (n+1)\zeta_{n+1}$ and $\zeta_{n+1}^{n+1} = 1$. It is a polynomial of degree $(n-1)$ whose weights are completely determined. Third, the quasi-trivial factor $Q_d(T)$ coming from a finite singularity has the form

$$Q_d(T) = \prod_{a+b=d,\,0\leq b\leq n} Q_{a,b}(T)^{(-1)^{b-1}(b-1)},$$

where $Q_{a,b}(T)$ is a polynomial whose degree $D_{n,a,b}$ and the weights of its roots are given. Finally, the trivial factor $A_d(T)$ is given by:

$A_d(T) = (1 - q^{\frac{d(n-1)}{2}}T)(1 - q^{\frac{d(n-1)}{2}+1}T)(1 - q^{\frac{d(n-2)}{2}+1}T)$ *if n and d are even.*

$A_d(T) = (1 - q^{\frac{d(n-2)}{2}+1}T)$ *if n is even and d is odd.*

$A_d(T) = (1 - q^{\frac{d(n-1)}{2}}T)$ *if n and d are odd.*

$A_d(T) = (1 - q^{\frac{d(n-1)}{2}+1}T)^{-1}$ *if n is odd and d is even.*

Corollary 5.5. *Let $n = 2$ and $f : \{x_1 + x_2 + \frac{1}{x_1 x_2} - y = 0\} \mapsto y$ with $p \nmid 3$. Then,*

$$Z_d(f,T)^{-1} = A_d(T)\frac{R_d(T)}{R_{d-2}(qT)},$$

where $A_d(T)$ is a trivial factor and $R_d(T) \in 1 + T\mathbb{Z}[T]$ is a non-trivial factor which is pure of weight $d+1$ and degree $2(d-1)$.

For all $d \leq 1$, $R_d(T) = 1$. $R_2(T)$ is a polynomial of degree 2 and weight 3. This suggests that $R_2(T)$ comes from a rigid Calabi-Yau variety. In general, $R_d(T)$ is of weight $d + 1$ and degree $2(d - 1)$.

As always, we may ask what are the slopes of $R_d(T)$?

The above one parameter family of Calabi-Yau hypersurfaces is the only higher dimensional example for which the moment zeta functions are determined so far. It shows that the calculation of the moment zeta function can be quite complicated in general. A related example is the one parameter family of higher dimensional Kloosterman sums, see [10], [11] for the L-function of higher symmetric power which gives the main piece of the moment zeta function.

l-adic Moment Zeta Function $(l \neq p)$

Fix a prime $l \neq p$. Given $\alpha \in \mathbb{Z}_l^*$ and $d_1 \equiv d_2 \bmod (l-1)l^{k-1}$ for some k, then $\alpha^{d_1} \equiv \alpha^{d_2} \bmod l^k$.

By rationality of $Z(f^{-1}(y), T)$ it follows that

$$\# f^{-1}(y)(\mathbb{F}_{q^d}) = \sum_i \alpha_i(y)^d - \sum_j \beta_j(y)^d$$

for some l-adic algebraic integers $\alpha_i(y)$ and $\beta_j(y)$. Consider

$$M_d(f) = \sum_{y \in Y(\mathbb{F}_q)} \# f^{-1}(y)(\mathbb{F}_{q^d}).$$

This can be rewritten as

$$= \sum_{y \in Y(\mathbb{F}_q)} \left(\sum_i \alpha_i(y)^d - \sum_j \beta_j(y)^d \right).$$

We may take some $D_l(f) \in \mathbb{Z}_{>0}$ such that if $d_1 \equiv d_2 \bmod D_l(f)l^{k-1}$ then

1. $M_{d_1}(f) \equiv M_{d_2}(f) \bmod l^k$.
2. $Z_{d_1}(f, T) \equiv Z_{d_2}(f, T) \bmod l^k \in 1 + T\mathbb{Z}[[T]]$.

Definition 5.6. The l-adic weight space is defined to be

$$W_l(f) = (\mathbb{Z}/D_l(f)\mathbb{Z}) \times \mathbb{Z}_l.$$

Let $s = (s_1, s_2) \in W_l(f)$. Take a sequence of $d_i \in \mathbb{Z}_{>0}$ such that

1. $d_i \to \infty$ in $\mathbb{C}$,
2. $d_i \equiv s_1 \bmod D_l(f)$,
3. $d_i \to s_2 \in \mathbb{Z}_l$.

With this we may define the l-adic moment zeta function

$$\zeta_s(f, T) = \lim_{i \to \infty} Z_{d_i}(f, T) \in 1 + T\mathbb{Z}_l[[T]].$$

This function is analytic in the l-adic open unit disk $|T|_l < 1$.

Question 5.7. Is $\zeta_s(f,T)$ analytic on $|T|_l \leq 1$? What about in $|T|_l < \infty$?

Embed $\mathbb{Z}$ in $W_l(f)$ in the following way:

$$\mathbb{Z} \hookrightarrow W_l(f),$$

$$d \mapsto (d,d).$$

Proposition 5.8. *If* $d \in \mathbb{Z}_{>0} \hookrightarrow W_l(f)$*, then* $\zeta_d(f,T) = Z_d(f,T) \in \mathbb{Q}(T)$.

Question 5.9. What if $s \in W_l(f) \setminus \mathbb{Z}$? This is open even when f is a non-trivial family of elliptic curves over $\mathbb{F}_p$.

p-adic Moment Zeta Functions $(l = p)$

As in the l-adic case, one has a p-adic continuous result.

If $d_1 \equiv d_2 \bmod D_p(f)p^{k-1}, d_1 \geq d_2 \geq c_f k$ for some k and sufficiently large constant c_f, then

$$M_{d_1}(f) \equiv M_{d_2}(f) \bmod p^k.$$

Also, define in the same way as above

$$\zeta_{s,p}(f,T) = \lim_{i \to \infty} Z_{d_i}(f,T) \in 1 + T\mathbb{Z}_p[[T]].$$

As before consider the embedding:

$$\mathbb{Z} \hookrightarrow W_p(f),$$

$$d \mapsto (d,d).$$

The following result was conjectured by Dwork [8].

Theorem 5.10 (Wan, [23][24][25]). *If* $s = d \in \mathbb{Z} \hookrightarrow W_p(f)$*, then* $\zeta_{d,p}(f,T)$ *is p-adic meromorphic in* $|T|_p < \infty$.

Furthermore, we have

Theorem 5.11 ([25]). *Assume the p-rank* ≤ 1*. Then for each* $s \in W_p(f)$*,* $\zeta_{s,p}(f,T)$ *is p-adic meromorphic in* $|T|_p < \infty$.

This can be extended a little further as suggested by Coleman.

Theorem 5.12 (Grosse-Klönne, [13]). *Assume the p-rank* ≤ 1*. For* $s = (s_1, s_2)$ *with* $s_1 \in \mathbb{Z}/D_p(f)$ *and* $s_2 \in \mathbb{Z}_p/p^\epsilon$ *(small denominator), then* $\zeta_{s,p}(f,T)$ *is p-adic meromorphic in* $|T|_p < \infty$.

Question 5.13. In the case $s \in W_p(f) - \mathbb{Z}$ and p-rank > 1, it is unknown if $\zeta_{s,p}(f,T)$ is p-adic meromorphic, even on the closed unit disk $|T|_p \leq 1$.

6. Moment Zeta Functions over $\mathbb{Z}$

Consider

$$f : X \mapsto Y/\mathbb{Z}[\frac{1}{N}].$$

The d-th moment zeta function of f is:

$$\zeta_d(f, s) = \prod_{p \nmid N} Z_d(f \otimes \mathbb{F}_p, p^{-s}).$$

Is this $\mathbb{C}$-meromorphic in $s \in \mathbb{C}$? Is $\zeta_d(f, s)$ or its special values p-adic continuous in some sense? If so, its p-adic limit $\zeta_s(f)(s \in \mathbb{Z}_p)$ is a p-adic zeta function of f.

Example 6.1. Consider the map

$$f : \{x_1 + x_2 + \frac{1}{x_1 x_2} - y = 0\} \mapsto y.$$

Then

$$Z_d(f \otimes \mathbb{F}_p, T)^{-1} = A_d(T) \frac{R_d(f \otimes \mathbb{F}_p, T)}{R_{d-2}(f \otimes \mathbb{F}_p, pT)}$$

where $A_d(T)$ is a trivial factor and R_d is a non-trivial factor of degree $2(d-1)$ and weight $d + 1$.

$$R_d(T) \leftrightarrow f^{\otimes d} = \{x_{11} + x_{12} + \frac{1}{x_{11} x_{12}} = \ldots = x_{d1} + x_{d2} + \frac{1}{x_{d1} x_{d2}}\}$$

Example 6.2. For $d = 2$, we have

$$x_1 + x_2 + \frac{1}{x_1 x_2} = y_1 + y_2 + \frac{1}{y_1 y_2}.$$

As Matthias Schuett observed during the workshop, $R_2(T) \leftrightarrow$ the unique new form of weight 4 and level 9.

Conjecture 6.3. $\prod_p R_d(f \otimes \mathbb{F}_p, p^{-s})$ *is meromorphic in* $s \in \mathbb{C}$ *for all* d.

This conjecture is known to be true if $d \leq 2$. It should be realistic to prove the conjecture for all positive integers d.

7. l-adic Partial Zeta Functions

We now consider the system of maps where $X \mapsto X_1 \times \ldots \times X_n$ is an embedding (See Figure 4).

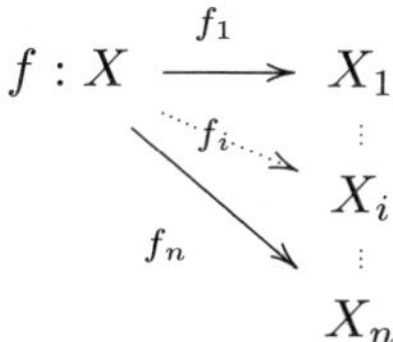

Figure 4. $f : X \mapsto X_1 \times \ldots \times X_n$

This allows us to define the partial zeta function

$$Z_{d_1,\ldots,d_n}(f,T) = \exp\left(\sum_{k=1}^{\infty}\frac{T^k}{k}\#\{x \in X(\overline{\mathbb{F}_q})\mid f_i(x) \in X_i(\mathbb{F}_{q^{d_i k}})\}\right) \in \mathbb{Q}(T).$$

Question 7.1. Is there any p-adic or l-adic continuity result as $\{d_1,\ldots,d_n\}$ varies p-adically or l-adically?

Example 7.2. Consider the surface and three projection maps:

$$f : x_1 + x_2 + \frac{1}{x_2 x_2} - r_3 = 0$$

Thus

$$M_{d_1,d_2,d_3}(f) = \#\{(x_1,x_2,x_3)\mid x_1 + x_2 + \frac{1}{x_1 x_2} - x_3 = 0, x_i \in \mathbb{F}_{q^{d_i}}, i = 1,2,3\}.$$

Is there a continuity result as $\{d_1,d_2,d_3\}$ vary?

8. Zeta Functions of Toric Affine Hypersurfaces

Let $\triangle \subset \mathbb{R}^n$ be an n-dimensional integral polytope. Let $f \in \mathbb{F}_q[x_1^{\pm 1},\ldots,x_n^{\pm 1}]$ with

$$f = \sum_{u \in \triangle \cap \mathbb{Z}^n} a_u X^u, a_u \in \mathbb{F}_q$$

such that $\triangle(f) = \triangle$. That is, $a_u \neq 0$ for each u which is a vertex of $\triangle$.

Question 8.1. Consider the toric affine hypersurface

$$U_f : \{f(x_1,\ldots,x_n) = 0\} \hookrightarrow \mathbb{G}_m^n.$$

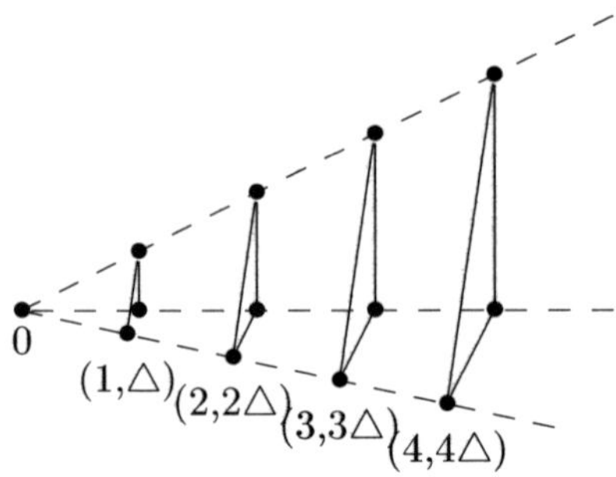

Figure 5. $C(\triangle)$

1. $\#U_f(\mathbb{F}_q) =?$
2. $Z(U_f, T) =?$

Definition 8.2.

1. If $\triangle' \subset \triangle$ is a face of $\triangle$, define

$$f^{\triangle'} = \sum_{u \in \triangle' \cap \mathbb{Z}^n} a_u X^u.$$

2. f is $\triangle$-regular if for every face $\triangle'$ (of any dimension) of $\triangle$, the system

$$f^{\triangle'} = x_1 \frac{\partial f^{\triangle'}}{\partial x_1} = \ldots = x_n \frac{\partial f^{\triangle'}}{\partial x_n} = 0$$

has no common zeros in $\mathbb{G}_m^n(\overline{\mathbb{F}_q})$.

Theorem 8.3 (GKZ, [12]).

1. *There is a nonzero polynomial* $\mathrm{disc}_\triangle \in \mathbb{Z}[a_u | u \in \triangle \cap \mathbb{Z}^n]$ *such that f is $\triangle$-regular if and only if* $\mathrm{disc}_\triangle(f) \neq 0$ *in* $\mathbb{F}_q$. *In other words,* $\mathrm{disc}_\triangle$ *is an integer coefficient polynomial that will determine $\triangle$-regularity.*
2. $\triangle(\mathrm{disc}_\triangle)$ *is determined. This is referred to as the secondary polytope.*

Question 8.4. For which p, $\mathrm{disc}_\triangle \otimes \mathbb{F}_p \neq 0$?

Definition 8.5. Let $C(\triangle)$ be the cone in $\mathbb{R}^{n+1}$ generated by 0 and $(1, \triangle)$

1. Define

$$W_\triangle(k) = \#\{(k, k\triangle) \cap \mathbb{Z}^{n+1}\}, k = 0, 1, \ldots$$

The Hodge numbers of $\triangle$ are defined by

$$h_\triangle(k) = W_\triangle(k) - \binom{n+1}{1} W_\triangle(k-1) + \binom{n+2}{2} W_\triangle(k-2) - \ldots,$$

$$h_\triangle(k) = 0, \text{if } k \geq n+1.$$

2. $\deg(\triangle) = d(\triangle) = n! Vol(\triangle) = \sum_{k=0}^{n} h_\triangle(k).$

Theorem 8.6 (Adolphson-Sperber [1], Denef-Loesser [4]). *Assume $f/\mathbb{F}_q$ is $\triangle$-regular. Then*

1. $Z(U_f, T) = \prod_{i=0}^{n-1}(1 - q^i T)^{(-1)^{n-i}\binom{n}{i+1}} P_f(T)^{(-1)^n}$ *with* $P_f(T) \in 1 + T\mathbb{Z}[T]$ *is of degree* $d(\triangle) - 1$.
2. $P_f(T) = \prod_{i=1}^{d(\triangle-1)}(1 - \alpha_i(f)T), |\alpha_i(f)| \leq \sqrt{q}^{n-1}$. *In particular,*

$$|\#U_f(\mathbb{F}_q) - \frac{(q-1)^n - (-1)^n}{q}| \leq (d(\triangle) - 1)\sqrt{q}^{n-1}.$$

The precise weights of the $\alpha_i(f)$'s were also determined by Denef-Loesser.

Question 8.7. For $i = 1, 2, \ldots, d(\triangle) - 1$, what is $\text{ord}_q(\alpha_i(f)) =?$

9. Newton and Hodge Polygons

Write

$$P_f(T) = 1 + c_1 T + c_2 T^2 + \ldots.$$

The q-adic Newton polygon of $P_f(T)$ is the lower convex closure in $\mathbb{R}^2$ of the points $(k, \text{ord}_q(c_k)), (k = 0, 1, \ldots, d(\triangle) - 1)$. Denote this Newton polygon by $NP(f)$. Note that $NP(f) = NP(f \otimes \mathbb{F}_{q^k})$ for all k.

Proposition 9.1. *Let h_s denote the horizontal length of the slope s side in $NP(f)$. Then, $P_f(T)$ has exactly h_s reciprocal zeros $\alpha_i(f)$ such that $\text{ord}_q(\alpha_i(f)) = s$ for each $s \in \mathbb{Q}_{\geq 0}$.*

Definition 9.2. The Hodge polygon of $\triangle$, denoted by $HP(\triangle)$, is the polygon in $\mathbb{R}^2$ with a side of slope $k - 1$ with horizontal length $h_\triangle(k)$ for $1 \leq k \leq n$ and vertices

$$(0, 0), \left(\sum_{m=1}^{k} h_\triangle(m), \sum_{m=1}^{k}(m-1)h_\triangle(m) \right), k = 1, 2, \ldots, n.$$

Theorem 9.3 (Adolphson-Sperber [1]). *The q-adic Newton polygon lies above the Hodge polygon, i.e., $NP(f) \geq HP(\triangle)$. In addition, the endpoints of the two coincide.*

Definition 9.4. If $NP(f) = HP(\triangle)$, then f is called ordinary.

Question 9.5. When is f ordinary? One hopes this is often.

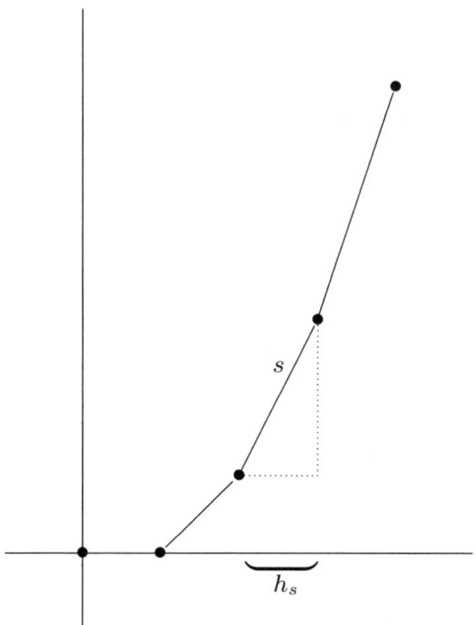

Figure 6. Newton Polygon

Let

$$M_p(\triangle) = \{f \in \overline{\mathbb{F}_p}[x_1^{\pm 1}, \cdots, x_n^{\pm 1}] | \triangle(f) = \triangle, \ f \ \triangle - \text{regular}\}.$$

Theorem 9.6 (Grothendieck, [18]). *There exists a generic Newton polygon, denoted by $GNP(\triangle, p)$, such that*

$$GNP(\triangle, p) = \inf\{NP(f) | f \in M_p(\triangle)\}$$

Hence for any $f \in M_p(\triangle)$,

$$NP(f) \geq GNP(\triangle, p) \geq HP(\triangle),$$

where the first inequality is an equality for most f (generic f).

Question 9.7. Given $\triangle$, for which p, is $GNP(\triangle, p) = HP(\triangle)$? In other words, when is f generically ordinary?

This suggests the following conjecture.

Conjecture 9.8 (Adolphson-Sperber [1]). *For each $p \gg 0$, $GNP(\triangle, p) = HP(\triangle)$.*

This is false in general. Some counterexamples can be found in [22].

Definition 9.9.

1. $S(\triangle)$ =the semigroup $C(\triangle) \cap \mathbb{Z}^{n+1}$.
   $S_1(\triangle)$ = the semigroup generated by $(1, \triangle) \cap \mathbb{Z}^{n+1}$.
2. Define the exponents of $\triangle$ as

$$I(\triangle) \quad = \inf\{D > 0 | Du \in S_1(\triangle), \forall u \in S(\triangle)\}$$
$$I_\infty(\triangle) = \inf\{D > 0 | Du \in S_1(\triangle), \forall u \in S(\triangle), u \gg 0\}$$

Conjecture 9.10. *If $p \equiv 1 \ mod \ I(\triangle)$ or if $p \equiv 1 \ mod \ I_\infty(\triangle)$ for $p \gg 0$, then*

1. $\operatorname{disc}_\triangle \otimes \mathbb{F}_p \neq 0$,
2. $GNP(\triangle, p) = HP(\triangle)$.

Part (2) is a weaker version of the conjecture in [22].

10. Generic Ordinarity

Toric Hypersurface

Let $\triangle \subset \mathbb{R}^n$ be a n-dimensional integral polytope and p a prime. Let $d(\triangle) = n! Vol(\triangle)$. Define

$$M_p(\triangle) = \{f \in \overline{\mathbb{F}_p}[x_1^{\pm 1}, \ldots, x_n^{\pm 1}] | \triangle(f) = \triangle, \ f \ \triangle - \text{regular}\}.$$

For each $f \in M_p(\triangle)$, let $NP(f)$ be the Newton polygon of the interesting factor $P_f(T)$ of the zeta function $Z(U_f, T)$. Note that changing the ground field will not change the Newton polygon. Recall that

$$NP(f) \geq GNP(\triangle, p) \geq HP(\triangle).$$

Note that $NP(f)$ is defined in a completely arithmetic fashion and is dependent on the coefficients of the polynomial f. On the other hand, $GNP(\triangle, p)$ is independent of coefficients while $HP(\triangle)$ is obtained combinatorially. If $GNP(\triangle, p) = HP(\triangle)$, we refer to p as ordinary for $\triangle$.

Conjecture 10.1 (Adolphson-Sperber). *For any $\triangle$, p is ordinary for all $p \gg 0$.*

This conjecture is too strong as Example 10.2 illustrates.

Example 10.2. Let $f = a_0 + a_1 x_1 + \ldots + a_n x_n + a_{n+1} x_1 x_2 \ldots x_n$ and

$$\triangle = Conv((0, \ldots, 0), (1, \ldots, 0), \ldots, (0, \ldots, 1), (1, 1, \ldots, 1)).$$

Therefore $d(\triangle) = n$ for $n \geq 2$. Furthermore, $\triangle$ is an empty simplex, i.e., a simplex with no lattice points other than vertices. It follows that

1. p is ordinary for $\triangle$ if and only if $p \equiv 1 \bmod (n - 1)$. This implies
2. If $n \geq 4$, then the Adolphson-Sperber conjecture is false.

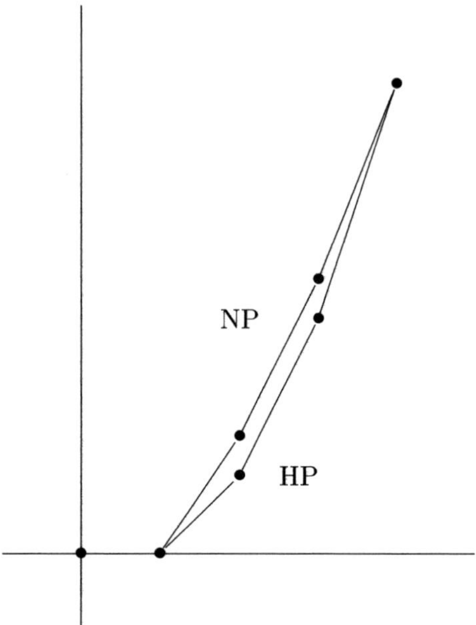

Figure 7. $NP \geq HP$

Convex Triangulation

Definition 10.3.

1. A triangulation of $\triangle$ is a decomposition

$$\triangle = \bigcup_{i=1}^{m} \triangle_i,$$

   such that each $\triangle_i$ is a simplex, $\triangle_i \cap \triangle_j$ is a common face for both $\triangle_i$ and $\triangle_j$.

2. The triangulation is called **convex** if there is a piecewise linear function $\phi : \triangle \mapsto \mathbb{R}$ such that

   (a) ϕ is convex i.e. $\phi(\frac{1}{2}x + \frac{1}{2}x') \leq \frac{1}{2}\phi(x) + \frac{1}{2}\phi(x')$, for all $x, x' \in \triangle$.
   (b) The domains of linearity of ϕ are precisely the n-dimensional simplices $\triangle_i$ for $1 \leq i \leq m$.

Examples of convex triangulations include the star decomposition, the hyperplane decomposition and the collapsing decomposition [27].

Basic Decomposition Theorem

The decomposition methods in [22], [27] generalize to prove the following decomposition theorem.

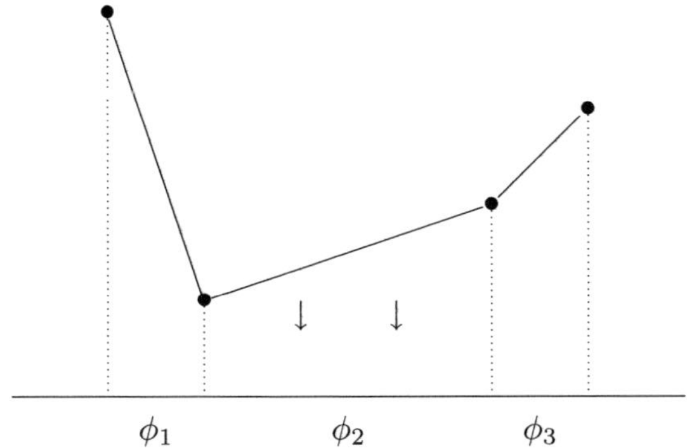

Figure 8. Piecewise projection down

Theorem 10.4.

1. *Let $\triangle = \cup_{i=1}^{m}\triangle_i$ be a convex integral triangulation of $\triangle$. If p is ordinary for each $\triangle_i$, $1 \le i \le m$, then p is ordinary for $\triangle$.*
2. *If $\triangle$ is a simplex and $p \equiv 1 \bmod d(\triangle)$, then p is ordinary.*

Corollary 10.5. *If $p \equiv 1 \bmod (lcm(d(\triangle_1), \ldots, d(\triangle_m)))$, then p is ordinary.*

Example 10.6. Let A be the convex closure of $(-1, -1)$, $(1, 0)$ and $(0, 1)$ in $\mathbb{R}^2$. The star decomposition in Figure 9 is convex and integral.

More generally,

Example 10.7. Consider $f : \{x_1 + x_2 + \ldots + x_n + 1/x_1 x_2 \ldots x_n - y = 0\}$ over $\mathbb{F}_p$. This is generically ordinary for all p. The proof uses the same star decomposition.

Example 10.8. Let $\triangle = \{(d, 0, \ldots, 0), (0, d, 0, \ldots, 0), \ldots, (0, \ldots, d), (0, \ldots, 0)\}$. We may make a parallel hyperplane cut as in Figure 10. This will make $d(\triangle_i) = 1$ for each piece $\triangle_i$ of the decomposition, see [22]. This proves that the universal family of affine (or projective) hypersurfaces of degree d and n variables over $\mathbb{F}_p$

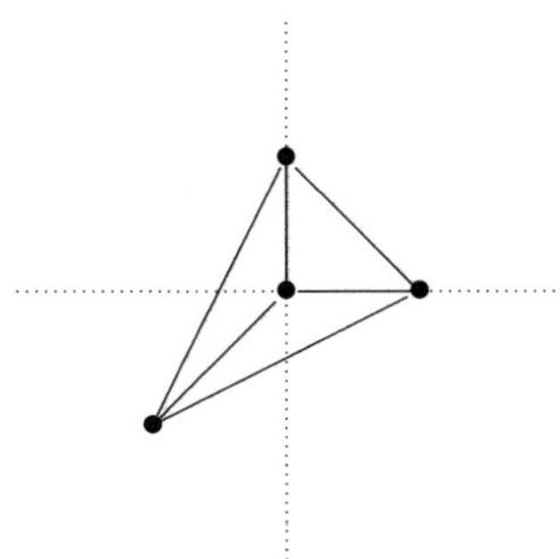

Figure 9. Star decomposition of A

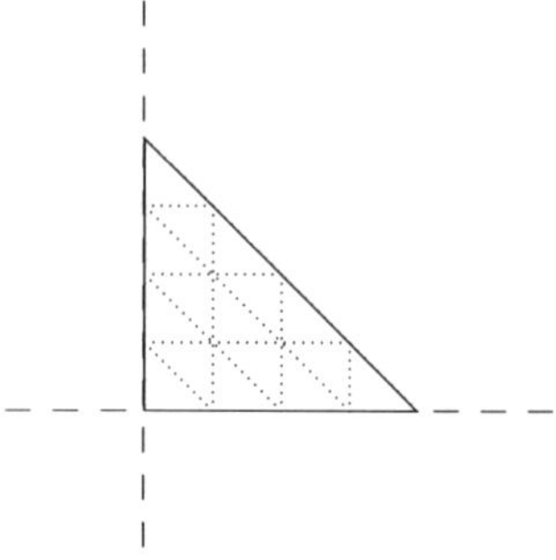

Figure 10. Parallel Hyperplane Decomposition into simplices

is also generically ordinary for every p. The projective hypersurface (complete intersection) case was first proved by Illusie [15].

Corollary 10.9. *If $n = \dim(\triangle) = 2$, then p is ordinary for $\triangle$ for all p.*

Corollary 10.10. *If $n = \dim(\triangle) = 3$, then p is ordinary for $p > 6Vol(\triangle)$.*

This corollary is proven by showing stability of the p-action on the weight. This is a different argument than by proving $d(\triangle_i) = 1$ argument.

Definition 10.11. Let $\triangle$ be an n-dimensional integral convex polytope in $\mathbb{R}^n$. Assume that 0 (origin) is in the interior of $\triangle$. Given such a situation, define $\triangle^* \subset \mathbb{R}^n$ by:

$$\triangle^* = \{(x_1, \ldots, x_n) \in \mathbb{R}^n \mid \sum_{i=1}^{n} x_i y_i \geq -1, \ \forall (y_1, \ldots, y_n) \in \triangle\}$$

Observe $\triangle^*$ is also a convex polytope in $\mathbb{R}^n$, though it may not have integral vertices. Also observe $(\triangle^*)^* = \triangle$.

Definition 10.12. $\triangle$ is called reflexive if $\triangle^*$ is also integral.

Corollary 10.13. *If $n = \dim(\triangle) = 4$ and if $\triangle$ is reflexive then p is ordinary for $\triangle$ for all $p > 12Vol(\triangle)$.*

Slope Zeta Function

The concept of slope zeta functions was developed for arithmetic mirror symmetry as we will describe here. More information can be found in [29], [30].

Let (X, Y) be a mirror pair over $\mathbb{F}_q$. Candelas, de la Ossa and Rodriques-Villegas in [3] desired a possible mirror relation of the type

$$Z(X, T) = \frac{1}{Z(Y, T)}$$

for 3 dimensional Calabi-Yau varieties. This is not true. If this were the case then

$$\sum \frac{T^k}{k} \#X(\mathbb{F}_q) = \sum \frac{T^k}{k}(-\#Y(\mathbb{F}_q)).$$

Therefore

$$\#X(\mathbb{F}_q) = -\#Y(\mathbb{F}_q),$$

which is impossible for large q on nonempty varieties.

The question is then to modify the zeta function suitably so that the desired mirror relation holds. The slope zeta function was introduced for this purpose.

Definition 10.14. Write $Z(X,T) = \prod_i (1 - \alpha_i T)^{\pm 1} \in \mathbb{C}_p(T)$.

1. The slope zeta function of X is defined to be the following two variable function:

$$S(X,U,T) = \prod_i (1 - U^{\mathrm{ord}_q(\alpha_i)}T)^{\pm 1}.$$

2. If $f : X \mapsto Y$ defined over $\mathbb{F}_q$ (a nice family) then the slope zeta function of f is the generic one among $S(f^{-1}(y),U,T)$ from all $y \in Y$, denoted by $S(f,U,T)$.

Conjecture 10.15. *Let X be a 3-dimensional Calabi-Yau variety over $\mathbb{Q}$. Assume that X has a mirror over $\mathbb{Q}$. Then the generic family containing X as a member is generically ordinary for all $p \gg 0$.*

This conjecture implies the following

Conjecture 10.16 (Arithmetic Mirror Conjecture). *Let $\{f,g\}$ be two generic mirror families of a 3-dimensional Calabi-Yau variety over $\mathbb{Q}$. Then for all $p \gg 0$,*

$$S(f \otimes \mathbb{F}_p, U, T) = \frac{1}{S(g \otimes \mathbb{F}_p, U, T)}.$$

References

[1]   A. Adolphson and S. Sperber, *Exponential sums and Newton polyhedra: Cohomology and estimates*, Ann. Math., 130 (1989), 367-406.

[2]   P. Berthelot and A. Ogus, *Notes on Crystalline Cohomology*, Princeton University Press, 1978.

[3]   P. Candelas, X. de la Ossa, F. Rodriques-Villegas, *Calabi-Yau manifolds over finitef fields II*, Fields Instit. of Commun., 38(2003).

[4]   J. Denef and F. Loeser, *Weights of exponential sums, intersection cohomology, and Newton polyhedra*, Invent. Math., 106(1991), no.2, 275-294.

[5]   P. Deligne, *Applications de la Formule des Traces aux Sommes Trigonométriques*, in Cohomologie Étale (SGA $4\frac{1}{2}$), 168-232, Lecture Notes in Math. 569, Springer-Verlag 1977.

[6]   P. Deligne, *La Conjecture de Weil II*, Publ. Math. IHES 52 (1980), 137-252.

[7]   B. Dwork, *On the rationality of the zeta function of an algebraic variety*, Amer. J. Math., 82(1960), 631-648.

[8]   B. Dwork, *Normalized period matrices II*, Ann. Math., 98(1973), 1-57.

[9]   L. Fu and D. Wan, *Moment L-functions, partial L-functions and partial exponential sums*, Math. Ann., 328(2004), 193-228.

[10] L. Fu and D. Wan, *L-functions for symmetric products of Kloosterman sums*, J. Reine Angew. Math., 589(2005), 79-103.

[11] L. Fu and D. Wan, *Trivial factors for L-functions of symmetric products of Kloosterman sheaves*, Finite Fields & Appl., to appear.

[12] I.M. Gelfand, M.M. Kapranov and A.V. Zelevinsky, *Discriminatns, Resultants and Multidimensional Determinants*, Birkhäuser Boston, Inc., Boston, MA, 1994.

[13] E. Grosse-Klönne, *On families of pure slope L-functions*, Documenta Math., 8(2003), 1-42.

[14] A. Grothendick, *Formule de Lefschetz et rationalité des fonctions L*, Séminaire Bourbaki, exposé 279, 1964/65.

[15] L. Illusie, *Ordinarité des intersections complétes générales*, Grothendieck Festschrift, Vol. II (1990). 375-405.

[16] N. Katz, *On a theorem of Ax*, Amer. J. Math., 93(1971), 485-499.

[17] N. Katz, *Slope filtration of F-crystals*, Astérisque 63(1979), 113-164.

[18] N. Katz, *Frobenius-Schur indicator and the ubiquity of Brock-Granville quadratic excess*, Finite Fields & Appl., 7(2001), 45-69.

[19] K. Kedlaya, *Fourier transforms and p-adic "Weil II"*, Compositio Mathematica, 142(2006), 1426-1450.

[20] B. Mazur, *Frobenius and the Hodge filtration*, Bull. Amer. Math. Soc., 78(1972), 653-667.

[21] A. Rojas-Leon and D. Wan, *Moment zeta functions for toric Calabi-Yau hypersurfaces*, Communications in Number Theory and Physics, Vol 1, No. 3, 2007.

[22] D. Wan, *Newton polygons of zeta functions and L-functions*, Ann. of Math, Vol. 2, No. 2(1993), 249-293.

[23] D. Wan, *Dwork's conjecture on unit root zeta functions*, Ann. Math., 150(1999), 867-927.

[24] D. Wan, *Higher rank case of Dwork's conjecture*, J. Amer. Math. Soc., 13(2000), 807-852.

[25] D. Wan, *Rank one case of Dwork's conjecture*, J. Amer. Math. Soc., 13(2000), 853-908.

[26] D. Wan, *Rationality of partial zeta functions*, Indagationes Math., New Ser., 14(2003), 285-292.

[27] D. Wan, *Variations of p-adic Newton polygons for L-functions of exponential sums*, Asian J. Math., Vol.8, 3(2004), 427-474.

[28] D. Wan, *Geometric moment zeta functions*, Geometric Aspects of Dwork Theory, Walter de Gruyter, 2004 , Vol II, 1113-1129.

[29] D. Wan, *Arithmetic mirror symmetry*, Pure Appl. Math. Q., 1(2005), 369-378.

[30] D. Wan, *Mirror symmetry for zeta functions*, Mirror Symmetry V, AMS/IP Studies in Advanced Mathematics, Vol.38, (2007), 159-184.

[31] J-D. Yu, *Variation of the unit roots along the Dwork family of Calabi-Yau varieties*, preprint, 2007.

De Rham cohomology of varieties over fields of positive characteristic

Torsten Wedhorn

Institut für Mathematik, Universität Paderborn, Germany
e-mail: wedhorn@math.uni-paderborn.de

Abstract. These are the elaborated notes of two talks given at the Summer School in Göttingen on Higher-Dimensional Geometry over Finite Fields. We study the De Rham cohomology of smooth and proper varieties over fields of positive characteristic in case that the Hodge spectral sequence degenerates. The De Rham cohomology carries the structure of a so-called F-zip. We explain two classifications of F-zips, one stems from representation theory of algebras and the other one uses algebraic groups and their compactifications. We show how this second classification can be extended if the De Rham cohomology is endowed with a symplectic or a symmetric pairing. Throughout we illustrate the theory via the examples of (polarized) abelian varieties and (polarized) K3-surfaces.

Introduction

In this article the De Rham cohomology of certain varieties over fields of positive characteristic is studied. The text is aimed mainly at students and non-specialists having some familiarity with the usual techniques from algebraic geometry. The emphasis will be on precise definitions and examples, in particular abelian varieties and K3-surfaces. For the proofs often only a reference to the literature is given.

To put the theory for varieties over fields of positive characteristic into perspective let us first look at complex varieties: Let X be a smooth and proper scheme over the complex numbers and denote by X^{an} the associated compact complex manifold. The De Rham cohomology $H^{\bullet}_{\mathrm{DR}}(X/\mathbb{C})$ is a finite complex vector space. It carries several additional structures (recalled in more detail in (1.3)).

(I) As a complex vector space $H^{\bullet}_{\mathrm{DR}}(X/\mathbb{C})$ is isomorphic to the singular cohomology $H^{\bullet}(X^{\mathrm{an}}, \mathbb{C}) = H^{\bullet}(X^{\mathrm{an}}, \mathbb{Z}) \otimes_{\mathbb{Z}} \mathbb{C}$ and therefore has a $\mathbb{Z}$-rational structure.

(II) The Hodge spectral sequence degenerates and $H^{\bullet}_{\mathrm{DR}}(X/\mathbb{C})$ is therefore endowed with the Hodge filtration whose successive quotients are the spaces $H^a(X, \Omega^b_{X/\mathbb{C}})$.

(III) The complex conjugate of the Hodge filtration is opposed to the Hodge filtration and therefore defines a canonical splitting of the Hodge filtration.

In other words, $H^\bullet_{\mathrm{DR}}(X/\mathbb{C})$ is endowed with an integral Hodge structure in the sense of [De1].

To know the $\mathbb{R}$-Hodge structure on $H^\bullet_{\mathrm{DR}}(X/\mathbb{C})$ is equivalent to the knowledge of the Hodge numbers $h^{ab}(X) = \dim_{\mathbb{C}}(H^a(X, \Omega^b_{X/\mathbb{C}}))$. This is therefore a discrete invariant. Moreover, the Hodge numbers are locally constant in families, i.e., if $X \to S$ is a smooth proper morphism of schemes of finite type over $\mathbb{C}$ the map $S(\mathbb{C}) \ni s \mapsto h^{ab}(X_s)$ is locally constant.

The integral Hodge structure is a much finer invariant. It varies in families and the study of "moduli spaces" of integral Hodge structures leads to Griffiths' period domains (see e.g. [BP]).

Consider the example of two abelian varieties X and Y over $\mathbb{C}$. Then $H^1_{\mathrm{DR}}(X/\mathbb{C})$ and $H^1_{\mathrm{DR}}(Y/\mathbb{C})$ with their real Hodge structures are isomorphic if and only if $\dim(X) = \dim(Y)$. If $H^1_{\mathrm{DR}}(X/\mathbb{C})$ and $H^1_{\mathrm{DR}}(Y/\mathbb{C})$ with their integral Hodge structures are isomorphic, X and Y are isomorphic (called to global Torelli property for abelian varieties).

Now let X be a smooth and proper scheme over an algebraically closed field k of characteristic $p > 0$. Here we do not have any analogue of the property (I) of the De Rham cohomology in the complex case (see however below). Moreover the Hodge spectral sequence does not degenerate in general, hence the analogue of property (II) does not hold. But it turns out that still many interesting varieties (e.g., abelian varieties, smooth complete intersections in the projective space, K3-surfaces, toric varieties) have property (II) (see (1.5)) and therefore we will just assume in this article that the Hodge spectral sequence degenerates. The general case has not been studied much. A notable exception is Ogus' paper [Og4].

Now consider property (III). First of all there is of course no complex conjugation and therefore no verbatim analogue of property (III). But there is another filtration on the De Rham cohomology, namely the filtration given by the second spectral sequence of hypercohomology. It plays a somewhat analogue role to the complex conjugate of the Hodge spectral sequence for varieties over $\mathbb{C}$ and therefore we call it the conjugate spectral sequence (following Katz [Ka]). The initial terms of the Hodge and the conjugate spectral sequences are linked by the Cartier isomorphism and it follows from our assumption that the conjugate spectral sequence degenerates as well (see (1.6)). Therefore we have a "conjugate" filtration on $H^\bullet_{\mathrm{DR}}(X/k)$. But it is not true in general that Hodge and conjugate filtration are opposed to each other. In fact this characterizes ordinary varieties (see (1.9)). In general we obtain the structure of an F-zip on $H^\bullet_{\mathrm{DR}}(X/k)$, a notion first defined in [MW] and recalled in (1.7). There are two ways to classify F-zips, both explained in Chapter 3. It turns out that the isomorphism class of the De Rham cohomology endowed with its F-zip structure is still a discrete invariant but it is not locally constant in families.

Again we illustrate this with the example of abelian varieties. For an abelian variety X over k of dimension g there are 2^g possible F-zip structures on $H^1_{\mathrm{DR}}(X/k)$ and in fact all of them occur (see (5.1)). In particular for an elliptic curve E there are two cases and this is the distinction whether E is ordinary or supersingular.

We conclude this introductory part with some remarks on the nonexistent analogue of property (I). To get an "integral" structure the best candidate is the crystalline cohomology $H^\bullet_{\mathrm{cris}}(X/W)$ which is W-module where $W = W(k)$ is the ring of Witt vectors of k. The relative Frobenius on X induces a semi-linear map ϕ on $H^\bullet_{\mathrm{cris}}(X/W)$. If $H^\bullet_{\mathrm{cris}}(X/W)$ is a free W-module, there is a functorial isomorphism $\alpha\colon H^\bullet_{\mathrm{cris}}(X/W)\otimes_W k \xrightarrow{\sim} H^\bullet_{\mathrm{DR}}(X/k)$. In this case ϕ defines on $H^\bullet_{\mathrm{cris}}(X/W) \otimes_W k$ the structure of an F-zip such that α is an isomorphism of F-zips. Therefore we could consider $H^\bullet_{\mathrm{cris}}(X/W)$ as an "integral" version of $H^\bullet_{\mathrm{DR}}(X/k)$. Again this is a much finer invariant and it would be desirable to construct moduli spaces of such structures.

We give now a more detailed description of the content of this article. In the first chapter we recall the definition of the De Rham cohomology of an algebraic variety and of the two canonical spectral sequences converging to it. Then we consider the case of complex varieties before turning towards the actual focus of this article, the De Rham cohomology of varieties over fields of positive characteristic and its F-zip structure. In the second chapter we detail two examples: Abelian varieties (and – more generally – at level 1 truncated Barsotti-Tate groups) and K3-surfaces.

The third chapter is dedicated to the two classifications of F-zips. The first stems from the field of representation theory of algebras and is essentially due to Gelfand and Ponomarev [GP] and Crawley-Boevey [Cr]. The second uses the theory of algebraic groups and has been given in [MW]. It relates the classification problem to the study of a Frobenius-linear version of the wonderful compactification of the projective linear group. We link both classifications (3.5) and illustrate both of them by classifying at level 1 truncated Barsotti-Tate groups (3.7) and (3.8).

Often the De Rham cohomology carries additional structures like symplectic or symmetric pairings (e.g., induced by polarizations or Poincaré duality). Therefore we define in Chapter 4 symplectic and orthogonal F-zips. Now the essential ingredient for the second classification has an analogue for arbitrary reductive groups G. Specializing G to the symplectic and the orthogonal group we obtain a description of isomorphism classes of symplectic and orthogonal F-zips. In Chapter 5 this is applied to the study of the De Rham cohomology of polarized abelian varieties and of polarized K3-surfaces.

The sixth chapter deals with families of F-zips over arbitrary schemes S of characteristic p. It is shown that every F-zip over S defines a decomposition in locally closed subschemes, namely in the loci where the "isomorphism class of the F-zip is constant" (see (6.1) for two (equivalent) precise definitions). We conclude with glancing at this decomposition for the special case that S is the moduli space of principally polarized abelian varieties or the moduli space of polarized K3-surfaces.

In an appendix we recall the necessary notions and results about reductive groups which are used for the second classification.

ACKNOWLEDGEMENT: I am grateful to D. Blottiere for useful comments.

Notation: Fix a prime number $p > 0$. We will always denote by k a field which we assume to be perfect of characteristic p from (1.5) on. If not otherwise specified, X will be a smooth proper scheme over k which usually will satisfy a further condition (deg) from (1.5) on.

1. De Rham cohomology of varieties over fields of positive characteristic

(1.1) Hypercohomology.

Let $\mathcal{A}$ and $\mathcal{B}$ be two abelian categories and assume that $\mathcal{A}$ has sufficiently many injective objects. Let $T\colon \mathcal{A} \to \mathcal{B}$ be a left exact functor. Recall that an object M in $\mathcal{A}$ is called T-*acyclic* if all derived functors $\mathrm{R}^i T(M) = 0$ for $i > 0$. An object in $\mathcal{A}$ is injective if and only if it is T-acyclic for all abelian categories $\mathcal{B}$ and for all left exact functors $T\colon \mathcal{A} \to \mathcal{B}$.

Let K be a complex of objects in $\mathcal{A}$ which is bounded below. The hypercohomology $RT(K)$ can be defined as follows. Choose a quasi-isomorphism $K \to I$ to a complex of T-acyclic objects I^j in $\mathcal{A}$ (e.g., one can choose a quasi-isomorphism of K to a complex of injective objects). Then set

$$\mathbb{R}^j T(K) := H^j(T(I)).$$

This definition is independent of the choice of $K \to I$. Clearly, each quasi-isomorphism $K \to K'$ induces an isomorphism $\mathbb{R}^j T(K) \xrightarrow{\sim} \mathbb{R}^j T(K')$. Note that if $K[a]$ ($a \in \mathbb{Z}$) is the complex $K[a]^i = K^{a+i}$ and $d_{K[a]} = (-1)^a d_K$, we have $\mathbb{R}^j T(K[a]) = \mathbb{R}^{j+a} T(M)$. If M is an object in $\mathcal{A}$, considered as a complex concentrated in degree zero, we have $\mathbb{R}^j T(M) = \mathrm{R}^j T(M)$.

Assume that K is endowed with a descending filtration of subcomplexes

$$K \supset \cdots \supset \mathrm{Fil}^i(K) \supset \mathrm{Fil}^{i+1}(K) \supset \ldots$$

which is biregular (i.e., the filtration $\mathrm{Fil}^\bullet K^j$ induced on each component satisfies $\mathrm{Fil}^i K^j = K^j$ for i sufficiently small and $\mathrm{Fil}^i K^j = 0$ for i sufficiently big). In that case we obtain a spectral sequence which converges to the hypercohomology of K (see e.g. [De1] 1.4)

$$E_1^{ab} = \mathbb{R}^{a+b} T(\mathrm{Fil}^a(K)/\mathrm{Fil}^{a+1}(K)) \implies \mathbb{R}^{a+b} T(K). \qquad (1.1.1)$$

On every complex there are (at least) two natural filtrations. First one can consider the so-called "naive" filtration $(\sigma^{\geq i}(K))_i$ given by

$$\sigma^{\geq i}(K)^j = \begin{cases} 0, & j < i; \\ K^j, & j \geq i. \end{cases}$$

In that case, $\mathrm{Fil}^a(K)/\mathrm{Fil}^{a+1}(K) = K^a[-a]$ is the complex consisting of the object K^a in degree a, and the spectral sequence (1.1.1) becomes

$$E_1^{ab} = \mathrm{R}^b T(K^a) \implies \mathbb{R}^{a+b} T(K), \tag{1.1.2}$$

which is sometimes also called the *first spectral sequence of hypercohomology*

The second natural filtration is the (ascending) canonical filtration $(\tau_{\leq i}(K))_i$ given by

$$\tau_{\leq i}(K)^j = \begin{cases} K^j, & j < i; \\ \mathrm{Ker}(K^i \xrightarrow{\ d\ } K^{i+1}), & j = i; \\ 0, & j > i. \end{cases}$$

To get a descending filtration we set $\mathrm{Fil}^a K := \tau_{\leq -a}(K)$. In this case we have $\mathrm{Fil}^a(K)/\mathrm{Fil}^{a+1}(K) = H^{-a}K[a]$ and by (1.1.1) we obtain a spectral sequence $E_1^{ab} = \mathbb{R}^{2a+b}T(H^{-a}K) \implies \mathbb{R}^{a+b}T(K)$. Replacing E_r^{ab} by $E_{r+1}^{-b,a+2b}$ we get a spectral sequence

$$E_2^{ab} = \mathrm{R}^a T(H^b(K)) \implies \mathbb{R}^{a+b} T(K), \tag{1.1.3}$$

which is sometimes also called the *second spectral sequence of hypercohomology*.

(1.2) De Rham cohomology.

We apply the general remarks on hypercohomology to the de Rham complex. Let k be a field and let X be any scheme of finite type over k. Denote by $\Omega^\bullet_{X/k}$ the de Rham complex

$$0 \to \mathscr{O}_X \to \Omega^1_{X/k} \to \Omega^2_{X/k} \to \cdots$$

of X over k [EGA] IV, (16.6). As X is locally of finite type, $\Omega^1_{X/k}$ is a coherent $\mathscr{O}_X$-module [EGA] IV, (16.3.9). As X is also quasi-compact, $\Omega^j_{X/k} = \bigwedge^j \Omega^1_{X/k}$ is zero for $j \gg 0$. The de Rham complex is a complex in the abelian category $\mathcal{A}$ of sheaves on X with values in k-vector spaces (it is not a complex in the category of $\mathscr{O}_X$-modules as the differentials are not $\mathscr{O}_X$-linear). Denote by $T = \Gamma$ the functor of global sections from $\mathcal{A}$ to the category of k-vector spaces. As usual we write $H^i(X, M)$ and $\mathbb{H}^i(X, K)$ instead of $\mathrm{R}^i\Gamma(M)$ and $\mathbb{R}^i\Gamma(K)$ (where M is an object in $\mathcal{A}$ and K is a complex in $\mathcal{A}$).

For each integer $i \geq 0$ the *i-th De Rham cohomology of X over k* is defined as

$$H^i_{\mathrm{DR}}(X/k) := \mathbb{H}^i(X, \Omega^\bullet_{X/k}).$$

Then (1.1.2) and (1.1.3) give spectral sequences

$${}'E_1^{ab} = H^b(X, \Omega^a_{X/k}) \implies H^{a+b}_{\mathrm{DR}}(X/k), \tag{1.2.1}$$

$${}''E_2^{ab} = H^a(X, \mathscr{H}^b(\Omega^\bullet_{X/k})) \implies H^{a+b}_{\mathrm{DR}}(X/k). \tag{1.2.2}$$

Here we denote by $\mathscr{H}^b(\Omega^\bullet_{X/k})$ the b-th cohomology sheaf of the complex $\Omega^\bullet_{X/k}$.

The first spectral sequence is usually called the *Hodge spectral sequence*. As explained above, there exists an integer $A > 0$ such that $\Omega^a_{X/k} = 0$ for $a > A$. Therefore we have

$$'E_1^{ab} = 0, \qquad \text{for } a > A,\ b > \dim(X). \tag{1.2.3}$$

By definition of a spectral sequence, the limit terms $H_{\mathrm{DR}}^{a+b}(X/k)$ are endowed with a descending filtration $'\mathrm{Fil}^\bullet$ such that $'\mathrm{Fil}^a / '\mathrm{Fil}^{a+1} = 'E_\infty^{ab}$. As $'E_\infty^{ab}$ is a subquotient of $'E_1^{ab}$, this filtration is finite and we see that $H_{\mathrm{DR}}^n(X/k) = 0$ for all $n > A + \dim(X)$.

If X is smooth of pure dimension d over k, $\Omega^1_{X/k}$ is a locally free $\mathscr{O}_X$-module of rank d and therefore $\Omega^a_{X/k} = 0$ for $a > d$.

If X is proper over k, the initial terms in the Hodge spectral sequence are finite-dimensional vector spaces and therefore also their subquotients $'E_\infty^{ab}$ are finite-dimensional. Therefore for all $n \geq 0$ we have

$$\dim_k(H_{\mathrm{DR}}^n(X/k)) < \infty,$$

if X is a proper k-scheme.

(1.3) De Rham cohomology of complex varieties.

We first study the case that X is a smooth and proper scheme over the field of complex numbers $\mathbb{C}$. In that case we use the GAGA principle ([Se1] and [SGA1] Exp. XII): There is a compact complex manifold X^{an} and a morphism $i_X \colon X^{\mathrm{an}} \to X$ of locally ringed spaces such that for every complex analytic space Y and every morphism $\phi \colon Y \to X$ of locally ringed spaces there exists a unique morphism of complex analytic spaces $\psi \colon Y \to X^{\mathrm{an}}$ such that $i_X \circ \psi = \phi$. Serre's GAGA theorems tells us that $\mathscr{F} \mapsto \mathscr{F}^{\mathrm{an}} := i_X^* \mathscr{F}$ is an equivalance of categories between the category of coherent $\mathscr{O}_X$-modules and the category of coherent $\mathscr{O}_{X^{\mathrm{an}}}$-modules. Moreover, i_X^* induces isomorphisms of finite-dimensional $\mathbb{C}$-vector spaces

$$H^a(X, \mathscr{F}) \xrightarrow{\sim} H^a(X^{\mathrm{an}}, \mathscr{F}^{\mathrm{an}}).$$

We apply this to the coherent module $\Omega^b_{X/\mathbb{C}}$. In that case $(\Omega^b_{X/\mathbb{C}})^{\mathrm{an}}$ is the sheaf $\Omega^b_{X^{\mathrm{an}}}$ of holomorphic b-forms on X^{an}. Clearly i_X^* is compatible with the naive filtration on the De Rham complex. We get a homomorphism of the Hodge spectral sequences which by the GAGA principle is an isomorphism on its initial terms (in the analytic setting the Hodge spectral sequence is sometimes also called the Hodge-Fröhlicher spectral sequence). Therefore i_X^* induces an isomorphism of filtered $\mathbb{C}$-vector spaces

$$H_{\mathrm{DR}}^n(X/k) \xrightarrow{\sim} H_{\mathrm{DR}}^n(X^{\mathrm{an}}) := \mathbb{H}^n(X^{\mathrm{an}}, \Omega^\bullet_{X^{\mathrm{an}}}).$$

There is a rich history of the study of De Rham cohomology of komplex analytic spaces, both analytic and algebraic. We refer to [Voi], [Dm], and [Ill2] and the references given there for an extensive discussion. Here we recall just some well-known facts from complex geometry.

Degeneration of the Hodge spectral sequence.

First Fact over the complex numbers 1.1. *The Hodge spectral sequence* $'E_1^{ab} = H^b(X^{\mathrm{an}}, \Omega_{X^{\mathrm{an}}}^a) \implies H_{\mathrm{DR}}^{a+b}(X^{\mathrm{an}})$ *degenerates at* $'E_1$.

In particular the (descending) Hodge filtration $C^\bullet$ on $H_{\mathrm{DR}}^n(X^{\mathrm{an}})$ given by the Hodge spectral sequence has as graded pieces

$$C^i/C^{i+1} = H^{n-i}(X^{\mathrm{an}}, \Omega_{X^{\mathrm{an}}}^i) = H^{n-i}(X, \Omega_{X/\mathbb{C}}^i).$$

Poincaré lemma.
The second fact is the Poincaré lemma. For this we consider the natural embedding $\mathbb{C}_{X^{\mathrm{an}}} \hookrightarrow \mathscr{O}_{X^{\mathrm{an}}}$, where $\mathbb{C}_{X^{\mathrm{an}}}$ denotes the sheaf of locally constant complex valued functions on X^{an}. The Poincaré lemma says:

Second Fact over the complex numbers 1.2. *The homomorphism* $\mathbb{C}_{X^{\mathrm{an}}} \to \Omega_{X^{\mathrm{an}}}^\bullet$ *is a quasi-isomorphism.*

Hence the initial terms of $''E_2^{ab} = H^a(X^{\mathrm{an}}, \mathscr{H}^b(\Omega_{X^{\mathrm{an}}}^\bullet))$ are zero for $b > 0$, and therefore this spectral sequence is trivial. The quasi-isomorphism $\mathbb{C}_{X^{\mathrm{an}}} \to \Omega_{X^{\mathrm{an}}}^\bullet$ induces an isomorphism

$$H_{\mathrm{sing}}^n(X^{\mathrm{an}}, \mathbb{C}) = H^n(X^{\mathrm{an}}, \mathbb{C}_{X^{\mathrm{an}}}) \xrightarrow{\sim} H_{\mathrm{DR}}^n(X^{\mathrm{an}}) = H_{\mathrm{DR}}^n(X/\mathbb{C}),$$

where the left hand side denotes the singular cohomology and the first equality is a standard fact from topology (e.g., [Br] III, §1).

The conjugate filtration.
Note that $H_{\mathrm{sing}}^n(X^{\mathrm{an}}, \mathbb{C}) = H_{\mathrm{sing}}^n(X^{\mathrm{an}}, \mathbb{R}) \otimes_{\mathbb{R}} \mathbb{C}$, and the complex conjugation on $\mathbb{C}$ induces an $\mathbb{R}$-linear map $\sigma \otimes z \mapsto \sigma \otimes \bar{z}$ on $H_{\mathrm{sing}}^n(X^{\mathrm{an}}, \mathbb{C})$. If $W \subset H_{\mathrm{sing}}^n(X^{\mathrm{an}}, \mathbb{C})$ is any complex subspace, its image $\overline{W}$ under this map is again a complex subspace. We define an ascending filtration $D_\bullet$ on $H_{\mathrm{sing}}^n(X^{\mathrm{an}}, \mathbb{C})$ by setting

$$D_i := \overline{C^{n-i}}.$$

We call this filtration the conjugate filtation. Hodge theory now provides the following fact.

Third Fact over the complex numbers 1.3. *For all* $i \in \mathbb{Z}$ *we have*

$$D_{i-1} \oplus C^i = H_{\mathrm{DR}}^n(X^{\mathrm{an}}).$$

(1.4) Varieties over fields of positive characteristic.

Let now k be a field of positive characteristic p and let X be a k-scheme.

For any $\mathbb{F}_p$-scheme S we denote by $\mathrm{Frob}_S\colon S \to S$ the absolute Frobenius of S (i.e., Frob_S is the identity on the underlying topological spaces and sends a local section x of $\mathcal{O}_S$ to x^p). To shorten notations we denote by $\sigma\colon k \to k$ the Frobenius $a \mapsto a^p$ on k and also the Frobenius morphism $\sigma = \mathrm{Frob}_{\mathrm{Spec}(k)}$. Consider the diagram

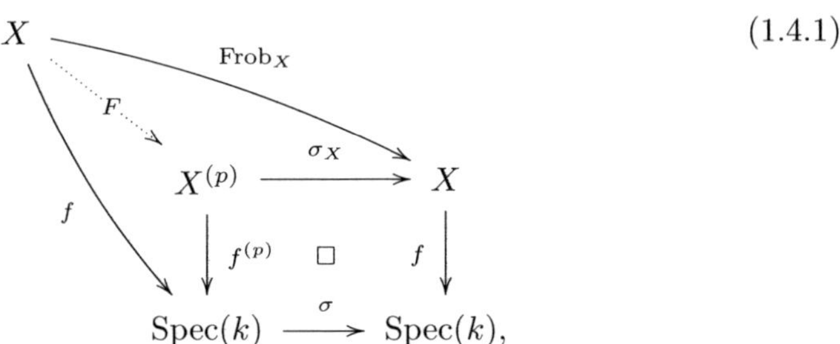

$$(1.4.1)$$

where $X^{(p)}$ is defined be the cartesian square and $F = F_{X/k}$ is the relative Frobenius of X over k, i.e., unique morphism making the above diagram commutative.

We describe this diagram locally: Assume that $X = \mathrm{Spec}(A)$ is affine. Via the choice of generators of A as a k-algebra, we can identify A with $k[\underline{X}]/(\underline{f})$ where $\underline{X} = (X_i)_{i \in I}$ is a tuple of indeterminates and $\underline{f} = (f_j)_{j \in J}$ is a tuple of polynomials in $k[\underline{X}]$. Then the diagram (1.4.1) is given by:

- $X^{(p)} = \mathrm{Spec}(A^{(p)})$ with $A^{(p)} = k[\underline{X}]/(f_j^{(p)}; j \in J)$, where for any polynomial $f = \sum_{\nu \in \mathbb{N}_0^{(I)}} a_\nu X^\nu \in k[\underline{X}]$ we set $f^{(p)} = \sum_{\nu \in \mathbb{N}_0^{(I)}} a_\nu^p X^\nu$.
- The morphism $\sigma_X^*\colon A \to A^{(p)}$ is induced by $k[\underline{X}] \to k[\underline{X}]$, $f \mapsto f^{(p)}$.
- The relative Frobenius $F^* = F_{X/k}^*$ is induced by the homomorphism of k-algebras $k[\underline{X}] \to k[\underline{X}]$ which sends an indeterminate X_i to X_i^p.

(1.5) De Rham cohomology of varieties over fields of positive characteristic.

From now on k will denote a perfect field of characteristic p, and $f\colon X \to \mathrm{Spec}(k)$ will be a smooth proper scheme over k.

We will study the analogies of the facts in (1.3).

First Fact in positive characteristic 1.4. *In general the Hodge spectral sequence* $'E_1^{ab} = H^b(X, \Omega_{X/k}^a) \implies H_{\mathrm{DR}}^{a+b}(X/k)$ *does not degenerate at* E_1.

Mumford [Mu1] has given examples of smooth projective surfaces such that the Hodge spectral sequence does not degenerate. To exclude such cases we will make from now on the following assumption.

(deg)   **Assumption**: The Hodge spectral sequence degenerates at E_1.

We list some examples where this assumption holds:

(1) Any abelian variety X over k satisfies (deg) ([Od] Prop. 5.1).

(2) Any smooth proper curve C over k satisfies (deg): To see this one can either use the previous example and deduce the degeneracy of the Hodge spectral sequence for C for that of the Hodge spectral sequence of the Jacobian of C, or one can use the result of Deligne and Illusie below.

(3) Any K3-surface X over k satisfies (deg): This follows from [De2] Prop. 2.2.

(4) Every smooth complete intersection in the projective space $\mathbb{P}_k^n$ satisfies (deg) as a scheme over k (see [SGA7] Exp. IX, Thm. 1.5).

(5) Every smooth proper toric variety satisfies (deg) (see [Bl]).

(6) Let X be a smooth proper k-scheme such that $F_*(\Omega_{X/k}^{\bullet})$ is decomposable (i.e., isomorphic in the derived category to a complex with zero differential). Then f satisfies (deg) by results of Deligne and Illusie, see [DI], Cor. 4.1.5. Moreover, this condition is satisfied if $\dim(X) \leq p$ and f admits a smooth lifting $\tilde{f} \colon \tilde{X} \to \operatorname{Spec}(W_2(k))$, where $W_2(k)$ is the ring of Witt vectors of length 2 of k. We refer to [Ill2] for an extensive discussion of this property.

(1.6) The Cartier isomorphism.

We now come to the analogue of the Poincaré lemma: Assume that X is any k-scheme (not necessarily smooth or proper). Again the differentials of the De Rham complex $\Omega_{X/k}^{\bullet}$ are in general not $\mathscr{O}_X$-linear. But it follows from the local description of the relative Frobenius in (1.4) that the differentials of $F_*(\Omega_{X/k}^{\bullet})$ are $\mathscr{O}_{X^{(p)}}$-linear (because $d(X_i^p f) = X_i^p d(f) + pd(X_i^{p-1})f = X_i^p d(f)$).

The Cartier isomorphism describes the cohomology sheaves of the complex $F_*(\Omega_{X/k}^{\bullet})$. It can be defined as follows. Let x be a local section of $\mathscr{O}_X$. Then $d\sigma_X^*(x)$ is a local section of $\Omega_{X^{(p)}/k}^1$ and there exists a unique homomorphism of $\mathscr{O}_{X^{(p)}}$-modules $\gamma \colon \Omega_{X^{(p)}/k}^1 \to \mathscr{H}^1(F_*(\Omega_{X/k}^{\bullet}))$ such that $\gamma(d\sigma_X^*(x))$ is the class of $x^{p-1}dx$.

Second Fact in positive characteristic 1.5. *For all $i \geq 0$ there exists a unique homomorphism of $\mathscr{O}_{X^{(p)}}$-modules*

$$\gamma^i \colon \Omega_{X^{(p)}/k}^i \to \mathscr{H}^i(F_*(\Omega_{X/k}^{\bullet}))$$

such that

$$\gamma^0(1) = 1,$$

$$\gamma^1 = \gamma,$$

$$\gamma^{i+j}(\omega \wedge \omega') = \gamma^i(\omega) \wedge \gamma^j(\omega').$$

Moreover, if X is smooth over k, γ^i is an isomorphism for all $i \geq 0$.

This family of isomorphisms is called the *(inverse) Cartier isomorphism*. It was initially defined by Cartier. The description here is due to Grothendieck and detailed in [Ka] 7. One can prove the fact that γ^i is an isomorphism along the following lines (see loc. cit. for the details): The assertion is clearly local on X. The formation of the De Rham complex and the definition of γ^i commutes with pull back via étale morphisms. Therefore we can assume that $X = \mathbb{A}_k^d$ is the d-dimensional affine space over k. The formation of the De Rham complex and the definition of γ^i also commutes with extension of scalars of the base field and hence we can assume that $k = \mathbb{F}_p$ and hence $\sigma = \mathrm{id}$. Finally using the Künneth formula we can assume that $X = \mathbb{A}_{\mathbb{F}_p}^1$. In this case it is an easy explicit calculation to check that γ^1 is an isomorphism.

Now we assume again that X is smooth and proper over k. Then $(\gamma^i)^{-1}$ induces for all $n \geq i$ a k-linear isomorphism

$$H^{n-i}(X^{(p)}, \Omega^i_{X^{(p)}/k}) \xrightarrow{\sim} H^{n-i}(X^{(p)}, \mathscr{H}^i(F_*(\Omega^\bullet_{X/k}))).$$

Using the natural isomorphisms

$$H^{n-i}(X^{(p)}, \Omega^i_{X^{(p)}/k}) \cong \sigma^* H^{n-i}(X, \Omega^i_{X/k}),$$

$$H^{n-i}(X^{(p)}, \mathscr{H}^i(F_*(\Omega^\bullet_{X/k})) \cong H^{n-i}(X^{(p)}, F_* \mathscr{H}^i(\Omega^\bullet_{X/k}))$$

$$\cong H^{n-i}(X, \mathscr{H}^i(\Omega^\bullet_{X/k})),$$

we obtain a k-linear isomorphism

$$\varphi_i \colon \sigma^*({}'E_1^{i,n-i}) = \sigma^* H^{n-i}(X, \Omega^i_{X/k})$$
$$\xrightarrow{\sim} H^{n-i}(X, \mathscr{H}^i(\Omega^\bullet_{X/k})) = {}''E_2^{n-i,i}. \tag{1.6.1}$$

Corollary 1.6. *If X satisfies the assumption* (deg), *the second spectral sequence* ${}''E_2^{ab} = H^a(X, \mathscr{H}^b(\Omega^\bullet_{X/k})) \implies H_{\mathrm{DR}}^{a+b}(X/k)$ *degenerates at* ${}''E_2$.

In characteristic p we will call this second spectral sequence the *conjugate spectral sequence* (following Katz [Ka]). Note that this is just a name. There is no complex conjugation here.

(1.7) F-zips.

We continue to assume that k is a perfect field and that X is a smooth proper k-scheme which satisfies (deg). We fix an integer $n \geq 0$ and set $M := H_{\mathrm{DR}}^n(X/k)$. We have seen that M carries the following structure.

(a) M is a finite dimensional k-vector space.

(b) The Hodge spectral sequence (1.2.1) provides a descending filtration ${}'\mathrm{Fil}^\bullet$ on M such that ${}'\mathrm{Fil}^i / {}'\mathrm{Fil}^{i+1} = H^{n-i}(X, \Omega^i_{X/k})$. If we define $C^i := \sigma^*({}'Fil^i)$, $C^\bullet$ is a descending filtration on $\sigma^*(M)$.

(c) The conjugate spectral sequence (1.2.2) provides a second descending filtration ${}''\mathrm{Fil}^\bullet$ on M. If we define $D_i := {}''\mathrm{Fil}^{n-i}$, $D_\bullet$ is an ascending filtration on M such that $D_i/D_{i-1} = H^{n-i}(X, \mathscr{H}^i(\Omega^\bullet_{X/k}))$.

(d) The isomorphism (1.6.1) is a k-linear isomorphism

$$\varphi_i \colon C^i/C^{i+1} \xrightarrow{\sim} D_i/D_{i-1}.$$

We now give an abstract definition for such a structure which we call an F-zip. This notion was first defined in [MW]. If M is any k-vector space, a *descending filtration* on M is by definition a family $(C^i)_{i\in\mathbb{Z}}$ of subspaces of M such that $C^i \supset C^{i+1}$ for all $i \in \mathbb{Z}$ and such that $\bigcup_i C^i = M$ and $\bigcap_i C^i = 0$. An *ascending filtration* is defined analogously.

Definition 1.7. *An F-zip over k is a tuple $\underline{M} = (M, C^\bullet, D_\bullet, \varphi_\bullet)$ such that*
(a) *M is a finite dimensional k-vector space,*
(b) *$C^\bullet$ is a descending filtration on $\sigma^*(M)$,*
(c) *$D_\bullet$ is an ascending filtration on M,*
(d) *$\varphi_i \colon C^i/C^{i+1} \xrightarrow{\sim} D_i/D_{i-1}$ is a k-linear isomorphism.*
Moreover, for an F-zip we call the function

$$\tau = \tau_{\underline{M}} \colon i \mapsto \dim_k(C^i/C^{i+1})$$

the filtration type of $C^\bullet$ or simply the type of $\underline{M}$. The elements in the support of τ, i.e., the $i \in \mathbb{Z}$ such that $\tau(i) \neq 0$, are called the weights of $\underline{M}$.

The notion of a morphism of F-zips over k is defined in an obvious way and we obtain an $\mathbb{F}_p$-linear category (which is not abelian; e.g., there do not exist kernels in general). If $\underline{M}$ is an F-zip over k and k' is a field extension of k, it is clear how to define the base change $\underline{M} \otimes_k k'$ which we will usually denote by $\underline{M}_{k'}$.

The Hodge spectral sequence, the conjugate spectral sequence, and the Cartier isomorphism are functorial in X, therefore we see:

Proposition 1.8. *Fix an integer $n \geq 0$. Then $X \mapsto H^n_{\mathrm{DR}}(X/k)$ defines a contravariant functor from the category of smooth proper k-schemes satisfying* (deg) *to the category of F-zips over k. Moreover, if τ is the type of the F-zip $H^n_{\mathrm{DR}}(X/k)$, the weights on $\underline{M}$ are contained in $\{0, \ldots, n\}$ and for $0 \leq i \leq n$ we have*

$$\tau(i) = \dim_k H^{n-i}(X, \Omega^i_{X/k}).$$

Example 1.9. For $d \in \mathbb{Z}$ the *Tate-F-zips $\underline{T}(d)$* is given as follows. The underlying vector space of $\underline{T}(d)$ is just k, we have $C^i = \sigma^*(k)$ for $i \leq d$ and $C^i = 0$ for $i > d$ which implies $D_i = 0$ for $i < d$ and $D_i = k$ for $i \geq d$. Finally $\varphi_d \colon k \to k$ is given by the Frobenius σ.

If X is a geometrically connected smooth proper k-scheme of dimension d, we have isomorphisms

$$H^0_{\mathrm{DR}}(X/k) \cong \underline{T}(0), \qquad H^{2d}_{\mathrm{DR}}(X/k) \cong \underline{T}(2d).$$

For the projective space of dimension d we have

$$H_{\mathrm{DR}}^i(\mathbb{P}_k^d/k) = \begin{cases} \underline{T}(i), & \text{if } 0 \le i \le 2d \text{ is even;} \\ 0, & \text{otherwise.} \end{cases}$$

(1.8) The category of F-zips.

For the proof of the following proposition we refer to [Wd3].

Proposition 1.10. *The category of F-zips over k has the following properties.*
(1) *There exist finite direct sums, and every F-zip $\underline{M}$ is a direct sum of a finite family of indecomposable F-zips (i.e., F-zips which are not isomorphic to a direct sum of nonzero F-zips). This finite family is uniquely determined by $\underline{M}$ up to order.*
(2) *There exists a natural $\otimes$-structure on this category and with this structure the category of F-zips is a rigid $\mathbb{F}_p$-linear $\otimes$-category.*
(3) *The Tate-F-zips are projective and injective in the following sense: For every F-zip $\underline{M}$ and for every surjective morphism $\underline{M} \to \underline{T}(d)$ (resp. for every injective morphism $\underline{T}(d) \to \underline{M}$) there exists a section (resp. a retraction).*

For every F-zip $\underline{M}$ and every $d \in \mathbb{Z}$ we set $\underline{M}(d) := \underline{M} \otimes \underline{T}(d)$. This is the F-zip obtained from $\underline{M}$ by shifting the indices of the filtrations $C^\bullet$ and $D_\bullet$ be d.

It follows by an easy descend argument from $\bar{k}$ to k that every F-zip $\underline{M}$ over k admits a unique (up to order) decomposition

$$\underline{M} = \underline{M}_{\mathrm{mw}} \oplus \bigoplus_{j=1}^{t} \underline{M}_{(d_j)}, \tag{1.8.1}$$

where $\underline{M}_{\mathrm{mw}}$ is an F-zip such that every indecomposable summand of $\underline{M}_{\mathrm{mw}}$ has mixed weights, i.e., more than a single weight, and where $(\underline{M}_{(d_j)})_{\bar{k}}$ is isomorphic to a direct sum of Tate-F-zips of weight d_j for pairwise different integers d_j.

(1.9) Ordinary varieties.

We now come to the analogue of the third fact 1.3. Let X be as above and fix $n \ge 0$. Consider the F-zip $(M := H_{\mathrm{DR}}^n(X/k), C^\bullet, D_\bullet, \varphi_\bullet)$ defined above.

Third Fact in positive characteristic 1.11. *In general, $\sigma^*(D_{i-1})$ and C^i are not complementary subspaces of $\sigma^*(M)$.*

This leads us to the following definition.

Definition 1.12. *An F-zip $\underline{M} = (M, C^\bullet, D_\bullet, \varphi_\bullet)$ is called* ordinary *if we have $\sigma^*(D_{i-1}) \oplus C^i = \sigma^*(M)$ for all $i \in \mathbb{Z}$.*

A smooth proper k-scheme satisfying (deg) *is called* ordinary, *if $H_{\mathrm{DR}}^n(X/k)$ with its natural F-zips structure is ordinary for all $n \ge 0$.*

In [IR] (4.12), Illusie and Raynaud define the notion of ordinariness for a smooth proper scheme X over k using the De Rham-Witt complex. It follows from loc. cit. (4.13) that the above definition is equivalent to the definition given by Illusie and Raynaud.

(1.10) Chern class of a line bundle.

Let X be any k-scheme. Let $\mathrm{dlog}\colon \mathscr{O}_X^\times \to \Omega^1_{X/k}$ the logarithmic derivation, given on local sections by $x \mapsto dx/x$. We consider dlog as a morphism

$$\mathrm{dlog}\colon \mathscr{O}_X^\times \to \sigma^{\geq 1}\Omega^\bullet_{X/k}[1]$$

of complexes of abelian sheaves on X. Taking the first hypercohomology we obtain a map

$$c_1'\colon \mathrm{Pic}(X) = H^1(X, \mathscr{O}_X^\times) \to \mathbb{H}^2(X, \sigma^{\geq 1}\Omega^\bullet_{X/k}).$$

The exact sequence

$$0 \longrightarrow \sigma^{\geq 1}\Omega^\bullet_{X/k} \longrightarrow \Omega^\bullet_{X/k} \longrightarrow \mathscr{O}_X \longrightarrow 0$$

provides a long exact sequence

$$\cdots \to \mathbb{H}^2(X, \sigma^{\geq 1}\Omega^\bullet_{X/k}) \overset{\alpha}{\longrightarrow} H^2_{\mathrm{DR}}(X/k) \overset{\beta}{\longrightarrow} H^2(X, \mathscr{O}_X) \to \cdots \qquad (1.10.1)$$

and we denote by $c_1\colon \mathrm{Pic}(X) \to H^2_{\mathrm{DR}}(X/k)$ the composition of c_1' with α. This is the Chern class map.

Now assume that X is a smooth proper k-scheme satisfying (deg) and endow $H^2_{\mathrm{DR}}(X/k)$ with its F-zip structure. As the Hodge spectral sequence degenerates, the long exact sequence (1.10.1) decomposes in short exact sequences and the map $\sigma^*(\beta)$ is nothing but the map $\sigma^*(H^2_{\mathrm{DR}}(X/k)) = C^0 \to C^0/C^1 = \sigma^*(H^2(X, \mathscr{O}_X))$. Therefore we see that for any line bundle $\mathscr{L}$ on X we have $\sigma^*(c_1(\mathscr{L})) \in C^1$.

(1.11) Smooth proper families and F-zips over schemes.

There is a relative version of the notion of an F-zip and an F-zip structure on the De Rham cohomology. Here we give only a brief outline and refer to [MW] for the details.

Let S be an arbitrary $\mathbb{F}_p$-scheme and let $f\colon X \to S$ be a smooth and proper morphism of schemes. We set $H^n_{\mathrm{DR}}(X/S) = \mathbb{R}^n f_*(\Omega^\bullet_{X/S})$. The general formalism explained in (1.1) (now applied to the functor f_*) provides two spectral sequences

$$'E_1^{ab} = R^b f_*(\Omega^a_{X/S}) \implies H^{a+b}_{\mathrm{DR}}(X/S),$$

$$''E_2^{ab} = R^a f_*(\mathscr{H}^b(\Omega^\bullet_{X/S})) \implies H^{a+b}_{\mathrm{DR}}(X/S),$$

called the *Hodge spectral sequence* and the *conjugate spectral sequence*. Again we have a Cartier isomorphism which induces an isomorphism of $\mathscr{O}_S$-modules

$$\mathrm{Frob}_S^*('E_1^{ba}) \xrightarrow{\sim} ''E_2^{ab}.$$

We say that f satisfies the condition (deg) if the following two conditions hold.

(a) The Hodge spectral sequence degenerates at E_1.

(b) The $\mathcal{O}_S$-modules $R^b f_*(\Omega^a_{X/S})$ are locally free of finite rank for all $a, b \geq 0$.

If f satisfies (deg), the formation of the Hodge spectral sequence (and in particular the formation of $R^b f_*(\Omega^a_{X/S})$ and of $H^n_{\mathrm{DR}}(X/S)$) commutes with base change $S' \to S$. It follows that the conjugate spectral sequence degenerates at E_2 and that its formation commutes with base change as well.

Examples for morphisms $f\colon X \to S$ satisfying (deg) are again abelian schemes, smooth proper (relative) curves, $K3$-surfaces, smooth complete intersections in $\mathbb{P}(\mathcal{E})$ for some vector bundle $\mathcal{E}$ on S.

Similar as in (1.7) we make the following definitions. If M is a locally free $\mathcal{O}_S$-module of finite rank, a descending filtration on M is a family $(C^i)_{i \in \mathbb{Z}}$ of $\mathcal{O}_S$-submodules of M which are locally direct summands such that $C^i \supset C^{i+1}$ for all $i \in \mathbb{Z}$ and such that $\bigcup_i C^i = M$ and $\bigcap_i C^i = 0$. An *ascending filtration* is defined analogously.

Definition 1.13. *An F-zip over S is a tuple $\underline{M} = (M, C^\bullet, D_\bullet, \varphi_\bullet)$ such that*

(a) *M is a locally free $\mathcal{O}_S$-module of finite rank,*

(b) *$C^\bullet$ is a descending filtration on $\mathrm{Frob}_S^*(M)$,*

(c) *$D_\bullet$ is an ascending filtration on M,*

(d) *$\varphi_i\colon C^i/C^{i+1} \xrightarrow{\sim} D_i/D_{i-1}$ is an $\mathcal{O}_S$-linear isomorphism.*

Again we can define the type *of $\underline{M}$ as $\mathbb{Z} \ni i \mapsto \mathrm{rk}_{\mathcal{O}_S}(C^i/C^{i+1})$, which is now a function with values in the set of locally constanst functions on S with values in $\mathbb{N}_0$.*

Note that this definition differs slightly from the definition given in [MW] although they are equivalent if S is the spectrum of a perfect ring.

The same construction of an F-zip structure on the De Rham cohomology for $S = \mathrm{Spec}(k)$ can be done for an arbitrary $\mathbb{F}_p$-scheme S and we obtain for each $n \geq 0$ a functor $(f\colon X \to S) \mapsto H^n_{\mathrm{DR}}(X/S)$ from the category of smooth proper S-schemes satisfying (deg) to the category of F-zips over S. This functor commutes with arbitrary base change $g\colon S' \to S$ (in the obvious sense).

2. Examples I

We continue to assume that k is a perfect field of characteristic $p > 0$.

(2.1) Abelian Varieties.

Let X be an abelian variety over k of dimension $g \geq 1$. Consider $H^1_{\mathrm{DR}}(X/k)$ with its F-zip structure. The $\otimes$-structure on the category of F-zips over k allows to form the F-zip $\bigwedge^n(H^1_{\mathrm{DR}}(X/k))$ for $n \geq 0$ (see [Wd3] for details) and the cup product defines an isomorphism of F-zips

$$\bigwedge^n (H^1_{\mathrm{DR}}(X/k) \xrightarrow{\ \sim\ } H^n_{\mathrm{DR}}(X/k).$$

In particular we see that the F-zip $H^1_{\mathrm{DR}}(X/k)$ already determines all F-zips $H^n_{\mathrm{DR}}(X/k)$ for $n \geq 0$.

Now let $H^1_{\mathrm{DR}}(X/k) = (M, C^\bullet, D_\bullet, \varphi_\bullet)$. Then M is a k-vector space of dimension $2g$ and the filtrations are given by

$$C^0 = \sigma^* M \supset C^1 = \sigma^* H^0(X, \Omega^1_{X/k}) \supset C^2 = 0,$$

$$D_{-1} = 0 \subset D_0 = H^1(X, \mathscr{H}^0(\Omega^\bullet_{X/k})) \subset D_1 = M$$

with $\dim(C^1) = \dim(D_0) = g$. The Cartier isomorphism induces two nontrivial isomorphisms

$$\varphi_0 \colon \sigma^*(M)/C^1 \xrightarrow{\ \sim\ } D_0,$$

$$\varphi_1 \colon C^1 \xrightarrow{\ \sim\ } M/D_0.$$

Moreover, the following assertions are equivalent.
(1) X is ordinary (i.e., the F-zips $H^n_{\mathrm{DR}}(X/k)$ are ordinary in the sense of Definition 1.12 for all $n \geq 0$).
(2) The F-zip $H^1_{\mathrm{DR}}(X/k)$ is ordinary (i.e., $\sigma^*(D_0) \oplus C^1 = \sigma^*(M)$).
(3) $X[p](\bar{k}) \cong (\mathbb{Z}/p\mathbb{Z})^g$.
(4) The Newton polygon of X has only slopes 0 and 1.

The last two (equivalent) conditions are often used to define the notion of ordinariness for an abelian variety. Here $X[p]$ denotes the kernel of the multiplication with p on X. In fact, the next three sections show that $X[p]$ determines the F-zip $H^1_{\mathrm{DR}}(X/k)$ (and vice versa).

(2.2) At level 1 truncated Barsotti-Tate groups.

Let X be an abelian variety over k. Then the kernel $G := X[p]$ of the multiplication with p is a so-called at level 1 truncated Barsotti-Tate group of height $2g$ and dimension g. We explain now what this means.

For any group scheme G over k, the relative Frobenius $F_{G/k} \colon G \to G^{(p)}$ defined in (1.4) is a homomorphism of group schemes. Moreover, if G is commutative, by [SGA3] Exp. VII$_A$, 4.3 there exists a natural homomorphism $V_{G/k} \colon G^{(p)} \to G$ of group schemes over k, called the *Verschiebung of G*, such that

$$V_{G/k} \circ F_{G/k} = p \operatorname{id}_G. \tag{2.2.1}$$

If $F_{G/k}$ is an epimorphism (i.e., if G is smooth over k), (2.2.1) determines $V_{G/k}$ uniquely.

An *at level 1 truncated Barsotti-Tate group over k* (or, shorter, a BT_1) is a finite commutative group scheme G over k which is annihilated by the multiplication with p such that the complex

$$G \xrightarrow{F_{G/k}} G^{(p)} \xrightarrow{V_{G/k}} G$$

of abelian fppf-sheaves is exact. The underlying scheme of G is of the form $\mathrm{Spec}(A)$ for a finite-dimensional k-algebra and $\dim_k(A)$ is a power of p and we define the *height h of G* by $p^h = \dim_k(A)$. As G is in general not smooth, the Lie algebra $\mathrm{Lie}(G)$ can be nonzero even though the underlying scheme of G is of dimension zero. We call $\dim(\mathrm{Lie}(G))$ the *dimension of G* and denote it by $\dim(G)$.

At level 1 truncated Barsotti-Tate groups can be described via Dieudonné theory. We call a *Dieudonné space over k* a triple (M, F, V), where M is a finite-dimensional k-vector space and $F\colon \sigma^*(M) \to M$ and $V\colon M \to \sigma^*(M)$ are k-linear maps such that $F \circ V = 0$, $V \circ F = 0$ and $\mathrm{Ker}(F) = \mathrm{Im}(F)$ (note that these conditions imply that $\mathrm{Ker}(V) = \mathrm{Im}(F)$). Then we have the following theorem (see e.g. [Dem] or [BBM] 3).

Theorem 2.1. *Crystalline Dieudonné theory provides a contravariant functor $\mathbb{D}$ from the category of BT_1 over k to the category of Dieudonné spaces over k, and this functor is an equivalence of categories. Moreover, if $\mathbb{D}(G) = (M, F, V)$, we have* $\mathrm{height}(G) = \dim_k(M)$ *and* $\dim(G) = \dim_k(M/\mathrm{Im}(F))$.

(2.3) Dieudonné spaces and F-Zips.

We can consider every Dieudonné space as a special case of an F-zip as follows. If (M, F, V) is a Dieudonné space, we set

$$C^i = \begin{cases} \sigma^*(M), & i \leq 0; \\ \mathrm{Ker}(F), & i = 1; \\ 0, & i \geq 2; \end{cases} \qquad D_i = \begin{cases} 0, & i \leq -1; \\ \mathrm{Ker}(V), & i = 0; \\ M, & i \geq 1. \end{cases}$$

Finally we define $\varphi_0\colon \sigma^*(M)/\mathrm{Ker}(F) \to \mathrm{Ker}(V) = \mathrm{Im}(F)$ as the isomorphism induced by F and $\varphi_1\colon \mathrm{Im}(V) = \mathrm{Ker}(F) \to M/\mathrm{Ker}(V)$ as the inverse of the isomorphism induced by V. Then $(M, C^\bullet, D_\bullet, \varphi_\bullet)$ is an F-zip.

This construction is clearly functorial in (M, F, V) and thus we obtain a functor from the category of Dieudonné spaces over k to the category of F-zips over k. Moreover, it is easy to see that this induces an equivalence of the category of Dieudonné spaces with the category of F-zips $\underline{M} = (M, C^\bullet, D_\bullet, \varphi_\bullet)$ such that $C^0 = \sigma^*(M)$ and $C^2 = 0$, i.e., the support of the type τ of $\underline{M}$ is contained in $\{0, 1\}$. We will call such F-zips *Dieudonné-zips*. From Theorem 2.1 we obtain.

Corollary 2.2. *Crystalline Dieudonné theory together with the construction above gives an equivalence $G \mapsto \underline{M}(G)$ between the category of BT_1 G over k and the category of Dieudonné zips $\underline{M}$ over k (considered as a full subcategory of the category of all F-zips over k). Via this equivalence we have* $\mathrm{height}(G) = \dim_k(M)$ *and* $\dim(G) = \tau(1)$, *where τ is the type of $\underline{M}(G)$.*

(2.4) Truncated Barsotti-Tate groups and de Rham cohomology of abelian varieties.

We now have constructed two functors from the category of abelian varieties over k to the category of F-zips over k. The first is the functor $X \mapsto H^1_{\mathrm{DR}}(X/k)$ constructed in (2.1). The second is the composition of the functors $X \mapsto X[p]$ and $G \mapsto \mathbb{D}(G)$ where we consider every Dieudonné space as an F-zip as explained in (2.3). We have the following result by [Od] Corollary 5.11.

Theorem 2.3. *The two functors $X \mapsto H^1_{\mathrm{DR}}(X/k)$ and $X \to \mathbb{D}(X[p])$ are isomorphic.*

Hence we see that $H^1_{\mathrm{DR}}(X/k)$ together with its F-zip structure and the p-torsion $X[p]$ determine each other. This is in effect only a special case ($n = 1$) of the more general result that for all $n \geq 1$ the p^n-torsion $X[p^n]$ and the crystalline cohomology $H^1_{\mathrm{cris}}(X/W_n(k))$ (which is a free $W_n(k)$-module M together with $W_n(k)$-linear maps $F \colon \sigma^*(M) \to M$ and $V \colon M \to \sigma^*(M)$ such that $F \circ V = p\,\mathrm{id}_M$ and $V \circ F = p\,\mathrm{id}_{\sigma^*(M)}$) determine each other (see loc. cit.).

(2.5) Decomposition of truncated Barsotti-Tate groups.

Let us first recall some facts on finite commutative group schemes (see e.g. [Dem]). To simplify the exposition, let us assume that k is algebraically closed.

Every finite commutative group scheme G over k admits a unique decomposition $G = G_p \times G^p$, where G_p is a group scheme killed by a power of p and where G^p is a finite étale group scheme of order prime to p.

Every such G_p admits itself a unique decomposition $G_p = G_{\mathrm{ét}} \times G_{\mathrm{mult}} \times G_{\mathrm{bi}}$, where

- $G_{\mathrm{ét}}$ is étale, i.e., $G_{\mathrm{ét}}$ is isomorphic to a product of constant group schemes associated to groups of the form $\mathbb{Z}/p^l\mathbb{Z}$.
- G_{mult} is of multiplicative type, i.e., G_{mult} is isomorphic to a product of group schemes of roots of unities of the form μ_{p^l}.
- G_{bi} is bi-infinitesimal, i.e., G and its Cartier dual $\underline{\mathrm{Hom}}(G, \mathbb{G}_m)$ are local as schemes and therefore the spectra of local Artinian k-algebras. Equivalently, there exist no nontrival homomorphisms of group schemes $\mathbb{Z}/p\mathbb{Z} \to G_{\mathrm{bi}}$ and $\mu_p \to G_{\mathrm{bi}}$.

In particular, every BT_1 admits such a decomposition $G = G_{\mathrm{ét}} \times G_{\mathrm{mult}} \times G_{\mathrm{bi}}$.

Let $\underline{M}(G)$ be the Dieudonné-zip over k corresponding to G via the equivalence in Corollary 2.2. Then (1.8.1) has the form $\underline{M} = \underline{T}(0)^{\oplus f} \oplus \underline{T}(1)^{\oplus e} \oplus \underline{M}_{\mathrm{mw}}$ and

$$\underline{M}(G_{\mathrm{ét}}) = \underline{T}(0)^{\oplus f}, \quad \underline{M}(G_{\mathrm{mult}}) = \underline{T}(1)^{\oplus e}, \quad \underline{M}(G_{\mathrm{bi}}) = \underline{M}_{\mathrm{mw}}.$$

(2.6) K3-surfaces.

We now study the De Rham cohomology of K3-surfaces. For this let k be a perfect field of characteristic $p > 0$. By definition, a K3-surface over k is a smooth proper connected k-scheme X of dimension 2 such that $H^1(X, \mathscr{O}_X) = 0$ and such that its canonical bundle $K_X = \Omega^2_{X/k}$ is trivial (i.e., isomorphic to $\mathcal{O}_X$). Note that every K3-surface is automatically projective.

Then it is well-known (see e.g. [De2]) that the Hodge spectral sequence degenerates at E_1 and that the Hodge numbers $(\dim_k(H^b(X, \Omega^a_{X/k})))_{0 \le a,b \le 2}$ are given by the matrix

$$\begin{pmatrix} 1 & 0 & 1 \\ 0 & 20 & 0 \\ 1 & 0 & 1 \end{pmatrix}. \tag{2.6.1}$$

Hence we see that

$$\dim_k(H^n_{\mathrm{DR}}(X/k)) = \begin{cases} 1, & n = 0, 4; \\ 22, & n = 2; \\ 0, & \text{otherwise.} \end{cases}$$

The F-zip structure on $M := H^2_{\mathrm{DR}}(X/k)$ is given by two filtrations $C^\bullet$ and $D_\bullet$ with

- $C^i = \sigma^*(M)$ for $i \le 0$, $\dim_k(C^1) = 21$, $\dim_k(C^2) = 1$, and $C^i = 0$ for $i \ge 3$,
- $D_i = 0$ for $i \le -1$, $\dim_k(D_0) = 1$, $\dim_k(D_1) = 21$, and $D_i = M$ for $i \ge 2$,
- two isomorphisms $\varphi_0: \sigma^*(M)/C^1 \overset{\sim}{\to} D_0$ and $\varphi_2: C^2 \overset{\sim}{\to} M/D_1$ of one-dimensional spaces, and an isomorphism $\varphi_1: C^1/C^2 \overset{\sim}{\to} D_1/D_0$ of spaces of dimension 20.

Moreover, the following assertions are equivalent.

(1) X is ordinary (i.e., the F-zips $H^n_{\mathrm{DR}}(X/k)$ are ordinary in the sense of Definition 1.12 for all $n \ge 0$).

(2) The F-zip $H^2_{\mathrm{DR}}(X/k)$ is ordinary.

(3) The Newton polygon of X has precisely the slopes 0, 1, and 2.

3. Classification of F-zips

In this section we will give two different classifications of F-zips over algebraically closed fields. The first classification (3.1) stems from representation theory of algebras and is a variation of a result of Crawley-Boevey [Cr] which goes back in this special case to work of Gelfand and Ponomarew [GP]. For the special case of Dieudonné zips this classification has obtained indepedently by Oort [Oo].

The second classification is shown by defining the variety X of rigidified F-zips and relate it to the wonderful compactification of the projective linear group by de Concini and Procesi ((3.2) and (3.3)). This classification was obtained in [MW] by relating X to a third variety and it was observed in [Wd2] how to relate this work to the wonderful compactification (see also below in (3.3)).

Notation: Let k be an algebraically closed field of characteristic p.

(3.1) Classification via representation theory.

Recall that by Proposition 1.10 every F-zip over k is a direct sum of indecomposable F-zips and that this decomposition is unique up to order. Argueing as in [Cr] it is possible to give the following description of isomorphism classes of indecomposable F-zips (see [Wd3]).

A *word* is by definition a formal sequences $\mathfrak{x} = x_1 x_2 \ldots x_n$ with $n \geq 0$ where $x_j \in \mathbb{Z}$. A *rotation* of $\mathfrak{x}$ is a word of the form $x_{j+1} \ldots x_n x_1 \ldots x_j$ for some j. The *product* $\mathfrak{x}\mathfrak{y}$ of two words is defined by placing them next to each other. We set

$$\mathscr{B} = \{\, \mathfrak{x} \text{ word} \mid \text{all nontrivial rotations of } \mathfrak{x} \text{ are different from } \mathfrak{x} \,\}/\sim$$

with $\mathfrak{x} \sim \mathfrak{y}$ if $\mathfrak{y}$ is a rotation of $\mathfrak{x}$. The elements of $\mathscr{B}$ are called *bands*.

To each representative $\mathfrak{x} = x_1 \ldots x_n$ of a band we associate an F-zip $\underline{M}(\mathfrak{x}) = (M, C^\bullet, D_\bullet, \varphi_\bullet)$ over $\mathbb{F}_p$ as follows. Let M_0 be an $\mathbb{F}_p$-vector spaces with basis $(e_1, \ldots, e_n)$. We set $x_0 := x_n$ and $x_{n+1} := x_1$ and define

$$C^i := \bigoplus_{\substack{1 \leq l \leq n, \\ x_l \geq i}} \mathbb{F}_p e_l,$$

$$D_i := \bigoplus_{\substack{1 \leq l \leq n, \\ x_l \ 1 \leq i}} \mathbb{F}_p e_l,$$

$$\varphi_i \colon C^i/C^{i+1} = \bigoplus_{\substack{1 \leq l \leq n, \\ x_l = i}} \mathbb{F}_p e_l$$

$$\xrightarrow{\ e_l \mapsto e_{l+1}\ } D_i/D_{i-1} = \bigoplus_{\substack{1 \leq l \leq n, \\ x_{l-1} = i}} \mathbb{F}_p e_l = \bigoplus_{\substack{1 \leq l \leq n, \\ x_l = i}} \mathbb{F}_p e_{l+1}.$$

$$(3.1.1)$$

Theorem 3.1. *The above construction* $[\mathfrak{x}] \mapsto \underline{M}(\mathfrak{x})_k$ *induces a bijection*

$$\mathscr{B} \leftrightarrow \left\{\begin{array}{c} \text{isomorphism classes of} \\ \text{indecomposable } F\text{-zips over } k \end{array}\right\}.$$

For example, the band corresponding to a Tate-F-zip $\underline{T}(d)$ is the word consisting of the single letter $\mathbf{d}$.

(3.2) The scheme of F-zips.

For the second classification we define the variety of rigidified F-zips. Here "rigidified" means that we fix the underlying module of the F-zip. Again let k be algebraically closed.

We first fix a filtration type, i.e., a function

$$\tau \colon \mathbb{Z} \to \mathbb{N}_0 \tag{3.2.1}$$

with finite support $i_1 > \cdots > i_r$. We set for $j = 1, \ldots, r$

$$n_j := \tau(i_j),$$

$$n := \sum_{j=1}^{r} n_j. \qquad (3.2.2)$$

We will now define an $\mathbb{F}_p$-scheme which parametrizes F-zips of type τ with fixed underlying module. Therefore let M_0 be a fixed $\mathbb{F}_p$-vector space of dimension n. Let X_τ be the $\mathbb{F}_p$-scheme whose S-valued points (for S an $\mathbb{F}_p$-scheme) are given by

$$X_\tau(S) = \{\, \underline{M} = (M, C^\bullet, D_\bullet, \varphi_\bullet) \mid \underline{M} \ F\text{-zip of type } \tau \text{ over } S$$

$$\text{with } M = M_0 \otimes_{\mathbb{F}_p} \mathscr{O}_S \,\}.$$

Let $G = \mathrm{GL}(M_0)$, which we consider as an algebraic group over $\mathbb{F}_p$. Then X_τ has a natural action of $G \times G$ defined as follows. If $\underline{M} = (M_0 \otimes_{\mathbb{F}_p} \mathscr{O}_S, C^\bullet, D_\bullet, \varphi_\bullet)$ is an S-valued of X_τ and $(h_1, h_2) \in G(S) \times G(S)$, we set

$$^{(h_1, h_2)}\underline{M} = (M_0 \otimes_{\mathbb{F}_p} \mathscr{O}_S, (F(h_2)(C^i))_i, (h_1(D_i))_i, (h_1 \varphi_i F(h_2)^{-1})_i), \qquad (3.2.3)$$

where $h_1 \varphi_i F(h_2)^{-1}$ denotes the composition

$$
\begin{array}{ccc}
F(h_2)(C^i)/F(h_2)(C^{i+1}) & & h_1(D_i)/h_1(D_{i-1}) \\[2mm]
\scriptstyle F(h_2)^{-1} \downarrow & & \uparrow \scriptstyle h_1 \\[2mm]
C^i/C^{i+1} & \xrightarrow{\ \varphi_i\ } & D_i/D_{i-1}.
\end{array}
$$

The $G \times G$-variety has the following properties ([MW] Lemma 5.1 and Lemma 4.2, [Wd2] (3.3)).

Proposition 3.2. *X_τ is a smooth connected $\mathbb{F}_p$-scheme of dimension equal to $\dim(G) = n^2$. The $G \times G$-action on X_τ is transitive.*

Embedding $G \hookrightarrow G \times G$ diagonally, the $G \times G$-action restricts to a G-action on X_τ. Every F-zip over k of type τ is isomorphic to an F-zip whose underlying k-vector space is equal to $M_0 \otimes_{\mathbb{F}_p} k$ and two such F-zips are isomorphic if and only if they are in the same $G(k)$-orbit. Hence:

Proposition 3.3. *Associating to each k-valued point of X_τ the isomorphism class of the corresponding F-zip defines a bijection*

$$\{G(k)\text{-orbits on } X_\tau(k)\} \longleftrightarrow \left\{ \begin{array}{c} \text{isomorphism classes of } F\text{-zips} \\ \text{over } k \text{ of type } \tau \end{array} \right\}.$$

To study the $G(k)$-orbits on $X_\tau(k)$ we relate X_τ to a part of the wonderful compactification $\bar{H}$ of $H := \mathrm{PGL}(M_0)$. We refer to the appendix for the necessary facts about reductive groups and the definition and some properties of $\bar{H}$.

(3.3) Classification via the wonderful compactification.

Recall that we fixed a filtration type τ (3.2.1) which gives a flag type $(n_1, \ldots, n_r)$ (3.2.2) in the sense of Example 7.2. Further we fixed an n-dimensional $\mathbb{F}_p$-vector space M_0, where $n = n_1 + \cdots + n_r$. Set $G = \mathrm{GL}(M_0)$ and $H = \mathrm{PGL}(M_0)$, considered as algebraic groups over $\mathbb{F}_p$.

To classify F-zips over the algebraically closed field k we proceed in two steps.

First step.
We construct a morphism

$$\gamma\colon X_\tau \longrightarrow \bar{H}_{(n_1,\ldots,n_r)},$$

where $\bar{H}_{(n_1,\ldots,n_r)}$ is the $H \times H$-orbit in $\bar{H}$ corresponding to $(n_1, \ldots, n_r)$ (Example 7.9). This morphism will be equivariant with respect to the canonical projection $G \times G \to H \times H$. Here the action of $G \times G$ on X_τ is the one defined in (3.2.3), and the action of $H \times H$ on $\bar{H}_{(n_1,\ldots,n_r)}$ is the composition of the homomorphism

$$F_2\colon H \times H \to H \times H, \quad (h_1, h_2) \mapsto (h_1, F(h_2)) \tag{3.3.1}$$

and the action defined in (7.7.2). In particular it will induce a map

$$[\gamma]\colon \{G(k)\text{-orbits of } X_\tau(k)\} \to \{H_F(k)\text{-orbits of } \bar{H}_{(n_1,\ldots,n_r)}(k)\},$$

where H_F is defined in (7.7.5). We will show that $[\gamma]$ is in fact a bijection.

Second step.
We invoke [MW] and [Wd2] to classify the H_F-orbits of $\overline{H_J^1}$ in the general setting of (7.7).

Making the first step.
We will define γ only on k-valued points although is is not difficult – albeit notationally a bit cumbersome – to define γ also on S-valued points where S is an arbitrary $\mathbb{F}_p$-scheme. For this we refer to [Wd3].

Set $M := M_0 \otimes_{\mathbb{F}_p} k$ and let $\underline{M} = (M, C^\bullet, D_\bullet, \varphi_\bullet)$ be a k-valued point of X_τ. Note that φ_i is an isomorphism of a zero-dimensional vector space except for $i \in \{i_1, \ldots, i_r\}$.

First we define a "projective version" PX_τ of X_τ. Set $T = \mathbb{G}_{m,\mathbb{F}_p}^r$. Then T acts on X_τ by

$$(C^\bullet, D_\bullet, (\varphi_{i_1}, \ldots, \varphi_{i_r})) \cdot (t_i)_{1 \leq i \leq r} = (C^\bullet, D_\bullet, (t_1^{-1}\varphi_{i_1}, \ldots, t_r^{-1}\varphi_{i_r})).$$

This is a free action. Embed $\mathbb{G}_{m,\mathbb{F}_p}$ diagonally into T. We set $PX_\tau = X_\tau / \mathbb{G}_{m,\mathbb{F}_p}$ and $\bar{X}_\tau = X_\tau / T$. We can define an action of $H \times H$ on PX_τ and $\bar{X}_\tau$ by the same recipe as the action of $G \times G$ on X_τ (3.2.3). The canonical morphisms

$$X_\tau \xrightarrow{\pi} PX_\tau \xrightarrow{\psi} \bar{X}_\tau$$

are clearly equivariant with respect to $G \times G \to H \times H$. It is easy to check that $\psi \circ \pi$ induces a bijection

$$\{G(k)\text{-orbits of } X_\tau(k)\} \xrightarrow{\sim} \{H(k)\text{-orbits of } \bar{X}_\tau(k)\}.$$

For a k-valued point $\underline{M} \in PX_\tau(k)$ denote by $H_{\underline{M}} \subset H \times H$ the stabilizer of $\underline{M}$.

Lemma 3.4. *$H_{\underline{M}}$ is a smooth algebraic group with $\dim(H_{\underline{M}}) = \dim(H)$.*

Proof. The smoothness of $H_{\underline{M}}$ is easy to check with the infinitesimal lifting criterion. From the transitivity of the $G \times G$-action on X_τ (Proposition 3.2) it follows that the $H \times H$-action on PX_τ (and hence on $\bar{X}_\tau$) is transitive as well. As $\dim(PX_\tau) = \dim(X_\tau) - 1 = \dim(G) - 1 = \dim(H)$, we see that $\dim(H_{\underline{M}}) = \dim(H)$. $\qquad\square$

We define an F_2-equivariant morphism (F_2 was defined in (3.3.1))

$$\rho \colon PX_\tau \to \mathcal{V}_H, \qquad \underline{M} \mapsto \mathrm{Lie}(H_{\underline{M}}),$$

where $\mathcal{V}$ is defined in (7.7). Note that for $\underline{M}_1, \underline{M}_2 \in PX_\tau(k)$ which have the same image in $\bar{X}_\tau$, the stabilizers $H_{\underline{M}_1}$ and $H_{\underline{M}_2}$ are equal. Therefore ρ factorizes over $PX_\tau \twoheadrightarrow \bar{X}_\tau$. Now it follows from [Lu] 12.3 (specialized to the case PGL_n) that ρ induces an F_2-equivariant isomorphism

$$\iota \colon \bar{X}_\tau \xrightarrow{\sim} \overline{H}_{(n_1,\ldots,n_r)}.$$

This concludes the first step.

Making the second step.
In this second step we will describe the H-orbits of $\overline{H_J^1}$ in the general setting of (7.7) and for an arbitrary subset $J \subset I$ where (W, I) is the Weyl system of H.

To describe this classification we have to introduce the following notation. Denote by W_J the subgroup of W generated by J. In any left coset $W_J w$ ($w \in W$) there exists a (necessarily unique) element $^J w$ such that $\ell(^J w) \le \ell(w')$ for all $w' \in W_J w$. We denote by $^J W \subset W$ the set of these representatives of minimal length of $W_J \backslash W$. We further set

$$d_J := \dim(H) - \max\{ \ell(w) \mid w \in {}^J W \} = \dim(H) - \dim(\mathrm{Par}_J),$$

where Par_J is the scheme of parabolic subgroups of H of type J (7.4) and (7.5) (the equality being a general fact from the theory of root systems of a reductive group).

Theorem 3.5. *There exists a natural bijection*

$$^J W \leftrightarrow \{ H_F(k)\text{-orbits of } \overline{H_J^1} \},$$

$$w \mapsto {}^w \overline{H_J^1} \tag{3.3.2}$$

such that

$$\dim({}^w \overline{H_J^1}) = \ell(w) + d_J. \tag{3.3.3}$$

Proof. Note that via the commutative diagram (7.7.4) it suffices to describe the H_F-orbits of $\overline{H}_J$. Now by [Wd2], $\overline{H}_J$ is $H \times H$-equivariantly isomorphic to another $H \times H$-scheme $Z_J = Z_{H,J}$ over $\mathbb{F}_p$. By [MW] Theorem 4.11 there is a natural bijection between the H-orbits of Z_J and $^J W$. $\square$

We will not explain in which sense (3.3.2) is natural but refer to [MW] for this. Instead we will make the Theorem 3.5 explicit in the case $H = H^1 = \mathrm{PGL}_n$ in the next section.

(3.4) Classification of F-zips.

By Example 7.4, for $H = \mathrm{PGL}_n$ we can identify

$$W = S_n, \qquad I = \{ \tau_i \mid i = 1, \ldots, n-1 \},$$

and $J \subset I$ corresponds to a flag type, i.e., to a tuple $(n_1, \ldots, n_r)$ with $\sum n_j = n$. By the definition of this correspondence in (7.4.2) and (7.5.2) we have

$$W_J = S_{n_1} \times S_{n_2} \times \ldots S_{n_r},$$

which is considered as a subgroup of $W = S_n$ in the obvious way.

For $j = 0, \ldots, r$ we set

$$m_j := n_1 + \cdots + n_j.$$

From the description of the length of a permutation in (7.5.2) it follows easily that

$$^J W = \{ w \in S_n \mid w^{-1}(m_{j-1}+1) < \cdots < w^{-1}(m_j),$$

$$\text{for all } j = 1, \ldots, r \}. \tag{3.4.1}$$

Let w_0 be the longest elements in S_n, i.e., $w_0(i) = n+1-i$ for all $1 \leq i \leq n$. Let $w_{0,J}$ be the longest element in W_J and set $w_0^J := w_0 w_{0,J}$, i.e.,

$$w_0^J(i) = n - m_j - m_{j-1} + i, \qquad \text{for } m_{j-1} < i \leq m_j.$$

To shorten notations we set

$$w^\circ := w w_0^J \tag{3.4.2}$$

for $w \in S_n$.

We will now associate to each $w \in S_n$ an F-zip ${}^w\underline{M} = (M_0, C^\bullet, D_\bullet, \varphi_\bullet)$ over $\mathbb{F}_p$ (where M_0 is our fixed n-dimensional $\mathbb{F}_p$-vector space), such that

$$ {}^J W \ni w \mapsto ({}^w\underline{M})_k \in X_\tau(k) $$

defines the bijection

$$ {}^J W \leftrightarrow \{H_F(k)\text{-orbits of } \overline{H}_J\} \leftrightarrow \{\mathrm{GL}_n(k)\text{-orbits of } X_\tau(k)\} $$

constructed in the two steps of (3.3).

Note that the Frobenius on $\mathbb{F}_p$ is the identity and therefore $\sigma^*(M_0) = M_0$. As $C^\bullet$ will be a filtration of type τ (3.2.1), we have $C^i = C^{i+1}$ and $D_i = D_{i-1}$ for $i \notin \{i_1, \ldots, i_r\}$. Therefore it suffices to define C^{i_j}, D_{i_j}, and φ_{i_j} for $j = 1, \ldots, r$. Recall the convention $i_1 > \cdots > i_r$.

We choose a basis $(e_1, \ldots, e_n)$ of M_0. Then ${}^w\underline{M}$ is defined by

$$
\begin{aligned}
C^{i_j} &:= \bigoplus_{l=1}^{m_j} \mathbb{F}_p e_l, \\[4pt]
D_{i_j} &:= \bigoplus_{l=m_{j-1}+1}^{n} \mathbb{F}_p e_{w^\circ(l)}, \\[4pt]
\varphi_{i_j} : C^{i_j}/C^{i_j+1} &= \bigoplus_{l=m_{j-1}+1}^{m_j} \mathbb{F}_p e_l \\[4pt]
&\xrightarrow{\;e_l \mapsto e_{w^\circ(l)}\;} D_{i_j}/D_{i_j-1} = \bigoplus_{l=m_{j-1}+1}^{m_j} \mathbb{F}_p e_{w^\circ(l)}.
\end{aligned}
\tag{3.4.3}
$$

Note that ${}^w\underline{M}$ also depends on τ. Altogether we obtain:

Theorem 3.6. *Let τ be a filtration type and let $(n_1, \ldots, n_r)$ be the associated flag type. Associating to $w \in S_n$ the F-zip ${}^w\underline{M}$ induces a bijection*

$$ {}^J W \leftrightarrow \left\{ \begin{array}{c} \text{isomorphism classes of } F\text{-zips} \\ \text{of type } \tau \text{ over } k \end{array} \right\}, $$

where ${}^J W$ is described in (3.4.1).

Of course, ${}^J W$ depends only on τ and we will often write ${}^\tau W$ instead of ${}^J W$.

Description of ordinary F-zips.
It follows from this explicit description that we have the following criterion for an F-zip to be ordinary in the sense of Definition 1.12.

Corollary 3.7. *Let $\underline{M}$ be an F-zip. The following assertions are equivalent.*
(1) $\underline{M}$ *is ordinary.*
(2) $\underline{M}$ *is a direct sum of Tate-F-zips (Example 1.9).*
(3) *The permutation corresponding to $\underline{M}$ via the bijection in Theorem 3.6 is the element of maximal length in ^{J}W, i.e., the element $w_{0,J}w_{0}$.*

Decomposition in Tate-F-zips and mixed weight F-zips.
Let $\underline{M}$ be an F-zip over k with set of weights $\{i_1 > \cdots > i_r\}$ and flag type $(n_1, \ldots, n_r)$. As k is algebraically closed now, the decomposition (1.8.1) of an F-zip $\underline{M}$ over k has the form

$$\underline{M} = \underline{M}_{\mathrm{mw}} \oplus \bigoplus_{j=1}^{r} \underline{T}(i_j)^{\oplus t_j}$$

with $0 \le t_j \le n_j$.

We use the notations of (3.4.1). Let $w \in {}^{J}W$ be the permutation corresponding to the isomorphism class of $\underline{M}$. It follows from the explicit description in Theorem 3.6 that

$$t_j = \#\{\, i \in \{m_{j-1}+1, \ldots, m_j\} \mid w^{\circ}(i) = i\,\}.$$

(3.5) Connecting the two classifications.

We now relate the representation theoretic classification of F-zips via bands in (3.1) with the group theoretic classification of (3.3). For this we again fix a type τ as in (3.2.1) and associate to each $w \in W = S_n$ a collection $\mathcal{X}$ of words $\mathfrak{x}_1, \ldots, \mathfrak{x}_c$ such that for $w \in {}^{\tau}W$ we have

$$^{w}(\underline{M}) \cong \underline{M}(\mathcal{X}) := \bigoplus_{a=1}^{c} \underline{M}(\mathfrak{x}_a),$$

where $\underline{M}(\mathfrak{x}_a)$ is the F-zip defined in (3.1). We first define a map

$$\iota \colon \{1, \ldots, n\} \to \{i_1, \ldots, i_r\} \subset \mathbb{Z},$$

$$\iota(\nu) = i_j, \qquad \text{for} \quad \sum_{h=1}^{j-1} n_h < \nu \le \sum_{h=1}^{j} n_h.$$

Now write w° as a product of cycles

$$\big[\nu_{11}, \nu_{12}, \ldots, \nu_{1d_1}\big] \cdots \big[\nu_{c1}, \nu_{c2}, \ldots, \nu_{cd_c}\big]$$

and then associate words $\mathfrak{x}_1, \ldots, \mathfrak{x}_c$ of integers by

$$\mathfrak{x}_a = \iota(\nu_{a1})\iota(\nu_{a2}) \cdots \iota(\nu_{ad_a}).$$

From this comparison we also see:

Proposition 3.8. *Let $\underline{M}$ be the F-zip of type τ corresponding to a permutation $w \in {}^{\tau}W$ via Theorem 3.6. Then the number of indecomposable summands of $\underline{M}$ is the same as the number of cycles of w°. In particular, $\underline{M}$ is indecomposable if and only if w° consists of one cycle only.*

(3.6) A simple example.

We fix an integer $n \geq 1$ and define a filtration type by

$$\tau \colon \mathbb{Z} \to \mathbb{N}_0, \qquad \tau(i) = \begin{cases} 1, & i = 0; \\ n - 1, & i = 1; \\ 0, & \text{otherwise.} \end{cases}$$

In particular F-zips of type τ will be given by Dieudonné zips (2.3). We have $i_1 = 1 > i_2 = 0$, $J = (n-1, 1)$ and ${}^{\tau}W = {}^{J}W$ consists of the permutations w_i for $i = 1, \ldots, n$ where

$$w_i := \begin{pmatrix} 1 & \ldots & i-1 & i & i+1 & \ldots & n \\ 1 & \ldots & i-1 & n & i & \ldots & n-1 \end{pmatrix}.$$

We have $w_1^{\circ} = \mathrm{id}$ and for $i \geq 2$:

$$w_i^{\circ} = \begin{pmatrix} 1 & \ldots & i-2 & i-1 & i & \ldots & n-1 & n \\ 2 & \ldots & i-1 & n & i & \ldots & n-1 & 1 \end{pmatrix}.$$

The corresponding F-zip ${}^{w_i}\underline{M}$ is given by $C^1 = \bigoplus_{l=1}^{n-1} \mathbb{F}_p e_l$, $D_0 = \mathbb{F}_p e_{w_i^{\circ}(n)}$ (i.e., $D_0 = \mathbb{F}_p e_1$ for $i > 1$ and $D_0 = \mathbb{F}_p e_n$ for $i = 1$, and $\varphi_0 \colon e_n \mapsto e_{n-1}$, $\varphi_1 \colon e_l \mapsto e_{w_i^{\circ}(l)}$ for $l = 1, \ldots, n-1$.

Writing w_i° as product of cycles we get

$$w_i^{\circ} = [1\ 2\ \ldots\ i-1\ n][i][i+1]\ldots[n-1].$$

The corresponding collection of words is then given by

$$\mathcal{X} : \mathfrak{x}_1 = \mathbf{1}^{i-1}\mathbf{0}, \ \mathfrak{x}_2 = \mathbf{1}, \ \mathfrak{x}_3 = \mathbf{1}, \ \ldots, \ \mathfrak{x}_{n-i+1} = \mathbf{1}.$$

Example of elliptic curves.
The special case $n = 2$ classifies the F-zip structure on $\underline{M}(E) = H^1_{\mathrm{DR}}(E/k)$ where E is an elliptic curve over the algebraically closed field k (2.1). We see that up to isomorphism there are only two F-zips of this type, isomorphic to ${}^{w}M_k$ for $w \in S_2$ or to $\underline{M}(\mathcal{X})_k$ where $\mathcal{X} = \mathbf{1}, \mathbf{0}$ or $\mathcal{X} = \mathbf{10}$. Then E is supersingular if and only if $\underline{M}(E) \cong {}^{w}M_k$ for $w = \mathrm{id}$ if and only if $\underline{M}(E) \cong \underline{M}(\mathcal{X})_k$ for $\mathcal{X} = \mathbf{10}$. Otherwise, E is ordinary.

(3.7) Classification of at level 1 truncated Barsotti-Tate groups.

Fix integers $0 \leq d \leq h$. To classify BT_1 of height h and dimension d over an algebraically clsed field k it suffices by Corollary 2.2 to parametrize the isomorphism classes of Dieudonné zips $\underline{M}$ of type τ, where $\tau(0) = h - d$ and $\tau(1) = d$. In this case we have (with the notations of (3.4))

$$^\tau W = \{\, w \in S_h \mid w^{-1}(1) < \cdots < w^{-1}(d), w^{-1}(d+1) < \cdots < w^{-1}(h) \,\}.$$

Therefore the two classifications give the following theorem.

Theorem 3.9. *The constructions in (3.4.3) and in (3.1.1) give bijections*

$$\left\{ \begin{array}{c} \textit{isomorphism classes of at level 1 truncated} \\ \textit{Barsotti-Tate groups over k of height h and dimension d} \end{array} \right\}$$

$$\leftrightarrow \; ^\tau W$$

$$\leftrightarrow \; \left\{ \begin{array}{c} \textit{unordered tuples of bands } \mathfrak{x}_1, \ldots, \mathfrak{x}_t \textit{ in the letters 0 and 1} \\ \textit{where 0 appears } h - d \textit{ times and 1 appears d times} \end{array} \right\}$$

(3.8) The p-rank and a-number.

Let k be algebraically closed. For every abelian variety X over k often two numerical invariants are considered, the p-rank and the a-number. Both depend only on the p-torsion $X[p]$ which is a BT_1 whose classification was obtained in (3.7). We will explain how to read off these invariants from the two classifications.

Let G be a BT_1 of height h and dimension d and let $\underline{M}$ be the corresponding Dieudonné zip (Corollary 2.2).

The p-rank.
As $G(k)$ is a finite abelian group killed by p, we have $G(k) \cong (\mathbb{Z}/p\mathbb{Z})^f$. This integer $f \geq 0$ is called the *p-rank of G*. By (2.5) the p-rank is simply the multiplicity of $\underline{T}(0)$ in $\underline{M}$. For the classifications this means the following.

If $\mathfrak{x}_1, \ldots, \mathfrak{x}_r$ are the bands corresponding to the indecompsable summands of the F-zip $\underline{M}$ via the classification (3.1), the p-rank of f equals the number of bands $\mathfrak{x}_i$ which consist of the single letter **0**.

If $w \in {}^\tau W$ is the permutation corresponding to the isomorphism class of $\underline{M}$ via Theorem 3.9, the p-rank of X is the number of $d + 1 \leq i \leq h$ such that $w^\circ(i) = i$.

The a-number.
Let α_p the finite commutative group scheme over k such that

$$\alpha_p(R) = \{\, a \in R \mid a^p = 0 \,\} \subset (R, +)$$

for all k-algebras R. This is a group scheme killed by p which is not a BT_1 (Frobenius and Verschiebung are both zero on α_p).

Consider the abelian group $\mathrm{Hom}(\alpha_p, G)$ of homomorphisms of group schemes over k. The obvious k-vector space structure on $\alpha_p(R)$ makes $\mathrm{Hom}(\alpha_p, G)$ into a k-vector space. Its dimension is called the *a-number of G* and denoted by $a(G)$.

Let (M, F, V) be the Dieudonné space associated to G via the equivalence of categories in Theorem 2.1 and $\underline{M}$ the corresponding Dieudonné zip. Then

$$a(\underline{M}) := a(G) = \dim_k(\mathrm{Ker}(F) \cap \sigma^*(\mathrm{Ker}(V))) = \dim_k(C^1 \cap \sigma^*(D_0)).$$

In terms of bands or permutations this number can be described as follows.

If $G = G_1 \times G_2$, we have $a(G) = a(G_1) + a(G_2)$ and hence it suffices to describe the a-number of an indecomposable Dieudonné zip $\underline{M}$ whose isomorphism class is therefore given by a single band $\mathfrak{x} = x_0 x_1 \ldots x_{n-1}$ where $x_i \in \{\mathbf{0}, \mathbf{1}\}$ for $i \in \mathbb{Z}/n\mathbb{Z}$. Then the description of the F-zip corresponding to $\mathfrak{x}$ in (3.1.1) shows

$$a(\underline{M}) = \{\, i \in \mathbb{Z}/n\mathbb{Z} \mid x_i = \mathbf{0}, \ x_{i+1} = \mathbf{1} \,\}.$$

We now assume that $d \leq h - d$ (otherwise we replace G by its Cartier dual, i.e., $\underline{M}$ by $\underline{M}^\vee(1)$, see Example 4.2 below). If $w \in {}^\tau W$ be the corresponding permutation (3.7), it follows immediately from the explicit description in (3.4.3) that

$$a(\underline{M}) = \#\{\, 1 \leq i \leq d \mid w^\circ(i) > d \,\}.$$

Via both descriptions (and also via the definition) we see that the a-number of an abelian variety X is zero if and only if the p-rank of X is equal to $\dim(X)$, i.e., if and only if X is ordinary.

4. F-zips with additional structures

Often the De Rham cohomology of an algebraic variety comes equipped with additional structures, e.g. pairings induced by Poincaré duality or by a polarization. Here we just consider two examples.

Notation: Let k be a perfect field of characteristic p.

(4.1) Symplectic and orthogonal F-Zips.

Let $d \in \mathbb{Z}$ and let S be an $\mathbb{F}_p$-scheme.

Definition 4.1. *A d-symplectic F-zip over S is an F-zip $\underline{M}$ over S together with a perfect pairing $\psi\colon \bigwedge^2(\underline{M}) \to \underline{T}(d)$, where $\underline{T}(d)$ is the Tate-zip of weight d ($\underline{T}(d)$ was in Example 1.9 only defined over a field, but the definition of $\underline{T}(d)$ over an arbitrary $\mathbb{F}_p$-scheme should be clear).*

Similarly, a d-orthogonal F-zip over S is an F-zip $\underline{M}$ over S together with a perfect pairing $\psi\colon \mathrm{Sym}^2(\underline{M}) \to \underline{T}(d)$. When we speak of orthogonal F-zips, we will always assume that $p \neq 2$.

For a more explicit definition of symplectic and orthogonal F-zips see also [MW] 6.1.

For a vector space M we denote by $M^\vee$ its dual. Note first that the existence of a perfect pairing $\psi\colon \underline{M} \otimes \underline{M} \to \underline{T}(d)$ implies in particular the existence of an isomorphism $M \xrightarrow{\sim} M^\vee$ which induces a k-linear isomorphism

$$C^i \xrightarrow{\sim} (C^{d+1-i})^\perp \subset \sigma^*(M)^\vee = \sigma^*(M^\vee).$$

This implies

$$\tau(i) = \tau(d - i) \tag{4.1.1}$$

for the type τ of $\underline{M}$.

We now fix a function $\tau\colon \mathbb{Z} \to \mathbb{N}_0$ with finite support $i_1 > \cdots > i_r$ and which satisfies (4.1.1). Again we set $n_j := \tau(i_j)$ and $n := n_1 + \cdots + n_r$. Then (4.1.1) implies

$$i_j + i_{r+1-j} = d, \tag{4.1.2}$$

$$n_j = n_{r+1-j} \tag{4.1.3}$$

for all $j = 1, \ldots, r$.

Example 4.2. Let G be a BT_1 over a perfect field k and let $\underline{M}$ be the corresponding Dieudonné zip (2.3). Then the Cartier dual $\underline{\mathrm{Hom}}(G, \mathbb{G}_m)$ of G is again a BT_1 and the corresponding Dieudonné zip is $\underline{M}^\vee(1)$.

Let k be algebraically closed and let $\underline{M}$ be an indecomposable F-zip over k with associated band $\mathfrak{x} = x_1 x_2 \ldots x_n$ (via (3.1)). Then $\underline{M}^\vee$ is given by the band $\mathfrak{x}^\vee := (-x_1)(-x_2) \ldots (-x_n)$ and $\underline{M}(d)$ (for $d \in \mathbb{Z}$) corresponds to the band $\mathfrak{x}(d) := (x_1 + d)(x_2 + d) \ldots (x_n + d)$.

Fix $d \in \mathbb{Z}$. We will now classify d-symplectic and d-orthogonal F-zips over k, where k is algebraically closed. For d-orthogonal F-zips we first have to define a further invariant, namely its discriminant.

(4.2) Classification of symplectic F-zips.

We fix an n-dimensional $\mathbb{F}_p$-vector space $M_0 \neq 0$ together with symplectic (i.e. alternating and non-degenerate) pairing $\psi_0\colon \bigwedge^2(M_0) \to \mathbb{F}_p$. This implies that n is even, say $n = 2g$. Set $G = \mathrm{Sp}(M_0, \psi_0)$, considered as an algebraic group over $\mathbb{F}_p$.

As in (3.2) we define an $\mathbb{F}_p$-scheme $X_\tau^{d\text{-symp}}$ whose S-valued points are those d-symplectic F-zips $((M, C^\bullet, D_\bullet, \varphi_\bullet), \psi)$ of type τ over S such that $(M, \psi) = (M_0, \psi_0) \otimes_{\mathbb{F}_p} \mathcal{O}_S$. The same definition as (3.2.3) defines a $G \times G$-action on $X_\tau^{d\text{-symp}}$ such that, if we embed $G \hookrightarrow G \times G$ diagonally, the $G(k)$-orbits of $X_\tau^{d\text{-symp}}(k)$ correspond to isomorphism classes of d-symplectic F-zips of type τ over k.

The center of G is μ_2, the group of second roots of unity, acting by multiplication of (M_0, ψ_0). Set $H := \mathrm{PSp}(M_0, \psi_0) = G/\mu_2$. The $H \times H$-orbits of the wonderful compactification $\overline{H}$ of H are given by conjugacy classes $\mathcal{C}_H$ of parabolic subgroups of H (7.8). As for PGL_n, $\mathcal{C}_H$ is in bijection to flag types $(n_1, \ldots, n_r)$ such that $n_1 + \cdots + n_r = 2g$ and such that (4.1.3) holds. In particular, our fixed type τ gives such a conjugacy class $(n_1, \ldots, n_r)$ of parabolic subgroups. Let (W, I) be the Weyl group of H together with its set of simple reflections and let J be the subset of I which corresponds to our fixed conjugacy class $(n_1, \ldots, n_r)$ via (7.5.1). Let $\overline{H}_J$ be the $H \times H$-orbit corresponding to J.

The same arguments as in (3.3) show that there exists a morphism

$$X_\tau^{d\text{-symp}} \to \overline{H}_J,$$

which induces a bijection from the set of $G(k)$-orbits of $X_\tau^{d\text{-symp}}(k)$ to the set of $H_F(k)$-orbits of $\overline{H}_J(k)$.

Now we can use the general Theorem 3.5 and obtain:

Theorem 4.3. *There exists a natural bijection*

$$\left\{ \begin{array}{c} \textit{isomorphism classes of d-symplectic} \\ \textit{F-zips over k of type } \tau \end{array} \right\} \leftrightarrow {}^J W$$

We will make ${}^J W$ more explicit in the symplectic case. The Weyl group W and its set of simple reflections I can again be described by the relative position of flags as in the case of PGL_n. In addition there is a symmetry condition imposed by the symplectic from. More precisely we can identify

$$\begin{aligned} W &= \{\, w \in S_{2g} \mid w(i) + w(2g + 1 - i) = 2g + 1 \text{ for all } i \,\}, \\ I &= \{\, s_i := \tau_i \tau_{2g-i} \mid i = 1, \ldots, g - 1 \,\} \cup \{ s_g := \tau_g \}. \end{aligned} \qquad (4.2.1)$$

We set $m_j := n_1 + \cdots + n_j$. The flag type $(n_1, \ldots, n_r)$ corresponds via (7.5.1) to the subset

$$J = \{\, s_i \mid i \notin \{m_1, m_2, \ldots, m_r\} \,\}$$

of I. Then ${}^J W$ consists of those elements in W such that

$$w^{-1}(m_{i-1} + 1) < \cdots < w^{-1}(m_i), \qquad \text{for } i = 1, \ldots, r.$$

(4.3) The discriminant of an orthogonal F-zip.

Fix $d \in \mathbb{Z}$ and let $(\underline{M}, \psi)$ be a d-orthogonal F-zip over an $\mathbb{F}_p$-scheme S. We will define the *discriminant of* $(\underline{M}, \psi)$, denoted by $\mathrm{disc}(\underline{M}, \psi)$, in two steps.

Assume first that M is locally free of rank one. Let n be the weight of $\underline{M}$. For every étale morphism $\iota \colon U \to S$ consider the set

$$B(U) := \{\, e \in \Gamma(U, \iota^*(M)) \mid e \text{ generates } M,\ \varphi_n(\sigma^*(e)) = e \,\}.$$

The constant étale sheaf $\mathbb{F}_p^{\times}$ acts by multiplication on B and this makes B into an $\mathbb{F}_p^{\times}$-torsor. For any $e \in B(U)$ we have $\psi(e \otimes e) \in \mathbb{F}_p^{\times}$ and its image in $\mathbb{F}_p^{\times}/(\mathbb{F}_p^{\times})^2$ is independent of e. This defines a global section $\mathrm{disc}(\underline{M}, \psi)$ of the constant étale sheaf $\mathbb{F}_p^{\times}/(\mathbb{F}_p^{\times})^2$.

Now let $(\underline{M}, \psi)$ be arbitrary. Then the maximal exterior power $\det(\underline{M})$ of $\underline{M}$ inherits a non-degenerate symmetric bilinear form $\det(\psi)$ by

$$\det(\psi)(m_1 \wedge \cdots \wedge m_h, m_1' \wedge \cdots \wedge m_h') = \sum_{\pi \in S_h} \mathrm{sgn}(\pi) \prod_{i=1}^{h} \psi(m_i, m_{\pi(i)}').$$

But now the underlying $\mathscr{O}_S$-module of $\det(\underline{M})$ is locally free of rank one and we set $\mathrm{disc}(\underline{M}, \psi) = \mathrm{disc}(\det(\underline{M}), \det(\psi))$.

If $(\underline{M}_1, \psi_1)$ and $(\underline{M}_2, \psi_2)$ are two d-orthogonal F-zips over S, we can build their orthogonal sum $(\underline{M}, \psi)$ in the obvious way. Then clearly we have $\det(\underline{M}, \psi) = \det(\underline{M}_1, \psi_1) \otimes \det(\underline{M}_2, \psi_2)$ and we see that

$$\mathrm{disc}(\underline{M}, \psi) = \mathrm{disc}(\underline{M}_1, \psi_1) \cdot \mathrm{disc}(\underline{M}_2, \psi_2). \tag{4.3.1}$$

(4.4) Classification of orthogonal F-zips.

To deal with the orhogonal case, recall that we assume that $p \neq 2$. Again let k be algebraically closed.

For an arbitrary field K, by a *symmetric space over K* we mean a finite-dimensional K-vector space M together with a non-degenerate symmetric pairing $\psi \colon \mathrm{Sym}^2(M) \to K$. The *dimension of (M, ψ)* is by definition the dimension of the underlying vector space M and $\mathrm{disc}(\psi) \in K^{\times}/(K^{\times})^2$ denotes the discriminant of (M, ψ).

We now fix an integer $n \geq 0$ and $\delta \in \mathbb{F}_p^{\times}/(\mathbb{F}_p^{\times})^2$. Any two symmetric spaces over $\mathbb{F}_p$ of dimension n and discriminant δ are isomorphic (see e.g. [Se2] chap. IV, 1.7). Let (M_0, ψ_0) such a symmetric space. Set $G = \mathrm{SO}(M_0, \psi_0)$ considered as an algebraic group over $\mathbb{F}_p$. Of course, G depends up to isomorphism only on n and δ.

To simplify the exposition we will from now on assume that $n = 2m + 1$ is odd. In this case the isomorphism class of G depends only on n and not on δ (any algebraic group over $\mathbb{F}_p$ which is isomorphic to G after base change to an algebraically closed field (i.e., any $\mathbb{F}_p$-form of G), is already isomorphic to G: as $\mathbb{F}_p$ has cohomological dimension 1, G has no nontrivial inner forms, and as the Dynkin diagram of G (which is of type (B_m)) has no automorphisms, G has no outer automorphisms and therefore every form of G is inner). Moreover G is adjoint in this case (Example 7.5). Both statements are not true if n is even.

Again we define an $\mathbb{F}_p$-scheme $X_{\tau,\delta}^{d\text{-orth}}$ whose S-valued points are those d−orthogonal F-zips $((M, C^{\bullet}, D_{\bullet}, \varphi_{\bullet}), \psi)$ of type τ over S whose discriminant in the sense of (4.3) is δ and such that $(M, \psi) = (M_0, \psi_0) \otimes_{\mathbb{F}_p} \mathscr{O}_S$. The same definition as (3.2.3) defines a $G \times G$-action on $X_{\tau,\delta}^{d\text{-orth}}$ such that, if we embed $G \hookrightarrow G \times G$ diagonally, the $G(k)$-orbits of $X_{\tau,\delta}^{d\text{-orth}}(k)$ correspond to isomorphism classes of d-orthogonal F-zips of type τ and discriminant δ over k.

As in the symplectic case, the $G \times G$-orbits of the wonderful compactification $\overline{G}$ of G correspond to flag types $(n_1, \ldots, n_r)$ such that $n_1 + \cdots + n_r = 2m + 1$ and such that (4.1.3) holds. (Note that the analogue statement would not hold if n were even!) In particular, our fixed type τ gives such a conjugacy class $(n_1, \ldots, n_r)$ of parabolic subgroups. Let (W, I) be the Weyl group of G together with its set of simple reflections and let J be the subset of I which corresponds to our fixed conjugacy class $(n_1, \ldots, n_r)$ via (7.5.1). Let $\overline{G}_J$ be the $G \times G$-orbit corresponding to J.

As in the linear and the symplectic case we obtain by the general classification of G_F-orbits on $\overline{G}_J$ (Theorem 3.5):

Theorem 4.4. *There exists a natural bijection*

$$\left\{ \begin{array}{c} \text{isomorphism classes of } d\text{-orthogonal} \\ F\text{-zips over } k \text{ of type } \tau \end{array} \right\} \leftrightarrow {}^J W$$

The Weyl group W, its set of simple reflections I, and ${}^J W$ have the same description as in the symplectic case (4.2) (replacing g by m).

5. Examples II

As examples we will study abelian varieties endowed with a polarization of degree prime to p and polarized K3-surfaces.

In this section we assume that k is a perfect field of characteristic p.

(5.1) Prime-to-p-polarized abelian varieties.

We now apply the general results of (4.2) to describe the isomorphism classes of the first De-Rham cohomology of abelian varieties endowed with a polarization of degree prime to p.

Let X be an abelian variety of dimension $g \geq 1$ over a perfect field k. We denote by $X^\vee$ the dual abelian variety and by $\xi: X \to X^\vee$ a polarization of degree d [Mu2]. We assume that d is prime to p.

Via the canonical perfect pairing $H^1_{\mathrm{DR}}(X/k) \otimes H^1_{\mathrm{DR}}(X^\vee/k) \to k$ (see e.g. [BBM] 5.1), we can identify $H^1_{\mathrm{DR}}(X^\vee/k)$ with $H^1_{\mathrm{DR}}(X/k)^\vee$ and ξ induces a k-linear map $\beta^{-1}: H^1_{\mathrm{DR}}(X/k)^\vee \to H^1_{\mathrm{DR}}(X/k)$ which is an isomorphism because d is prime to p. We denote by $\psi: H^1_{\mathrm{DR}}(X/k) \otimes H^1_{\mathrm{DR}}(X/k) \to k$ the pairing associated to $\beta := (\beta^{-1})^{-1}$. As ξ is a polarization, we have $\xi^\vee = \xi$ and this implies $\beta^\vee = -\beta$ [BBM] 5.2.13. Therefore ψ is skew-symmetric and hence alternating if $p > 2$. For $p = 2$, ψ is also alternating, but one has to work harder to see this (e.g. [dJ] 2).

By (2.1), the type τ of the F-zip $H^1_{\mathrm{DR}}(X/k)$ is given by $\tau(i) = g$ for $i = 0, 1$ and $\tau(i) = 0$ otherwise. The filtration $C^\bullet$ and $D_\bullet$ are given by $C^1 = \sigma^* H^0(X, \Omega^1_{X/k})$ and $D_0 = H^1(X, \mathcal{H}^0(\Omega^\bullet_{X/k}))$.

Proposition 5.1. *$H_{\mathrm{DR}}^1(X/k)$ together with its F-zip structure and the alternating pairing ψ is a 1-symplectic F-zip of type τ.*

Proof. It remains to show that the k-linear isomorphism β is an isomorphisms of F-zips $H_{\mathrm{DR}}^1(X/k) \xrightarrow{\sim} H_{\mathrm{DR}}^1(X/k)^\vee(1)$. This is well-known in the language of Dieudonné spaces (2.3). Therefore define

$$F \colon \sigma^*(H_{\mathrm{DR}}^1(X/k)) \twoheadrightarrow \sigma^*(H_{\mathrm{DR}}^1(X/k))/C^1 \xrightarrow{\varphi_0} D_0 \hookrightarrow H_{\mathrm{DR}}^1(X/k),$$

$$V \colon H_{\mathrm{DR}}^1(X/k) \twoheadrightarrow H_{\mathrm{DR}}^1(X/k)/D_0 \xrightarrow{\varphi_1^{-1}} C^1 \hookrightarrow \sigma^*(H_{\mathrm{DR}}^1(X/k)).$$

Going through the definition of the dual and the Tate twist of an F-zip, the assertion of the lemma is now equivalent to the equality

$$\psi(F(m_1), m_2) = \sigma^*(\psi)(m_1, V(m_2)) \tag{5.1.1}$$

for all $m_1 \in \sigma^*(H_{\mathrm{DR}}^1(X/k))$ and $m_2 \in H_{\mathrm{DR}}^1(X/k)$. Note that (5.1.1) implies (but is not equivalent to) that $H^0(X, \Omega_{X/k}^1)$ and $H^1(X, \mathscr{H}^0(\Omega_{X/k}^\bullet))$ are Lagrangian subspaces of $H_{\mathrm{DR}}^1(X/k)$. $\qquad\square$

Altogether we have associated to every abelian variety endowed with a polarization of degree prime to p a 1-symplectic F-zip $(H_{\mathrm{DR}}^1(X/k), \psi)$ of type τ. This construction is (contravariantly) functorial in (X, ξ). Conversely for fixed d prime to p, every 1-symplectic F-zip of type τ is isomorphic to $(H_{\mathrm{DR}}^1(X/k), \psi)$ for some abelian variety endowed with a polarization of degree d. For $p > 2$ this is a special case ($F = \mathbb{Q}$) of [Wd1] (7.2).

The flag type associated to τ is given by (g, g) and the subset J of simple reflections in the Weyl group is $\{s_1, \ldots, s_{g-1}\}$ (with the notations of (4.2)).

Now assume that k is algebraically closed. By Theorem 4.3 isomorphism classes of 1-symplectic F-zips over k of type τ are given by $^J W$, where

$$^J W = \{\, w \in W \mid w^{-1}(1) < \cdots < w^{-1}(g) \,\}. \tag{5.1.2}$$

Note that every $w \in {}^J W$ automatically satisfies $w^{-1}(g+1) < \cdots < w^{-1}(2g)$.

(5.2) Connection with Oort's classification.

Oort has given a different classification of symplectic Dieudonné spaces (i.e., Dieudonné spaces together with a symplectic pairing satisfying (5.1.1)) in [Oo]. We now explain the connection between the two classifications.

We can describe the set $^J W$ in (5.1.2) also as follows. For $w \in {}^J W$ we define a map

$$\varphi_w \colon \{0, \ldots, g\} \to \{1, \ldots, g\},$$
$$\varphi_w(i) = \#\{\, a \subset \{1, \ldots, g\} \mid w(a) > i \,\}. \tag{5.2.1}$$

We obtain a map $\varphi_w \colon \{1, \ldots, g\} \to \mathbb{N}_0$ such that $\varphi_w(j) \le \varphi_w(j+1) \le \varphi_w(j)+1$, i.e., an elementary sequence in the language of [Oo]. Then the isomorphism class of the 1-symplectic F-zip of type τ corresponding to $w \in {}^J W$ is given by the isomorphism class of the symplectic Dieudonné space corresponding to φ_w by the construction in [Oo] (9.1).

(5.3) Polarized $K3$-surfaces.

We now recall first the notion of a polarization on a K3-surface and some facts about K3-surfaces in general. Then we show that any K3-surface with polarization of degree prime to p gives rise to a 2-orthogonal F-zip of type τ, where

$$\tau(i) = \begin{cases} 1, & i = 0, 2; \\ 19, & i = 1; \\ 0, & \text{otherwise.} \end{cases} \tag{5.3.1}$$

These are classified by the general classification of orthogonal F-zips in (4.4) which we make more explicit in this case.

Some facts about K3-surfaces.
Let X be a K3-surface over the perfect field k (2.6). A *polarization of X* is an ample line bundle $\mathcal{L}$. Its self-intersection $(\mathcal{L}, \mathcal{L})$ is called its *degree*. Hirzebruch-Riemann-Roch tells us

$$\chi(\mathcal{L}) = \chi(\mathcal{O}_X) + \frac{1}{2}\big((\mathcal{L}, \mathcal{L}) - (\mathcal{L}, K_X)\big).$$

By (2.6.1) we have $\chi(\mathcal{O}_X) = 2$ and by the definition of a K3-surface, K_X is trivial. We obtain $\chi(\mathcal{L}) = 2 + (\mathcal{L}, \mathcal{L})/2$ and in particular the degree is an even number.

The Picard functor $\mathrm{Pic}_{X/k}$ is representable by a group scheme locally of finite type and its Lie algebra is isomorphic to $H^1(X, \mathcal{O}_X)$ by [BLR] chapter 8. As $H^1(X, \mathcal{O}_X) = 0$ by (2.6.1), $\mathrm{Pic}_{X/k}$ is étale and its identity component $\mathrm{Pic}^0_{X/k}$ is trivial. Therefore $\mathrm{Pic}(X)$ is equal to the Néron-Severi group $\mathrm{NS}(X) := \mathrm{Pic}(X)/\mathrm{Pic}^0(X)$ and this is a finitely generated free abelian group.

Next we recall some facts about cup-product and chern classes for K3-surfaces. As a reference we use [De2], where crystalline versions of these constructions are explained. But the crystalline cohomology $H^i_{\mathrm{cris}}(X/W(k))$ is a free $W(k)$-module, where $W(k)$ denotes the ring of Witt vectors of k by loc. cit. Prop. 1.1. Therefore we have

$$H^i_{\mathrm{cris}}(X/W(k)) \otimes_{W(k)} k = H^i_{\mathrm{DR}}(X/k)$$

for all $i \geq 0$ (see e.g. [Ill1] 1.3(b)) and all our constructions are just "(mod p)-versions" of the crystalline theory explained in [De2].

Composing the cup-product with the trace map we get a non-degenerate symmetric pairing

$$\psi \colon H^2_{\mathrm{DR}}(X/k) \otimes H^2_{\mathrm{DR}}(X/k) \xrightarrow{\cup} H^4_{\mathrm{DR}}(X/k) \xrightarrow{\mathrm{tr}} k. \tag{5.3.2}$$

For all $\mathscr{L} \in \mathrm{Pic}(X)$ we have

$$(\mathscr{L}, \mathscr{L}) = \psi(c_1(\mathscr{L}), c_1(\mathscr{L})), \qquad (5.3.3)$$

where $c_1 \colon \mathrm{Pic}(X) = \mathrm{NS}(X) \to H^2_{\mathrm{DR}}(X/k)$ is the Chern class map (1.10). Here we consider the integer on the left hand side as an element of k.

Orthogonal F-zips associated to polarized K3-surfaces.
Let X be a K3-surface over k. We have seen in (2.6) that $H^2_{\mathrm{DR}}(X/k)$ carries the structure of an F-zip of type $(1, 20, 1) \in \mathbb{N}_0^{\{0,1,2\}} \subset \mathbb{N}_0^{\mathbb{Z}}$. The following lemma follows from work of Ogus (namely that in the language of [Og2] §1 the second crystalline cohomology is a K3-crystal).

Lemma 5.2.   *The pairing ψ (5.3.2) defines on $H^2_{\mathrm{DR}}(X/k)$ the structure of a 2-orthogonal F-zip of type $\tilde{\tau}$.*

From now on we fix an integer $d \geq 1$. Let $\mathscr{L}$ be a polarization on X of degree $2d$. We assume that p does not divide $2d$. Then (5.3.3) implies that the subspace $\langle c_1(\mathscr{L}) \rangle$ of $H^2_{\mathrm{DR}}(X/k)$ generated by $c_1(\mathscr{L})$ is nonzero and that the restriction of ψ to $\langle c_1(\mathscr{L}) \rangle$ is non-degenerate. We denote by $H_{\mathrm{prim}}(X, \mathscr{L})$ the orthogonal complement of $\langle c_1(\mathscr{L}) \rangle$. It follows from [Og1] that $\langle c_1(\mathscr{L}) \rangle$ carries an induced F-zip structure and that there exists an isomorphism $\langle c_1(\mathscr{L}) \rangle \xrightarrow{\sim} \underline{T}(1)$ such that the restriction of ψ to $\langle c_1(\mathscr{L}) \rangle$ corresponds via this isomorphism to the natural pairing $\underline{T}(1) \otimes \underline{T}(1) \to \underline{T}(2)$. Therefore also the orthogonal complement $H_{\mathrm{prim}}(X, \mathscr{L})$ carries the structure of a 2-orthogonal F-zip over k.

Altogether we associated to each polarized K3-surface $(X, \mathscr{L})$ of degree $2d$ a 2-orthogonal F-zip $(H_{\mathrm{prim}}(X, \mathscr{L}), \psi)$ whose type is the function τ defined in (5.3.1). We will call such an F-zip a K3-zip.

Lemma 5.3. *Denote by δ the image of $-2d$ in $\mathbb{F}_p^\times / (\mathbb{F}_p^\times)^2$. Then the discriminant of the K3-zip $(H_{\mathrm{prim}}(X, \mathscr{L}), \psi)$ (in the sense of (4.3)) is δ.*

Proof. By [Og2] §1 (1.4) we have $\mathrm{disc}(H^2_{\mathrm{DR}}(X/k)) = -1$ and by the definition of $2d$, we have $\mathrm{disc}(\langle c_1(\mathscr{L}) \rangle) = 2d$. Therefore the lemma follows from (4.3.1). $\square$

Now let k be algebraically closed. By the general classification Theorem 4.4, isomorphism classes of K3-zips over k with discriminant δ are in bijection with $^J W$, where

$$W = \{\, w \in S_{20} \mid w(i) + w(21 - i) = 21 \text{ for all } i \,\},$$
$$^J W = \{\, w \in W \mid w^{-1}(2) < \cdots < w^{-1}(19) \,\}. \qquad (5.3.4)$$

Therefore we have a bijection

$$^J W \leftrightarrow \{1, \ldots, 20\}, \quad w \mapsto w^{-1}(1). \qquad (5.3.5)$$

Note that the element in $^J W$ corresponding to $r \in \{1, \ldots, 20\}$ has length $r - 1$.

Description of K3-zips.

We will now give an explicit description of the K3-zip over the algebraically closed field k corresponding to w_r. Therefore fix an $r \in \{1, \ldots, 20\}$. Let M_0 be a 21-dimensional $\mathbb{F}_p$-vector space and choose a basis $e_1, \ldots, e_{21}$ of M_0. Let ψ_0 be the symmetric bilinear form on M associated to the quadratic from

$$\sum x_i e_i \mapsto x_1 x_{21} + \cdots + x_{10} x_{12} + x_{11}^2.$$

Note that we can identify W with

$$W' := \{ w \in S_{21} \mid w(i) + w(22 - i) = 22 \text{ for all } i \}$$

by sending $w \in W$ to $w' \in W'$ defined by $w'(i) := w(i)$ for $i \leq 10$, $w'(11) := 11$, and $w'(i) = w(i - 1)$ for $i \geq 12$. Under this bijection ${}^J W$ is send to

$$^J W' := \{ w \in W' \mid w^{-1}(2) < \cdots < w^{-1}(20) \}.$$

As in (5.3.5) we have a bijection

$$\{1, \ldots, 10, 12, \ldots, 21\} \leftrightarrow {}^J W', \qquad r \mapsto w_r. \qquad (5.3.6)$$

We denote the K3-zip corresponding to w_r by ${}^r \underline{M}$. Note that we have $\ell(w_r) = r - 1$.

The underlying symmetric space of ${}^r \underline{M}$ is $(M_0, \psi_0)_k$ and the underlying F-zip is $(^{w_r} \underline{M})_k$, where $^{w_r} \underline{M}$ is the F-zip of type τ constructed in Example 3.4.3. We distinguish three cases.

$r = 21$: In this case w_r is the element of maximal length,

$$^1 \underline{M} \cong \underline{T}(0) \oplus \underline{T}(1)^{\oplus 19} \oplus \underline{T}(2),$$

and ψ induces a perfect duality between $\underline{T}(0)$ and $\underline{T}(2)$. This is the case where $\underline{M}$ is ordinary (Corollary 3.7).

$12 \leq r \leq 20$: In this case

$$^r \underline{M} \cong \underline{M}_1 \oplus \underline{T}(1)^{\oplus 2r - 23} \oplus \underline{M}_2,$$

where $\underline{M}_1 \cong {}^{w_n} \underline{M}$ is the indecomposable F-zip defined in Example 3.6 for $n = 22 - r$, and $\underline{M}_2$ is in perfect duality via ψ to $\underline{M}_1$ and hence $\underline{M}_2 \cong \underline{M}_1^{\vee}(2)$. In other words, via the classification (3.1), $\underline{M}_1$ is given by the word $\mathbf{1}^{(21-r)}\mathbf{0}$, and $\underline{M}_2$ is given by the word $\mathbf{1}^{(21-r)}\mathbf{2}$, where $\mathbf{i}^{(d)}$ means that $\mathbf{i}$ is repeated d-times.

$1 \leq r \leq 10$: Then

$$^r \underline{M} \cong \underline{N} \oplus \underline{T}(1)^{\oplus 21 - 2r},$$

where $\underline{N}$ is an indecomposable F-zip such that ψ induces on $\underline{N}$ the structure of a 2-othogonal F-zip. The band corresponding to $\underline{N}$ via the classification in (3.1) is given by $\mathbf{21}^{(r-1)}\mathbf{01}^{(r-1)}$.

6. Families of F-zips

(6.1) F-zip stratification.

Let τ be a filtration type (3.2.1) and let $\underline{M}$ be an F-zip over an $\mathbb{F}_p$-scheme S. For $w \in {}^{\tau}W$ we define a locally closed subschemes S_w of S which is the "locus in S where $\underline{M}$ is of isomorphism class w". We will do this in two equivalent ways.

Definition of S_w via the stack of F-zips.
The first way is via the moduli stack of F-zips of type τ. For this we define the algebraic stack (in the sense of [LM])

$$\mathscr{X}_{\tau} := [G \backslash X_{\tau}],$$

where X_{τ} is the scheme of F-zips $\underline{M}$ of type τ such that $M = M_0 \otimes_{\mathbb{F}_p} \mathscr{O}_S$ for fixed $\mathbb{F}_p$-vector space M_0 (defined in (3.2)) and where $G = \mathrm{GL}(M_0)$. The action of G on X_{τ} is given via (3.2.3), where $G \hookrightarrow G \times G$ is diagonally embedded. Then for any $\mathbb{F}_p$-scheme S the category of 1-morphisms $S \to \mathscr{X}_{\tau}$ is equivalent to the category of F-zips over S of type τ where as morphisms we take only isomorphisms of F-zips. By Proposition 3.2 this is a smooth algebraic stack of dimension 0.

Let k be an algebraically closed extension of $\mathbb{F}_p$. The G_k-orbits on $(X_{\tau})_k$ are parametrized by ${}^{\tau}W$ (3.3) and are already defined over $\mathbb{F}_p$, because the standard F-zips of (3.4.3) are defined over $\mathbb{F}_p$. For $w \in {}^{\tau}W$ let ${}^w X_{\tau} \subset X_{\tau}$ this orbit. Then ${}^w\mathscr{X}_{\tau} := [G \backslash {}^w X_{\tau}]$ is an algebraic substack of $\mathscr{X}_{\tau}$ which is smooth of dimension $\ell(w) - \dim(\mathrm{Par}_J)$ by (3.3.3).

Now an F-zip $\underline{M}$ over S of type τ defines a 1-morphism $\gamma_{\underline{M}} : S \to \mathscr{X}_{\tau}$ and we define $S_w \subset S$ as the inverse image of ${}^w\mathscr{X}_{\tau}$ under $\gamma_{\underline{M}}$.

Definition of S_w via isotrivial F-zips.
We now translate the definition of S_w given above in a language which avoids algebraic stacks. For this we use the explicitly defined "standard" F-zips ${}^w\underline{M}$ for $w \in {}^{\tau}W$ over $\mathbb{F}_p$. We say that an F-zip $\underline{M}$ over an $\mathbb{F}_p$-scheme T is *isotrivial of isomorphism class w* if for any $t \in T$ there exists an open affine neighborhood U of t and a faithfully flat morphism $\alpha : V \to U$ of finite presentation such that $\alpha^*(\underline{M}_{|U}) \cong {}^w\underline{M} \otimes_{\mathbb{F}_p} \mathscr{O}_V$. Note that in this definition we could have replaced ${}^w\underline{M}$ by any other F-zip over $\mathbb{F}_p$ which becomes isomorphic to ${}^w\underline{M}$ over some field extension (or, equivalently, over some finite field extensions) of $\mathbb{F}_p$.

Now let $\underline{M}$ be an arbitary F-zip of type τ over S. Then S_w is the (necessarily unique) subscheme of S which satisfies the following universal property. A morphism $\alpha : T \to S$ factors through S_w if and only if $\alpha^*(\underline{M})$ is isotrivial of isomorphism class w.

Set-theoretically S is the disjoint union of the subschemes S_w. But even set-theoretically it is not true in general that the closure of a subscheme S_w is the union of other subschemes of the form $S_{w'}$.

Clearly there is a variant for this for d-symplectic or d-orthogonal F-zips over a scheme S, where again we obtain subschemes S_w which are now indexed by $w \in {}^J W$, where ${}^J W$ is described in (4.2) and in (4.4), respectively.

(6.2) Example: The Ekedahl-Oort stratification for the moduli space of principally polarized abelian varieties.

Fix $g \geq 1$. Let $\mathcal{A}_g$ be the moduli stack of principally polarized abelian schemes of dimension g in characteristic p, i.e., for each $\mathbb{F}_p$-scheme S the category of 1-morphisms $S \to \mathcal{A}_g$ is the category of principally polarized g-dimensional abelian schemes over S, where the morphisms are isomorphisms between abelian schemes preserving the principal polarization. This is a Deligne-Mumford stack by [FC] chapter 1. As we did not define the notion of an F-zip over an algebraic stack (which is not difficult), we consider the following variant of $\mathcal{A}_g$. Fix an integer $N \geq 3$ prime to p and consider the moduli stack $\mathcal{A}_{g,N}$ of principally polarized abelian schemes of dimension g in characteristic p together with a level-N-structure (loc. cit.). This is in fact a smooth quasi-projective scheme over $\mathbb{F}_p$ of dimension $g(g+1)/2$.

Let (X, ξ, η) be the universal principally polarized abelian scheme with level-N-structure over $\mathcal{A}_{g,N}$. As explained in (1.11), $H^1_{\mathrm{DR}}(X/\mathcal{A}_{g,N})$ carries the structure of an F-zip. Again it is not difficult to see that the principal polarization induces on $H^1_{\mathrm{DR}}(X/\mathcal{A}_{g,N})$ the structure of a 1-symplectic F-zip over $\mathcal{A}_{g,N}$. By (6.1) we obtain locally closed subschemes $(\mathcal{A}_{g,N})_w$ for $w \in {}^J W$, where ${}^J W$ is described in (5.1). The collection of these subschemes is called the *Ekedahl-Oort stratification of* $\mathcal{A}_{g,N}$. They have been defined as reduced subschemes in [Oo].

We recall briefly some properties of the Ekedahl-Oort stratification from [Oo], [Wd1], [MW], and [Wd2].

(1) For all $w \in {}^J W$ the Ekedahl-Oort stratum $(\mathcal{A}_{g,N})_w$ is quasi-affine and smooth of dimension $\ell(w)$. In particular, they are all nonempty.

(2) The closure of $(\mathcal{A}_{g,N})_w$ is the union of those Ekedahl-Oort strata $(\mathcal{A}_{g,N})_{w'}$ such that $w' \preceq w$ where $\preceq$ can be described explicitly. The partial order is a refinement of the Bruhat order on ${}^J W$ (it is strictly finer if and only if $g \geq 5$).

(3) There is a unique closed Ekedahl-Oort stratum (namely $(\mathcal{A}_{g,N})_{\mathrm{id}}$) and a unique open (and therefore dense) stratum (namely $(\mathcal{A}_{g,N})_{w_0, J w_0}$) which equals the ordinary locus in $\mathcal{A}_{g,N}$. For $w \neq \mathrm{id}$ the closure of $(\mathcal{A}_{g,N})_w$ is connected.

(6.3) Example: The moduli space of polarized K3-surfaces.

Fix an integer $d \geq 1$ and assume that p does not divide $2d$. We now consider the moduli stack $\mathcal{F}_{2d}$ of K3-surfaces together with a polarization of degree $2d$. This is a smooth Deligne-Mumford stack by work of Rizow [Ri] (see also [Ol]). Again the universal polarized K3-surface defines a morphism from $\mathcal{F}_{2d}$ into the algebraic stack classifying 2-orthogonal F-zips of the type τ defined in (5.3.1).

As above we obtain locally closed substacks $(\mathcal{F}_{2d})_s$ which are indexed by $s \in \{1, \ldots, 20\}$ via the bijection (5.3.5). This is a refinement of the height stratification (this is essentially shown in [Og3]; we omit the details).

7. Appendix: On reductive groups and the wonderful compactification

Notation: In the appendix we denote by κ an arbitrary field and by k an algebraically closed extension of κ. If X is a scheme over κ, we set $X_k := X \otimes_\kappa k$. Moreover, if G is any group, $g \in G$, and H is a subgroup of G, we set ${}^g H := gHg^{-1}$.

We first recall some notions and facts about reductive groups ((7.1) – (7.6)). We refer to [Sp] or [SGA3] for the proofs.

(7.1) Examples of reductive groups.

Let G be a connected reductive group over κ. Here we will need only the following examples of reductive groups.

(A) For $n \geq 1$ let V be an n-dimensional κ-vector space. Then $\mathrm{GL}(V)$, $\mathrm{SL}(V)$, and $\mathrm{PGL}(V)$, considered as algebraic groups over κ, are reductive groups.

(B) Let $V = \kappa^{2m+1}$ and q be the quadratic form $q(x_0, x_1, \ldots, x_{2m}) = x_0^2 + x_1 x_{2m} + \ldots x_m x_{m+1}$. Then $\mathrm{SO}(V, q)$ is a reductive group.

(C) For $g \geq 1$, the algebraic group Sp_{2g} is a reductive group over κ.

(D) Let $V = \kappa^{2m}$ and q be the quadratic form $q(x_1, \ldots, x_{2m}) = x_1 x_{2m} + \ldots x_m x_{m+1}$. Then $\mathrm{SO}(V, q)$ is a reductive group.

(7.2) Maximal tori.

Denote by $\mathbb{G}_m$ the multiplicative group over κ, i.e., $\mathbb{G}_m(R) = R^\times$ for every κ-algebra R. Recall that a *torus over κ* is an algebraic group T over κ such that $T_k \cong (\mathbb{G}_m)_k^n$ for some integer $n \geq 1$. A torus is *split* if $T \cong \mathbb{G}_m^n$. If G is a an algebraic group, an algebraic subgroup T of G is called a *maximal torus of G* if it is a torus such that T_k is maximal in the set of subtori of G_k. Maximal tori always exist.

Example 7.1. If $G = \mathrm{GL}_n$, the subgroup D of diagonal matrices in GL_n is a maximal torus which is also split. More generally, an algebraic subgroup T of G is a maximal torus if and only if T_k is conjugate in G_k to D_k. If κ is not separably closed, there exist also non-split maximal tori. In particular such tori are not conjugate to D over κ.

For every maximal torus T in SL_n (resp. in PGL_n) there exists a unique maximal torus T' in GL_n such that $T = T' \cap \mathrm{SL}_n$ (resp. such that T is the image of T' under the canonical homomorphism $\mathrm{GL}_n \to \mathrm{PGL}_n$).

(7.3) Parabolic subgroups and Borel subgroups.

An algebraic subgroup P of the reductive group G over κ is called a *parabolic subgroup* if the quotient G/P is a proper κ-scheme. Such subgroups are automatically connected and their own normalizers. A parabolic subgroup B of G is called *Borel subgroup* if B_k is a minimal parabolic subgroup of G_k.

Example 7.2. For $G = \mathrm{GL}(V)$, $G = \mathrm{PGL}(V)$, or $G = \mathrm{SL}(V)$ (notations of Example 7.1(A)), a subgroup P of G is a parabolic subgroup if and only if $P = P_{\mathcal{F}}$ is the stabilizer of a flag $\mathcal{F}$ of subspaces

$$\mathcal{F}: \quad 0 = V_0 \subset V_{n_1} \subset V_{n_1+n_2} \subset \cdots \subset V_{n_1+\cdots+n_r} = V_n = V$$

of V where $\dim(V_i) = i$ and $(n_1, \ldots, n_r)$ is a tuple of integers $n_i \geq 1$ such that $n_1 + \cdots + n_r = n$. We call this tuple the *flag type of* $\mathcal{F}$.

Note that $\mathcal{F} \mapsto P_{\mathcal{F}}$ defines a bijection between flags in V and parabolic subgroup of G. Two parabolic subgroups of G are conjugate if and only if the correspondoing flags have the same flag type. A parabolic subgroup $P_{\mathcal{F}}$ is a Borel subgroup if and only if $\mathcal{F}$ is a complete flag, i.e., if its flag type is equal to $(1, \ldots, 1)$.

If G is one of the reductive groups defined in Example 7.1(B) – (D), we can consider the natural embedding $G \hookrightarrow \mathrm{GL}_n$ for $n = 2m + 1$, $n = 2g$, $n = 2m$ respectively. Then every parabolic subgroup P and any Borel subgroup B of G is of the form $P = P' \cap G$ and $B = B' \cap G$ where P' is a parabolic subgroup and B' is a Borel subgroup of GL_n.

(7.4) Scheme of parabolic subgroups.

To simplify the exposition we will assume from now on that G is *split*, i.e., that G has a split maximal torus (7.2). This implies that G admits a Borel subgroup. G is split if κ is algebraically closed by definition. Moreover, all the examples given in (A) – (D) are split over arbitrary fields κ.

We denote by $\mathcal{C}_G$ the set of conjugacy classes of parabolic subgroups of G. We fix a Borel subgroup B of G and a conjugacy class $J \in \mathcal{C}_G$. Then there exists a unique parabolic subgroup $P_C \in J$ of G which contains B. In particular, any two Borel subgroups of G are conjugate. We define a partial order on $\mathcal{C}_G$ by setting

$$J_1 \leq J_2 \quad \text{if and only if} \quad P_{J_1} \subset P_{J_2}. \tag{7.4.1}$$

As a parabolic subgroup is its own normalizer, the map $G \ni g \mapsto {}^g P_J$ defines a bijection $G/P_J \xrightarrow{\sim} C$. This allows to define the scheme Par_J of parabolic subgroups in J. It is a smooth and proper scheme over κ. In particular we get the scheme of Borel subgroups of G which is denoted by Bor.

Example 7.3. Let G be $\mathrm{GL}(V)$, $\mathrm{SL}(V)$, or $\mathrm{PGL}(V)$. Two parabolic subgroups $P_{\mathcal{F}_1}$ and $P_{\mathcal{F}_2}$ of G are conjugated if and only if the types of $\mathcal{F}_1$ and $\mathcal{F}_2$ are equal. By associating to a flag type $(n_1, \ldots, n_r)$ the set

$$J_{(n_1,\ldots,n_r)} := \{1, \ldots, n-1\} \setminus \{n_1, n_1 + n_2, \ldots, n_1 + \cdots + n_{r-1}\}, \tag{7.4.2}$$

we identify $\mathcal{C}_G$ with the set of subsets of $I := \{1, \ldots, n-1\}$ (as a partially ordered set). The conjugacy class of a Borel subgroup corresponds to the empty set, and the conjugacy class of G (consisting only of G itself) corresponds to I.

(7.5) Weyl groups.

The scheme of pairs of Borel subgroups $\mathrm{Bor} \times \mathrm{Bor}$ carries a G-action by

$$g \cdot (B_1, B_2) := ({}^{g}B_1, {}^{g}B_2).$$

We denote by W the set of $G(k)$-orbits on $\mathrm{Bor}(k) \times \mathrm{Bor}(k)$. For $w \in W$ we denote the corresponding orbit by $\mathcal{O}(w)$. Moreover we set

$$\ell(w) := \dim(\mathcal{O}(w)) - \dim(\mathrm{Bor}), \qquad \text{for } w \in W;$$

$$I := \{\, w \in W \mid \ell(w) = 1 \,\}.$$

We define a group structure on W as follows. Let $w_1, w_2 \in W$. We can choose Borel subgroups B_1, B_2, and B_3 of G_k containig a common maximal torus of G_k such that $(B_1, B_2) \in \mathcal{O}(w_1)$ and $(B_2, B_3) \in \mathcal{O}(w_2)$. Then we define $w := w_1 w_2$ as the orbit $\mathcal{O}(w)$ such that $(B_1, B_3) \in \mathcal{O}(w)$. This defines a group structure on W. The identity of W is given by the orbit of Borel subgroups (B_1, B_2) such that $B_1 = B_2$ and the inverse on W is induced by $(B_1, B_2) \mapsto (B_2, B_1)$. Then (W, I) is a Coxeter system ([BouLie] chap. IV, §1) with length function ℓ. In particular, I generates W and $\ell(w)$ is the minimal number of elements $s_1, \dots, s_d \in I$ such that $w = s_1 s_2 \dots s_d$.

For $i \subset I$ we say that a parabolic subgroup P of G has *type* $\{i\}$ if for any two Borel subgroups $B_1, B_2 \subset P_k$ with $B_1 \neq B_2$ we have $(B_1, B_2) \in \mathcal{O}(i)$. If P is any parabolic subgroup of G, we define the *type of P* as the subset J of I which consists of those $i \in I$ such that P contains a parabolic subgroup of type $\{i\}$. Two parabolic subgroups of G have the same type if and only if they are conjugate. Therefore we obtain a bijection

$$\alpha \colon \mathcal{C}_G \leftrightarrow 2^{I}. \tag{7.5.1}$$

This bijection is an isomorphism of partially ordered sets for the order on $\mathcal{C}_G$ defined in (7.4.1) and the inclusion order on the set of subsets of I. If J is a subset of I we also write Par_J instead of $\mathrm{Par}_{\alpha^{-1}(J)}$ for the scheme of parabolic subgroups of type J.

Example 7.4. Let G be $\mathrm{GL}(V)$, $\mathrm{PGL}(V)$, or $\mathrm{SL}(V)$. By Example 7.2 each Borel subgroup of G_k is the stabilizer of a unique flag

$$\mathcal{F} \colon \quad 0 = V_0 \subset V_1 \subset \cdots \subset V_{n-1} \subset V_n = V \otimes_{\kappa} k$$

with $\dim(V_i) = i$. Consider two such flags $\mathcal{F}$ and $\mathcal{F}'$ corresponding to Borel subgroups B and B'. Set $\mathrm{gr}^i_{\mathcal{F}} = V^i/V^{i-1}$. This is a one-dimensional k-vector space and therefore there exists a unique $\pi(i) = \pi_{B,B'}(i) \in \{1, \dots, n\}$ such that the graded piece $\mathrm{gr}^{\pi(i)}_{\mathcal{F}'} \mathrm{gr}^i_{\mathcal{F}}$ of the flag induced by $\mathcal{F}'$ on $\mathrm{gr}^i_{\mathcal{F}}$ is nonzero. The map $\mathrm{Bor}(k) \times \mathrm{Bor}(k) \ni (B, B') \mapsto \pi^{-1}_{B,B'}$ induces an isomorphism of groups

$$\iota \colon W \xrightarrow{\sim} S_n.$$

We use ι to identify W with S_n. Via this identification we have for $w \in S_n$

$$\ell(w) = \#\{\, (i,j) \mid i,j \in \{1,\ldots,n\}, i < j, w(i) > w(j) \,\};$$
$$I = \{\tau_1, \tau_2, \ldots, \tau_{n-1}\},$$

$$(7.5.2)$$

where $\tau_i \in S_n$ denotes transposition of i and $i+1$.

In (7.4.2) we already identified conjugacy classes of parabolic subgroups of G with subsets of $\{1,\ldots,n-1\}$. Sending $i \in \{1,\ldots,n-1\}$ to $\tau_i \in I$ defines the bijection α in (7.5.1).

(7.6) Adjoint groups.

Recall that a connected linear algebraic group H over $\mathbb{F}_p$ is called *adjoint* if the scheme-theoretical center is trivial. Such an algebraic group is automatically reductive.

Example 7.5. Examples for adjoint groups are PGL_n for $n \geq 1$ and $\mathrm{SO}(V,q)$ where (V,q) is the odd-dimensional quadratic space defined in (7.1) (B). In the latter case $\hat{H} = \mathrm{O}(V,q)$ is a linear algebraic group whose connected component of the identity is $\mathrm{SO}(V,q)$. If $p \neq 2$, $\mathrm{O}(V,q)$ has two connected components.

On the other hand, SL_n (for $n > 1$), GL_n (for $n \geq 1$), Sp_{2g} (for $g \geq 1$), and $\mathrm{SO}(V,q)$, where (V,q) is the even-dimensional quadratic space defined in (7.1) (D), are examples of algebraic groups which are not adjoint.

(7.7) The wonderful compactification.

We will now define the wonderful compactification of an adjoint group.

The algebraic group.
Let $\hat{H}$ be a linear algebraic group over κ and denote by H the connected component of the identity. We assume that H is an adjoint group over κ. We further fix a connected component H^1 of $\hat{H}$. To simplify the exposition we assume that H is split.

We define an $(H \times H)$-action on H^1 by setting

$$(h_1, h_2) \cdot h := h_1 h h_2^{-1}.$$

$$(7.7.1)$$

The wonderful compactification.
Let $\mathcal{V}_H$ be the scheme over κ whose points are the $\dim(H)$-dimensional Lie-subalgebras of $\mathrm{Lie}(H \times H)$. This is a closed subscheme of the Grassmannian of $\dim(H)$-dimensional subspaces of $\mathrm{Lie}(H \times H)$ and therefore it is projective. It is endowed with an action by $H \times H$ via

$$(h_1, h_2) \cdot \mathfrak{g} := \mathrm{ad}(h_1, h_2)(\mathfrak{g}).$$

$$(7.7.2)$$

For every $h \in H^1$ we define a subgroup $H_h \subset H \times H$ as

$$H_h := \{\, (h_1, h_2) \in H \times H \mid h_1 = h h_2 h^{-1} \,\}.$$

Then $\dim(H_h) = \dim(H)$ and $H_{(h_1, h_2) \cdot h} = \mathrm{int}(h_1, h_2)(H_h)$ and therefore

$$i \colon H^1 \hookrightarrow \mathcal{V}_H, \qquad h \mapsto \mathrm{Lie}(H_h)$$

is a well defined $(H \times H)$-equivariant morphism with respect to the actions (7.7.1) and (7.7.2). As H is adjoint, ι is an immersion.

We define $\overline{H^1}$ as the closure of $i(H^1)$ in $\mathcal{V}_H$. This is a closed $(H \times H)$-invariant subscheme of $\mathcal{V}_H$ and therefore it is an integral projective scheme over κ with an $(H \times H)$-action.

Definition 7.6. *The projective scheme* $\overline{H^1}$ *is called the* wonderful compactification *of* H^1.

We collect some facts about the wonderful compactification.

Theorem 7.7. $\overline{H^1}$ *is a smooth projective κ-scheme and $\overline{H^1} \setminus H^1$ is a divisor with normal crossings in $\overline{H^1}$.*

Recall that we identified the set $\mathcal{C}_H$ of conjugacy classes of parabolic subgroups of H with the set of subsets of I, where I is the set of simple reflections of the Coxeter system (W, I) (7.5.1).

Theorem 7.8. *There is a bijection*

$$\mathcal{C}_H = 2^I \leftrightarrow \left\{ \begin{array}{c} (H(k) \times H(k)) \text{-}orbits \\ of\ \overline{H^1}(k) \end{array} \right\}, \tag{7.7.3}$$

$$J \mapsto \overline{H^1_J},$$

such that
(1) $\overline{H^1_I} = H^1$ *and* $\overline{H^1_J}$ *is contained in the closure of* $\overline{H^1_{J'}}$ *if and only if* $J \subset J'$.
(2) *The codimension of* $\overline{H^1_J}$ *in* $\overline{H^1}$ *is* $\#(I \setminus J)$.
(3) *The intersection of the closure of* $\overline{H^1_J}$ *and the closure of* $\overline{H^1_{J'}}$ *is* $\overline{H^1_{J \cap J'}}$.

Proof of Theorems 7.7 and 7.8. If $H^1 = H$, theorems 7.7 and 7.8 are known (see [dCP] and [St]). We will reduce to this case.

We will use the following diction. If X and Y are varieties with an action by algebraic groups G and H, respectively, and if $\alpha \colon G \to H$ is a homomorphism of algebraic groups, we say that a morphism $\varphi \colon X \to Y$ is α-equivariant if $\varphi(g \cdot x) = \alpha(g) \cdot \varphi(x)$ for all $g \in G$ and $x \in X$.

Now choose $h^1 \in H^1(\kappa)$. Then

$$
\begin{array}{ccc}
H & \overset{\sim}{\underset{\varphi \colon\, h \mapsto h^1 h}{\longrightarrow}} & H^1 \\[2pt]
{\scriptstyle i}\big\downarrow & & \big\downarrow{\scriptstyle i} \\[2pt]
\mathcal{V}_H & \overset{\sim}{\underset{\bar{\varphi}\colon\, \mathfrak{g} \mapsto \mathrm{ad}(h^1,1)(\mathfrak{g})}{\longrightarrow}} & \mathcal{V}_H
\end{array}
\tag{7.7.4}
$$

is a commutative diagram. Moreover, if we define

$$\alpha\colon H \times H \to H \times H, \qquad (h_1, h_2) \mapsto (h^1 h_1 (h^1)^{-1}, h_2),$$

the morphisms φ and $\bar{\varphi}$ are both α-equivariant. Therefore $\bar{\varphi}$ defines an α-equivariant isomorphism $\overline{H} \to \overline{H^1}$. As α is an automorphism of $H \times H$, we can replace H^1 by H for the proof. $\qquad\square$

Example 7.9. For $H = \mathrm{PGL}_n$ it follows that the $H \times H$-orbits of $\overline{H}$ are in bijection with subsets of $J \subset \{\tau_1, \ldots, \tau_{n-1}\}$ (7.5.2) or also with flag types $(n_1, \ldots, n_r)$ (7.4.2). We denote the corresponding orbit either by $\overline{H}_J$ or by $\overline{H}_{(n_1, \ldots, n_r)}$.

The diagonal action.
Wo consider now a Frobenius-linear version of the diagonal action of H on $\overline{H^1}$. Therefore assume that $\kappa = \mathbb{F}_p$. Let $F\colon \hat{H} \to \hat{H}$ be the Frobenius endomorphism. If $\hat{H}$ is an algebraic subgroup of GL_n (and every affine algebraic group is isomorphic to an algebraic subgroup of GL_n), F is given on k-valued points by $(a_{ij}) \mapsto (a_{ij}^p)$ for a matrix $(a_{ij}) \in \hat{H}(k) \subset \mathrm{GL}_n(k)$.

Consider the closed embedding

$$H \hookrightarrow H \times H, \quad h \mapsto (h, F(h)) \tag{7.7.5}$$

and denote by H_F the image of H in $H \times H$. Then by restricting the $(H \times H)$-action to H_F we obtain an action of H_F on $\overline{H^1}$.

References

[BBM]   P. Berthelot, L. Breen, W. Messing, *Théorie de Dieudonné cristalline II*, LNM **930**, Springer-Verlag (1982).

[Bl]   M. Blickle, *Cartier isomorphism for toric varieties*, J. Algebra **237** (2001), 342–357.

[BLR]   S. Bosch, W. Lütkebohmert, M. Raynaud, *Néron Models*, Springer Verlag (1990).

[BouLie]   N. Bourbaki, *Groupes et Algèbres de Lie*, chap. I-III, Springer (1989); chap. IV, V, VI, Masson (1968); chap. VII, VIII, Masson (1990); chap. IX, Masson (1982).

[BP]   J. Bertin, C. Peters, *Variations de structures de Hodge, variétés de Calabi-Yau et symétrie miroir*, Introduction à la théorie de Hodge, Panor. Synthses **3**, Soc. Math. France (1996), 169–256.

[Br]   G.E. Bredon, *Sheaf Theory*, Second edition, Graduate Texts in Mathematics **170**, Springer-Verlag (1997).

[Cr]   W. W. Crawley-Boevey, *Functorial Filtrations II: Clans and the Gelfand Problem*, J. London Math. Soc. (2) **40** (1989), no. 1, 9–30.

[dCP]   C. De Concini, C. Procesi, *Complete symmetric varieties*, Invariant theory (Montecatini, 1982), Lecture Notes in Math. **996** (1983), Springer 1983, 1–44.

[De1]   P. Deligne, *Théorie de Hodge II*, Publ. IHES **40** (1971), 5–57.

[De2]   P. Deligne, *Relèvement des surfaces K3 en caractéristique nulle (rédige par L. Illusie)*, in: Surfaces Algébriques, J. Giraud et al., eds., LNM **868**, Springer-Verlag, Berlin, 1981, 58–79.

[Dem]   M. Demazure, *Lectures on p-divisible groups*, LNM **302**, Springer (1970).

[Dm]   J.-P. Demailly, *Théorie de Hodge L^2 et théorèmes de annulation*, Introduction la théorie de Hodge, Panor. Synthèses **3** (1996), 3–111.

[DI]   P. Deligne, L. Illusie, *Relèvements modulo p^2 et décomposition du complexe de de Rham*, Invent. Math. **89** (1987), 247–270.

[dJ]   A.J. de Jong, *The moduli spaces of principally polarized abelian varieties with* $\Gamma_0(p)$-*level structure*, J. Algebraic Geometry **2** (1993), no. 4, 667–688.

[EGA]   A. Grothendieck, J. Dieudonné, *Eléments de Géométrie Algébrique*, I Grundlehren der Mathematik **166** (1971) Springer, II-IV Publ. Math. IHES **8** (1961), **11** (1961), **17** (1963), **20** (1964), **24** (1965), **28** (1966), **32** (1967).

[FC]   G. Faltings, C.-L. Chai, *Degeneration of abelian varieties*, Ergebnisse der Mathematik und ihrer Grenzgebiete (3. Folge) **22**, Springer (1990).

[GP]   I.M. Gelfand, V.A. Ponomarev, *Indecomposable representations of the Lorentz group*, Uspehi Mat. Nauk **23** 1968 no. 2 (140), 3–60.

[Ill1]   L. Illusie, *Crystalline cohomology*, in *Motives*, Proc. Sympos. Pure Math. **55** (1991), Part 1, 43–70.

[Ill2]   L. Illusie, *Frobenius et dégénerescence de Hodge*, Introduction la théorie de Hodge, Panor. Synthèses **3** (1996), 113–168.

[IR]   L. Illusie, M. Raynaud, *Les suites spectrales associées au complex de de Rham-Witt*, Publ. Math. IHES **57** (1983), 73–212.

[Ka]   N. Katz, *Nilpotent Connections and the Monodromy Theorem*, Publ. Math. IHES **39** (1970), 175–232.

[LM]   G. Laumon, L. Moret-Bailly, *Champs algébriques*, Ergebnisse der Mathematik und ihrer Grenzgebiete (3. Folge) **39**, Springer (2000).

[Lu]   G. Lusztig, *Parabolic character sheaves II*, Mosc. Math. J. **4** (2004), no. 4, 869–896.

[Mu1]   D. Mumford, *Pathologies of modular surfaces*, American J. Math. **83** (1961), 339–342.

[Mu2]   D. Mumford, *Abelian Varieties*, Oxford University Press, 2nd edition (1974).

[MW]   B. Moonen, T. Wedhorn, *Discrete Invariants of Varieties in Positive Charakteristic*, Int. Math. Res. Not. **2004:72** (2004), 3855–3903.

[Od]   T. Oda, *The first de Rham cohomology group and Dieudonné modules*, Ann. scient. Éc. Norm. Sup. (4), **2**, (1969), 63–135.

[Og1]   A. Ogus, *Supersingular K3-crystals*, Astérisque **64** (1979), 3–86.

[Og2]   A. Ogus, *A crystalline Torelli theorem for supersingular K3-surfaces*, in Arithmetic and geometry Vol. II, Progr. Math. **36**, Birkhuser 1983, 361–394.

[Og3]   A. Ogus, *Singularities of the height strata in the moduli of K3-surfaces*, Moduli of Abelian Varieties (ed. by C. Faber, G. van der Geer, F. Oort), Progress in Mathematics **195**, Birkhäuser (2001).

[Og4]   A. Ogus, *Frobenius and the Hodge spectral sequence*, Adv. Math. **162** (2001), no. 2, 141–172.

[Ol]   M. Olsson, *Semistable degenerations and period spaces for polarized K3-surfaces*, Duke Math. J. **125** (2004), no. 1, 121–203.

[Oo]   F. Oort, *A stratification of a moduli space of abelian varieties*, Moduli of Abelian Varieties (ed. by C. Faber, G. van der Geer, F. Oort), Progress in Mathematics **195**, Birkhäuser (2001).

[Ri]   J. Rizov, *Moduli stacks of polarized K3-surfaces in mixed characteristic*, Serdica Math. J. **32** (2006), no. 2-3, 131–178 (see also arXiv:math/0506120).

[Se1]   J.P. Serre, *Géométrie algébrique et géométrie analytique*, Ann. Inst. Fourier **6** (1956), 1–42.

[Se2]   J.P. Serre, *A Course in Arithmetic*, Graduate Texts in Mathematics **7**, Springer-Verlag 1973.

[SGA1]   A. Grothendieck et al., *Revêtements étales et Groupe Fondamental (SGA 1)*, Documents Mathématiques **3**, SMF 2003.

[SGA3]   A. Grothendieck et al., *Schémas en groupes*, LNM **151**, **152**, **153**, Springer 1970.

[SGA7]   A. Grothendieck, M. Raynaud, et. al., *Séminaire de Géometrie Algébrique du Bois-Marie, Groupes de Monodromie en Géometrie Algébrique* (1967–68), Lecture Notes in Mathematics **288**, **340**, Springer (1972–73).

[Sp]   T.A. Springer, *Linear algebraic groups*, 2nd ed., Progress in Mathematics **9** (1998), Birkhuser Boston.

[St]   E. Strickland, *A vanishing theorem for group compactifications*, Math. Ann. **277** (1987), no. 1, 165–171.

[Voi]   C. Voisin, *Théorie de Hodge et géométrie algébrique complexe*, Cours Spécialisés **10**,

SMF (2002).

[Wd1]   T. Wedhorn, *The dimension of Oort strata of Shimura varieties of PEL-type*, Moduli of Abelian Varieties (ed. by C. Faber, G. van der Geer, F. Oort), Progress in Mathematics **195**, Birkhäuser (2001).

[Wd2]   T. Wedhorn, *Specialiation of F-zips*, preprint 2005, arXiv:math/0507175.

[Wd3]   T. Wedhorn, *Zips, representations of a clan, and the wonderful compactification*, in preparation.

Higher-Dimensional Geometry over Finite Fields
D. Kaledin and Y. Tschinkel (Eds.)
IOS Press, 2008

Homomorphisms of abelian varieties over finite fields

Yuri G. Zarhin

Department of Mathematics, Pennsylvania State University, University Park,
PA 16802, USA
Institute for Mathematical Problems in Biology, Russian Academy of Sciences,
Pushchino, Moscow Region, Russia
e-mail: zarhin@math.psu.edu

Abstract. We give a proof of Tate's theorems on homomorphisms of
abelian varieties over finite fields and the corresponding ℓ-divisible
groups.

The aim of this note is to give a proof of Tate's theorems on homomorphisms
of abelian varieties over finite fields and the corresponding ℓ-divisible groups
[27,12], using ideas of [32,33]. We give a unified treatment for both $\ell \neq p$ and
$\ell = p$ cases. In fact, we prove a slightly stronger version of those theorems with
"finite coefficients". We use neither the existence (and properties) of the Frobenius
endomorphism (for $\ell \neq p$) nor Dieudonne modules (for $\ell = p$).

The paper is organized as follows. (A rather long) Section 1 contains auxil-
iary results about finite commutative group schemes and abelian varieties with
special reference to isogenies and polarizations. We discuss ℓ-divisible groups (aka
Barsotti–Tate groups) in Section 2. Section 3 contains useful results that play a
crucial role in the proof of main results that are stated in Section 4.

The next five Sections contain proofs of results that were stated in Section
3. In Section 5 we discuss abelian subvarieties of a given abelian variety. Section
6 deals with the finiteness of the set of abelian varieties of given dimension and
"bounded degree" over a finite field. In Section 7 we present a so called *quaternion trick*. In Section 8 we prove a crucial result about arbitrary finite group
subschemes of abelian varieties over finite fields. In Section 9 we try to divide
endomorphisms of a given abelian variety modulo n.

The main results of this paper are proven in Section 10. Their variants for
Tate modules are discussed in Section 11. An example of non-isomorphic elliptic
curves over a finite field with isomorphic ℓ-divisible groups (for all primes ℓ) is
discussed in Section 12.

I am grateful to Frans Oort and Bill Waterhouse for useful discussions and to
the referee, whose comments helped to improve the exposition. My special thanks
go to Dr. Boris Veytsman for his help with TeXnical problems.

1. Definitions and statements

Throughout this paper K is a field and $\bar{K}$ its algebraic closure. If X (resp. W) is an algebraic variety (resp. group scheme) over K then we write $\bar{X}$ (resp. $\bar{W}$) for the corresponding algebraic variety $X \times_{\mathrm{Spec}(K)} \mathrm{Spec}(\bar{K})$ (resp. group scheme $W \times_{\mathrm{Spec}(K)} \mathrm{Spec}(\bar{K})$) over $\bar{K}$. If $f : X \to Y$ is a a regular map of algebraic varieties over K then we write $\bar{f}$ for the corresponding map $\bar{X} \to \bar{Y}$.

1.1. Finite commutative group schemes over fields. We refer the reader to the books of Oort [17], Waterhouse [31] and Demazure–Gabriel [3] for basic properties of commutative group schemes; see also [25,21].

Recall that a group scheme V over K is called finite if the structure morphism $V \to \mathrm{Spec}(K)$ is finite. Since $\mathrm{Spec}(K)$ is a one-point set, it follows from the definition of finite morphism [7, Ch. II, Sect. 3] that V is an affine scheme and $\Gamma(V, \mathcal{O}_V)$ is a finite-dimensional commutative K-algebra. The K-dimension of the $\Gamma(V, \mathcal{O}_V)$ is called the *order* of V and denoted by $\#(V)$. An analogue of Lagrange theorem [19] asserts that multiplication by $\#(V)$ kills commutative V.

Let V and W be finite commutative group schemes over K and let $u : V \to W$ be a morphism of group K-schemes. Both V and W are affine schemes, $A = \Gamma(V, \mathcal{O}_V)$ and $B = \Gamma(W, \mathcal{O}_W)$ are finite-dimensional (commutative) K-algebras (with 1), $V = \mathrm{Spec}(A), W = \mathrm{Spec}(B)$ and u is induced by a certain K-algebra homomorphism

$$u^* : B \to A.$$

Since V and W are commutative group schemes, A and B are cocommutative Hopf K-algebras. Since u is a morphism of group schemes, u^* is a morphism of Hopf algebras. It follows that $C := u^*(B)$ is a K-subalgebra and also a Hopf subalgebra in A. It follows that $U := \mathrm{Spec}(C)$ carries the natural structure of a finite group scheme over K such that the natural scheme morphism $U \to V$ induced by $u^* : B \twoheadrightarrow u^*(B) = C$ is a morphism of group schemes. In addition, the inclusion $C \subset A$ induces the morphism of schemes $V \to U$, which is also a morphism of group schemes. The latter morphism is an epimorphism in the category of finite commutative group schemes over K, because the corresponding map

$$C = \Gamma(U, \mathcal{O}_U) \to \Gamma(V, \mathcal{O}_V) = A$$

is nothing else but the inclusion map $C \subset A$ and therefore is injective [18] (see also [5]).

On the other hand, the surjection $B \twoheadrightarrow C$ provides us with a canonical isomorphism $U \cong \mathrm{Spec}(B/\ker(u^*))$; in addition, we observe that $\mathrm{Spec}(B/\ker(u^*))$ is a (closed) group subscheme of $\mathrm{Spec}(B) = W$. We denote $\mathrm{Spec}(B/\ker(u^*))$ by $u(V)$ and call it the image of u or the image of V with respect to u and denote by $u(V)$. Notice that the set theoretic image of u is closed and our definition of the image of u coincides with the one given in [4, Sect. 5.1.1].

One may easily check that the closed embedding $j : u(V) \hookrightarrow V$ induced by $B \twoheadrightarrow B/\ker(u^*)$ is an image in the category of (affine) schemes over K. This means that if $\alpha, \beta : W \to S$ are two morphisms of schemes over K such that their *restrictions* to $u(V)$ do coincide, i.e., $\alpha j = \beta j$ (as morphisms from $u(V)$ to S) then $\alpha u = \beta u$ (as morphisms from U to S). It follows that j is also an image in the category of finite commutative group schemes. group [21, Sect. 10].

Theorem 1.2 (Theorem of Gabriel [18,5]). *The category of finite commutative group schemes over a field is abelian.*

Remark 1.3. Let V be a finite commutative group scheme over K and let W be its finite closed group subscheme. If $V \to U$ is a *surjective* morphism of finite commutative group schemes over K then [5]

$$\#(V) = \#(W) \cdot \#(U).$$

Recall that $\Gamma(W, \mathcal{O}_W)$ is the quotient of $\Gamma(V, \mathcal{O}_V)$. In particular, if the orders of V and W do coincide then $V = W$.

1.4. Abelian varieties over fields. We refer the reader to the books of Mumford [16], Shimura [26] for basic properties of abelian varieties (see also Lang's book [8] and papers of Waterhouse [30], Deligne [2], Milne [13] and Oort [20]). If X is an abelian variety over K then we write $\mathrm{End}(X)$ for the ring of all K-endomorphisms of X. If m is an integer then write m_X for the multiplication by m in X; in particular, 1_X is the identity map. (Sometimes we will use notation m instead of m_X.)

If Y is an abelian variety over K then we write $\mathrm{Hom}(X, Y)$ for the group of all K-endomorphisms $X \to Y$.

Remark 1.5. Warning: sometimes in the literature, including my own papers, the notation $\mathrm{End}(X)$ is used for the ring of $\bar{K}$-endomorphisms.

It is well known [16, Sect. 19, Theorem 3] that $\mathrm{Hom}(X, Y)$ is a free commutative group of finite rank. We write X^t for the dual of X (See [13, Sect. 9–10] for the definition and basic properties of the dual of an abelian variety.) In particular, X^t is also an abelian variety over K that is isogenous to X (over K). If $u \in \mathrm{Hom}(X, Y)$ then we write u^t for its dual in $\mathrm{Hom}(Y, X)$. We have

$$\bar{X}^t = \overline{X^t}.$$

If n is a positive integer then we write X_n for the kernel of n_X; it is a finite commutative (sub)group scheme (of X) over K of rank $2\dim(X)$. By definition, $X_n(\bar{K})$ is the kernel of multiplication by n in $X(\bar{K})$.

If n is not divisible by $\mathrm{char}(K)$ then X_n is an étale group scheme and it is well-known [16, Sect. 4] that $X_n(\bar{K})$ is a free $\mathbb{Z}/n\mathbb{Z}$-module of rank $2\dim(X)$ and all $\bar{K}$-points of X_n are defined over a finite separable extension of K. In particular, $X_n(\bar{K})$ carries a natural structure of Galois module.

1.6. Isogenies. Let $W \subset X$ be a finite group subscheme over K. It follows from the analogue of Lagrange theorem that $W \subset X_d$ for $d = \#(W)$. The quotient $Y := X/W$ is an abelian variety over K and the canonical isogeny $\pi : X \to X/W = Y$

has kernel W and degree $\#(W)$ ([16, Sect. 12, Corollary 1 to Theorem 1], [3, Sect. 2, pp. 307-314]). In particular, every homomorphism of abelian varieties $u : X \to Z$ over K with $W \subset \ker(u)$ factors through π, i.e., there exists a unique homomorphism of abelian varieties $v : Y \to Z$ over K such that

$$u = v\pi.$$

If m is a positive integer then

$$\pi m_X = m_Y \pi \in \mathrm{Hom}(X, Y).$$

Let us put

$$m^{-1}W := \ker(\pi m_X) = \ker(m_Y \pi) \subset X.$$

For every commutative K-algebra R the group of R-points $m^{-1}W(R)$ is the set of all $x \in X(R)$ with

$$mx \in W(R) \subset X(R).$$

For example, if $W = X_n$ then

$$Y = X, \pi = n_X, m^{-1}X_n = X_{nm}.$$

In general, if $W \subset X_n$ then $m^{-1}W$ is a closed group subscheme in $X_n m$. E.g., W is always a closed group subscheme of X_{dm} and therefore is a finite group subscheme of X over K. The order

$$\#(m^{-1}W) = \deg(\pi m_X) = \deg(\pi)\deg(m_X) = \#(W) \cdot m^{2\dim(X)}.$$

We have

$$X_m \subset m^{-1}W, \ m_X(m^{-1}W) \subset W$$

and the kernel of $m_X : m^{-1}W \to W$ coincides with X_m.

Lemma 1.7. *The image $m_X(m^{-1}W) = W$.*

Proof. Let us denote the image by G. By Remark 1.3, $\#(G)$ is the ratio

$$\#(m^{-1}W)/\#(X_m) = \dim(W),$$

i.e., the orders of G and W do coincide. Since $G \subset W$, we have (by the same Remark) $G = W$. $\square$

Example 1.8. If $W = X_n$ then $m^{-1}X_n = X_{nm}$ and therefore $m(X_{nm}) = X_n$.

Lemma 1.9. *If r is a positive integer then $r(X_n) = X_{n_1}$ where $n_1 = n/(n, r)$.*

Proof. We have $r = (n, r) \cdot r_1$ where r_1 is a positive integer such that n_1 and r_1 are relatively prime. This implies that $r_1(X_{n_1}) = X_{n_1}$. By Lemma 1.9, $(n, r)(X_n) = X_{n_1}$. This implies that

$$r(X_n) = r_1(n, r)(X_n) = r_1((n, r)(X_n)) = r_1(X_{n_1}) = X_{n_1}.$$

$\square$

Lemma 1.10. *Let X and Y be abelian varieties over a field K. Let $u : X \to Y$ be a K-homomorphism of abelian varieties. Let $n > 1$ be an integer and $u_n : X_n \to Y_n$ the morphism of commutative group schemes over K induced by u.*

(i) Suppose that u is an isogeny and $\deg(u)$ and n are relatively prime. Then $u_n : X_n \to Y_n$ is an isomorphism.

(ii) Suppose that $u_n : X_n \to Y_n$ is an isomorphism. Then u is an isogeny and $\deg(u)$ and n are relatively prime.

Proof. Let u be an isogeny such that $m := \deg(u)$ and n are relatively prime. Then $\ker(u) \subset X_m$. It follows that there exists a K-isogeny $v : Y \to X$ such that

$$vu = m_X, \quad uv = m_Y.$$

(i). Since multiplication by m is an automorphism of both X_n and Y_m, we conclude that $u_n : X_n \to Y_n$ and $v_n : Y_n \to X_n$ are isomorphisms.

(ii). Suppose that u_n is an isomorphism. This implies that the orders of X_n and Y_n coincide and therefore $\dim(X) = \dim(Y)$. We need to prove that u is isogeny and $\deg(u)$ and n are relatively prime. In order to do that, we may assume that K is algebraically closed (replacing K, X, Y, u by $\bar{K}, \bar{X}, \bar{Y}, \bar{u}$ respectively). Let us put $Z := u(Y) \subset X$: clearly, Z is a (closed) abelian subvariety of Y and therefore $\dim(Z) \leq \dim(Y)$. It is also clear that $u : X \to Y$ coincides with the composition of the natural surjection $X \to u(X) = Z$ and the inclusion map $j : Z \hookrightarrow X$. This implies that $u_n(X_n)$ is a (closed) group subscheme of $j_n(Z_n) \subset Y_n$. It follows that

$$\#(u_n(X_n)) \leq \#(j_n(Z_n)) \leq \#(Z_n) = n^{2\dim(Z)}.$$

Since u_n is an isomorphism, $u_n(X_n) = Y_n$ and therefore

$$\#(u_n(X_n)) = \#(Y_n) = n^{2\dim(Y)}.$$

It follows that

$$n^{2\dim(Y)} \leq n^{2\dim(Z)}$$

and therefore $\dim(Y) \leq \dim(Z)$. (Here we use that $n > 1$.) Since Z is a closed subvariety in Y, we conclude that $\dim(Z) = \dim(Y)$ and $Y = Z$. In other words, u is surjective. Taking into account that $\dim(X) = \dim(Y)$, we conclude that u ia an isogeny.

Now let $m = dr$ where d is the largest common divisor of n and m. Then r and n are relatively prime; in particular, multiplication by r is an automorphism of X_n. Let us denote $\ker(u)$ by W: it is a finite commutative group scheme over K of order m and therefore

$$W \subset X_m.$$

This implies that for every commutative K-algebra R we have

$$m \cdot W(R) = \{0\}.$$

On the other hand, since u_n is an isomorphism, the kernel of $W(R) \xrightarrow{n} W(R)$ is $\{0\}$. Since $d \mid n$, the kernel of $W(R) \xrightarrow{d} W(R)$ is also $\{0\}$. This implies that $r \cdot W(R) = \{0\}$ for all R. Hence $W \subset X_r$. It follows that $\deg(u) = \#(W)$ divides $\#(X_r) = r^{2\dim(X)}$ and therefore is coprime to n. $\qquad\square$

The next statement will be used only in Section 12.

Proposition 1.11. *Let X and Y be abelian varieties over a field K. Suppose that for every prime ℓ there exists an isogeny $X \to Y$, whose degree is not divisible by ℓ. Then for every positive integer n there exists an isogeny $X \to Y$, whose degree is coprime to n. In particular, $X_n \cong Y_n$.*

Proof. Recall that the additive group $\mathrm{Hom}(X, Y)$ is isomorphic to $\mathbb{Z}^\rho$ for some nonnegative integer ρ. In our case, X and Y are isogenous over K and therefore $\rho > 0$.

Let n be a positive integer and let $P(n)$ be the (finite) set of its prime divisors. For each $\ell \in P(n)$ pick an isogeny $v^{(\ell)} : X \to Y$, whose degree is not divisible by ℓ. By Lemma 1.10(i), $v^{(\ell)}$ induces an isomorphism $X_\ell \cong Y_\ell$. Now, by the Chinese Remainder Theorem, there exists $u \in \mathrm{Hom}(X, Y) \cong \mathbb{Z}^\rho$ such that

$$u - v^{(\ell)} \in \ell \cdot \mathrm{Hom}(X, Y) \ \forall \ \ell \in P.$$

This implies that for each $\ell \in P$ the homomorphisms u and $v^{(\ell)}$ induce the same morphism $X_\ell \cong Y_\ell$, which, as we know, is an isomorphism. It follows from Lemma By Lemma 1.10(ii) that u is an isogeny, whose degree is not divisible by ℓ. Hence $\deg(u)$ and n are coprime. Applying again Lemma 1.10(i), we conclude that u induces an isomorphism $X_n \cong Y_n$. $\qquad\square$

1.12. Polarizations. A homomorphism $\lambda : X \to X^t$ is a *polarization* if there exists an ample invertible sheaf $\mathcal{L}$ on $\bar{X}$ such that $\bar{\lambda}$ coincides with

$$\Lambda_{\mathcal{L}} : \bar{X}^t \to \bar{X}^t, \ z \mapsto \mathrm{cl}(T_z^* L \otimes L^{-1})$$

where $T_z : \bar{X} \to \bar{X}$ is the translation map

$$x \mapsto x + z$$

and cl stands for the isomorphism class of an invertible sheaf. Recall [16, Sect, 6, Proposition 1; Sect. 8, Theorem 1; Sect. 13, Corollary 5] that a polarization is an *isogeny*. If λ is an isomorphism, i.e., $\deg(\lambda) = 1$, we call λ a *principal polarization* and the pair (X, λ) is called a principally polarized abelian variety (over K).

If $n := \deg(\lambda) = \#(\ker(\lambda))$ then $\ker(\lambda)$ is killed by multiplication by n, i.e., $\ker(\lambda) \subset X_n$. For every positive integer m we write λ^n for the polarization

$$X^m \to (X^m)^t = (X^t)^m, \ (x_1, \ldots, x_m) \mapsto (\lambda(x_1), \ldots, \lambda(x_m))$$

that corresponds to the ample invertible sheaf $\otimes_{i=1}^{m} \mathrm{pr}_i^* \mathcal{L}$ where $\mathrm{pr}_i : X^m \to X$ is the ith projection map. We have

$$\dim(X^m) = m \cdot \dim(X), \ \deg(\lambda^m) = \deg(\lambda)^m$$

and $\ker(\lambda^m) = \ker(\lambda)^m \subset (X^m)_n$ if $\ker(\lambda) \subset X_n$.

There exists a *Riemann form* - a skew-symmetric pairing of group schemes over $\bar{K}$ [16, Sect. 23]

$$e_\lambda : \ker(\bar{\lambda}) \times \ker(\bar{\lambda}) \to \mathbf{G}_\mathrm{m}$$

where $\mathbf{G}_\mathrm{m}$ is the multiplicative group scheme over $\bar{K}$.

If

$$e_{\lambda^m} : \ker(\bar{\lambda}^m) \times \ker(\bar{\lambda}^m) \to \mathbf{G}_\mathrm{m}$$

is the Riemann form for λ^m then in obvious notation

$$e_{\lambda^m}(x, y) = \prod_{i=1}^{m} e_\lambda(x_i, y_i)$$

where

$$x = (x_1, \ldots, x_m), \ y = (y_1, \ldots, y_m) \in \ker(\bar{\lambda})^m = \ker(\bar{\lambda}^m).$$

We have

$$\mathrm{Mat}_m(\mathbb{Z}) \subset \mathrm{Mat}_m(\mathrm{End}(\bar{X})) = \mathrm{End}(X^m).$$

One may easily check that every $u \in \mathrm{Mat}_m(\mathbb{Z})$ leaves the group subscheme $\ker(\bar{\lambda}^m)$ invariant and

$$e_{\lambda^m}(ux, y) = e_{\lambda^m}(x, u^* y)$$

where u^* is the transpose of the matrix u. Notice that u^* viewed as an element of

$$\mathrm{Mat}_m(\mathbb{Z}) \subset \mathrm{Mat}_m(\mathrm{End}(X^t)) = \mathrm{End}((X^t)^m)$$

coincides with $u^t \in \mathrm{End}((X^m)^t)$.

1.13. Polarizations and isogenies. Let $W \subset \ker(\lambda)$ be a finite group subscheme over K. Recall that $Y := X/W$ is an abelian variety over K and the canonical isogeny $\pi : X \to X/W = Y$ has kernel W and degree $\#(W)$.

Suppose that $\bar{W}$ is isotropic with respect to e_λ, i.e., the restriction of e_λ to $\bar{W} \times \bar{W}$ is trivial. Then there exists an ample invertible sheaf $\mathcal{M}$ on $\bar{Y}$ such that $\mathcal{L} \cong \bar{\pi}^* \mathcal{M}$ [16, Sect. 23, Corollary to Theorem 2, p. 231] and the $\bar{K}$-polarization $\Lambda_{\bar{\mathcal{M}}} : \bar{Y} \to \bar{Y}^t$ satisfies

$$\bar{\lambda} = \overline{\pi^t} \Lambda_{\bar{\mathcal{M}}} \bar{\pi}.$$

Since $\bar{\pi}^t$ and $\bar{\pi}$ are isogenies that are defined over K, the polarization $\Lambda_{\bar{\mathcal{M}}}$ is also defined over K, i.e., there exists a K-isogeny $\mu : Y \to Y^t$ such that $\Lambda_{\bar{\mathcal{M}}} = \bar{\mu}$ and

$$\lambda = \pi^t \mu \pi.$$

It follows that

$$\deg(\lambda) = \deg(\pi) \deg(\mu) \deg(\pi^t) = \deg(\pi)^2 \deg(\mu) = (\#(W))^2 \deg(\mu).$$

Therefore μ is a principal polarization (i.e., $\deg(\mu) = 1$) if and only if

$$\deg(\lambda) = (\#(W))^2.$$

2. ℓ-divisible groups, abelian varieties and Tate modules

Let h be a non-negative integer and ℓ a prime. The following notion was introduced by Tate [28,25].

Definition 2.1. An ℓ-divisible group G over K of height h is a sequence $\{G_\nu, i_\nu\}_{\nu=1}^{\infty}$ in which:

- G_ν is a finite commutative group scheme over K of order $\ell^{h\nu}$.
- i_ν is a closed embedding $G_\nu \hookrightarrow G_{\nu+1}$ that is a morphism of group schemes. In addition, $i_\nu(G_\nu)$ is the kernel of multiplication by ℓ^ν in $G_{\nu+1}$.

Example 2.2. Let X be an abelian variety over K of dimension d. Then it is known [28,25] that the sequence $\{X_{\ell^\nu}\}_{\nu=1}^{\infty}$ is an ℓ-divisible group over K of height $2d$. Here i_ν is the *inclusion map* $X_{\ell^\nu} \hookrightarrow X_{\ell^{\nu+1}}$. We denote this ℓ-divisible group by $X(\ell)$.

2.3. Homomorphisms of ℓ-divisible groups and abelian varieties. If $H = \{H_\nu, j_\nu\}_{\nu=1}^{\infty}$ is an ℓ-divisible group over K then a morphism $u : G \to H$ is a sequence $\{u_{(\nu)}\}_{\nu=1}^{\infty}$ of morphisms of group schemes over K

$$u_{(\nu)} : G_\nu \to H_\nu$$

such that the composition

$$u_{(\nu+1)}i_\nu : G_\nu \hookrightarrow G_{\nu+1} \to H_{\nu+1}$$

coincides with

$$j_\nu u_{(\nu)} : G_\nu \to H_\nu \hookrightarrow H_{\nu+1},$$

i.e., the diagram

$$
\begin{array}{ccc}
G_\nu & \xrightarrow{\ u_{(\nu)}\ } & H_\nu \\
\downarrow{\scriptstyle i_\nu} & & \downarrow{\scriptstyle j_\nu} \\
G_{\nu+1} & \xrightarrow{\ u_{(\nu+1)}\ } & H_{\nu+1}
\end{array}
$$

is commutative.

Remark 2.4. A morphism u is an isomorphism of ℓ-divisible groups if and only if all $u_{(\nu)}$ are isomorphisms of the corresponding finite group schemes.

The group $\mathrm{Hom}(G, H)$ of morphisms from G to H carries a natural structure of $\mathbb{Z}_\ell$-module induced by the natural structures of $\mathbb{Z}/\ell^\nu = \mathbb{Z}_\ell/\ell^\nu$-module on $\mathrm{Hom}(G_\nu, H_\nu)$. Namely, if $u = \{u_{(\nu)}\}_{\nu=1}^\infty \in \mathrm{Hom}(G, H)$ and $a \in \mathbb{Z}_\ell$ then $au = \{(au)_{(\nu)}\}_{\nu=1}^\infty$ may be defined as follows. For each ν pick $a_\nu \in \mathbb{Z}$ with $a - a_\nu \in \ell^\nu \mathbb{Z}_\ell$ and put

$$(au)_{(\nu)} := a_\nu u_{(\nu)} : G_\nu \to H_\nu.$$

Since multiplication by ℓ^ν kills G_ν, the definition of $(au)_{(\nu)}$ does not depend on the choice of a_ν.

Let X and Y be abelian varieties over K. There is a natural homomorphism of commutative groups $\mathrm{Hom}(X, Y) \to \mathrm{Hom}(X(\ell), Y(\ell))$. Namely, if $u \in \mathrm{Hom}(X, Y)$ then $u(X_{\ell^\nu})$ lies in the kernel of multiplication by ℓ^ν, i.e. $u(X_{\ell^\nu}) \subset Y_{\ell^\nu}$. In fact, we get the natural homomorphism

$$\mathrm{Hom}(X, Y) \otimes \mathbb{Z}/\ell^\nu \to \mathrm{Hom}(X_{\ell^\nu}, Y_{\ell^\nu}),$$

which is known to be an embedding. (See also Lemma 9.1 below.)

Since $\mathrm{Hom}(X(\ell), Y(\ell))$ is a $\mathbb{Z}_\ell$-module, we get the natural homomorphism of $\mathbb{Z}_\ell$-modules

$$\mathrm{Hom}(X, Y) \otimes \mathbb{Z}_\ell \to \mathrm{Hom}(X(\ell), Y(\ell)).$$

Explicitly, if $u \in \mathrm{Hom}(X, Y) \otimes \mathbb{Z}_\ell$ then for each ν we may pick

$$w(\nu) \in \mathrm{Hom}(X, Y) = \mathrm{Hom}(X, Y) \otimes 1 \subset \mathrm{Hom}(X, Y) \otimes \mathbb{Z}_\ell$$

such that

$$u - w(\nu) \in \ell^\nu \cdot \{\mathrm{Hom}(X,Y) \otimes \mathbb{Z}_\ell\} = \{\ell^\nu \cdot \mathrm{Hom}(X,Y)\} \otimes \mathbb{Z}_\ell = \mathrm{Hom}(X,Y) \otimes \ell^\nu \mathbb{Z}_\ell.$$

Then the corresponding morphism of group schemes $u_{(\nu)} := w(\nu) : X_{\ell^\nu} \to Y$ does not depend on the choice of $w(\nu)$ and defines the corresponding morphism of ℓ-divisible groups

$$u_{(\nu)} : X_{\ell^\nu} \to Y_{\ell^\nu}; \ \nu = 1, 2, \dots.$$

Remark 2.5. Since $\mathrm{Hom}(X,Y)$ is a free commutative group of finite rank, the $\mathbb{Z}_\ell$-module $\mathrm{Hom}(X,Y) \otimes \mathbb{Z}_\ell$ is a free module of finite rank.

The following assertion seems to be well known (at least, when $\ell \neq \mathrm{char}(K)$).

Lemma 2.6. *The natural homomorphism of $\mathbb{Z}_\ell$-modules*

$$\mathrm{Hom}(X,Y) \otimes \mathbb{Z}_\ell \to \mathrm{Hom}(X(\ell), Y(\ell))$$

is injective.

Proof. If it is not injective and u lies in the kernel then $u_{(\nu)} \in \ell^\nu \cdot \mathrm{Hom}(X,Y)$ for all ν. Since $u - u_{(\nu)} \in \ell^\nu \cdot \{\mathrm{Hom}(X,Y) \otimes \mathbb{Z}_\ell\}$, we conclude that $u \in \ell^\nu \cdot \{\mathrm{Hom}(X,Y) \otimes \mathbb{Z}_\ell\}$ for all ν. Since $\mathrm{Hom}(X,Y) \otimes \mathbb{Z}_\ell$ is a free $\mathbb{Z}_\ell$-module of finite rank, it follows that $u = 0$. $\qquad\square$

Corollary 2.7. *The following conditions are equivalent:*

> *(i) There exists an isogeny $u : X \to Y$, whose degree is not divisible by ℓ.*
> *(ii) There exists $w \in \mathrm{Hom}(X,Y) \otimes \mathbb{Z}_\ell$ that induces an isomorphism of ℓ-divisible groups $X(\ell) \to Y(\ell)$.*

Proof. Let $u : X \to Y$ be an isogeny, whose degree is not divisible by ℓ. Applying Lemma 1.10(i) to all $n = \ell^\nu$, we conclude that u induces an isomorphism $X(\ell) \cong Y(\ell)$.

Now suppose that $w \in \mathrm{Hom}(X,Y) \otimes \mathbb{Z}_\ell$ that induces an isomorphism of ℓ-divisible groups $X(\ell) \to Y(\ell)$. In particular, w induces an isomorphism of finite group schemes $w_{(1)} : X_\ell \cong Y_\ell$. On the other hand, there exists $u \in \mathrm{Hom}(X,Y)$ such that

$$w - u \in \ell \cdot \{\mathrm{Hom}(X,Y) \otimes \mathbb{Z}_\ell\} = \mathrm{Hom}(X,Y) \otimes \ell\mathbb{Z}_\ell.$$

This implies that u and w induce the same morphism of finite group schemes $X_\ell \to Y_\ell$. It follows that the morphism

$$u_\ell = u_{(1)} : X_\ell \to Y_\ell$$

induced by u coincides with $w_{(1)}$ and therefore is an isomorphism. Now Lemma 1.10(ii) implies that u is an isogeny, whose degree is not divisible by ℓ. $\qquad\square$

2.8. Tate modules. In this subsection we assume that ℓ is a prime different from $\mathrm{char}(K)$. If $n = \ell^\nu$ then X_n is an étale finite group scheme of order $n^{2\dim(X)}$ and we will identify its with the Galois module of its $\bar{K}$-points. (Actually, all points of X_n are defined over a separable algebraic extension of K). The Tate ℓ-module $T_\ell(X)$ is defined as the projective limit of Galois modules X_{ℓ^ν} where the transition map $X_{\ell^{\nu+1}} \to X_{\ell^\nu}$ is multiplication by ℓ. The Tate module carries a natural structure of free $\mathbb{Z}_\ell$-module of rank $2\dim(X)$; it is also provided with a natural structure of Galois module in such a way that natural homomorphisms $T_\ell(X) \to X_{\ell^\nu}$ induce isomorphisms of Galois modules

$$T_\ell(X) \otimes \mathbb{Z}/\ell^\nu \cong X_{\ell^\nu}.$$

Explicitly, $T_\ell(X)$ is the set of all collections $x = \{x_\nu\}_{\nu=1}^\infty$ with

$$x_\nu \in X_{\ell^\nu}, \quad x_{\nu+1} = \ell x_\nu \ \forall \nu.$$

The map $x \mapsto x_\nu$ defines the surjective homomorphism of Galois modules $T_\ell(X) \to X_{\ell^\nu}$, whose kernel coincides with $\ell^\nu \cdot T_\ell(X)$ and therefore induces the isomorphism of Galois modules $T_\ell(X)/\ell^\nu \cong X_{\ell^\nu}$ mentioned above.

If Y is an abelian variety over K then we write $\mathrm{Hom}_{\mathrm{Gal}}(T_\ell(X), T_\ell(Y))$ for the $\mathbb{Z}_\ell$-module of all homomorphisms of $\mathbb{Z}_\ell$-modules $T_\ell(X) \to T_\ell(Y)$ that commute with the Galois action(s), i.e., are also homomorphisms of Galois modules.

The $\mathbb{Z}_\ell$-module $\mathrm{Hom}_{\mathrm{Gal}}(T_\ell(X), T_\ell(Y))$ is the set of collections $w = \{w_\nu\}_{\nu=1}^\infty$ of homomorphisms of Galois modules

$$w_\nu : T_\ell(X)/\ell^\nu = X_{\ell^\nu} \to Y_{\ell^\nu} = T_\ell(Y)/\ell^\nu$$

such that

$$w_\nu(x_\nu) = \ell \cdot u w_{\nu+1}(x_{\nu+1}) \quad \forall x = \{x_\nu\}_{\nu=1}^\infty \in T_\ell(X).$$

Now if $z \in X_{\ell^\nu}$ then there exists $x \in T_\ell(X)$ with $x_\nu = z$. We have $\ell x_{\nu+1} = x_\nu = z$ and

$$w_\nu(z) = w_\nu(x_\nu) = \ell \cdot w_{\nu+1}(x_{\nu+1}) = w_{\nu+1}(\ell x_{\nu+1}) = w_{\nu+1}(x_\nu) = w_{\nu+1}(z),$$

i.e., the restriction of $w_{\nu+1}$ to X_{ℓ^ν} coincides with w_ν. This means that the collection $\{w_\nu\}_{\nu=1}^\infty$ defines a morphism of ℓ-divisible groups over K

$$X(\ell) \to Y(\ell).$$

Conversely, if $u = \{u_{(\nu)}\}_{\nu=1}^\infty$ is a morphism $X(\ell) \to Y(\ell)$ over K then

$$u_{(\nu)} : X_{\ell^\nu} \to Y_{\ell^\nu}$$

is a homomorphism of Galois modules; in addition, the restriction of $u_{(\nu+1)}$ to X_{ℓ^ν} coincides with $u_{(\nu)}$. This implies that for each $\{x_\nu\}_{\nu=1}^\infty \in T_\ell(X)$

$$u_{(\nu)}(x_\nu) = u_{(\nu+1)}(x_\nu) = u_{(\nu+1)}(\ell x_{\nu+1}) = \ell u_{(\nu+1)}(x_{\nu+1})$$

for all ν. This means that the collection $\{u_{(\nu)}\}_{\nu=1}^{\infty}$ defines a homomorphism of Galois modules $T_\ell(X) \to T_\ell(Y)$. Those observations give us the natural isomorphism of $\mathbb{Z}_\ell$-modules

$$\mathrm{Hom}(X(\ell), Y(\ell)) = \mathrm{Hom}_{\mathrm{Gal}}(T_\ell(X), T_\ell(Y)).$$

3. Useful results

Theorem 3.1 ([32,34,14]). *Let X be an abelian variety of positive dimension over a field K and X^t its dual. Then $(X \times X^t)^4$ admits a principal K-polarization.*

We prove Theorem 3.1 in Section 7.

Theorem 3.2 ([11]). *Let X be an abelian variety over K. The set of abelian K-subvarieties of X is finite, up to the action of the group $\mathrm{Aut}(X)$ of K-automorphisms of X.*

We sketch the proof of Theorem 3.2 in Section 5.

Lemma 3.3 (Tate ([27], Sect. 2, p. 136)). *Let K be a finite field, and let g and d be positive integers. The set of K-isomorphism classes of g-dimensional abelian varieties over K that admit a K-polarization of degree d is finite.*

Lemma 3.3 will be proven in Section 6.

Theorem 3.4 ([32], Th. 4.1). *Let K be a finite field, g a positive integer. Then the set of K-isomorphism classes of g-dimensional abelian varieties over K is finite.*

Proof of Theorem 3.4 (modulo Theorem 3.1 and Lemma 3.3). Suppose that X is a g-dimensional abelian variety over K. By Lemma 3.3, the set of $4g$-dimensional abelian varieties over K of the form $(X \times X^t)^4$ is finite, up to K-isomorphism. The abelian variety X is isomorphic over K to an abelian subvariety of $(X \times X^t)^4$. In order to finish the proof, one has only to recall that thanks to Theorem 3.2, the set of abelian subvarieties of a given abelian variety is finite, up to a K-isomorphism. $\square$

We need Theorem 1.2 in order to state the following assertion.

Corollary 3.5 (Corollary to Theorem 3.4). *Let X be an abelian variety of positive dimension over a finite field K. There exists a positive integer $r = r(X, K)$ that enjoys the following properties:*

(i) *If Y is an abelian variety over K that is K-isogenous to X then there exists a K-isogeny $\beta : X \to Y$ such that $\ker(\beta) \subset X_r$.*

(ii) *If n is a positive integer and $W \subset X_n$ is a group subscheme over K then there exists an endomorphism $u \in \mathrm{End}(X)$ such that*

$$rW \subset uX_n \subset W.$$

Remark 3.6. The assertion 3.5(i) follows readily from Theorem 3.4.

We prove Corollary 3.5(ii) in Section 8.

4. Main results

Theorem 4.1. *Let X be an abelian variety of positive dimension over a finite field K. There exists a positive integer $r_1 = r_1(X, K)$ that enjoys the following properties:*

Let n be a positive integer and $u_n \in \mathrm{End}(X_n)$. Let us put $m = n/(n, r_1)$. Then there exists $u \in \mathrm{End}(X)$ such that the images of u and u_n in $\mathrm{End}(X_m)$ do coincide.

We prove Theorem 4.1 in Section 10.

Applying Theorem 4.1 to a product $X = A \times B$ of abelian varieties A and B, we obtain the following statement.

Theorem 4.2. *Let A, B be abelian varieties of positive dimension over a finite field K. There exists a positive integer $r_2 = r_2(A, B)$ that enjoys the following properties:*

Suppose that n is a positive integer and $u_n : A_n \to B_n$ is a morphism of group schemes over K. Let us put $m = n/(n, r_2)$. Then there exists a homomorphism $u : A \to B$ of abelian varieties over K such that the images of u and u_n in $\mathrm{Hom}(A_m, B_m)$ do coincide.

The following assertions follow readily from Theorem 4.2.

Corollary 4.3 (First Corollary to Theorem 4.2). *If n and r_2 are relatively prime (e.g., n is a prime that does not divide r_2) then the natural injection*

$$\mathrm{Hom}(A, B) \otimes \mathbb{Z}/n \hookrightarrow \mathrm{Hom}(A_n, B_n)$$

is bijective.

Corollary 4.4 (Second Corollary to Theorem 4.2). *Let ℓ be a prime and $\ell^{r(\ell)}$ is the exact power of ℓ dividing r_2. Then for each positive integer i the image of*

$$\mathrm{Hom}(A_{\ell^{i+r(\ell)}}, B_{\ell^{i+r(\ell)}}) \to \mathrm{Hom}(A_{\ell^i}, B_{\ell^i})$$

coincides with the image of

$$\mathrm{Hom}(A, B) \otimes \mathbb{Z}/\ell^i \hookrightarrow \mathrm{Hom}(A_{\ell^i}, B_{\ell^i}).$$

5. Abelian subvarieties

We follow the exposition in [11].

The next statement is a corollary of a finiteness result of Borel and Harish-Chandra [1, Theorem 6.9]; it may also be deduced from the Jordan–Zassenhaus theorem [23, Theorem 26.4].

Proposition 5.1 ([11], p. 514). *Let F be a finite-dimensional semisimple $\mathbb{Q}$-algebra, M a finitely generated right F-module, L a $\mathbb{Z}$-lattice in M. Let G be the group of those automorphisms σ of the F-module M for which $\sigma(L) = L$. Then the number of G-orbits of the set of F-submodules of M is finite.*

Now let X be an abelian variety over K. We are going to apply Proposition 5.1 to

$$F = \operatorname{End}(X) \otimes \mathbb{Q}, \ M = \operatorname{End}(X) \otimes \mathbb{Q}, \ L = \operatorname{End}(X).$$

One may identify G with the group $\operatorname{Aut}(X) = \operatorname{End}(X)^*$ of automorphisms of X: here elements of $\operatorname{End}(X)^*$ act as left multiplications on $\operatorname{End}(X) \otimes \mathbb{Q} = M$.

On the other hand, to each abelian K-subvariety $Y \subset X$ corresponds the right ideal

$$I(Y) = \{u \in \operatorname{End}(X) \mid u(X) \subset Y\}$$

and the F-submodule

$$I(Y)_{\mathbb{Q}} = I(Y) \otimes \mathbb{Q} \subset \operatorname{End}(X) \otimes \mathbb{Q} = M.$$

Using the theorem of Poincaré–Weil [13, Proposition 12.1], one may prove ([11, p. 515] that $I(Y)_{\mathbb{Q}}$ uniquely determines Y. Even better, if Y' is an abelian K-subvariety of X and

$$uI(Y)_{\mathbb{Q}} = I(Y')_{\mathbb{Q}}$$

for $u \in \operatorname{Aut}(X) = \operatorname{End}(X)^*$ then $Y' = u(Y)$. Now Proposition 5.1 implies the finiteness of the number of orbits of the set of abelian K-subvarieties of X under the natural action of $\operatorname{Aut}(X)$. This proves Theorem 3.2. (See [10] for variants and complements.)

6. Polarized abelian varieties

Lemma 6.1 (Mumford's lemma [15]). *Let X be an abelian variety of positive dimension over a field K. If $\lambda : X \to X^t$ is a polarization then there exists an ample invertible sheaf $\mathcal{L}$ on X such that*

$$\Lambda_{\bar{\mathcal{L}}} = 2\bar{\lambda}$$

where $\bar{\mathcal{L}}$ is the invertible sheaf on $\bar{X}$ induced by $\mathcal{L}$.

Proof. See [15, Ch. 6, Sect. 2, pp. 120–121] where a much more general case of abelian schemes is considered. (In notation of [15], S is the spectrum of K.) Let me just recall an explicit construction of $\mathcal{L}$. Let $\mathbb{P}$ be the universal Poincaré invertible sheaf on $X \times X^t$ [13, Sect. 9]. Then $\mathcal{L} := (1_X, \lambda)^*\mathbb{P}$ where $(1_X, \lambda) : X \to X \times X^t$ is defined by the formula

$$x \mapsto (x, \lambda(x)).$$

$\square$

Proof of Lemma 3.3. So, let X be a g-dimensional abelian variety over a finite field K and let $\lambda : X \to X^t$ be a polarization of degree d. We follow the exposition in [22, p. 243]. By Lemma 6.1, there exists an invertible ample sheaf $\mathcal{L}$ on X such that the self-intersection index of $\bar{\mathcal{L}}$ equals $2^g dg!$ [16, Sect. 16]. The invertible sheaf $\bar{\mathcal{L}}^3$ is very ample, its space of global section has dimension $6^g d$; the self-intersection index of $\mathcal{L}$ equals $6^g dg!$ [16, Sect. 16]. This implies that $\mathcal{L}^3$ is also very ample and gives us an embedding (over K) of X into the $6^g d - 1$-dimensional projective space as a closed K-subvariety of degree $6^g dg!$. All those subvarieties are uniquely determined by their Chow forms ([29, Ch. 1, Sect. 6.5], [6, Lecture 21, pp. 268–273]), whose coefficients are elements of K. Since K is finite and the number of coefficients depends only on the degree and dimension, we get the desired finiteness result. $\square$

7. Quaternion trick

Let X be an abelian variety of positive dimension over a field K and $\lambda : X \to X^l$ a K-polarization. Pick a positive integer n such that

$$\ker(\lambda) \subset X_n.$$

Lemma 7.1. *Suppose that there exists an integer a such that $a^2 + 1$ is divisible by n. Then $X \times X^t$ admits a principal polarization that is defined over K.*

Proof. Let

$$V \subset \ker(\lambda) \times \ker(\lambda) \subset X_n \times X_n \subset X \times X$$

be the graph of multiplication by a in $\ker(\lambda)$. Clearly, V is a finite group subscheme over K that is isomorphic to $\ker(\lambda)$ and therefore its order is equal to $\deg(\lambda)$. Notice that $\deg(\lambda)$ is the square root of $\deg(\lambda^2)$.

For each commutative $\bar{K}$-algebra R the group $\bar{V}(R)$ of R-points coincides with the set of all the pairs (x, ax) with $x \in \ker(\bar{\lambda}) \subset \bar{X}_n$. This implies that for all $(x, ax), (y, ay) \in \bar{V}(R)$ we have

$$e_{\lambda^2}((x, ax), (y, ay)) = e_\lambda(x, y) \cdot e_\lambda(ax, ay) = e_\lambda(x, y) \cdot e_\lambda(a^2 x, y) =$$

$$e_\lambda(x, y) \cdot e_\lambda(-x, y) = e_\lambda(x, y)/e_\lambda(x, y) = 1.$$

In other words, $\bar{V}$ is isotropic with respect to e_{λ^2}; in addition,

$$\#(\bar{V})^2 = \deg(\lambda)^2 = \deg(\lambda^2).$$

This implies that X^2/V is a principally polarized abelian variety over K. On the other hand, we have an isomorphism of abelian varieties over K

$$f : X \times X \to X \times X = X^2, \; (x, y) \mapsto (x, ax) + (0, y) = (x, ax + y)$$

and

$$V = f(\ker \lambda \times \{0\}) \subset f(X \times \{0\}).$$

Thus, we obtain K-isomorphisms

$$X^2/V \cong X/\ker(\lambda) \times X = X^t \times X = X \times X^t.$$

In particular, $X \times X^t$ admits a principal K-polarization and we are done. $\quad\square$

Proof of Theorem 3.1. Choose a quadruple of integers a, b, c, d such that

$$0 \neq s := a^2 + b^2 + c^2 + d^2$$

is congruent to -1 modulo n. We denote by $\mathcal{I}$ the "quaternion"

$$\mathcal{I} = \begin{pmatrix} a & -b & -c & -d \\ b & a & d & c \\ c & -d & a & b \\ d & c & -b & a \end{pmatrix} \in \mathrm{Mat}_4(\mathbb{Z}) \subset \mathrm{Mat}_4(\mathrm{End}(X) = \mathrm{End}(X*^4)).$$

We have

$$\mathcal{I}^*\mathcal{I} = a^2 + b^2 + c^2 + d^2 = s \in \mathbb{Z} \subset \mathrm{Mat}_4(\mathbb{Z}) \subset \mathrm{Mat}_4(\mathrm{End}(X) = \mathrm{End}(X^4)).$$

Let

$$V \subset \ker(\lambda^4) \times \ker(\lambda^4) \subset (X^4)_n \times (X^4)_n \subset X^4 \times X^4 = X^8$$

be the graph of

$$\mathcal{I} : \ker(\lambda^4) \to \ker(\lambda^4).$$

Clearly, V is a finite group subscheme over K and its order is equal to $\deg(\lambda^4)$. Notice that $\deg(\lambda^4)$ is the square root of $\deg(\lambda^8)$.

For each commutative $\bar{K}$-algebra R the group $\bar{V}(R)$ of R-points consists of all the pairs $(x, \mathcal{I}x)$ with $x \in \ker(\bar{\lambda}^4) \subset (\bar{X}^4)_n$. This implies that for all $(x, \mathcal{I}x), (y, \mathcal{I}y) \in \bar{V}(R)$ we have

$$e_{\lambda^4}((x, \mathcal{I}x), (y, \mathcal{I}y)) = e_{\lambda^4}(x, y) \cdot e_{\lambda^4}(\mathcal{I}x, \mathcal{I}y) = e_{\lambda^4}(x, y) \cdot e_\lambda(x, \mathcal{I}^t\mathcal{I}y) =$$

$$e_\lambda(x, y) \cdot e_\lambda(x, sy) = e_\lambda(x, y) \cdot e_\lambda(x, -y) = e_\lambda(x, y)/e_\lambda(x, y) = 1.$$

In other words, $\bar{V}$ is isotropic with respect to e_{λ^4}; in addition,

$$\#(\bar{V})^2 = \deg(\lambda^4)^2 = \deg(\lambda^8).$$

This implies that X^8/V is a principally polarized abelian variety over K. On the other hand, we have an isomorphism of abelian varieties over K

$$f : X^4 \times X^4 \to X^4 \times X^4 = X^8, \ (x,y) \mapsto (x, \mathcal{I}x) + (0,y) = (x, \mathcal{I}x + y)$$

and

$$V = f(\ker(\lambda^4) \times \{0\}) \subset f(X^4 \times \{0\}).$$

Thus, we obtain K-isomorphisms

$$X^4/V \cong X^4/\ker \lambda^4 \times X^4 = (X^4)^t \times X^4 = (X \times X^t)^4.$$

In particular, $(X \times X^t)^4$ admits a principal K-polarization and we are done. $\quad\square$

Remark 7.2. We followed the exposition in [32, Lemma 2.5], [34, Sect. 5]. See [14, Ch. IX, Sect. 1] where Deligne's proof is given.

8. Finite group subschemes of abelian varieties

Proof of Corollary 3.5(ii). Let r be as in 3.5(i). Let us consider the abelian variety $Y := X/W$ and the canonical K-isogeny $\pi : X \to X/W = Y$. Clearly,

$$W = \ker(\pi).$$

Since $W \subset X_n$, there exists a K-isogeny $v : Y \to X/X_n = X$ such that the composition $v\pi$ coincides with multiplication by n in X; in addition,

$$\pi n_X = n_Y \pi : X \to Y$$

is a K-isogeny, whose degree is $\#(W) \times n^{2\dim(X)}$. Here n_X (resp. n_Y) stands for multiplication by n in X (resp. in Y). Let us put

$$U = \ker(\pi n_X) = \ker(n_Y \pi) \subset X;$$

it is a finite commutative group K-(sub)scheme and

$$\#(U) = \#(W) \times n^{2\dim(X)}.$$

Then

$$X_n \subset U, \ W \subset U; \ \pi(U) \subset Y_n, \ n_X(U) \subset W.$$

The order arguments imply that the natural morphisms of group K-schemes

$$\pi : U \to Y_n, \; n_X : U \to W$$

are surjective, i.e.,

$$\pi(U) = Y_n, \; nU = W.$$

We have

$$v(Y_n) = v(\pi(U)) = v\pi(U) = nU = W,$$

i.e.,

$$v(Y_n) = W.$$

By 3.5(i), there exists a K-isogeny $\beta : X \to Y$ with $\ker(\beta) \subset X_r$. Then there exists a K-isogeny $\gamma : Y \to X$ such that $\gamma\beta = r_X$. This implies that

$$\gamma r_Y = r_X \gamma = \gamma\beta\gamma = \gamma(\beta\gamma),$$

i.e.,

$$\gamma r_Y = \gamma(\beta\gamma).$$

It follows that $r_Y = \beta\gamma$, because $\ker(\gamma)$ is finite while $(r_Y - \beta\gamma)Y$ is an abelian subvariety. This implies that

$$\beta(X_n) \supset \beta(\gamma(Y_n)) = \beta\gamma(Y_n) = rY_n.$$

Let us put

$$u = v\beta \in \mathrm{End}(X).$$

We have

$$Y_n \supset \beta(X_n) \supset rY_n.$$

This implies that

$$W = v(Y_n) \supset v(\beta)(X_n) = u(X_n),$$

$$u(X_n) = v(\beta(X_n)) \supset v(rY_n) = r(W)$$

and therefore

$$W \supset u(X_n) \supset r(W).$$

$\square$

9. Dividing homomorphisms of abelian varieties

Results of this Section will be used in the proof of Theorem 4.1 in Section 10.

Throughout this Section, Y is an abelian variety over a field K. The following statement is well known.

Lemma 9.1. *let* $u : Y \to Y$ *be a* K*-isogeny. Suppose that* Z *is an abelian variety over* K*. Let* $v \in \mathrm{Hom}(Y, Z)$ *and* $\ker(u) \subset \ker(v)$ *(as a group subscheme in* Y*). Then there exists exactly one* $w \in \mathrm{Hom}(Y, T)$ *such that* $v = wu$*, i.e., the diagram*

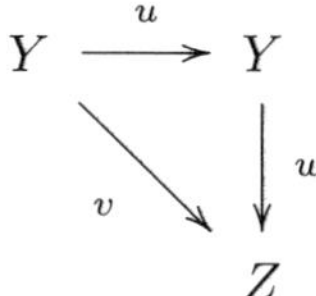

is commutative. In addition, w *is an isogeny if and only if* v *is an isogeny.*

Proof. We have $Y \cong Y/\ker(u)$. Now the result follows from the universality property of quotient maps.

$\square$

Let n be a positive integer and u an endomorphism of Y. Let us consider the homomorphism of abelian varieties over K

$$(n_Y, u) : Y \to Y \times Y, \quad y \mapsto (ny, uy).$$

Then

$$\ker((n_Y, u)) = \ker(Y_n \xrightarrow{u} Y_n) \subset Y_n \subset Y.$$

Slightly abusing notation, we denote the finite commutative group K-(sub)scheme $\ker((n_Y, u))$ by $\{\ker(u) \bigcap Y_n\}$.

Lemma 9.2. *Let* Y *be an abelian variety of positive dimension over a field* K*. Then there exists a positive integer* $h = h(Y, K)$ *that enjoys the following properties: If* n *is a positive integer,* $u, v \in \mathrm{End}(Y)$ *are endomorphisms such that*

$$\{\ker(u) \bigcap Y_n\} \subset \{\ker(v) \bigcap Y_n\}$$

then there exists a K*-isogeny* $w : Y \to Y$ *such that*

$$hv - wu \in n \cdot \mathrm{End}(Y).$$

In particular, the images of hv *and* wu *in* $\mathrm{End}(Y_n)$ *do coincide.*

Proof. Since $\mathcal{O} := \mathrm{End}(Y)$ is an order in the semisimple finite-dimensional $\mathbb{Q}$-algebra $\mathrm{End}(Y) \otimes \mathbb{Q}$, the Jordan–Zassenhaus theorem [23, Th. 26.4] implies that there exists a positive integer M that enjoys the following properties:

if I is a left ideal in $\mathcal{O}$ that is also a subgroup of finite index then there exists $a_I \in \mathcal{O}$ such that the principal left ideal $a \cdot \mathcal{O}$ is a subgroup in I of finite index dividing M; in particular,

$$M \cdot I \subset a_I \cdot \mathcal{O} \subset I.$$

Clearly, such a_I is invertible in $\mathrm{End}(Y) \otimes \mathbb{Q}$ and therefore is an isogeny. Let us put

$$h := M^3.$$

Let us consider the left ideals

$$I = n\mathcal{O} + u\mathcal{O}, \quad J = n\mathcal{O} + v\mathcal{O}$$

in $\mathcal{O}$. Then both I and J are subgroups of finite index in $\mathcal{O}$. So, there exist K-isogenies

$$a_I : Y \to Y, \quad a_J : Y \to Y$$

such that

$$M \cdot I \subset a_I \cdot \mathcal{O} \subset I, \quad M \cdot I \subset a_J \cdot \mathcal{O} \subset J.$$

In particular, there exist $b, c \in \mathcal{O}$ such that

$$M a_I - bu \in n \cdot \mathcal{O}, \quad Mv = c a_J.$$

In obvious notation

$$\{\ker(v) \bigcap Y_n\} \subset \ker(a_J) \subset \{\ker(Mv) \bigcap Y_{Mn}\} = M^{-1}\{\ker(v) \bigcap Y_n\} \subset Y,$$

$$\{\ker(u) \bigcap Y_n\} \subset \ker(a_I) \subset \{\ker(Mu) \bigcap Y_{Mn}\} = M^{-1}\{\ker(u) \bigcap Y_n\} \subset Y.$$

This implies that

$$\ker(a_I) \subset M^{-1}\{\ker(u) \bigcap Y_n\} \subset M^{-1}\{\ker(v) \bigcap Y_n\} \subset M^{-1}\ker(a_J) = \ker(Ma_J)$$

and therefore

$$\ker(a_I) \subset \ker(Ma_J).$$

By Lemma 9.1, there exists a K-isogeny $z : Y \to Y$ such that $Ma_J = za_I$ and therefore $M^2 a_J = Mza_I$. This implies that

$$M^3 v = M^2 ca_J = Mc(Ma_J) = Mc(za_I) = cz(Ma_I) =$$

$$cz[bu + (Ma_I - bu)] = (czb)u + cz(Ma_I - bu).$$

Since $h = M^3$ and $bu - Ma_I$ is divisible by n in $\mathcal{O} = \mathrm{End}(Y)$,

$$hv - (czb)u \in n \cdot \mathrm{End}(Y).$$

So, we may put $w = czb$. $\qquad\qquad\square$

10. Endomorphisms of group schemes

Proof of Theorem 4.1. Let X be an abelian variety of positive dimension over a finite field K. Let us put $Y := X \times X$. Let $h = h(Y)$ be as in Lemma 9.2 and $r = r(Y, K)$ be as in Corollary 3.5. Let us put

$$r_1 = r_1(X, K) := r(Y, K)h(Y, K).$$

Let n be a positive integer and $u_n \in \mathrm{End}(X_n)$. Let W be the graph of u_n in $X_n \times X_n = (X \times X)_n = Y_n$, i.e., the image of

$$(\mathbf{1}_n, u_n) : X_n \hookrightarrow X_n \times X_n = (X \times X)_n = Y_n.$$

Here $\mathbf{1}_n$ is the identity automorphism of X_n.

By Corollary 3.5, there exists $v \in \mathrm{End}(Y)$ such that

$$rW \subset u(Y_n) \subset W.$$

Let $\mathrm{pr}_1, \mathrm{pr}_2 : Y = X \times X \to X$ be the projection maps and

$$q_1 : X = X \times \{0\} \subset X \times X = Y, \ q_2 : X = \{0\} \times X \subset X \times X = Y$$

be the inclusion maps. Let us consider the homomorphisms

$$\mathrm{pr}_1 v, \mathrm{pr}_2 v : Y \to X$$

and the endomorphisms

$$v_1 = q_1 \mathrm{pr}_1 v, \ v_2 = q_1 \mathrm{pr}_2 v \in \mathrm{End}(X \times X) = \mathrm{End}(Y).$$

Clearly,

$$v : Y \to Y = X \times X$$

is "defined" by pair

$$(\mathrm{pr}_1 v, \mathrm{pr}_2 v) : Y \to X \times X = Y.$$

Since W is a graph,

$$\mathrm{pr}_1(W) = X_n, \quad v(Y_n) \subset W$$

and

$$\{\ker(\mathrm{pr}_1 v) \bigcap Y_n\} \subset \{\ker(\mathrm{pr}_2 v) \bigcap Y_n\}.$$

Since q_1 and q_2 are embeddings,

$$\{\ker(v_1) \bigcap Y_n\} \subset \{\ker(v_2) \bigcap Y_n\}.$$

By Lemma 9.2, there exists a K-isogeny $w : Y \to Y$ such that the restrictions of hv_2 and wv_1 to Y_n do coincide. Taking into account that

$$v_1(X \times X) \subset X \times \{0\}, \quad v_2(X \times X) \subset \{0\} \times X,$$

we conclude that if we put

$$w_{12} = \mathrm{pr}_2 w q_1 \in \mathrm{End}(X)$$

then the images of $h\,\mathrm{pr}_2 v$ and $w_{12}\mathrm{pr}_1 v$ in $\mathrm{Hom}(Y_n, X_n) = \mathrm{Hom}(X_n \times X_n, X_n)$ do coincide.

Since W is the graph of u_n and $u(Y_n) \subset W$,

$$\mathrm{pr}_2 v = u_n \mathrm{pr}_1 v \in \mathrm{Hom}(Y_n, X_n);$$

here both sides are viewed as morphisms of group schemes $Y_n \to X_n$. This implies that in $\mathrm{Hom}(Y_n, X_n)$ we have

$$w_{12}\mathrm{pr}_1 v = h\,\mathrm{pr}_2 v = h\,u_n \mathrm{pr}_1 v.$$

This implies that $w_{12} = h\,u_n$ on

$$\mathrm{pr}_1 v(Y_n) \subset X_n.$$

We have

$$\mathrm{pr}_1 v(Y_n) \supset r\,\mathrm{pr}_1(r(W)) = r(X_n)$$

and therefore $w_{12} = h\,u_n$ on $r(X_n)$. By Lemma 1.8,

$$r(X_n) = X_{n_1},$$

where $n_1 = n/(n, r)$. So, $w_{12} = h\,u_n$ on X_{n_1}. Let us put $d := (n_1, h)$. Clearly, $X_d \subset X_{n_1}$ and $w_{12} = hu_n$ kills X_d, because d divides h. This implies that there exists $u \in \text{End}(X)$ such that $w_{12} = d\,u$. If we put $m = n_1/d$ then h/d is a positive integer relatively prime to m and $(h/d)\,u\,d = (h/d)\,u_n\,d$ on X_{n_1} and therefore $(h/d)\,u = (h/d)\,u_n$ on $d(X_{n_1}) = X_m$. Since multiplication by (h/d) is an automorphism of X_m, we conclude that $u = u_n$ on X_m. $\square$

Corollary 10.1. *Let K be a finite field, X and Y abelian varieties over K. Let S be the set of positive integers n such that the finite commutative group K-schemes X_n and Y_n are isomorphic. If S is infinite then X and Y are isogenous over K. In addition, if S is the set of powers of a prime ℓ then there exists a K-isogeny $X \to Y$, whose degree is not divisible by ℓ.*

Proof. Pick $n \in S$ such that $n > r_2 := r_2(X, Y)$ where r_2 is as in Theorem 4.2. Then $m := n/(n, r_2)$ is strictly greater than 1. (In addition, if n is a power of ℓ then m is also a power of ℓ.(Fix an isomorphism $w_n : X_n \cong Y_n$. By Theorem 4.2, there exists $u \in \text{Hom}(X, Y)$ such that the induced morphism $u_m : X_m \to Y_m$ coincides with the restriction (image) of w_n to (in) $\text{Hom}(X_m, Y_m)$. But this restriction is an isomorphism, since w_n is an isomorphism. It follows that u_m is an isomorphism. Now the desired result follows from Lemma 1.10(ii). $\square$

Theorem 10.2 (Tate's theorem on homomorphisms). *Let K be a finite field, ℓ an arbitrary prime, X and Y abelian varieties over K of positive dimension. Let $X(\ell)$ and $Y(\ell)$ be the ℓ divisible groups attached to X and Y respectively. Then the natural embedding*

$$\text{Hom}(X, Y) \otimes \mathbb{Z}_\ell \hookrightarrow \text{Hom}(X(\ell), Y(\ell))$$

is bijective.

Remark 10.3. Our proof will work for both cases $\ell \neq \text{char}(K)$ and $\ell = \text{char}(K)$.

Proof of Theorem 10.2. Any element of $\text{Hom}(X(\ell), Y(\ell))$ is a collection

$$\{w_{(\nu)} \in \text{Hom}(X_{\ell^\nu}, Y_{\ell^\nu})\}_{\nu=1}^{\infty}$$

such that every $w_{(\nu)}$ coincides with the "restriction" of $w_{(\nu+1)}$ to X_{ℓ^ν}. It follows from Corollary 4.4 that there exists $u_\nu \in \text{Hom}(X, Y) \otimes \mathbb{Z}/\ell^\nu$ such that $w_{(\nu)} = u_\nu$. This implies that the image of $u_{\nu+1}$ in $\text{Hom}(X, Y) \otimes \mathbb{Z}/\ell^\nu$ coincides with u_ν for all ν. This means that if u is the projective limit of u_ν in $\text{Hom}(X, Y) \otimes \mathbb{Z}_\ell$ then u induces (for all ν) the morphism from X_{ℓ^ν} to Y_{ℓ^ν} that coincides with u_ν and therefore with $w_{(\nu)}$. $\square$

Corollary 10.4. *Let K be a finite field, ℓ an arbitrary prime, X and Y abelian varieties over K of positive dimension. Then the following conditions are equivalent:*

- *There exists a K-isogeny $X \to Y$, whose degree is not divisible by ℓ.*
- *The ℓ-divisible groups $X(\ell)$ and $Y(\ell)$ are isomorphic.*

Proof. It follows readily from Theorem 10.2 and Corollary 2.7. $\square$

11. Homomorphisms of Tate modules and isogenies

Throughout this Section, K is a finite field and ℓ is a prime $\neq \mathrm{char}(K)$.

Combining Theorem 10.2 with results of Section 2.8, we obtain the following statement.

Theorem 11.1 (Tate [27]). *Let X and Y be abelian varieties over K. Then*

$$\mathrm{Hom}(X, Y) \otimes \mathbb{Z}_\ell = \mathrm{Hom}_{\mathrm{Gal}}(T_\ell(X), T_\ell(Y)).$$

Let X be an abelian variety over K. Let us consider the $\mathbb{Q}_\ell$-vector space

$$V_\ell(X) = T_\ell(X) \otimes_{\mathbb{Z}_\ell} \mathbb{Q}_\ell$$

provided with the natural structure of Galois module. We have

$$\dim_{\mathbb{Q}_\ell}(V_\ell(X)) = 2\dim(X)$$

and the map

$$T_\ell(X) \hookrightarrow V_\ell(X), \ z \mapsto z \otimes 1$$

identifies $T_\ell(X)$ with a Galois-invariant $\mathbb{Z}_\ell$-lattice. This implies that the natural map

$$\mathrm{Hom}_{\mathrm{Gal}}(T_\ell(X), T_\ell(Y)) \otimes_{\mathbb{Z}_\ell} \mathbb{Q}_\ell \to \mathrm{Hom}_{\mathrm{Gal}}(V_\ell(X), V_\ell(Y))$$

is bijective. Here $\mathrm{Hom}_{\mathrm{Gal}}(V_\ell(X), V_\ell(Y))$ is the $\mathbb{Q}_\ell$-vector space of $\mathbb{Q}_\ell$-linear homomorphisms of Galois modules $V_\ell(X) \to V_\ell(Y)$.

Applying Theorem 11.1, we obtain the following statement.

Theorem 11.2 (Tate [27]). *Let X and Y be abelian varieties over K. Then the natural map*

$$\mathrm{Hom}(X, Y) \otimes \mathbb{Q}_\ell = \mathrm{Hom}_{\mathrm{Gal}}(V_\ell(X), V_\ell(Y))$$

is bijective.

The following assertion is very useful.

Corollary 11.3 (Tate's isogeny theorem [27]). *Let X and Y be abelian varieties over K. Then X and Y are isogenous over K if and only if the Galois modules $V_\ell(X)$ and $V_\ell(Y)$ are isomorphic.*

Proof. If X and Y are isogenous over K then there exist a positive integer N and isogenies

$$\alpha : X \to Y, \ \beta : Y \to X$$

such that

$$\beta\alpha = N_X, \ \alpha\beta = N_Y.$$

By functoriality, α and β induce homomorphisms of Galois modules

$$\alpha(\ell) : V_\ell(X) \to V_\ell(Y), \ \beta(\ell) : V_\ell(Y) \to V_\ell(X)$$

such that the compositions $\beta(\ell)\alpha(\ell)$ and $\alpha(\ell)\beta(\ell)$ coincide with multiplication by N in $V_\ell(X)$ and $V_\ell(Y)$ respectively. It follows that $\alpha(\ell)$ is an isomorphism of Galois modules $V_\ell(X)$ and $V_\ell(Y)$.

Suppose now that the Galois modules $V_\ell(X)$ and $V_\ell(Y)$ are isomorphic. Then their $\mathbb{Q}_\ell$-dimensions coincide and therefore

$$\dim(X) = \dim(Y).$$

Choose an isomorphism

$$w : V_\ell(X) \cong V_\ell(Y)$$

of Galois modules. Replacing (if necessary) w by $\ell^M w$ for sufficiently large positive integer M, we may and will assume that

$$w(T_\ell(X)) \subset T_\ell(Y).$$

The image $w(T_\ell(X))$ is a $\mathbb{Z}_\ell$-lattice in $V_\ell(Y)$. This implies that $w(T_\ell(X))$ is a subgroup of finite index in $T_\ell(Y)$. So, we may view w as an *injective* homomorphism $T_\ell(X) \to T_\ell(Y)$ of Galois modules. There exists a positive integer M such that if

$$w' \in \mathrm{Hom}_{\mathrm{Gal}}(T_\ell(X), T_\ell(Y)), \ w' - w \in \ell^M \cdot \mathrm{Hom}_{\mathrm{Gal}}(T_\ell(X), T_\ell(Y))$$

then

$$w' : T_\ell(X) \to T_\ell(Y)$$

is also injective. Since $\mathrm{Hom}(X, Y)$ is everywhere dense with respect to ℓ-adic topology in

$$\mathrm{Hom}(X, Y) \otimes \mathbb{Z}_\ell = \mathrm{Hom}_{\mathrm{Gal}}(T_\ell(X), T_\ell(Y)),$$

there exists $u \in \mathrm{Hom}(X, Y)$ such that the induced (by u) homomorphism of Galois modules

$$u(\ell) : T_\ell(X) \to T_\ell(Y)$$

is injective. This implies that

$$\mathrm{rk}_{\mathbb{Z}_\ell}(u(\ell)(T_\ell(X))) = \mathrm{rk}_{\mathbb{Z}_\ell}(T_\ell(X)) = 2\dim(X) = 2\dim(Y).$$

I claim that u is an isogeny. Indeed, let us put $Z := u(X)$: it is a (closed) abelian subvariety of Y that is defined over K. The homomorphism $u : X \to Y$ coincides with the composition of the natural surjection $X \to Z$ and the inclusion map $j : Z \hookrightarrow X$. This implies that $u(\ell)(T_\ell(X))$ is contained in $j(\ell)(T_\ell(Z))$ where

$$j(\ell) : T_\ell(Z) \to T_\ell(Y)$$

is the homomorphism of Tate modules induced by j. It follows that

$$2\dim(Z) = \mathrm{rk}(T_\ell(Z)) \geq \mathrm{rk}(j(\ell)(T_\ell(Z))) \geq$$

$$\mathrm{rk}(u(\ell)(T_\ell(X))) = 2\dim(X) = 2\dim(Y)$$

and therefore $\dim(Z) \geq \dim(Y)$. (Hereafter rk stands for the rank of a free $\mathbb{Z}_\ell$-module.)

Since Z is a closed subvariety of Y, we conclude that $\dim(Z) = \dim(Y)$ and therefore $Z = Y$. This implies that $u : X \to Y$ is surjective. Since $\dim(X) = \dim(Y)$, we conclude that u is an isogeny. $\qquad\square$

Corollary 11.3 admits the following "refinement".

Corollary 11.4. *Let X and Y be abelian varieties over K. The following assertions are equivalent.*

- *There exists an isogeny $X \to Y$, whose degree is not divisible by ℓ.*
- *The Galois modules $T_\ell(X)$ and $T_\ell(Y)$ are isomorphic.*

Proof. It follows readily from Corollary 10.4 and the last displayed formula in Subsection 2.8. $\qquad\square$

12. An example

Corollaries 10.1 and Corollary 10.4 suggest the following question: if X and Y are abelian varieties over a finite field K such that $X_n \cong Y_n$ for all n and $X(\ell) \cong Y(\ell)$ for all ℓ then is it true that X and Y are isomorphic? The aim of this Section is to give a negative answer to this question. Our construction is based on the theory of elliptic curves with complex multiplication [24,9].

We start to work over the field $\mathbb{C}$ of complex numbers. Let $F \subset \mathbb{C}$ be an imaginary quadratic field with the ring of integers $\mathcal{O}_F$. For every non-zero ideal $\mathfrak{b} \subset \mathcal{O}_F$ there exists an elliptic curve $E^{(\mathfrak{b})}$ over $\mathbb{C}$ such that that its group of complex points $E^{(\mathfrak{b})}(\mathbb{C})$ (viewed as a complex Lie group) is $\mathbb{C}/\mathfrak{b}$. There is a natural ring isomorphism $\mathcal{O}_F \cong \mathrm{End}(E^{(\mathfrak{b})})$ where any $a \in \mathcal{O}_F$ acts on $E^{(\mathfrak{b})}(\mathbb{C})$ as

$$z + \mathfrak{b} \mapsto az + \mathfrak{b}.$$

In particular, $E^{(\mathfrak{b})}$ is an elliptic curve with complex multiplication and $j(E^{(\mathfrak{b})}) \in \mathbb{C}$ is an *algebraic integer*.

Let us put $E := E^{(\mathcal{O}_F)}$. There is a natural bijection of groups

$$\mathfrak{b} \cong \operatorname{Hom}(E, E^{(\mathfrak{b})}), \ \ c \mapsto u(c),$$

where homomorphism $u(c)$ acts on complex points as

$$u(c) : \mathbb{C}/\mathcal{O}_F \to \mathbb{C}/\mathfrak{b}, \ \ z + \mathcal{O}_F \mapsto cz + \mathfrak{b}.$$

In addition, for every non-zero c the homomorphism $u(c) : E \to E^{(\mathfrak{b})}$ is an isogeny, whose degree is the order of the (finite) quotient $\mathfrak{b}/c\mathcal{O}_F$. In particular, E and $E^{(\mathfrak{b})}$ are isomorphic if and only if $\mathfrak{b}$ is a principal ideal. This implies that if $\mathfrak{b}$ is not principal then

$$\mathrm{j}(E^{(\mathfrak{b})}) \neq \mathrm{j}(E).$$

Lemma 12.1. *For every prime ℓ there exists a non-zero $c \in \mathfrak{b}$ such that the order of $\mathfrak{b}/c\mathcal{O}_F$ is not divisible by ℓ.*

Proof. We may assume that $\mathfrak{b}$ is not principal. If $\ell\mathcal{O}_F$ is a prime ideal in $\mathcal{O}_F$, pick any $c \subset \mathfrak{b} \setminus \ell\mathfrak{b}$. If $\ell\mathcal{O}_F$ is a square $\mathfrak{L}^2$ of a prime ideal $\mathfrak{L}$, pick any $c \in \mathfrak{b} \setminus \mathfrak{L} \cdot \mathfrak{b}$. If $\ell\mathcal{O}_F$ is a product $\mathfrak{L}_1\mathfrak{L}_2$ of two distinct prime ideals $\mathfrak{L}_1, \mathfrak{L}_2 \subset \mathcal{O}_F$, pick

$$c_1 \in \mathfrak{L}_1 \cdot \mathfrak{b} \setminus \mathfrak{L}_2 \cdot \mathfrak{b}, \ \ c_2 \in \mathfrak{L}_2 \cdot \mathfrak{b} \setminus \mathfrak{L}_1 \cdot \mathfrak{b}$$

and put $c = c_1 + c_2$; clearly,

$$c \notin \mathfrak{L}_1 \cdot \mathfrak{b}, \ \ c \notin \mathfrak{L}_2 \cdot \mathfrak{b}.$$

In all three cases

$$c\mathcal{O}_F = \mathfrak{M} \cdot \mathfrak{b}$$

where the ideal $\mathfrak{M} = \prod_{\mathfrak{P}} \mathfrak{P}^{m_{\mathfrak{P}}}$ is a (finite) product of powers of (non-zero) prime ideals $\mathfrak{P}$, none of which divides ℓ. It follows that $\mathfrak{b}/c\mathcal{O}_F$ is a (finite) $\mathcal{O}_F/\mathfrak{M}$-module. By the Chinese Remainder Theorem,

$$\mathcal{O}_F/\mathfrak{M} = \oplus_{\mathfrak{P}} \mathcal{O}_F/\mathfrak{P}^{m_{\mathfrak{P}}}.$$

Therefore $\mathfrak{b}/c\mathcal{O}_F$ is a product of finite $\mathcal{O}_F/\mathfrak{P}^{m_{\mathfrak{P}}}$-modules. Since the multiplication by the residual characteristic of $\mathfrak{P}$ kills $\mathcal{O}_F/\mathfrak{P}$, it follows that the $m_{\mathfrak{P}}$th power of this characteristic kills every $\mathcal{O}_F/\mathfrak{P}^{m_{\mathfrak{P}}}$-module. This implies that the order of $\mathfrak{b}/c\mathcal{O}_F$ is a product of powers of residual characteristics of $\mathfrak{P}$'s and therefore is not divisible by ℓ. $\qquad\square$

Corollary 12.2. *For every prime ℓ there exists an isogeny $E \to E^{(\mathfrak{b})}$, whose degree is not divisible by ℓ.*

12.3. The construction. Choose an imaginary quadratic field F with class number > 1 and pick a *non*-principal ideal $\mathfrak{b} \subset \mathcal{O}_F$. We have

$$\mathrm{j}(E^{(\mathfrak{b})}) \neq \mathrm{j}(E).$$

There exists an algebraic number field $L \subset \mathbb{C}$ such that:

- L contains F, $\mathrm{j}(E)$ and $\mathrm{j}(E^{(\mathfrak{b})})$.
- The elliptic curves E and $E^{(\mathfrak{b})}$ are defined over L.
- All homomorphisms between E and $E^{(\mathfrak{b})}$ are defined over L.

Let us choose a maximal ideal $\mathfrak{q} \subset \mathcal{O}_F$ such that both E and $E^{(\mathfrak{b})}$ have good reduction at $\mathfrak{q}$ and $j(E) - j(E^{(\mathfrak{b})})$ does *not* lie in $\mathfrak{q}$. (Those conditions are satisfied by all but finitely many $\mathfrak{q}$.) Let K be the (finite) residue field at $\mathfrak{q}$, let $\mathbf{E}$ and $\mathbf{E}^{(\mathfrak{b})}$ be the reductions at $\mathfrak{q}$ of E and $E^{(\mathfrak{b})}$ respectively: they are elliptic curves over K. Then $\mathrm{j}(\mathbf{E})$ and $\mathrm{j}(\mathbf{E}^{(\mathfrak{b})})$ are the reductions modulo $\mathfrak{q}$ of $\mathrm{j}(E)$ and $\mathrm{j}(E^{(\mathfrak{b})})$ respectively. Our assumptions on $\mathfrak{q}$ imply that

$$\mathrm{j}(\mathbf{E}) \neq \mathrm{j}(\mathbf{E}^{(\mathfrak{b})}).$$

Therefore $\mathbf{E}$ and $\mathbf{E}^{(\mathfrak{b})}$ are not isomorphic over K and even over $\bar{K}$!

On the other hand, it is known [9, Ch. 9, Sect. 3] that there is a natural embedding

$$\mathrm{Hom}(E, E^{(\mathfrak{b})}) \hookrightarrow \mathrm{Hom}(\mathbf{E}, \mathbf{E}^{(\mathfrak{b})})$$

that respects the degrees of isogenies. It follows from Corollary 12.2 that for every prime ℓ there exists an isogeny $\mathbf{E} \to \mathbf{E}^{(\mathfrak{b})}$, whose degree is not divisible by ℓ. Now Proposition 1.11 implies that $\mathbf{E}_n \cong \mathbf{E}^{(\mathfrak{b})}{}_n$ for all positive integers n. It follows from Corollary 10.4 that the ℓ-divisible groups $\mathbf{E}(\ell)$ and $\mathbf{E}^{(\mathfrak{b})}(\ell)$ are isomorphic for all ℓ, including $\ell = \mathrm{char}(K)$. Since both $\mathbf{E}(\bar{K})$ and $\mathbf{E}^{(\mathfrak{b})}(\bar{K})$ are torsion groups, they are isomorphic as Galois modules. This implies that their subgroups of all Galois invariants are isomorphic, i.e., the finite groups $\mathbf{E}(K)$ and $\mathbf{E}^{(\mathfrak{b})}(K)$ are isomorphic.

References

[1]   A. Borel, Harish-Chandra, *Arithmetic subgroups of algebraic groups*, Ann. of Math. **75** (1962), 485–535.

[2]   P. Deligne, *Variétés abéliennes ordinaire sur un corps fini*, Invent. Math. **8** (1969), 238–243.

[3]   M. Demazure, P. Gabriel, Groupes algébriques, Tome I, North Holland, Amsterdam 1970.

[4]   D. Eisenbud, J. Harris, The geometry of schemes, GTM **197**, Springer-Verlag, New York 2000.

[5]   R. Hoobler and A. Magid, *Finite group schemes over fields*, Proc. Amer. Math. Soc. **33** (1972), 310–312.

[6]   J. Harris, Algebraic geometry, Corrected 3rd printing, Springer Verlag New York, 1995.

[7]   R. Hartshorne, Algebraic geometry, GTM **52**, Springer Verlag, New York Heidelberg Berlin, 1977.

[8]   S. Lang, Abelian varieties, 2nd edition, Springer Verlag, New York, 1983.

[9] S. Lang, Elliptic functions, Addison-Wesley, 1973.

[10] H.W. Lenstra, Jr., F. Oort, Yu. G. Zarhin, *Abelian subvarieties*, University of Utrecht, Department of Mathematics, Preprint 842, March 1994; 19 pp.

[11] H.W. Lenstra, Jr., F. Oort, Yu. G. Zarhin, *Abelian subvarieties*, J. Algebra **80** (1996), 513–516.

[12] J.S. Milne, W. C. Waterhouse, *Abelian varieties over finite fields*, Proc. Symp. Pure Math. 20 (1971), 53-64.

[13] J.S. Milne, *Abelian varieties*, Chapter V in: Arithmetic geometry (G. Cornell, J.H. Silverman, eds.). Springer-Verlag, New York 1986.

[14] L. Moret-Bailly, Pinceaux de variétés abéliennes, Astérisque, vol. 129 (1985).

[15] D. Mumford, J. Fogarty, F. Kirwan, Geometric invariant theory, 3rd enlarged edition, Springer Verlag 1994.

[16] D. Mumford, Abelian varieties, 2nd edition, Oxford University Press, 1974.

[17] F. Oort, Commutative group schemes, Springer Lecture Notes in Math. **15** (1966).

[18] F. Oort and J.R. Strooker, *The category of finite groups over a field*, Indag. Math. **29** (1967), 163–169.

[19] F. Oort and J. Tate, *Group schemes of prime order*, Ann. Sci. École Norm. Sup. (4) **3** (1970), 1–21.

[20] F. Oort, *Abelian varieties over finite fields*, This volume, `www.math.uu.nl/people/oort/`.

[21] R. Pink, *Finite group schemes*, Lecture course in WS 2004/05 ETH Zürich, `www.math.ethz.ch/ pink/ftp/FGS/CompleteNotes.pdf`.

[22] C.P. Ramanujam, *The theorem of Tate*, Appendix I to [16].

[23] I. Reiner, Maximal orders, First edition, Academic Press, London, 1975; Second edition, Clarendon Press, Oxford, 2003

[24] J.-P. Serre, *Complex multiplication*, Chapter 13 in: Algebraic Number Theory (J. Cassels A. Frölich, eds), Academic Press, London, 1967.

[25] S.S. Shatz, *Group schemes, Formal groups and p-divisible groups*, Chapter III in: Arithmetic Geometry (G. Cornell, J.H. Silverman, eds.). Springer-Verlag, New York 1986.

[26] G. Shimura, Abelian varieties with complex multiplication and modular functions, Princeton University Press, Princeton, 1997.

[27] J.T. Tate, *Endomorphisms of abelian varieties over finite fields*, Invent. Math. **2** (1966), 134–144.

[28] J.T. Tate, *p-divisible groups*, In: Proceedings of a Conference on Local Fields, Driebergen, 1966. Springer-Verlag, Berlin Heidelberg New York, 1967, pp. 158–183.

[29] I.R. Shafarevich, Basic algebraic geometry, First edition, Springer Verlag, Berlin Heidelberg New York 1977.

[30] W.C. Waterhouse, *Abelian varieties over finite fields*, Ann. Sci. Écol. Norm. Supér. (4) **2**, (1969), 521-560.

[31] W.C. Waterhouse, Introduction to affine group schemes, Springer-Velag, New York 1979.

[32] Yu. G. Zarhin, *Endomorphisms of abelian varieties and points of finite order in characteristic P*, Mat. Zametki, **21** (1977), 737-744; Mathematical Notes **21** (1978) 415–419.

[33] Yu. G. Zarhin, *Homomorphisms of Abelian varieties and points of finite order over fields of finite characteristic* (in Russian), pp. 146-147, In: Problems in Group Theory and Homological Algebra (A. L. Onishchik, editor), Yaroslavl Gos. Univ., Yaroslavl, 1981; MR0709632 (84m:14051).

[34] Yu. G. Zarhin, *A finiteness theorem for unpolarized Abelian varieties over number fields with prescribed places of bad reduction*, Invent. Math. **79** (1985), 309–321.

Higher-Dimensional Geometry over Finite Fields
D. Kaledin and Y. Tschinkel (Eds.)
IOS Press, 2008

Author Index